Rhabdoviruses

Molecular Taxonomy, Evolution, Genomics, Ecology, Host–Vector Interactions, Cytopathology and Control

Edited by

Ralf G. Dietzgen

Queensland Alliance for Agriculture and Food Innovation
The University of Queensland
St. Lucia, QLD
Australia

and

Ivan V. Kuzmin

Division of High-Consequence Pathogens and Pathology
National Center for Emerging and Zoonotic Infectious Diseases
Centers for Disease Control and Prevention
Atlanta, GA
USA

Caister Academic Press

Caister Academic Press
Norfolk, UK

www.caister.com

British Library Cataloguing-in-Publication Data
A catalogue record for this book is available from the British Library

ISBN: 978-1-908230-11-9

Cover design adapted from CDC Image Library (Frederick Murphy)

Printed and bound in Great Britain

Contents

Contributors

El-Desouky Ammar
United States Department of Agriculture
ARS Subtropical Insects Research
Fort Pierce, FL
USA

eldammar@hotmail.com

Kim R. Blasdell
CSIRO Livestock Industries
Australian Animal Health Laboratory
Geelong, VIC
Australia

kim.blasdell@csiro.au

Ralf G. Dietzgen
Queensland Alliance for Agriculture and Food Innovation
The University of Queensland
St. Lucia, QLD
Australia

r.dietzgen@uq.edu.au

Michael M. Goodin
Department of Plant Pathology
University of Kentucky
Lexington, KY
USA

mgoodin@uky.edu

Aneta Gubala
Human Protection and Performance Division
Defence Science and Technology Organisation
Port Melbourne, VIC
Australia

ania.gubala@dsto.defence.gov.au

Rongliang Hu
Laboratory of Epidemiology
Veterinary Research Institute
Academy of Military Medical Sciences
Changchun
China

ronglianghu@hotmail.com

Francis M. Jiggins
Department of Genetics
University of Cambridge
Cambridge
UK

f.jiggins@gen.cam.ac.uk

D. Albert Joubert
CSIRO Livestock Industries
Australian Animal Health Laboratory
Geelong, VIC
Australia

albert.joubert@csiro.au

Gael Kurath
U.S. Geological Survey
Western Fisheries Research Center
Seattle, WA
USA

gkurath@usgs.gov

Ivan V. Kuzmin
Division of High-Consequence Pathogens and Pathology
National Center for Emerging and Zoonotic Infectious Diseases
Centers for Disease Control and Prevention
Atlanta, GA
USA

ikuzmin@cdc.gov

Ben Longdon
Department of Genetics
University of Cambridge
Cambridge
UK

b.longdon@ed.ac.uk

Byoung-Eun Min
Department of Plant Pathology
University of Kentucky
Lexington, KY
USA

byoungeunmin@yahoo.com

Isabel S. Novella
Department of Medical Microbiology and Immunology
University of Toledo
Toledo, OH
USA

isabel.novella@utoledo.edu

Steven J. Pauszek
Agricultural Research Service
United States Department of Agriculture
Foreign Animal Disease Research Unit
Plum Island Animal Disease Center
Orient, NY
USA

steve.pauszek@ars.usda.gov

John B. Presloid
Department of Medical Microbiology and Immunology
University of Toledo
Toledo, OH
USA

john.presloid@utoledo.edu

Margaret G. Redinbaugh
United States Department of Agriculture
ARS Corn and Soybean Research and Department of Plant Pathology
Ohio State University
Wooster, OH
USA

peg.redinbaugh@ars.usda.gov

Luis L. Rodriguez
Agricultural Research Service
United States Department of Agriculture
Foreign Animal Disease Research Unit
Plum Island Animal Disease Center
Orient, NY
USA

luis.rodriguez@ars.usda.gov

Charles E. Rupprecht
Centers for Disease Control and Prevention
Atlanta, GA
USA

cyr5@cdc.gov

Dennis Slate
United States Department of Agriculture
Animal and Plant Health Inspection Service
Wildlife Services
Concord, NH
USA

dennis.slate@aphis.usda.gov

Todd G. Smith
Centers for Disease Control and Prevention
Atlanta, GA
USA

tgsmith1@cdc.gov

Noel Tordo
Antiviral Strategies Unit
Institut Pasteur
Paris
France

ntordo@pasteur.fr

Peter J. Walker
CSIRO Livestock Industries
Australian Animal Health Laboratory
Geelong, VIC
Australia

peter.walker@csiro.au

Anna E. Whitfield
Department of Plant Pathology
Kansas State University
Manhattan, KS
USA

aewtospo@ksu.edu

Lena Wilfert
Institute of Evolutionary Biology
Ashworth Labs
University of Edinburgh
Edinburgh
UK

lena.wilfert@ed.ac.uk

Xianfu Wu
Centers for Disease Control and Prevention
Atlanta, GA
USA

xaw6@cdc.gov

1 Introduction

Ralf G. Dietzgen and Ivan V. Kuzmin

Rhabdoviruses constitute a diverse group of viruses with non-segmented, single-stranded, negative-sense RNA genomes (*Mononegavirales*), which share a common morphology, phylogeny, and similar genome organization. The outstanding characteristic of rhabdoviruses is their large, bullet-shaped (Ge *et al.*, 2010) or bacilliform (Jackson *et al.*, 2005) virions after which the family *Rhabdoviridae* is named. The *Big Picture Book of Viruses* provides a collection of stunning and beautiful images of rhabdovirus electron micrographs and cartoons (www.virology.net/big_virology/BVRNArhabdo.html). Despite (or because of) these common characteristics, rhabdoviruses successfully utilize a plethora of ecological niches, infecting a variety of plants, invertebrates, and vertebrates.

Recent research demonstrated the presence of rhabdoviral sequences in the genomes of arthropods (Fort *et al.*, 2012) and plants (Chiba *et al.*, 2011), indicating the ancient evolutionary origin and long-standing association of rhabdoviruses with their hosts. Although the majority of today's rhabdoviruses are transmitted by arthropods, groups, such as lyssaviruses and novirhabdoviruses, have evolved to circulate between vertebrates without participation of arthropod vectors. Rhabdoviruses include pathogens of tremendous public health, veterinary, and agricultural significance. This explains the attention, paid to different aspects of pathobiology and epidemiology of diseases, caused by rhabdoviruses, as well as to the structural, genetic and evolutionary properties of rhabdoviruses themselves. Moreover, because of their relative simplicity, rhabdoviruses have been used as vectors for foreign genes for the creation of recombinant vaccines against other infections (McGettigan *et al.*, 2001; McKenna *et al.*, 2004; Kapadia *et al.*, 2008), in basic investigations of viral evolution (Novella, 2003; Novella *et al.*, 2010; Longdon *et al.*, 2011), virus replication and protein interactions (Min *et al.*, 2010; Albertini *et al.*, 2011), and even as neuroscience tools for studying neuron functionality (Osakada *et al.*, 2011).

This book on *Rhabdoviruses: Molecular Taxonomy, Evolution, Genomics, Ecology, Host–Vector Interactions, Cytopathology and Control* aims to provide a contemporary view of rhabdoviruses that includes the prototypical rabies and vesicular stomatitis viruses, but that extends to rhabdoviruses infecting fish, insects and plants. The first chapters provide an overview of general rhabdovirus characteristics and taxonomy, while subsequent chapters look in more detail at recent advances in knowledge of the different viruses that infect terrestrial and aquatic vertebrates, invertebrates and plants, rhabdovirus evolution and control.

Chapter 2 describes rhabdovirus particle morphology, genome organization and viral life cycle including transcription, translation, replication, encapsidation and budding from cell membranes. Chapter 3 covers the taxonomic classification of rhabdoviruses into the established six genera, *Vesiculovirus, Lyssavirus, Ephemerovirus, Novirhabdovirus, Cytorhabdovirus* and *Nucleorhabovirus*, and the species currently placed in those genera. It also discusses the ongoing evolution of taxonomic concepts through our growing knowledge of rhabdoviruses that assists more comprehensive species demarcation. The chapter concludes with taxonomic suggestions for rhabdovirus species that are currently not assigned to a genus and points out challenges and potential solutions in dealing with the significant number of 'tentative' rhabdoviruses that are currently not sufficiently characterized to be included in the taxonomic system.

The following chapters describe the current knowledge and the most recent advances in research of the four established genera of animal-infecting rhabdoviruses. In Chapter 4 Rodriguez and Pauszek describe vesiculoviruses – the most basically organized and well-studied members of the family. In additional to the mammalian vesiculoviruses, authors provide an insight into 'vesiculovirus-related' fish rhabdoviruses, which in the future will be likely segregated in a new, separate genus. Chapter 5, written by Kuzmin and Tordo, is dedicated to lyssaviruses, which cause such deadly and threatening disease as rabies. Authors describe multiple facets of lyssavirus taxonomy, pathobiology and epidemiology, acquired during multiple years of rabies research. In Chapter 5 Walker and co-authors describe ephemeroviruses, with special attention to their complex genome organization and functions, which only recently have been put under the scope of rigorous investigations. Fish novirhabdoviruses, which have increasing economic significance because of their worldwide distribution, are described by Kurath in Chapter 7. The special emphasis of this chapter is made on evolutionary patterns, distribution and molecular epidemiology of novirhabdoviruses. In Chapter 8, Longdon and co-authors describe the history of sigma virus discovery, and provide characteristics of six different, newly identified sigma viruses, including their spread through insect populations and development of virus resistance. Chapters 9 and 10 deal with aspects of plant-adapted rhabdoviruses with special emphasis on their interactions with plant hosts and insect vectors. Nucleorhabdoviruses replicate in both plant and invertebrate cells. Beyond their role as plant pathogens, their nucleotropic lifestyle has facilitated novel insights into plant nuclear biology. Goodin and Min (Chapter 9) describe protein–protein interactions and membrane dynamics in virus-infected plant cells especially in relation to cell-to-cell movement which is unique and essential for plant-adapted viruses. The interactions of plant-adapted rhabdoviruses with insect tissues leading to replication in and transmission by planthopper and leafhopper vectors to cereal crops are discussed by Redinbaugh and co-authors. Chapter 10 covers current knowledge of virus acquisition, replication, movement and transmission by these vector insects. Leaps in our understanding of virus-plant, virus-vector and plant–vector interactions can be expected in coming years from the emerging data on both host plant and insect vector genomes and interactomes. Exciting advances in the molecular characterization of recently discovered animal rhabdoviruses with complex genomes are presented in Chapter 11 and the use of rhabdoviruses as experimental model for the evolution of RNA viruses is discussed in Chapter 12. The final two chapters of this book deal in depth with advances in rabies prevention and control and new approaches in lyssavirus vaccine development.

When the offer to compile a book about recent advances on rhabdoviruses came from Caister Academic Press, the task was more complex than difficult. First, we selected the most important topics of rhabdovirus research, which have been extensively developed during recent years – obviously a somewhat personal choice. Second, we attempted to recruit a group of experts in these areas, clearly understanding that they are tremendously busy with other tasks and duties. Indeed, some of our recruitment efforts failed, but others were successful. The final concept of the book was shaped based on the expertise of the authors, who kindly agreed to dedicate their time, efforts and talent, providing not only the most up-to-date reviews, but also outlining perspectives and expressing their expert opinions in the topics of interest. As editors, we acquired from the authors a significant body of novel knowledge and ideas, and we hope that readers will approve our collective efforts and find this book insightful, stimulating and a source of useful information on the rhabdoviruses.

References

Albertini, A., Bressanelli, S., Lepault, J., and Gaudin, Y. (2011). Structure and working of viral fusion machinery. Curr. Top. Membr. *68*, 49–80.

Chiba, S., Kondo, H., Tani, A., Saisho, D., Sakamoto, W., Kanematsu, S., and Suzuki, N. (2011). Widespread endogenization of genome sequences of non-retroviral RNA viruses into plant genomes. PLoS Path. *7*, e1002146.

Fort, P., Albertini, A., Van-Hua, A., Berthomieu, A., Roche, S., Delsuc, F., Pasteur, N., Capy, P., Gaudin, Y., and Weill, M. (2012). Fossil rhabdoviral sequences integrated into arthropod genomes: ontogeny, evolution, and potential functionality. Mol. Biol. Evol. *29*, 381–390.

Ge, P., Tsao, J., Schein, S., Green, T.J., Luo, M., and Zhou, Z.H. (2010). Cryo-EM model of the bullet-shaped vesicular stomatitis virus. Science *327*, 689–693.

Jackson, A.O., Dietzgen, R.G., Goodin, M.M., Bragg, J.N., and Deng, M. (2005). Biology of plant rhabdoviruses. Annu. Rev. Phytopathol. *43*, 623–660.

Kapadia, S.U., Simon, I.D., and Rose, J.K. (2008). SARS vaccine based on a replication-defective recombinant vesicular stomatitis virus is more potent than one based on a replication-competent vector. Virology *376*, 165–172.

McGettigan, J.P., Sarma,. S., Orenstein, J.M., Pomerantz, R.J., and Schnell, M.J. (2001). Expression and immunogenicity of human immunodeficiency virus type 1 Gag expressed by a replication-competent rhabdovirus-based vaccine vector. J. Virol. *75*, 8724–8732.

McKenna, P.M., Aye, P.P., Dietzschold, B., Montefiori, D.C., Martin, L.N., Marx, P.A., Pomerantz, R.J., Lackner, A., and Schnell, M.J. (2004). Immunogenicity study of glycoprotein-deficient rabies virus expressing simian/human immunodeficiency virus SHIV89.6P envelope in a rhesus macaque. J. Virol. *78*, 13455–13459.

Longdon, B., Hadfield, J.D., Webster, C.L., Obbard, D.J., and Jiggins, F.M. (2011). Host phylogeny determines viral persistence and replication in novel hosts. PLoS Pathog. *7*, e1002260.

Min, B.E., Martin, K., Wang, R., Tafelmeyer, P., Bridges, M., and Goodin, M. (2010). A host–factor interaction and localization map for a plant-adapted rhabdovirus implicates cytoplasm-tethered transcription activators in cell-to-cell movement. Mol. PlantMicrobe Interact. *23*, 1420–1432.

Novella, I.S. (2003). Contributions of vesicular stomatitis virus to the understanding of RNA virus evolution. Curr. Opin. Microbiol. *6*, 399–405.

Novella, I.S., Presloid, J.B., Zhou, T., Smith-Tsurkan, S.D., Ebendick-Corpus, B.E., Dutta, R.N., Lust, K.L., and Wilke, C.O. (2010). Genomic evolution of vesicular stomatitis virus strains with differences in adaptability. J. Virol. *84*, 4960–4968.

Osakada, F., Mori, T., Cetin, A.H., Marshel, J.H., Virgen, B., and Callaway, E.M. (2011). New rabies virus variants for monitoring and manipulating activity and gene expression in defined neural circuits. Neuron *71*, 617–631.

Morphology, Genome Organization, Transcription and Replication of Rhabdoviruses

2

Ralf G. Dietzgen

Abstract

Rhabdoviruses have large enveloped virions with bullet-shaped or bacilliform morphology. They are composed of an infectious nucleocapsid enveloped by a host-derived lipid membrane containing glycoprotein spikes. Their genomes consist of a non-segmented, negative-sense ssRNA which encodes at least five structural proteins. The replication cycle of all rhabdoviruses involves the same general steps. Following cell entry and uncoating, the nucleocapsid that contains all the components necessary for early transcription is released into the cytoplasm of the infected cell. The different viral genes are transcribed and proteins translated, followed by genome replication, encapsidation and budding from cell membranes.

Virion morphology

Shape and composition

Rhabdoviruses have large enveloped virions, 100–430 nm in length and 45–100 nm in diameter. The lipid bilayer envelope is acquired from host cell membranes during budding. Those viruses that infect vertebrates are bullet- or cone-shaped, while plant-adapted rhabdoviruses may appear bacilliform, bullet-shaped or pleomorphic depending on fixation and staining used for electron microscopy. The outer surface of virions (except for the quasi-planar end of bullet-shaped viruses) is covered with glycoprotein spikes, protruding through the virion lipid envelope, which are 5–10 nm long and about 3 nm in diameter. Internally, the nucleocapsid core, about 30–70 nm in diameter, exhibits helical symmetry and can be seen as cross-striations (spacing about 5 nm) in negatively stained and thin-sectioned virions. The nucleocapsid consists of a ribonucleoprotein (RNP) complex comprising the genomic RNA tightly bound to the nucleoprotein (N), together with an RNA-dependent RNA polymerase (L) and phosphoprotein (P). The nucleocapsid is active for transcription and replication: the N-RNA template is processed by the L protein, which contains most enzymatic activities, and its cofactor the P protein. The nucleocapsid has a helical symmetry, is uncoiled and filamentous, about 700 × 20 nm in size. In the virion, the matrix protein (M) condenses the nucleocapsid; it interacts with the N–RNA complex and associates with the host-derived lipid bilayer containing the transmembrane G protein. This is beautifully illustrated by the recent high resolution cryo-electron microscopy model of the vesiculovirus (VSV) particle (Ge *et al.*, 2010).

Genome

The rhabdovirus genome consists of a non-segmented linear, negative-sense single stranded RNA, about 11–16 kb in length which represents about 1–2% of virion weight. The RNA has a 3′-terminal free hydroxyl group and a 5′-triphosphate and is not polyadenylated. The 3′ and 5′ ends have inverted complementary sequences with transcription and replication initiation signals. Full-length positive-sense RNA, which is an intermediate during the replication process, as well as the full-length negative-sense RNA genome are permanently associated with N protein.

Proteins

All rhabdoviruses feature five structural proteins designated L, G, N, P and M (Dietzgen *et al.*, 2011).

The large *L protein* (220–240 kDa) has multiple domains and performs the functions required for genome transcription and replication, including RNA-dependent RNA polymerase, mRNA 5′ capping enzyme, cap methyltransferase, 3′ poly (A) polymerase and protein kinase activities.

The *glycoprotein G* (65–90 kDa) is assembled into trimers to form the virion surface spikes. G protein binds to vertebrate and invertebrate host cell receptors, induces virus endocytosis and mediates fusion of viral and endosomal membranes. G is variously N-glycosylated and palmitoylated; it lacks O-linked glycans and has haemagglutinin activity. G protein induces production of virus-neutralizing antibodies and elicits cell-mediated immune responses.

The *nucleoprotein N* (47–62 kDa) is the major component of the nucleocapsid and actively interacts with RNA, L and P proteins. N associates with full-length negative- and positive-sense RNAs and defective RNAs, but not with viral mRNAs. Newly synthesized N protein is thought to modulate the balance between genome transcription and replication by influencing recognition of transcription signals. N elicits cell-mediated immune responses and humoral antibodies.

The *phosphoprotein P* (20–30 kDa) plays multiple roles during transcription and replication. P is a non-catalytic cofactor of the viral polymerase and is variously phosphorylated. P mediates the physical link and proper positioning of the L protein on the N-RNA template. P also acts as a chaperone during synthesis of N, by forming N–P complexes that prevent N from self-aggregation and binding to cellular RNA. During genome replication, N is transferred from the N–P complexes to the nascent viral RNA to ensure its specific encapsidation into new nucleocapsids.

The structure, host interactions and functions of rabies virus (RABV) P protein have recently been reviewed (Leyrat *et al.*, 2011). P protein was shown to have multiple binding sites for different viral or host proteins to allow their assembly into multi–molecular complexes. P forms elongated and flexible dimers and has a modular organization composed of three structured domains separated by intrinsically disordered regions. The highly acidic N-terminal domain has binding sites for the viral L and N proteins, host cell kinases and importins. The central domain has an L protein binding site and dimerization domain, and the C-terminal basic domain is involved in N-RNA binding. Phosphorylation of RABV and VSV P protein occurs in the N- and C-terminal structured domains and is mediated sequentially by two different kinases. P protein may also interact with the host's cellular transport systems such as the dynein motor complex, nucleocytoplasmic transporters and microtubules (Das *et al.*, 2006; Albertini *et al.*, 2008; Min *et al.*, 2010) to facilitate intracellular movement of viral components. Furthermore, P interferes with the innate immune response by inhibiting different steps of the host cell interferon response. In infected cells, RABV P is heterogeneous and present in several truncated forms and in different phosphorylation states. Truncations are produced from internal in-frame AUG initiation codons by a ribosomal leaky scanning mechanism. The different truncated forms of P protein localize to the cytoplasm or the nucleus dependent on the presence of nuclear localization and nuclear export signals in the N- and C-terminal regions. The biological roles of these P protein forms have not been resolved yet.

The basic *matrix protein M* is an inner component of the virion (20–30 kDa). It appears to be involved in regulation of the genome RNA transcription. M protein binds to nucleocapsids and the cytoplasmic domain of G protein, thereby facilitating the budding process. It may be phosphorylated or palmitoylated. M protein is found in the nucleus of infected cells where it is thought to inhibit host cell transcription. M also mediates other pathological effects such as cell rounding, intracellular membrane redistribution and apoptosis.

In addition to these structural proteins that are shared by all rhabdoviruses, several rhabdovirus genomes encode diverse accessory proteins some of which were suggested to be significant for pathogenesis and apoptosis in animals, and cell-to-cell movement in plants (Walker *et al.*,

2011). Others appear to be secreted or have the characteristics of membrane-anchored glycoproteins or viroporins. However, the majority of these accessory proteins appear to be novel and their functions remain unknown.

Lipids and carbohydrates

Virions contain 15–25% of lipids; their composition reflects that of the host cell membrane where virions acquire their lipid envelope by budding. Generally phospholipids represent about 55–60%, and sterols and glycolipids about 35–40% of the total lipids. G protein may have covalently associated fatty acids proximal to the lipid envelope. Carbohydrates represent about 3% of virion weight; they occur as N-linked glycan chains on G protein and as glycolipids. In mammalian cells, the oligosaccharide chains are generally of the complex type while in insect cells they are of non-complex types.

Genome organization

The negative-sense single-stranded RNA genome of rhabdoviruses includes five major genes that are arranged in the conserved linear order 3′-N–P–M–G–L-5′. One or more additional interposed genes have been identified (Walker *et al.*, 2011). These accessory proteins may be encoded either in the same or overlapping open reading frame (ORF) within an existing gene or in a novel ORF between the structural protein genes (Fig. 2.1). The ends of the genomic RNA, termed 3′ leader and 5′ trailer sequences, exhibit terminal complementarity and contain promoter sequences that initiate replication of the genome and antigenome, respectively. They are approximately 50–100 nt in animal rhabdoviruses, but considerably longer in plant-adapted rhabdoviruses with 84–206 nt and 145–389 nt, respectively (Jackson *et al.*, 2005). Each of the individual genes is flanked by transcription

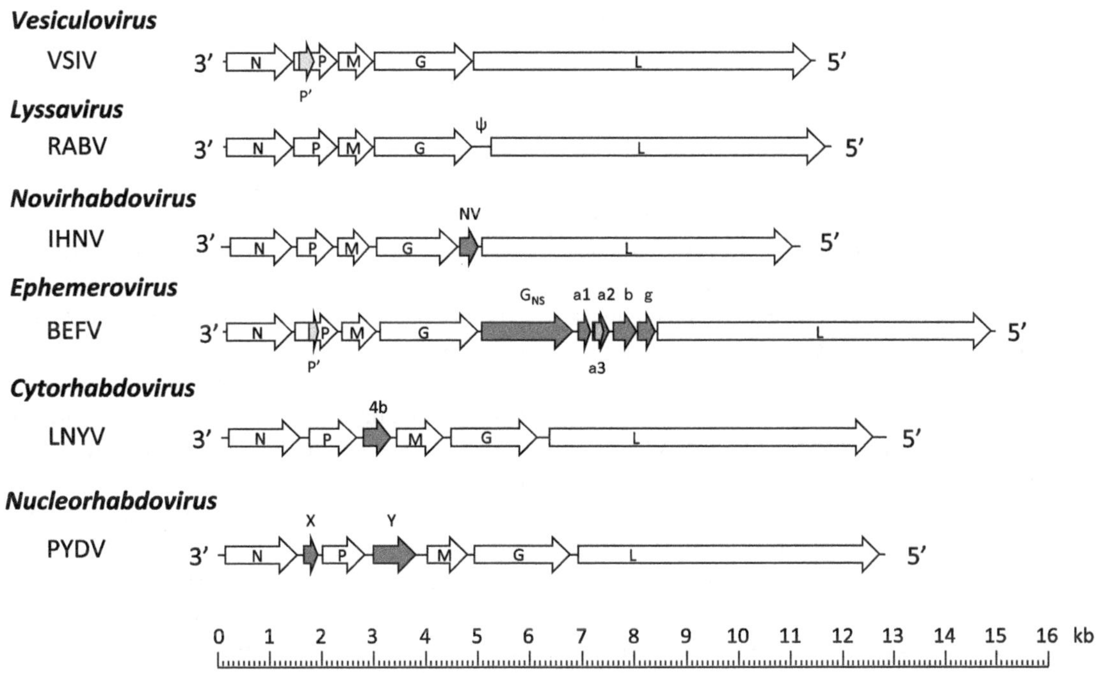

Figure 2.1 Genome organization of rhabdoviruses representing the six recognized genera of the *Rhabdoviridae*. The positions of open reading frames are indicated and those encoding accessory proteins are shaded. Virus names and abbreviations used are vesicular stomatitis Indiana virus (VSIV), rabies virus (RABV), infectious haematopoietic necrosis virus (IHNV), bovine ephemeral fever virus (BEFV), lettuce necrotic yellows virus (LNYV) and potato yellow dwarf virus (PYDV). The scale bar indicates approximate genome size (kb). Adapted from Walker *et al.* (2011).

initiation and termination/polyadenylation signals that are largely conserved among members of the same species and genus. Transcription units are separated by short untranscribed intergenic regions.

Transcription and replication

The replication cycle of all rhabdoviruses involves principally the same general steps. Following cell entry and uncoating, the nucleocapsid that contains all the components necessary for early transcription is released into the cytoplasm of the infected cell. The different viral genes are transcribed and proteins translated, followed by genome replication, encapsidation and budding from cell membranes.

Cell entry

Except for plant-adapted rhabdoviruses, which generally penetrate the plant cell wall through mechanical damage caused by insect vectors, the first steps of rhabdovirus infection involve attachment to host cells and penetration via endocytosis, which is mediated by the glycoprotein. In addition, carbohydrate moieties, phospholipids and gangliosides may play a complementary role for virus binding. After penetration, low pH within the endosome provokes fusion between endosomal and viral membranes, liberating the viral nucleocapsid into the cytoplasm. pH-induced conformational changes of the G protein catalyse this fusion process (Le Blanc *et al.*, 2005).

Nucleocapsid complexes: N-RNA, N-P and L-P

A typical feature of all negative-sense RNA viruses, including rhabdoviruses is that the RNA genome (or antigenome) never exists as free RNA, but instead is always encapsidated with many copies of the N protein. This stable N–RNA complex is the real template for both transcription and replication that is recognized by the viral polymerase (L–P) complex. The association is highly specific since in infected cells N protein does not associate with any other RNAs, including viral mRNAs. Structural and mechanistic comparative studies for negative sense RNA viruses were recently reviewed by Ruigrok *et al.* (2011). Based on the atomic crystal structure of the N–RNA complex, two domains of the N protein form a positively charged RNA binding groove into which the RNA binds by interactions with the sugar-phosphate backbone (Albertini *et al.*, 2008; Rugrok *et al.*, 2011). Every N protein molecule binds to nine or ten nucleotides and completely enwraps the genomic RNA to form a long helical N–RNA complex. The viral polymerase complex attaches to this RNP complex. The viral polymerase complex consists of two subunits, the enzymatically active L protein and the non-catalytic cofactor P. The L–P complex recognizes the polymerase entry site at the RNA 3′ end. L is positioned on the N–RNA complex through P. P also acts as a chaperone of nascent RNA-free N by forming N–P complexes that prevent N from binding to cellular RNAs and from polymerizing in the absence of RNA. Detailed structural analyses of the rhabdovirus transcription/replication complex have recently been compiled by Ivanov *et al.* (2011).

Transcription

Once the nucleocapsid is released into the cytoplasm, the genome RNA is repetitively transcribed by the virion transcriptase. This primary transcription does not depend on *de novo* host protein synthesis. N protein removal does not occur since the transcriptase only recognizes the RNA-N protein complex as template. The viral genes are transcribed progressively in decreasing molar abundance based on their distance from the 3′ end, i.e. N > P > M > G > L (Banerjee and Barik, 1992). At each transcription stop, a fraction of the transcriptase dissociates from the N-RNA template, regulating the relative abundance of each viral protein. The mRNAs have a 5′ cap and are 3′ polyadenylated and generally monocistronic (Fig. 2.2). Polycistronic mRNAs have been identified for viruses in some species; they result from a read-through or the absence of a stop transcription signal thereby allowing transcription extension across the adjacent 5′ cistron.

The viral mRNAs generally have similar 5′-terminal sequences (e.g. m^7Gppp(m) AmA(m)CA for vesiculoviruses and lyssaviruses) corresponding to the cap structure fused

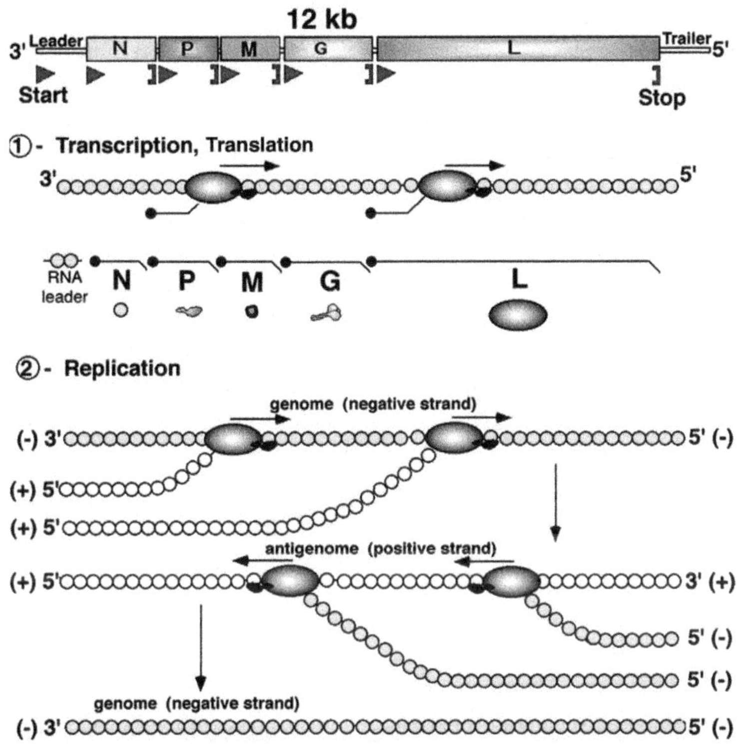

Figure 2.2 Rhabdovirus genome organization, transcription (1) and replication (2). Panel 1 shows the genome organization of vesicular stomatitis Indiana virus and the process of consecutive transcription of leader RNA and monocistronic mRNAs. Panel 2 illustrates replication of the negative-sense genome via a positive antigenome intermediate. The switch from transcription to replication appears to be regulated by the N protein. This figure is reprinted from Dietzgen *et al.* (2011) with permission from Elsevier (Copyright 2011).

to the first nucleotides copied from the start transcription signal. The mRNA cap formation pathway for vesiculoviruses is unconventional and differs from that employed by eukaryotic cells as reviewed by Ogino and Banerjee (2011). VSV L protein employs a novel 'RNA transfer mechanism' to produce the cap structure using a putative polyribonucleotidyl transferase domain.

The mRNAs' 3′-poly(A) tail is produced by the viral transcriptase upon copying in a reiterative mode the 7 U residues terminating each stop transcription signal. In certain cases the 5′-end of an mRNA may overlap the 3′-end of the preceding gene (Dietzgen *et al.*, 2011).

A short not capped or polyadenylated and untranslated leader RNA, corresponding to the positive-sense complement of the 3′-terminus of the viral RNA (i.e. preceding the N mRNA), is also transcribed. Unlike the mRNAs, leader RNA has a 5′ triphosphate terminus. Leader RNA transcripts of some viruses have been identified in the nucleus of infected cells.

Recently, the first case of RNA splicing among rhabdoviruses has been reported for a new rhabdovirus isolated from *Culex* mosquito (Kuwata *et al.*, 2011). This unusual virus has extremely elongated bacilliform virions, replicates in the nucleus of infected cells and requires RNA splicing for viral mRNA maturation. Its genome organization mirrors that of other rhabdoviruses, except for the presence of a putative 76-nt intron in the L protein coding region.

Translation

The viral mRNAs are generally translated on cytoplasmic polysomes except for the G mRNA which is translated on membrane-bound polysomes. Cellular chaperones are engaged to transport the M protein to the plasma membrane and G protein from the endoplasmic reticulum to the Golgi apparatus and then to the plasma membrane. Post-translational modification of G protein involves removal of the amino-proximal signal peptide and stepwise glycosylation in compartments of the Golgi apparatus.

Replication

After accumulation of newly synthesized viral proteins, the viral polymerase switches to a processive mode and ignores gene junctions to synthesize full-length positive-sense, antigenome RNA copies. The amount of N protein mediates this switch from transcription to replication. When it is sufficiently high to allow encapsidation of the nascent RNA, the polymerase switches to replication mode. This involves a change in the stoichiometry of L and P proteins in the polymerase complex. The antigenome RNA in turn serves as a replicative intermediate to direct synthesis of encapsidated negative-sense genome. This RNA is available for secondary transcription, translation and replication. Genomic N–RNA complexes can also be incorporated into progeny virions during the budding process.

Plant-infecting nucleorhabdoviruses and cytorhabdoviruses replicate in viroplasms in the cell nucleus and cytoplasm, respectively. As for other rhabdoviruses, replication occurs on the RNA–N protein complex and requires the newly synthesized N, P and L proteins to concomitantly encapsidate the nascent RNA into a nucleocapsid structure. Apart from freshly translated N–P–L proteins, replication may require host factors. However, vesiculoviruses can replicate in enucleated cells, indicating that newly synthesized host gene products are not required.

Assembly and budding

Viral nucleocapsid structures are assembled in association with M protein and lipid membranes containing viral G protein. The site of formation of particles depends on the virus and host cell. For vesiculoviruses, lyssaviruses, ephemeroviruses and novirhabdoviruses, nucleocapsids are synthesized in the cytoplasm and viruses bud from the plasma membrane in most, but not all cells. Some lyssaviruses can bud from intracytoplasmic membranes and in such cases virions can be seen in the cytoplasm of infected cells. In addition, there are aggregations of nucleocapsids in the Golgi complex and endoplasmic reticulum such as Negri bodies that are of diagnostic value. Cytorhabdoviruses bud from intracytoplasmic membranes associated with viroplasms; none have been observed to bud from plasma membranes. Nucleorhabdoviruses bud from the inner nuclear membrane and accumulate in perinuclear spaces.

References

Albertini, A.A.V., Schoehn, G., Weissenhorn, W., and Ruigrok, R.W.H. (2008). Structural aspects of rabies virus replication. Cell. Mol. Life Sci. *65*, 282–294.

Banerjee, A.K., and Barik, S. (1992). Gene expression of vesicular stomatitis virus genome RNA. Virology *188*, 417–428.

Das, S.C., Nayak, D., Zhou, Y., and Pattnaik, A.K. (2006). Visualization of intracellular transport of vesicular stomatitis virus nucleocapsids in living cells. J. Virol. *80*, 6368–6377.

Dietzgen, R.G., Calisher, C.H., Kurath, G., Kuzmin, I.V., Rodriguez, L.L., Stone, D.M., Tesh, R.B., Tordo, N., Walker, P.J., Wetzel, T., *et al.* (2011). Family *Rhabdoviridae*. In Virus Taxonomy. Ninth Report of the International Committee on Taxonomy of Viruses, King, A.M.Q., Adams, M.J., Carstens, E.B., and Lefkowitz, E.J., eds (Elsevier Academic Press, San Diego, CA), pp. 686–714.

Ge, P., Tsao, J., Schein, S., Green, T.J., Luo, M., and Zhou, Z.H. (2010). Cryo-EM model of the bullet-shaped vesicular stomatitis virus. Science *327*, 689–693.

Jackson, A.O., Dietzgen, R.G., Goodin, M.M., Bragg, J.N., and Deng, M. (2005). Biology of plant rhabdoviruses. Annu. Rev. Phytopathol. *43*, 623–660.

Kuwata, R., Isawa, H., Hoshino, K., Tsuda, Y., Yanase, T., Sasaki, T., Kobyashi, M., and Sawabe, K. (2011). RNA splicing in a new rhabdovirus from *Culex* mosquitoes. J. Virol. *85*, 6185–6196.

Ivanov, I., Yabukarski, F., Ruigrok, R.W.H., and Jamin, M. (2011). Structural insights into the rhabdovirus transcription/replication complex. Virus Res. *162*, 126–137.

Le Blanc, I., Luyet, P.P., Pons, V., Ferguson, C., Emans, N., Petiot, A., Mayran, N., Demaurex, N., Faure, J., Sadoul, R., *et al.* (2005). Endosome-to-cytosol transport of viral nucleocapsids. Nat. Cell. Biol. *150*, 601–612.

Leyrat, C., Ribeiro, E.A. Jr, Gérard, F.C.A., Ivanbov, I., Ruigrok, R.W.H., and Jamin, M. (2011). Structure, interactions with host cell and functions of rhabdovirus phosphoprotein. Future Virol. *6*, 465–481.

Min, B.E., Martin, K., Wang, R., Tafelmeyer, P., Bridges, M., and Goodin, M. (2010). A host–factor interaction and localization map for a plant-adapted rhabdovirus implicates cytoplasm-tethered transcription activators in cell-to-cell movement. Mol. Plant Microbe Interact. *23*, 1420–1432.

Ogino, T., and Banerjee, A.K. (2011). An unconventional pathway of mRNA cap formation by vesiculoviruses. Virus Res. *162*, 100–109.

Ruigrok, R.W.H., Crepin, T., and Kolakofsky, D. (2011). Nucleoproteins and nucleocapsids of negative-strand RNA viruses. Curr. Opin. Microbiol. *14*, 504–510.

Walker, P.J., Dietzgen, R.G., Joubert, D.A., and Blasdell, K.R. (2011). Rhabdovirus accessory genes. Virus Res. *162*, 110–125.

Taxonomy of Rhabdoviruses

3

Ralf G. Dietzgen and Ivan V. Kuzmin

Abstract

Taxonomy helps us to classify viruses according to guidelines developed by the International Committee on Taxonomy of Viruses (ICTV). Historical approaches, based on physico-chemical characteristics, serological relationships, genetic distances, or pathobiology, have been replaced with more complex concepts of viral species, that includes all the above, coupled with the genetic structure, evolutionary relationships, and ecologic properties of viruses. The species are segregated into genera, families, and orders, similar to the hierarchical systems developed for other forms of life. Currently, the family *Rhabdoviridae* includes six genera and several species, not assigned to any of the established genera. Taxonomic concepts evolve along with our growing scientific knowledge, development of novel tools and approaches, and appreciation of a variety of characters for comprehensive delineation of taxonomic entities. In this chapter we describe the most recent advances of rhabdovirus taxonomy, reflected in the ninth report of the ICTV, and those that arose after the report was prepared for publication.

Introduction

Life represents a continuum of evolving living organisms. Bacteria, archaea and eukarya (fungi, plants and animals) are hosts of viruses, and viral genomes are an important part of the phylogenetic 'tree of life' (Ludmir and Enquist, 2009). Putting aside the question if viruses are alive, we have to recognize that they follow the same general patterns as other life forms, perhaps with the exception of independent reproduction. Moreover, because of their relative simplicity, high multiplication rates and small genomes, viruses are subject to more rapid evolutionary changes. Even a single point-mutation can lead to dramatic changes in virus phenotype, which may positively or negatively affect progeny, and alter the direction of viral evolution (Dietzschold *et al.*, 1983). Any taxonomic classification system incorporates somewhat artificial elements, in an attempt to subdivide the continuum of life into discrete logical entities, particularly at the level of species and subspecies. However, taxonomy is a useful tool, not only to classify known living forms, but, more importantly, to demonstrate their evolutionary, morphological, and ecological relationships. Indeed, taxonomy evolves along with our growing scientific knowledge, development of novel tools and approaches, and appreciation of a variety of characters for comprehensive delineation of taxonomic entities.

The early classification of viruses was based on the diseases they caused. Electron microscopy enabled us to observe virus particles and facilitated virus classification based on their morphology. For example, more than 100 unclassified putative rhabdoviruses are currently listed as tentative members of the family, based largely on their unique shape and size, as observed by electron microscopy (Shope *et al.*, 1970; Travassos da Rosa *et al.*, 2002). Another approach of virus classification was based on serological cross-reactivity, with delineation of serotypes, and antigenic typing with poly- or monoclonal antibodies (Federer *et al.*, 1967; Tesh *et al.*, 1975; Wiktor and Koprowski, 1978; Calisher *et al.*, 1989). This technique was later expanded by genotyping, based on calculation of genetic distances between viral gene sequences (Bourhy *et al.*, 1993; Troyer and Kurath, 2003). Such physico-chemical and molecular approaches have discovered many important characteristics of viruses, including elements of their evolution and phylogeny. However, the concept of viral species is more complex.

A universal taxonomic system that reflects evolutionary relationships and ecology has been developed by the International Committee on Taxonomy of Viruses (ICTV), on behalf of the global community of virologists in an attempt to classify all known viruses. This approach has resulted in a hierarchical classification system of taxa including virus species, genera, families and some orders. This taxonomic system serves a conceptual framework for scientists and continues to evolve as new viruses are discovered and others are better characterized. Additional biological and genetic information is available at an accelerating rate. The rapid developments in next generation sequencing technologies and bioinformatics analyses provide a flood of new genetic data that, in association with traditional characterization tools, will assist greatly the taxonomic classification of unassigned viruses, as well as delineation of new species and genera. Currently, rhabdoviruses are classified as species in six genera that comprise the family *Rhabdoviridae* in the order *Mononegavirales*. This chapter provides an overview of the taxonomic structure of the family and discusses potential opportunities and challenges for further improvements to rhabdovirus taxonomy.

Negative-sense, single-stranded RNA viruses

The family *Rhabdoviridae*, together with the *Bornaviridae*, *Filoviridae* and *Paramyxoviridae* families are included in the order *Mononegavirales*. Viruses classified in this order are enveloped and have non-segmented, linear, negative-sense, single-stranded RNA genomes. Other features that the *Rhabdoviridae* have in common with the *Filoviridae* and *Paramyxoviridae* include a similar genome organization (3′-untranslated region (UTR) – core protein genes – envelope protein genes – RNA-dependent RNA polymerase gene – 5′UTR), helical nucleocapsids, a single 3′ promoter to initiate transcription, and complementary 3′ and 5′ termini. The ribonucleoprotein cores are infectious, but not the naked RNA. Maturation of virions occurs by budding through cellular membranes. The respective polymerase proteins contain regions of conserved amino acid homology, especially in the catalytic domain, which can be used to demonstrate phylogenetic relationships between the families (Fig. 3.1).

Rhabdoviridae taxonomic structure

The name rhabdovirus is derived from the Greek 'rhabdos' which means 'rod' or 'stick'. Virions are bullet-shaped, cone-shaped or bacilliform. The family *Rhabdoviridae* consists of six genera that were initially created based on serological and antigenic cross-reactivity, and further supplemented by genetic and other relevant information (such as evolutionary relatedness, pathobiology and ecology) of their member species: *Vesiculovirus*, *Lyssavirus*, *Ephemerovirus*, *Novirhabdovirus*, *Cytorhabdovirus* and *Nucleorhabdovirus* (Fig. 3.2). Serological and phylogenetic relationships between viruses usually correlate. However, some confusion occurred historically when the taxonomy was based solely on a limited number of characteristics. For example, based on antigenic cross-reactivity, Obodhiang and Kotonkan viruses were initially considered 'rabies-related', before establishment of the *Lyssavirus* genus (Shope, 1982). More rigorous serological comparisons (Calisher *et al.*, 1989), supplemented with phylogenetic data and ecological patterns of these viruses, demonstrated that both are ephemeroviruses (Bourhy *et al.*, 2005; Kuzmin *et al.*, 2006; Dietzgen *et al.*, 2011). Furthermore, in lyssaviruses, genetic distances are significantly shorter than in rhabdoviruses of other genera, and the majority of lyssaviruses cross-react serologically. However, from an ecological standpoint, all currently established species are clearly discriminated from each other (see Chapter 5). Therefore, viral species definitions based on several different sets of data provide better taxonomic resolution than serotyping or genotyping alone, despite the time and efforts needed for collection of the relevant information.

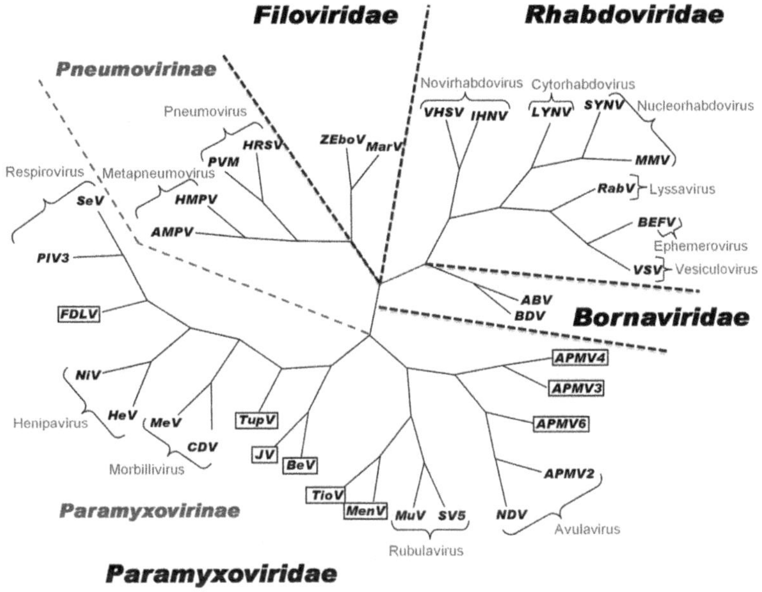

Figure 3.1 Unrooted phylogenetic tree of members of the order *Mononegavirales*. The tree was constructed using the CLUSTALX and PHYLIP programs with the sequences of the conserved domain III of the polymerase proteins. Nine paramyxoviruses, not yet assigned to genera, are included (boxed): avian parainfluenza virus types 3 (APMV3), avian parainfluenza virus types 4 (APMV4), avian parainfluenza virus types 6 (APMV6), Beilong virus (BeV), Fer de Lance virus (FDLV), J virus (JV), Menangle virus (MenV), Tioman virus (TioV) and Tupaia paramyxovirus (TupV). Abbreviations: ABV, avian bornavirus; AMPV, avian metapneumovirus; APMV2, avian parainfluenza virus types 2; BDV, Borna disease virus; BEFV, bovine ephemeral fever virus; CDV, canine distemper virus; HeV, hendra virus; HMPV, human metapneumovirus; HRSV, human respiratory syncytial virus; IHNV, infectious hemorraghic necrosis virus; LNYV, lettuce necrotic yellows virus; MarV, Marburg virus; MeV, measles virus; MMV, maize mosaic virus; MuV, mumps virus; NDV, Newcastle disease virus; NiV, Nipah virus; PIV3, parainfluenza virus type 3; PVM, pneumonia virus of mice; RabV, rabies virus; SeV, sendai virus; SV5, simian virus 5; SYNV, Sonchus yellow net virus; VSV, vesicular stomatitis Indiana virus; VHSV, viral haemorrhangic septicaemia virus; ZeboV, Zaire ebolavirus. This figure is reproduced from Easton and Pringle (2011) with permission from Elsevier (Copyright 2011).

Molecular phylogenies inferred from rhabdovirus gene sequences suggest a monophyletic origin of the family. Furthermore, they support the assignment of the six established genera and the species within these genera (Fig. 3.2). The P and G genes (and deduced proteins) demonstrate relatively low sequence identities compared to the N and L genes. In plant-adapted rhabdoviruses in the genera *Cytorhabdovirus* and *Nucleorhabdovirus*, only the L protein sequences are well conserved. Phylogenetic analyses indicate that vesiculoviruses and ephemeroviruses are the most closely related of the established genera. Together with Flanders virus (FLAV), Ngaingan virus (NGAV), Wongabel virus (WONV), Tupaia virus (TUPV) and several other, currently unclassified viruses, they have been proposed as members of a dipteran-mammal associated rhabdovirus 'supergroup', termed 'Dimarhabdovirus' (Bourhy *et al.*, 2005). However, the terminology for this 'supergroup' (which is not a recognized taxonomic entity) should be re-evaluated, as it includes viruses isolated from a wide range of hosts, such as birds, for Kolongo, Sanjimba and other related Central-African rhabdoviruses (Kuzmin *et al.*, 2006; Dacheux *et al.*, 2010), and waterborne fish vesiculoviruses, such as spring viraemia of carp virus (SVCV). Moreover, members of this 'supergroup' such as Almpiwar and Charleville viruses were isolated from reptiles (Tesh *et al.*, 1975; Monath *et al.*, 1979).

With the exception of nucleorhabdoviruses that replicate in the nucleus, all other members of the *Rhabdoviridae* multiply in the cytoplasm

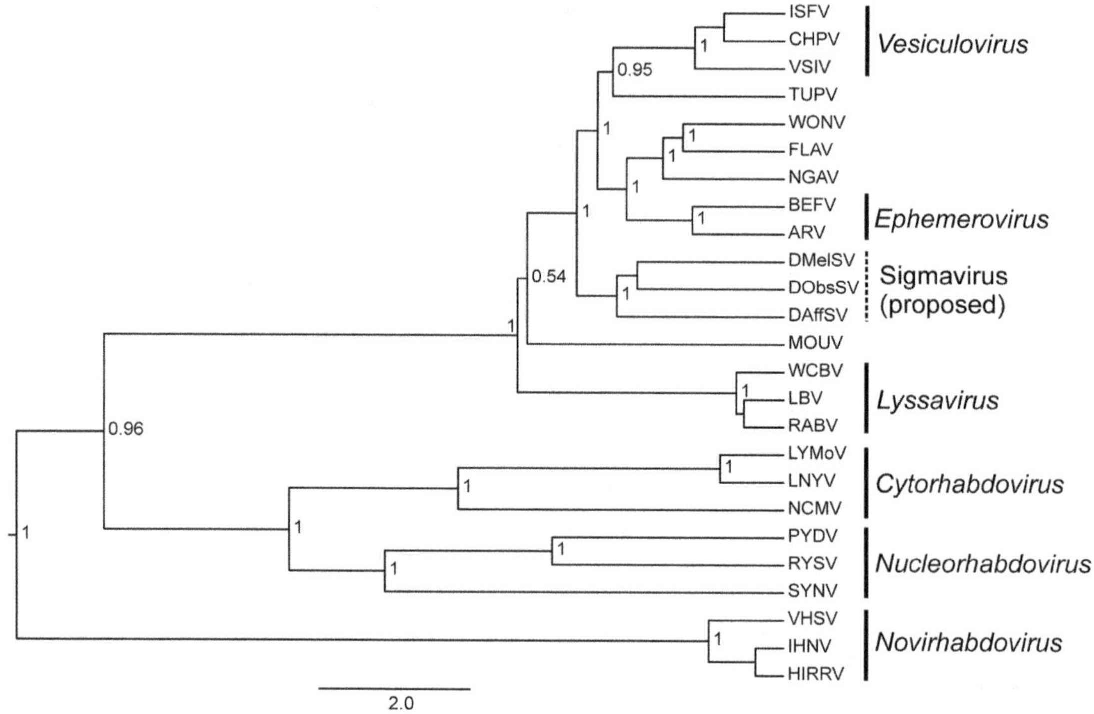

Figure 3.2 Bayesian tree of rhabdoviruses, based on L gene sequences. A limited number of representatives of each genus, as well as members of viral species not assigned into any established genus, are shown. Posterior probabilities for the key nodes are indicated.

of infected cells. Vesiculo-, lyssa-, ephemero- and novirhabdoviruses infect vertebrates while two genera of plant-adapted rhabdoviruses have been established, based on the sites of virus replication and morphogenesis – cytoplasm: *Cytorhabdovirus,* nucleus: *Nucleorhabdovirus.* The majority of rhabdoviruses are associated with arthropods, predominantly with insects (Hogenhout *et al.*, 2003; Kuzmin *et al.*, 2009). They include viruses that circulate solely within insect populations (such as Sigma viruses in fruit flies), plant rhabdoviruses transmitted in a persistent and propagative manner by insects (such as cytorhabdoviruses and nucleorhabdoviruses), and animal rhabdoviruses transmitted by insects (such as vesiculoviruses, ephemeroviruses, FLAV, and others). For many rhabdoviruses isolated from plants, invertebrates and vertebrates, ecological properties and circulation patterns remain unresolved (Dietzgen *et al.*, 2011). Clearly the evolution of rhabdoviruses has been driven by insects. Lyssaviruses, that have adapted to circulation among mammals without participation of insect vectors, represent one of the few exceptions to this rule. Nevertheless, the most parsimonious reconstruction of their evolution would likely include a host shift from insects to mammals. Today, we efficiently use oral vaccination of mammalian mesocarnivores with live attenuated rabies viruses (Slate *et al.*, 2009). Perhaps, by conducting such intervention we partly mimic the evolutionary pathway, which brought ancient *Lyssavirus* ancestors to circulate among insectivorous mammals, such as bats, which are considered the drivers of *Lyssavirus* evolution (Kuzmin and Rupprecht, 2007).

In addition to the five structural genes, common to all rhabdoviruses (N, P, M, G and L), the majority of insect-borne viruses harbour additional structural and non-structural genes, such as 1–5 accessory genes in cyto- and nucleorhabdoviruses, gene X of Sigma virus, G_{NS} and minor genes of ephemeroviruses, and multiple additional genes in WONV and NGAV viruses (Walker *et al.*, 2011). Functions of the majority of these genes are unknown. Lyssaviruses, in addition to the five structural genes, have only a long (~450–700 nucleotides) non-coding region between the G and L genes. In some vaccine strains of rabies virus, this region contains nucleotide sequences, similar to incomplete signals for transcription initiation and termination. Therefore, this region was suggested to be a remnant gene, or pseudogene (Tordo *et al.*, 1986). Perhaps, this gene (and relevant protein) facilitated the proposed historical switch of an ancestor virus from insects to bats, but disappeared because it no longer served a function, once lyssaviruses had adapted to circulation among mammalian hosts without the participation of insect vectors.

Vesiculoviruses, which share with lyssaviruses the same basic genome organization consisting of five structural genes, are capable of efficient infection of insects and mammals. Vesiculoviruses can replicate in insects of many taxa. However, populations of vesiculovirus virions demonstrate significant genetic variability. For example, vesiculovirus strains adapted to persistent infection of sand fly cells in culture are maladapted to mammalian BHK-21 cells. Recovery in the latter for only ten passages results in the loss of all mutations accumulated during the adaptation to persistence (including insertions) and the acquisition of a new set of mutations. The most parsimonious explanation for these sequence changes is that the mammalian-adapted mutants were present at a low frequency in the persistence-adapted populations, as confirmed experimentally (Novella *et al.*, 2007). Perhaps such mechanisms could historically have led the evolution of lyssavirus ancestors as well. However, such high pre-existing variability was never documented in lyssaviruses. Moreover, attempts to inoculate rabies virus into arthropods and insect cell lines failed. The only lyssavirus that demonstrated a limited ability to propagate in insect cells and in mosquitoes after adaptation was Mokola virus (Buckley, 1975; Aitken *et al.*, 1984). Perhaps, the observed genetic and antigenic similarity of the contemporary *Lyssavirus* species is the consequence of later constraints, applied by their adaptation to circulate within mammalian hosts without participation of insect vectors and unique neurotropic pathobiology (Kuzmin *et al.*, 2006).

The other rhabdoviruses that circulate without participation of arthropod vectors are fish rhabdoviruses, currently included in the genera *Vesiculovirus* and *Novirhabdovirus*. They are transmitted among susceptible fish in a waterborne manner, as described in Chapters 4 and 7. However, SVCV was detected in such water arthropods as penaeid shrimp, which suggests that arthropods might participate in SVCV circulation at least historically (Johnson *et al.*, 1999). Arthropod transmission has not been reported for novirhabdoviruses, to date.

The longest genetic distances separate plant-adapted rhabdoviruses within the *Nucleorhabdovirus* and *Cythorabdovirus* genera (Fig. 3.2). The approach for delineation of these genera is different from that accepted for animal rhabdoviruses (i.e. the intracellular location of virus replication in host cells is the major characteristic, placing a virus in one of the two genera). Very likely, future research will compile additional pieces of the taxonomic puzzle, and provide support for delineation of a greater number of genera of plant-adapted rhabdoviruses.

A brief overview of genera demarcation in the *Rhabdoviridae*

Genus *Vesiculovirus*

The genus name is derived from the Latin *vesicula*, meaning 'small bladder or blister'. The type species is *Vesicular stomatitis Indiana virus*. Vesiculoviruses cause diseases in mammals or fish (as long as we consider SVCV and related fish rhabdoviruses are members of this genus). Those viruses infecting mammals are transmitted by insects. Vesiculoviruses possess the shortest (~11 kb) and simplest genome organization (3′-nucleoprotein–phospoprotein–matrix protein glycoprotein–polymerase-5′) of all the rhabdoviruses.

Species demarcation is primarily based on serology combined with phylogenetic sequence analysis. Biological characteristics used to distinguish virus species within the genus include host range and means of transmission. Species recognized by the ICTV in its ninth report are *Carajas virus*, *Chandipura virus*, *Cocal virus*, *Isfahan virus*, *Maraba virus*, *Piry virus*, *Spring viraemia of carp virus*, *Vesicular stomatitis Alagoas virus*, *Vesicular stomatitis Indiana virus* and *Vesicular stomatitis New Jersey virus* (Dietzgen *et al.*, 2011). This report also lists 19 related viruses which may be members of this genus, but have not been approved as species (see Chapter 4).

Genus *Lyssavirus*

The name of the genus originates from Greek mythology: Lyssa (Λυσσα) was a goddess, or spirit of rage, fury, raging madness, and frenzy. The type species is *Rabies virus*. Lyssaviruses are essential neurotropic pathogens of mammals, and circulate without participation of arthropod vectors. They are excreted in the saliva, and delivered into wounds by bites of infected animals. Species demarcation is largely based on antigenic cross-reactivity using polyclonal antisera and monoclonal antibodies; on genetic characteristics such as genetic distances; topology and consistency of phylogenetic trees; and ecological characteristics. Genetically, lyssaviruses form a very tight monophyletic cluster within the *Rhabdoviridae*, well distanced from other family members. We do not know any 'intermediate' link between the lyssaviruses and other rhabdovirus representatives. Species recognized by the ICTV in its ninth report are *Aravan virus, Australian bat lyssavirus, Duvenhage virus, European bat lyssavirus 1, European bat lyssavirus 2, Irkut virus, Khujand virus, Lagos bat virus, Mokola virus, Rabies virus* and *West Caucasian bat virus* (Dietzgen *et al.*, 2011). *Shimoni bat virus* (Kuzmin *et al.*, 2010) has been recently ratified by the ICTV as a new species of this genus. Based on phylogenetic reconstructions and serological cross-reactivity, members of the genus have been subdivided into at least two phylogroups. Mokola, Lagos bat and Shimoni bat viruses belong to Phylogroup II, while all other lyssaviruses belong to Phylogroup I, except West Caucasian bat virus, which is most divergent and does not fit into any of the two phylogroups, and may be considered as a representative of a Phylogroup III.

Genus *Ephemerovirus*

The genus name is derived from the Greek *ephemeros,* which means 'short-lived'. The type species is *Bovine ephemeral fever virus*. Ephemeroviruses are animal rhabdoviruses that are transmitted by arthropods. Their genomes are more than 14 kb in size and complex. Multiple additional accessory protein genes are located between the G and L genes, including as a distinguishing feature a second, non-structural glycoprotein gene (G_{NS}) that appears to have arisen by gene duplication.

Viruses in the genus cross-react strongly in complement-fixation and indirect immunofluorescence assays, but show limited cross-neutralization. Different species may share up to 91% nucleoprotein amino acid sequence identity. Species recognized by the ICTV are *Adelaide River virus, Berrimah virus* and *Bovine ephemeral fever virus*. Recently, *Kotonkan virus* has been approved by the EC as a new species, awaiting ratification by the ICTV. There are several other related viruses including Obodhiang and Kimberley viruses which are likely members of this genus, but these are awaiting additional characterization.

Genus *Novirhabdovirus*

The genus name indicates the presence of a short 'non-virion' (NV) protein coded by the viral genome, and refers to this distinguishing feature of rhabdoviruses classified in this genus. The NV protein is expressed in infected cells, but not detectable in purified virions. In addition to this feature, novirhabdoviruses share phylogenetic relatedness and ecology, causing diseases in fish. The type species is *Infectious haematopoietic necrosis virus*. Waterborne and egg-associated transmission has been demonstrated.

Species demarcation is based on serological cross-neutralization and increasingly available sequence data. Species recognized by the ICTV in its ninth report are *Hirame rhabdovirus, Infectious haematopoietic necrosis virus, Snakehead virus* and *Viral haemorrhagic septicaemia virus* (Dietzgen *et al.*, 2011). Two viruses (Eel virus B12 and Eel virus C26) are listed as potential members, but have so far not been approved as new species.

Genus *Cytorhabdovirus*

The genus name is derived from the Greek *kytos* meaning 'cell'. Cytorhabdoviruses replicate in the cytoplasm of infected plant cells and virions bud in association with endoplasmic reticulum membranes. The type species of the genus is *Lettuce necrotic yellows virus*. Viruses in this genus infect monocot and dicot plants and are transmitted by leafhoppers, planthoppers or aphids, in which they also replicate.

Species are primarily differentiated by plant host range and vector specificity. Nucleic acid hybridization, serology and more recently genome sequence have also been used as species demarcation criteria. Species recognized by the ICTV in its ninth report are *Barley yellow striate mosaic virus, Broccoli necrotic yellows virus, Festuca leaf streak virus, Lettuce necrotic yellows virus, Lettuce yellow mottle virus, Northern cereal mosaic virus, Sonchus virus, Strawberry crinkle virus* and *Wheat American striate mosaic virus* (Dietzgen *et al.*, 2011). A small number of viruses are regarded as related and potential members of this genus, but they await more detailed characterization before taxonomic assignment.

Genus *Nucleorhabdovirus*

The genus name is derived from the Latin *nux, nucis*, meaning 'nut'. The unique characteristic that differentiates members of this genus from cytorhabdoviruses and indeed all other rhabdoviruses is that nucleorhabdoviruses replicate in the nuclei of infected plant cells. Morphogenesis occurs at the inner nuclear membrane and virions accumulate in perinuclear spaces. The type species of the genus is *Potato yellow dwarf virus*. Viruses in this genus infect monocot and dicot plants and are transmitted by leafhoppers, planthoppers or aphids in which they also replicate.

Species are primarily differentiated by plant host range and vector specificity. Nucleic acid hybridization, serology and more recently genome sequence have also been used as species demarcation criteria. Species recognized by the ICTV are *Datura yellow vein virus, Eggplant mottled dwarf virus, Maize fine streak virus, Maize mosaic virus, Potato yellow dwarf virus, Rice yellow stunt virus, Sonchus yellow net virus, Sowthistle yellow vein virus* and *Taro vein chlorosis virus* (Dietzgen *et al.*, 2011). *Maize Iranian mosaic virus* has been recently approved as a new species. A small number of viruses are considered as related, potential members of this genus, but these have so far not been proposed as new species and require more detailed molecular and biological characterization.

More detailed descriptions of each genus are provided in the following chapters of the book. Below we address several viral species that are currently not assigned to an established genus.

Rhabdovirus species not assigned to a genus

There are currently five rhabdovirus species recognized by the ICTV that have not been assigned to any of the six genera. These are *Flanders virus, Ngaingan virus, Wongabel virus, Tupaia virus* and *Sigma virus*. The former three viruses and *Moussa virus* (which recently obtained species approval from the ICTV), are associated with mosquitoes. Sequence comparisons between Flanders, Ngaingan and Wongabel viruses revealed limited relatedness between them and a more distant relationship with ephemeroviruses. The Tupaia virus genome contains a unique gene encoding a small hydrophobic transmembrane protein (SH) between the M and G genes and it is related to the recently characterized Durham virus, that also encodes a similar SH protein (Allison *et al.*, 2011). To date, Tupaia virus was isolated from a tree shrew (*Tupaia belangeri*), and could be propagated in tree shrew cell cultures only (Springfeld *et al.*, 2005). In contrast, Durham virus was isolated from an American coot (*Fulica americana*). The ecology of these viruses has not been sufficiently resolved. The genome of *Moussa virus* is quite intriguing. Although it is short and codes for the basic rhabdoviral five non-overlapping structural proteins, only the N, G and L proteins demonstrate considerable sequence identity to the corresponding genes of other rhabdoviruses. In contrast, the other two ORFs, with positions corresponding to the P and M genes in other rhabdoviruses, do not show any significant similarity with these, either at the nucleotide or amino acid sequence levels (Quan *et al.*, 2010). Sigma viruses naturally infect fruit flies and are vertically transmitted through the germ cells. Recent extensive studies demonstrated that sigma viruses include several distinct lineages and it has recently been proposed to the ICTV that a separate genus

Sigmavirus be created, encompassing seven sigma viruses as distinct viral species (Longdon *et al.*, 2010; Chapter 8).

Taxonomic challenges

Along with the established genera and species, several phylogenetic and antigenic groupings at different levels have been described within the *Rhabdoviridae*. There are more than 100 animal and plant viruses listed in the ninth report of the ICTV which may be members of the family *Rhabdoviridae*, but that have not been sufficiently characterized to include them in the official taxonomy (Dietzgen *et al.*, 2011). Comparative relationships between viral phylogeny and taxonomy remain incomplete until more biological and sequence data become available for these viruses. Relationships of viruses in the so-called Dimarhabdovirus 'supergroup' require further investigation. Here we still have either antigenic cross-reactivity only, or partial sequences of different genes, or suggestions based on morphology only from electron microscopy.

Nevertheless, some inferences can be made for more precise consideration in the future, and may help to focus research in certain directions. For example, N gene sequences demonstrated that Mount Elgon Almpiwar virus (MEBV), Kern Canyon virus (KCV) and Oita virus (OITAV) constitute a well supported monophyletic cluster (Kuzmin *et al.*, 2006). All three viruses were isolated from insectivorous bats, but in three different continents: Africa, North America, and Asia, respectively. Calisher *et al.* (1989) demonstrated that antigenically KCV is related to Fukuoka virus (FUKV) from Japan, whereas comparisons of partial L gene sequences demonstrated that FUKV is related to Keuraliba virus (KEUV), Le Dantek virus (LDV), and Nkolbisson virus (NKOV; all three viruses originated from western Africa) (Dacheux *et al.*, 2010). Unfortunately, the latter study did not include MEBV, KCV and OITAV, but we can expect that all these viruses are likely related, and may represent a new genus within the family.

Similarly, a group of seven rhabdoviruses (six of which were isolated from birds and one from dipterans in Central Africa), segregated into another monophyletic clade which may be potentially described as a new genus once additional relevant information is collected (Kuzmin *et al.*, 2006; Dacheux *et al.*, 2010). The recently characterized Australian Oak Vale virus demonstrates phylogenetic similarity to these viruses as well (Quan, 2011).

Several other rhabdovirus groups have been delineated based on their antigenic cross-reactivity and similarity of partial L gene sequences, available for comparisons (Calisher *et al.*, 1989; Bourhy *et al.*, 2005; Dacheux *et al.*, 2010). For example, the Hart Park group includes two 'orphan' viral species, Wongabel and Ngaingan, which currently are not included in any established genus, and at least five other viruses with only a limited set of characteristics available to date. The Tibrogargan group includes Tibrogargan virus (TIBV) along with 'orphan' species, TUPV, and recently characterized Durham virus (Allison *et al.*, 2011). The TIBV and Coastal Plains virus have recently been proposed as new species in a proposed new genus, *Tibrovirus*. The bat group includes at least three partly characterized viruses, isolated from lizards and dipterans. All these 'groups' should be assessed taxonomically in the future, when additional scientific information will balance clarity versus complexity, and help us properly evaluate relationships and connections within the family.

The ICTV *Rhabdoviridae* study group is considering ways to classify taxonomically several related fish-infecting rhabdoviruses that appear sufficiently different from both vesiculoviruses and novirhabdoviruses. It may be possible to group these viruses into new genera based on biological data and full-length genome sequences that have become available recently (Sheppard *et al.*, 2007; Chen *et al.*, 2009).

Among the plant-infecting negative sense-RNA viruses, the unclassified mite-transmitted orchid fleck virus (OFV) has a bipartite genome with significant sequence similarities to nucleorhabdoviruses (Kondo *et al.*, 2006). The gene order is reminiscent of rhabdoviruses with RNA 2 encoding the L gene. However, this virus cannot be classified taxonomically in the proposed new genus 'Dichorhabdovirus' in the *Rhabdoviridae*, since with a bipartite genome it clearly does not fit into the *Mononegavirales*. The coffee ringspot virus and the nuclear type of citrus leprosis virus are unclassified viruses similar to OFV that are transmitted by mites, have short bacilliform virions and accumulate in the nucleus. There are conflicting reports whether these viruses are enveloped or not. Once more physicochemical and genetic data for these viruses become available, it should be possible to resolve their relationships. The lettuce big-vein associated virus (genus *Varicosavirus*) also has a bipartite negative-sense RNA genome and its conserved L gene polymerase motif places it phylogeneticlly close to plant rhabdoviruses (Sasaya *et al.*, 2002). However, segmented and non-segmented negative-sense RNA viruses are phylogenetically clearly partitioned. Tordo *et al.* (1992) considered it 'plausible that segmented negative-sense RNA viruses descended from a distant mononegavirus ancestor'. Research in support of such suggestion remains to be conducted.

Disclaimer

The findings and conclusions in this report are those of the authors and do not necessarily represent the views of the Centers for Disease Control and Prevention.

References

Aitken, T.H., Kowalski, R.W., Beaty, B.J., Buckley, S.M., Wright, J.D., Shope, R.E., and Miller, B.R. (1984). Arthropod studies with rabies-related Mokola virus. Am. J. Trop. Med. Hyg. *33*, 945–952.

Allison, A.B., Palacios, G., Travassos da Rosa, A., Popov, V.L., Lu, L., Xiao, S.Y., DeToy, K., Briese, T., Lipkin, W.I., Keel, M.K., *et al.* (2011). Characterization of Durham virus, a novel rhabdovirus that encodes both a C and SH protein. Virus Res. *155*, 112–122.

Bourhy, H., Kissi, B., and Tordo, N. (1993). Molecular diversity of the *Lyssavirus* genus. Virology *194*, 70–81.

Bourhy, H., Cowley, J.A., Larrous, F., Holmes, E.C., and Walker, P.G. (2005). Phylogenetic relationships among rhabdoviruses inferred using the L polymerase gene. J. Gen. Virol. *86*, 2849–2858.

Buckley, S.M. (1975). Arbovirus infection of vertebrate and insect cell cultures, with special emphasis on Mokola, Obodhiang, and Kotonkan viruses of the rabies serogroup. Ann. N.Y. Acad. Sci. *266*, 241–250.

Calisher, C.H., Karabatsos, N., Zeller, H., Digoutte, J.P., Tesh, R.B., Shope, R.E., Travassos da Rosa, A.P., and St George, T.D. (1989). Antigenic relationships among rhabdoviruses from vertebrates and hematophagous arthropods. Intervirol. *30*, 241–257.

Chen, H.L., Liu, Z.X., He, J.Q., Gao, L.Y., Shi, X.J., and Jiang, Y.L. (2009). Characterization of the complete genome sequence of pike fry rhabdovirus. Arch. Virol. *154*, 1489–1494.

Dacheux, L., Berthet, N., Dissard, G., Holmes, E.C., Delmas, O., Larrous, F., Guigon, G., Dickinson, P., Faye, O., Sall, A.A., *et al.* (2010). Application of broad-spectrum resequencing microarray for genotyping rhabdoviruses. J. Virol. *84*, 9557–9574.

Dietzgen, R.G., Calisher, C.H., Kurath, G., Kuzmin, I.V., Rodriguez, L.L., Stone, D.M., Tesh, R.B., Tordo, N., Walker, P.J., Wetzel, T., *et al.* (2011). Family *Rhabdoviridae*. In Virus Taxonomy. Ninth Report of the International Committee on Taxonomy of Viruses, King, A.M.Q., Adams, M.J., Carstens, E.B., and Lefkowitz, E.J., eds (Elsevier Academic Press, San Diego, CA), pp. 686–714.

Dietzschold, B., Wunner, W.H., Wiktor, T.J., Lopes, A.D., Lafon, M., Smith, C.L., and Koprowski, H. (1983). Characterization of an antigenic determinant of the glycoprotein that correlates with pathogenicity of rabies virus. Proc. Natl. Acad. Sci. U.S.A. *80*, 70–74.

Easton, A.J., and Pringle, C.R. (2001). Order Mononegavirales. In Virus Taxonomy. Ninth Report of the International Committee on Taxonomy of Viruses, King, A.M.Q., Adams, M.J., Carstens, E.B., and Lefkowitz, E.J., eds (Elsevier Academic Press, San Diego, CA), pp. 653–657.

Federer, K.E., Burrows, R., and Brooksby, J.B. (1967). Vesicular stomatitis virus – the relation between some strains of the Indiana serotype. Res. Vet. Sci. *8*, 103–117.

Hogenhout, S.A., Redinbaugh, M.G., and Ammar, E.-D. (2003). Plant and animal rhabdovirus host range: a bug's view. Trends Microbiol. *11*, 264–271.

Johnson, M.C., Maxwell, J.M., Loh, P.C., and Leong, J.A. (1999). Molecular characterization of the glycoproteins from two warm water rhabdoviruses: snakehead rhabdovirus (SHRV) and rhabdovirus of penaeid shrimp (RPS)/spring viremia of carp virus (SVCV). Virus Res. *64*, 95–106.

Kondo, H., Maeda, T., Shirako, Y., and Tamada, T. (2006). Orchid fleck virus is a rhabdovirus with an unusual bipartite genome. J. Gen. Virol. *87*, 2413–2421.

Kuzmin, I.V., and Rupprecht, C.E. (2007). Bat rabies. In Rabies, Jackson, A., and Wunner, W., eds (Elsevier, London, UK), pp. 259–307.

Kuzmin, I.V., Hughes, G.J., and Rupprecht, C.E. (2006). Phylogenetic relationships of seven previously unclassified viruses within the family Rhabdoviridae using partial nucleoprotein gene sequences. J. Gen. Virol. *87*, 2323–2331.

Kuzmin, I.V., Novella, I.S., Dietzgen, R.G., Pahdi, and Rupprecht, C.E. (2009). The rhabdoviruses: Biodiversity, phylogenetics and evolution. Infect. Genet. Evol. *9*, 541–553.

Longdon, B., Obbard, D.J., and Jiggins, F.M. (2010). Sigma viruses from three species of *Drosophila* form a major new clade in the rhabdovirus phylogeny. Proc. Biol. Sci. *277*, 35–44.

Ludmir, E.B., and Enquist, L.W. (2009). Viral genomes are part of the phylogenetic tree of life. Nat. Rev. Microbiol. *7*, 615–616.

Monath, T.P., Cropp, C.B., Frazier, C.L., Murphy, F.A., and Whitfield, S.G. (1979). Viruses isolated from reptiles: identification of three new members of the family Rhabdoviridae. Arch. Virol. *60*, 1–12.

Novella, I.S., Ebendick-Corpus, B.E., Zarate, S., and Miller, E.L. (2007). Emergence of mammalian-adapted vesicular stomatitis virus from persistent infections of insect-vector cells. J. Virol. *81*, 6664–6668.

Quan, P.L., Junglen, S., Tashmukhamedova, A., Conlan, S., Hutchison, S.K., Kurth, A., Ellerbrok, H., Egholm, M., Briese, T., Leendertz, F.H., *et al.* (2010). Moussa virus: a new member of the Rhabdoviridae family isolated from *Culex decens* mosquitoes in Côte d'Ivoire. Virus Res. *147*, 17–24.

Quan, P.L., Williams, D.T., Johansen, C.A., Jain, K., Petrosov, A., Diviney, S.M., Tashmukhamedova, A., Hutchison, S.K., Tesh, R.B., Mackenzie, J.S., *et al.* (2011). Genetic characterization of K13965, a strain of Oak Vale virus from Western Australia. Virus Res. *160*, 206–213.

Sasaya, T., Ishikawa, K., and Koganezawa, H. (2002). The nucleotides sequence of RNA1 of *Lettuce big-vein virus*, genus *Varicosavirus*. J. Gen. Virol. *297*, 289–297.

Sheppard, A.M., Le Deuff, R.M., Martin, P.D., Woolford, G., Way, K., and Stone, D.M. (2007). Genotyping spring viraemia of carp virus and oter piscine vesiculo-like viruses using reverse hybridisation. Dis. Aquat. Org. *76*, 163–168.

Shope, R.E. (1982). Rabies-related viruses. Yale J. Biol. Med. *55*, 271–275.

Shope, R.E., Murphy, F.A., Harrison, A.K., Causey, O.R., Kemp, G.E., Simpson, D.I., and Moore, D.L. (1970). Two African viruses serologically and morphologically related to rabies virus. J. Virol. *6*, 690–692.

Slate, D., Algeo, T.P., Nelson, K.M., Chipman, R.B., Donovan, D., Blanton, J.D., Niezgoda, M., and Rupprecht, C.E. (2009). Oral rabies vaccination in North America: opportunities, complexities, and challenges. PLoS Negl. Trop. Dis. *3*, e549.

Springfeld, C., Darai, G., and Cattaneo, R. (2005). Characterization of the Tupaia rhabdovirus genome reveals a long open reading frame overlapping with P and a novel gene encoding a small hydrophobic protein. J. Virol. *79*, 6781–6790.

Tesh, R.B., Peralta, P.H., Shope, R.E., Chaniotis, B.N., and Johnson, K.M. (1975). Antigenic relationships among phlebotomus fever group arboviruses and their implication for the epidemiology of sandfly fever. Am. J. Trop. Med. Hyg. *24*, 135–144.

Tordo, N., Poch, O., Ermine, A., Keith, G., and Rougeon, F. (1986). Walking along the rabies genome: is the large G-L intergenic region a remnant gene? Proc. Natl. Acad. Sci. U.S.A. *83*, 3914–3918.

Tordo, N., De Haan, P., Goldbach, R., and Poch, O. (1992). Evolution of negative-stranded RNA genomes. Semin. Virol. *3*, 341–357.

Travassos da Rosa, A.P., Mather, T.N., Takeda, T., Whitehouse, C.A., Shope, R.E., Popov, V.L., Guzman, H., Coffey, L., Araujo, T.P., and Tesh, R.B. (2002). Two new rhabdoviruses (Rhabdoviridae) isolated from birds during surveillance for arboviral encephalitis, northeastern United States. Emerg. Infect. Dis. *8*, 614–618.

Troyer, R.M., and Kurath, G. (2003). Molecular epidemiology of infectious hematopoietic necrosis virus reveals complex virus traffic and evolution within southern Idaho aquaculture. Dis. Aquat. Organ. *55*, 175–185.

Walker, P.J., Dietzgen, R.G., Joubert, D.A., and Blasdell, K.R. (2011). Rhabdovirus accessory genes. Virus Res. *162*, 110–125.

Wiktor, T.J., and Koprowski, H. (1978). Monoclonal antibodies against rabies virus produced by somatic cell hybridization: detection of antigenic variants. Proc. Natl. Acad. Sci. U.S.A. *75*, 3938–3942.

Genus *Vesiculovirus*

4

Luis L. Rodriguez and Steven J. Pauszek

Abstract

The *Vesiculovirus* genus of the *Rhabdoviridae* includes viruses infecting mammals, insects and fish that have been taxonomically classified using a combination of serological relatedness, host range, genome organization, pathobiology and phylogenetic analysis of sequence data. There are 10 viruses assigned to the genus *Vesiculovirus* and a number of serologically related viruses that have been tentatively assigned to the genus. The best known members of the genus are vesicular stomatitis viruses which cause vesicular disease in farm animals including horses, cattle and pigs. Some vesiculoviruses cause disease in humans ranging from febrile syndromes (e.g. vesicular stomatitis virus) to fatal encephalitis (e.g. Chandipura virus). This chapter reviews the state-of-the-art knowledge about vesiculoviruses as well as some related viruses tentatively assigned to the genus.

Introduction

The first *Vesiculovirus* described was vesicular stomatitis virus (VSV), the prototype virus for the genus and one of the most studied RNA viruses. Although the first description of the virus was published in 1927 (Cotton, 1927) the vesicular stomatitis disease (initially known as sore mouth in cattle and horses) had been described affecting horses, pigs and cattle as early as 1801, along the eastern coast of the United States (Hanson, 1952). Outbreaks of 'sore tongue' were reported causing severe lameness in horses needed on the battlefield during the Civil War in 1862 (McClellan, 1864). Also VSV was carried from the US to Europe (France) with infected horses during World War I, and a vesicular disease was reported causing outbreaks in horses in South Africa in 1897 (Hanson, 1952). These two instances are the only reports of VSV causing disease outside the Americas. Reports of a vesicular disease in horses and cattle in Central and South America date back to the 1930s and continue to be reported every year in most endemic regions. There are well documented large epidemics of VSV occurring in the south-western US every 8–12 years from 1916 through 2006 (Rainwater-Lovett *et al.*, 2007).

Another vesiculovirus, Chandipura virus, was originally found as the cause of a febrile disease in humans in India in 1966. It was also found in sand flies, and it was shown that mosquitoes can be experimentally infected (Bhatt and Rodrigues, 1967; Rao *et al.*, 1967). More recently, Chandipura virus was associated with serious outbreaks of encephalitis, often fatal, in children from various parts of India (Hassler *et al.*, 2003; John, 2004; Rao *et al.*, 2004; Chadha *et al.*, 2005; Narasimha Rao *et al.*, 2008; Tandale *et al.*, 2008).

In 1977, Isfahan virus was found in sand flies in Iran and also in humans with febrile disease (Saidi *et al.*, 1977; Tesh *et al.*, 1977). This virus has also been isolated from sand flies collected in gerbil colonies in the former USSR (Gaidamovich *et al.*, 1980).

In the Americas, the majority of vesiculoviruses have been found in insects and mammals. The exception to this insect/mammal association is a virus that causes a disease in fish. This serious disease, referred to as 'infectious dropsy of carp,' appeared in 1930 and the aetiological agent was identified in the 1980s as spring viraemia of carp virus (SVCV) and has been currently classified as a *Vesiculovirus* (Dietzgen *et al.*, 2011). This is the only member of the genus not infecting insects and/or mammals. It is very likely that many others, yet to be discovered vesiculoviruses exist in nature, circulating in insects, mammals or fish, perhaps without causing a visible disease.

Taxonomy and phylogeny

Vesiculoviruses have been taxonomically classified using a combination of morphological and serological relatedness, host range, genome organization, pathobiology and via phylogenetic analysis of gene sequences (when available). As defined in the ninth report of the International Committee on Taxonomy of Viruses, there are ten established *Vesiculovirus* species and nine related viruses tentatively assigned to the genus (Dietzgen, 2011) (Table 4.1). Based on a recent published report (Pauszek *et al.*, 2011), virus CoAr 171638 (Tesh *et al.*, 1987) has been identified as a vesiculovirus-related virus. Currently, only ten full length genomes of vesiculoviruses (MARAV, VSIV, COCV, VSAV, VSNJV, CHPV, ISFV, and SVCV) or related proposed vesiculoviruses (PFRV, EXEV) are present in GenBank (Table 4.2). Phylogenetic analyses based on the concatenated amino acid sequence of the five structural proteins show strongly supported grouping of the insect/mammal associated vesiculoviruses separate from those infecting fish (Fig. 4.1a). This monophyletic grouping was maintained when the amino acid sequence of the five structural proteins were analysed individually.

While full-length genomic sequences or even complete sequence of individual viral genes, are not available for the majority of viruses tentatively assigned to the *Vesiculovirus* genus, phylogenies derived from the amino acid sequence of block III of the L polymerase have been used recently to determine evolutionary relationships among a large number of unclassified rhabdoviruses (Bourhy *et al.*, 2005; Dacheux *et al.*, 2010). A phylogenetic tree of all available limited L gene sequences of classified or tentative vesiculoviruses shows three clearly demarcated groups: group I containing vesicular stomatitis viruses and viruses isolated from insects; group II containing vesiculoviruses infecting humans, and group III containing viruses of fish (Fig. 4.1b). This analysis provides support for the classification of the unassigned viruses CoAr 171638, PERV, JURV and MSPV as vesiculoviruses. Although viruses GLOV, PORV, KLAV and BTKV fall within the Dimarhabdovirus supergroup (rhabdoviruses infecting diptera and mammals) (Bourhy *et al.*, 2005), phylogenetic analysis based on the block III of the L polymerase does not support their classification as vesiculoviruses.

Antigenic properties

Vesiculoviruses have been identified and classified into serotypes based on their serological

Table 4.1 List of accepted and related/tentative vesiculoviruses discussed in this chapter. Assigned abbreviations are given in parenthesis

Vesiculovirus species	Related/tentative *Vesiculovirus* species
Carajas virus (CJSV)	Boteke virus (BTKV)
Chandipura virus (CHPV)	CoAr 171638 (CoArV 171638)
Cocal virus (COCV)	Eel virus European X (EXEV)
Isfahan virus (ISFV)	Grey Lodge virus (GLOV)
Maraba virus (MARAV)	Jurona virus (JURV)
Piry virus (PIRYV)	Klamath virus (KLAV)
Spring viraemia of carp virus (SVCV)	Malpais Spring virus (MSPV)
Vesicular stomatitis Alagoas virus (VSAV)	Perinet virus (PERV)
Vesicular stomatitis Indiana virus (VSIV)	Pike fry rhabdovirus (PFRV)
Vesicular stomatitis New Jersey virus (VSNJV)	Porton virus (PORV)

Table 4.2 Vesiculovirus genome organization and length

Species	Genome (nt)	N (aa)	P (aa)	M (aa)	G (aa)	L (aa)
VSIV	11,161–11,336	422	265	229	511	2109
COCV	11,003	422	264	229	512	2108
VSAV	11,070	422	260–261	229	511	2108
CoAr 171638	ND	ND	270	229[a]	513	ND
MARAV	11,135	422	265	229	512	2109
VSNJV	11,119–11,123	422	274	229	517	2109
CJSV	ND	ND	271	228	523	2109
CHPV	11,119–11,120	422	293	229	530	2092
ISFV	11088	423	289	226	523	2093
MSPV	ND	ND	273[a]	225[a]	521[a]	ND
PIRYV	ND	ND	327	229	529	ND
SVCV	11,019–11,047	418	309	223	509	2095–2101
PFRV	11097	419	309	222	508	2095
EXEV	11806	428	275	222	531	2075

ND, not determined.
[a]Previously unpublished.

relationships (Cartwright and Brown, 1972). Cross-neutralization, which is determined by antibodies recognizing epitopes located in the viral G protein, has been used to distinguish serotypes and subtypes of VSV. The G protein also determines protection in animals vaccinated with either inactivated whole virus or subunit vaccines (Mackett *et al.*, 1985). Furthermore, recombinant viruses containing the G protein of VSNJV, but all other proteins from VSIV, induced protective immune responses against challenge with VSNJV (Martinez *et al.*, 2004). The N protein is a more cross-reactive antigen than G and it is the base for complement fixation tests that helped define members of the genus (Brooksby, 1948; Jenney *et al.*, 1958). One of the criteria used for *Vesiculovirus* classification is cross-reactivity by complement fixation.

In the case of vesicular stomatitis viruses there are two major serotypes: New Jersey and Indiana (Federer *et al.*, 1967). The serotype Indiana has been subdivided into three distinct serological complexes. Indiana 1 comprises the classical Indiana viruses (VSIV) found from northern South America to southern US. The Indiana 2 subtype has Cocal virus (COCV) as the prototype virus which was originally isolated from mites collected from rice rats in Trinidad in 1961. Indiana 2 viruses cause disease in cattle and horses in Brazil and Argentina. The Indiana 3 subtype is represented by Alagoas virus (VSAV) which was first isolated from a mule in Alagoas, Brazil, in 1964. This subtype is the most common cause of vesicular stomatitis in livestock in Brazil (Pauszek *et al.*, 2011). In a recent study the phylogenetic relationships among South American vesicular stomatitis viruses were found to follow very closely the serological classification of these viruses. However, other vesiculoviruses found in South American insects but not causing clinical disease grouped into distinct genetic lineages from vesicular stomatitis viruses (Pauszek *et al.*, 2011).

In the case of the fish *Vesiculovirus* SVCV, there is a single serotype. Antibodies directed against SVCV cross-react to various degrees with other fish rhabdoviruses; pike fry rhabdovirus (PFRV), Grass carp rhabdovirus (GrCRV) and tench rhabdovirus (TenRV), indicating that these viruses are closely related which is reflected in the phylogenetic trees (Fig. 4.1). SVCV and PFRV have been shown to share common antigenic determinants on the G, N and M proteins, but can be differentiated by neutralization assays.

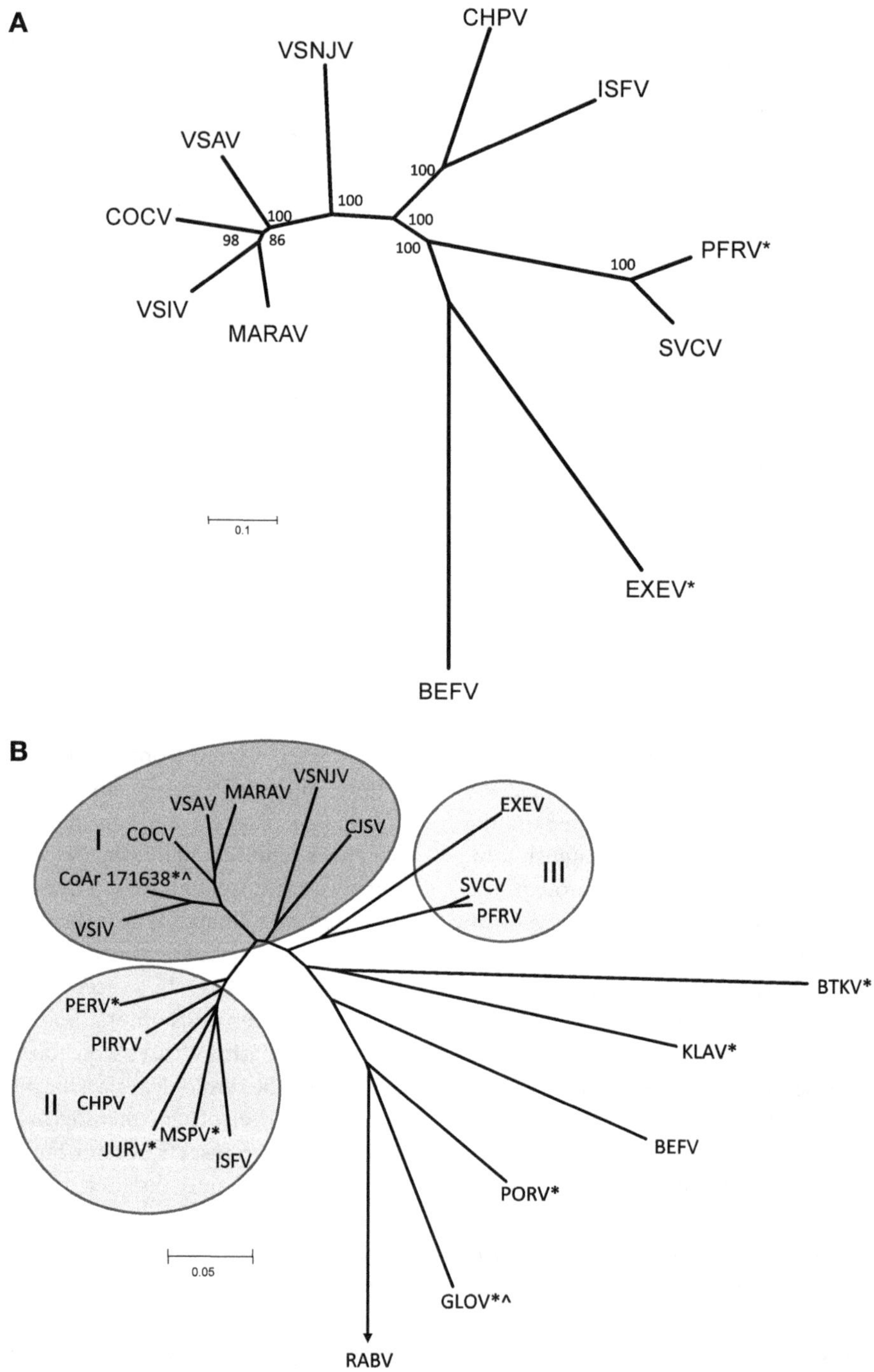

Figure 4.1 Phylogeny of the vesiculoviruses and related dimarhabdoviruses from concatenated full-length amino acid sequences of the five structural proteins (a) or block III of the L polymerase (b). Trees show three clearly demarcated groups: group I containing vesicular stomatitis viruses and viruses isolated from insects; group II containing vesiculoviruses infecting humans, and group III containing viruses of fish. The trees were inferred using the neighbour-joining method and numerical values indicate bootstrap support values (2000 replicates) for (a). The trees were calculated using the Poisson correction method and evolutionary distances represented as the number of amino acid substitutions per site. Gaps in the alignment were analysed with pairwise deletion. *Viruses related or tentatively assigned to the vesiculoviruses as defined in the ninth report of the International Committee on Taxonomy of Viruses; ^previously unpublished sequences.

Viral proteins, structure and morphology

Vesiculoviruses have five structural proteins (designated N, P, M, G and L) (Cartwright *et al.*, 1970; Kang and Prevec, 1970). Most vesiculoviruses also have two small, highly basic proteins coded in a second ORF within P (termed C′ and C) (Spiropoulou and Nichol, 1993). Three viral proteins (N, P and L) combine to form the transcription and replication complex which has RNA-dependent-RNA polymerase (RdRP) activity (Baltimore *et al.*, 1970; Davis and Wertz, 1982). The N protein encapsidates the viral RNA and functions in close association with the P protein. The P protein is a highly phosphorylated protein associated with viral polymerase activity. It mediates the binding of the L protein to the nucleocapsid core and facilitates access of the polymerase to the RNA template during transcription and replication (Wertz, 1983). Phosphorylation of P protein seems to be necessary for optimal transcriptase activity. The exact roles of the non-structural C and C′ proteins are unclear. Engineered viruses that do not express C proteins are indistinguishable from wild type virus in protein synthesis, virus production and host-protein synthesis shut off in tissue culture cells (Kretzschmar *et al.*, 1996). The L protein is multifunctional and performs most of the polymerase-associated functions including RNA synthesis, capping, methylation and poly (A) addition (Davis and Wertz, 1982). It also has protein kinase activity which preferentially phosphorylates serine residues on the P protein. The matrix protein (M) is the most abundant protein of virions. The M binds specifically to the G protein monomers and promotes their trimerization. The G protein is a typical class I membrane associated glycoprotein, with approximately 90% of the N-terminal region of the molecule projecting from the surface of the virion or infected cell, a hydrophobic transmembrane domain anchoring the protein in the membrane, and a C-terminal 28 amino acid cytoplasmic domain projecting to the interior of the infected cells (Rose *et al.*, 1980). The G protein forms trimers which constitute the approximately 400 spikes that are embedded on the virion bilayer phospholipid envelope. The G protein plays a major role in attachment and penetration of vesiculoviruses into susceptible cells and budding of virions from infected cells. It is the major target of serotype-specific neutralizing antibodies and is capable of inducing cell membrane fusion at low pH (Martinez and Wertz, 2005).

Virions are enveloped and bullet-shaped. The three-dimensional structure of VSV has been recently determined utilizing cryo-electron microscopy (Ge *et al.*, 2010). The virion structure is the result of the nucleocapsid forming a conical-shaped tip and a helical structure in the trunk. Matrix protein interacts with an inner helix of N protein and RNA as well as with neighbouring M and N subunits, providing rigidity to the ribonucleoprotein. Interactions among the N and M proteins provide scaffolding and stability to the complex and are critical in maintaining the bullet shape (Ge *et al.*, 2010). Two layers of lipids obtained from the host cell membrane with embedded trimers of the envelope glycoprotein form the outer layers of the virion and mediate the interaction with cellular receptor(s).

Genome organization and viral replication

Vesiculoviruses contain a single molecule of linear, negative-sense RNA of about 11 kb in length. The shortest genome is 11,003 nt (COCV) and the longest is 11,336 nt (VSIV) (Rodriguez *et al.*, 2002; Pauszek *et al.*, 2008). The genome organization for all vesiculoviruses consists of a short (~47 nt) leader sequence followed by the five structural protein genes in the order 3′ N–P–M–G–L, and a trailer sequence at the 5′ end. Most differences in length are found in non-translated regions, particularly in the M and G mRNAs. Despite the variability in lengths of the five mRNAs of the Indiana subtypes, four of the five predicted structural proteins (N, M, G and L) are within one amino acid residue in length from each other (Pauszek *et al.*, 2008).

Vesiculovirus replication is well studied and has served as a model for replication of rhabdoviruses. Transcription begins at a single entry site at the 3′ end of the genome with each gene expressed as a capped and polyadenylated monocistronic mRNA. The first transcript is the 3′ leader sequence, which is neither capped nor polyadenylated. It is transported to the nucleus where it inhibits host cell transcription. The leader transcript is followed by the N mRNA, which is capped during synthesis by the virion polymerase complex (composed of N, P and L). The intergenic sequence (5′-AGUUUUUUUCAUA-3′) serves as a signal for polyadenylation, termination and re-initiation of transcription in decreasing amounts as the polymerase complex moves away from the single promoter. Therefore, the gene order (i.e. 3′ N > P > M > G > L 5′) provides an efficient way of regulation of gene expression, by which proteins that are necessary in larger amounts (such as N) are located near the 3′ end and are transcribed in larger amounts, and those needed in lesser amounts are located towards the 5′ end and are transcribed less frequently (Ball and White, 1976).

After translation of the mRNAs to produce the viral proteins, genome replication starts. The RdRP initiates the replication process at the 3′ end of the genome, ignores all signals for stop and polyadenylation of individual mRNAs, and instead synthesizes a full-length complementary antigenome. This antigenome serves as a template for synthesis of a negative-sense leader RNA, and also for synthesis of full-length progeny genomes. Full-length genomes can either serve as templates for secondary transcription, or can be assembled into infectious particles. Factors determining the RdRP functions in transcription or replication modes are not fully understood but the proportion of N protein and available RNP templates are thought to be critical for determining these functions. Following replication, further rounds of transcription (secondary transcription), translation and replication ensues (Barr *et al.*, 2002).

Host range and geographical distribution.

The host range of vesiculoviruses include a variety of mammals, birds, fish and insects, found to be infected either by direct virus isolation or by evidence of exposure to virus from serological tests. The viruses infecting mammals are all transmitted by insects and therefore are considered arthropod borne viruses (arboviruses). Vesicular stomatitis viruses have been found causing disease in horses, cattle, pigs and can also infect humans causing flu-like symptoms. Other vesiculoviruses cause febrile disease and encephalitis in humans (e.g. Piry virus and Chandipura virus).

The natural cycle of vesiculoviruses infecting mammals remains largely undefined. These viruses are commonly found in a number of insect species including sand flies (flebotomine flies), black flies (simulid flies), several species of *Culicoides* and mosquitoes. Serological evidence suggests they are capable of infecting not only a number of wild mammals, but also birds and even reptiles living in endemic areas (Tesh *et al.*, 1969, 1970). This wide range of hosts might explain the ability of vesiculoviruses to infect and replicate in a very diverse range of vertebrate and invertebrate cell cultures.

The incredibly broad host range of the vesiculoviruses most likely contributes to their expansive geographical distribution. Viruses of two species, VSNJV and VSIV, are found endemically from northern South America (Peru, Colombia, Ecuador, Venezuela), through Central America into southern Mexico with sporadic incursions into northern Mexico and south-western United States. Vesicular stomatitis virus subtypes Indiana-2 or Indiana-3 sporadically cause disease in Brazil while Indiana 2 has also occurred in Argentina (Pauszek *et al.*, 2011).

The geographic distribution of Chandipura virus seems to be mainly in India, with initial reports in northern India (Chandipur, Orissa) and later reports in Central and Southern India (Rao *et al.*, 2004). However, it has also been reported in Nigeria (Berge, 1975). Isfahan virus has been reported in Central Iran (Esfahan) (Saidi *et al.*, 1977) and has also been reported in neighbouring Turkmenistan (Gaidamovich *et al.*, 1980). Spring viraemia of carp virus has been reported in Europe, Asia and the Americas (Walker and Winton, 2010). Two vesiculoviruses (Carajas virus and Maraba virus) were isolated from phlebotomine sand flies in Brazil (Travassos da Rosa *et al.*, 1984) but have not been associated with disease in livestock or humans. Reports of naturally occurring Piry virus infection have been limited to Brazil but human disease associated to laboratory infections have been reported (Wilks and House, 1984).

Viral ecology

Vesicular stomatitis viruses are transmitted to cattle, horses and pigs by various blood-sucking insects (sandflies, blackflies, and culicoides) found to be infected during epidemics, and can also be transmitted between mammals by direct contact (Mead *et al.*, 2004). Additionally, experimental mechanical transmission has been achieved by feeding vesicular stomatitis virus laboratory infected grasshoppers to cattle (Nunamaker *et al.*, 2003; Drolet *et al.*, 2009). However, grasshoppers have never been shown to carry vesiculoviruses in natural infections. Vesiculoviruses have been shown to be transmitted transovarially and, interestingly, also horizontally between infected and non-infected black flies while co-feeding on mammalian hosts (Mead *et al.*, 2000). The latter mechanism of transmission might explain the noticeable absence of viremic mammalian hosts for vesiculoviruses, an unusual feature for an arbovirus (Smith *et al.*, 2011). Black flies have been shown to transmit VSNJV both biologically and mechanically to horses, pigs and cattle, and those species have also been shown to serve as sources of virus to uninfected flies feeding at or near lesion sites (Mead *et al.*, 2004; Howerth *et al.*, 2006; Mead *et al.*, 2009; Smith *et al.*, 2009). These multiple mechanisms of transmission offer a variety of possible life cycles by VSV in nature.

There is evidence of a correlation between ecological conditions and VSV RNA evolution (Rodriguez *et al.*, 1996). Studies in endemic areas have shown that specific genetic lineages of vesiculoviruses exist in specific ecological zones over long periods of time, causing local outbreaks on a regular basis. Some of these endemic areas exist in southern and central Mexico but specific viral lineages from these regions sporadically cause outbreaks in the western US (Rainwater-Lovett *et al.*, 2007). Although the mechanisms of introduction are not clear, these outbreaks usually occur in the spring and summer months and subside after the first frost, presumably when insect vector populations decrease (Rodriguez, 2002). These viral lineages are capable of overwintering and re-emerge in the US in the spring-summer but usually become extinct after one to two years. New outbreaks are associated with new genetic lineages closely related to those circulating in endemic areas (Perez *et al.*, 2010).

The fish *Vesiculovirus* SVCV affects cyprinid (carp, minnows) fish. Naturally occurring SVC infections have been recorded in the common carp (*Cyprinus carpio*), koi carp (*Cyprinus carpio koi*), crucian carp (*Carassius carassius*), sheatfish, (also known as European catfish or wels) (*Silurus glanis*), silver carp (*Hypophthalmichthys molitrix*), bighead carp (*Aristichthys nobilis*), grass carp (white amur) (*Ctenopharyngodon idella*), goldfish (*Carassius auratus*), orfe (*Leuciscus idus*), and tench (*Tinca tinca*).

The virus can be transmitted by ectoparasites such as carp louse (*Argulus foliaceus*) and the leech (*Pisicola geometra*), but waterborne transmission without any vector organism is also effective (Ahne *et al.*, 2002). Like other viruses from fish, SVCV is typically isolated in fish cell cultures at 15–25°C, a lower replication temperature than mammalian rhabdoviruses. The disease patterns are influenced by water temperature, age and condition of the fish, population density and stress factors. The immune status of the fish is also an important factor with both non-specific (interferon) and specific immunity (serum antibodies, cellular immunity). Clinical disease is usually observed at water temperature between 5–18°C and is most severe at temperatures below 10°C, when it is believed the host immune response is suppressed or delayed.

Pathobiology

Among vesiculoviruses the best characterized are those causing VSV in livestock, CHPV in humans, and SVCV in fish. There is also a great deal of information published on various vesiculoviruses and their effect in laboratory animals. We will only cover the pathobiology occurring after the natural infection. Vesicular stomatitis (VS) is a disease of cattle, horses and pigs that occurs in North, Central and South America. Serological evidence of the infection is also found in wild animals and in humans living in endemic regions (Tesh *et al.*, 1969). The clinical signs include the appearance of vesicular lesions in the mouth (tongue, gums) and hooves of animals, as well as on the udders of lactating cattle. In cattle and pigs the clinical signs resemble those of foot-and-mouth disease (FMD), a devastating reportable disease of livestock that causes massive economic losses. When VS outbreaks occur, animal health authorities impose strict animal movement quarantines until FMD is ruled out with laboratory tests. In humans, VSV infection has been documented both naturally during outbreaks in livestock as well as a result of a laboratory exposure (Johnson *et al.*, 1966; Reif *et al.*, 1987). Vesicular lesions are not commonly observed, but rather a febrile, flu-like disease with signs of mild non-fatal encephalitis occurs (Quiroz *et al.*, 1988).

The basic mechanisms of infection leading to vesicular lesions remain unclear. Until recently, most pathogenesis studies have been done in laboratory rodents in which clinical signs are not vesicular in nature, but the infection manifests as a fatal encephalitis depending on host factors such as age, route of inoculation and virus strain. In cattle and pigs the development of lesions is dependent upon the site and the dose of inoculation. Vesicular lesions are observed only when cattle or swine are inoculated intradermally or by scarification at sites where lesions are found (snout, mouth, lip or coronary band of the hoof). Vesicular lesions are not observed after inoculation of the skin of the neck or abdomen, nor by the intranasal or intravenous routes (Scherer *et al.*, 2007). Likewise, clinical disease has been demonstrated in cattle and swine experimentally infected via black flies (*Simulim vittatum*) biting at lesion sites, but not when biting at non-lesion sites (Mead *et al.*, 2009). The viral and host determinants of virulence that define the clinical outcome are not clear. Few genetic determinants of virulence relevant to clinical disease in farm animals have been identified, except for the viral glycoprotein (Martinez *et al.*, 2003). The mechanisms of pathogenesis are not well elucidated, but recent data suggest that viral replication is limited at the non-lesion sites probably due to strong innate responses present at these sites (Reis *et al.*, 2010).

Early pathogenesis studies in cattle indicated that large amounts of virus inoculated in submucosa of the tongue resulted in local vesicular lesions (Cotton, 1926; Ribelin, 1958; Seibold and Sharp, 1960). However, these studies are unlikely representative of natural infection. Most natural infections in cattle living in endemic areas are subclinical and detected only by seroconversion without associated clinical signs (Rodriguez *et al.*, 1990; Vanleeuwen *et al.*, 1995). When clinical disease occurs in cattle in endemic regions, animals show vesicular lesions at specific sites: coronary bands, teats, mouth, or tongue, but rarely at more than one site (Heiny, 1945; Vanleeuwen *et al.*, 1995). The host and viral determinants for this tropism remain unclear. Little is known about the primary site of virus replication, cell types involved in supporting viral growth, or those involved in controlling the infection. In recent studies it was shown that bites by infected insects at lesion sites consistently resulted in lesion formation at much lower doses than the doses necessary to cause lesions via needle injection or scarification (Reis *et al.*, 2010). In pigs, generalized lesions are more common, and direct contact transmission between animals is readily observed in swine infected with VSNJV but not with VSIV (Stallknecht *et al.*, 2004). This is consistent with findings that the glycoprotein is a virulence determinant for VSV, indicating that the VSNJV G significantly increases VSIV virulence in swine (Martinez *et al.*, 2003).

Role of insects in VSV natural transmission cycle

The natural cycle of VSV remains obscure, although there is strong evidence that insects carry the virus in nature. In endemic regions VSV has been found infecting sand flies (*Lutzomyia* spp.), biting gnats (*Culicoides* spp.) and even mosquitoes (*Aedes* spp.) (Tesh *et al.*, 1970; Liu and Zee, 1976; Tesh *et al.*, 1987). During outbreaks in the southwestern United States, VSV has been isolated from *Culicoides* (*Culicoides* spp.), black flies (*Simulium* spp.) (Walton *et al.*, 1987; Francy *et al.*, 1988), mosquitoes and in non-biting insects such as eye gnats and house flies (Sudia *et al.*, 1967; Francy *et al.*, 1988). Members of three insect groups: sand flies (Diptera: Psychodidae), black flies (Diptera: Simuliidae) and *Culicoides* (Diptera: Ceratopogonidae) have experimentally been shown to be biological vectors of VSV (Tesh *et al.*, 1971; Cupp *et al.*, 1992). Biological vectors are those capable not only of supporting viral replication but also of transmitting the virus to susceptible hosts. To date, only black flies (*S. vittatum*) have been shown to be capable of VSV transmission resulting in clinical disease in livestock species including pigs, cattle and horses (Mead *et al.*, 2004; Howerth *et al.*, 2006; Mead *et al.*, 2009). *Culicoides* sonorensis transmitted VSNJV to cattle but the infection was subclinical and resulted only in seroconversion (Perez de Leon and Tabachnick, 2006). However, in that work insect feeding was done at non-lesion sites which might explain the result. Other insects such as sand flies (*Lutzomyia shannoni*) have been shown to be capable of replicating VSV and transmitting the virus to laboratory mice (Comer *et al.*, 1991). A recent report showed that grasshoppers infected either by needle injection or by feeding on grass painted with VSNJV could transmit VSV to cattle fed with these infected insects. Although it is possible that they might play a role in VSV transmission, there are no reports of infected grasshoppers during VSV outbreaks (Nunamaker *et al.*, 2003; Drolet *et al.*, 2009). An interesting mechanism of VSNJV transmission between infected and non-infected black flies while co-feeding in naïve mice and cattle has been recently reported (Mead *et al.*, 2000; Smith *et al.*, 2009). This is the only example of horizontal transmission of an animal arbovirus by insects. Transmission did not require direct contact between infected and non infected flies and in some cases concurrent co-feeding was not necessary for transmission to occur. These authors also showed that vesicular lesions can serve as a source of insect infection (Smith *et al.*, 2009).

It has been previously documented that bites from arthropods can modulate vertebrate host functions by several mechanisms including modulation of the immune response, vasodilation and other physiological effects (Schoeler and Wikel, 2001). Arthropod salivary gland extracts can augment the growth of VSV in vitro in primary cell cultures and in laboratory mice, presumably by interfering with the host innate immune response (Limesand *et al.*, 2000, 2003). Although few studies exist in animals, tissues from cattle infected by scarification showed higher number of immune cells positive for MHC-II antigens than cattle infected by fly bite suggesting a down regulation of inflammatory response by insect factors. These data, combined with the fact that fly bite induced vesicular lesions with lower VSV titres than those required to see vesicular lesions after scarification, suggest that insect bites can potentiate VSV infection (Reis *et al.*, 2010).

Epidemiology and control

VS is endemic from southern Mexico to northern South America with outbreaks caused primarily by VSNJV, reported every year, mostly in cattle, but also in horses and pigs (Rodriguez, 2002). In Mexico disease outbreaks are more common in the central and southern states with less frequent incursions into northern states (Arroyo *et al.*, 2011). In the United States, VS outbreaks occur at approximately 8- to 10-year intervals in cycles lasting one to three years in the south-western states (Texas, Arizona, New Mexico, Utah, Colorado), with the last three outbreak cycles occurring in 1984/85, 1995/98 and 2004/06 (Rainwater-Lovett *et al.*, 2007). Phylogenetic and spatial analyses of the viruses causing the 2004–2006 outbreak cycles showed that the same viral lineage overwintered in the US and that the virus was not re-introduced each subsequent year (Perez *et al.*, 2010). Two additional small outbreaks of VSNJV confirmed only by serological tests were reported in Texas and New Mexico in 2009 and in Arizona in 2010 (http://www.aphis.usda.gov/vs/nahss/equine/vsv/vsv2010.htm). It is interesting to point out that large VSV outbreaks were reported in 2008 and 2009 in northern Mexico near the US border (Arroyo *et al.*, 2011).

Phylogenetic analyses show that each outbreak cycle in the US is caused by a single viral lineage that is genetically closer to viruses circulating concurrently in endemic areas of Mexico than to viruses causing previous outbreaks in the US (Rainwater-Lovett *et al.*, 2007). New viral outbreaks in the US typically start in the Mexico border states of Texas, Arizona or New Mexico in early spring and progress north following valleys and rivers along the Rocky Mountain range reaching as far north as Montana or Nebraska and usually end after the first frosts of the winter (Rodriguez, 2002). This pattern of VSV occurrence in the US is consistent with vector transmission rather than animal movement.

Control measures of vesiculoviruses depend on the knowledge of the viral ecology, transmission and epidemiology. In the case of VSV, data indicates that the virus persists in endemic regions in a stable manner, causing disease in a small proportion of the infected animals. The problem arises when virus incursions occur into non-endemic areas, spreading rapidly and affecting a large number of animals. Control measures include stop of animal movement (quarantine), separation of infected and non infected animals and recommendations to control exposure to biting insects (Hurd *et al.*, 1999). Vaccines to VSV have been produced and tested and show that immunization with a killed virus vaccine protects against homologous challenge with a virus of the same serotype (House *et al.*, 2003). However, implementing vaccination programmes for a sporadic disease has had limited acceptance and currently is not utilized even in endemic regions.

In the case of CHPV, several vaccines have been recently developed including subunit and inactivated vaccines (Venkateswarlu and Arankalle, 2009, 2010; Jadi *et al.*, 2011). However, routine vaccination is not commonly practiced. Likewise, SVCV experimental vaccines have been reported by routine vaccination is not practiced (Emmenegger and Kurath, 2008).

References

Ahne, W., Bjorklund, H.V., Essbauer, S., Fijan, N., Kurath, G., and Winton, J.R. (2002). Spring viremia of carp (SVC). Dis. Aquat. Organ. *52*, 261–272.

Arroyo, M., Perez, A.M., and Rodriguez, L.L. (2011). Characterization of the temporal and spatial distribution and reproductive ratio of vesicular stomatitis outbreaks in Mexico in 2008. Am. J. Vet. Res. *72*, 233–238.

Ball, L.A., and White, C.N. (1976). Order of transcription of genes of vesicular stomatitis virus. Proc. Natl. Acad. Sci. U.S.A. *73*, 442–446.

Baltimore, D., Huang, A.S., and Stampfer, M. (1970). Ribonucleic acid synthesis of vesicular stomatitis virus, II. An RNA polymerase in the virion. Proc. Natl. Acad. Sci. U.S.A. *66*, 572–576.

Barr, J.N., Whelan, S.P., and Wertz, G.W. (2002). Transcriptional control of the RNA-dependent RNA polymerase of vesicular stomatitis virus. Biochim. Biophys. Acta. *1577*, 337–353.

Berge, T., ed. (1975). International Catalogue of Arboviruses Including Certain Other Viruses of Vertebrates (Public Health Service, Athens, GA).

Bhatt, P.N., and Rodrigues, F.M. (1967). Chandipura: a new Arbovirus isolated in India from patients with febrile illness. Indian J. Med. Res. *55*, 1295–1305.

Bourhy, H., Cowley, J.A., Larrous, F., Holmes, E.C., and Walker, P.J. (2005). Phylogenetic relationships among rhabdoviruses inferred using the L polymerase gene. J. Gen. Virol. *86*, 2849–2858.

Brooksby, J.B. (1948). Vesicular stomatitis and foot-and-mouth disease differentiation by complement-fixation. Proc. Soc. Exp. Biol. Med. *67*, 254–258.

Cartwright, B., and Brown, F. (1972). Serological relationships between different strains of vesicular stomatis virus. J. Gen. Virol. *16*, 391–398.

Cartwright, B., Talbot, P., and Brown, F. (1970). The proteins of biologically active sub-units of vesicular stomatitis virus. J. Gen. Virol. *7*, 267–272.

Chadha, M.S., Arankalle, V.A., Jadi, R.S., Joshi, M.V., Thakare, J.P., Mahadev, P.V., and Mishra, A.C. (2005). An outbreak of Chandipura virus encephalitis in the eastern districts of Gujarat state, India. Am. J. Trop. Med. Hyg. *73*, 566–570.

Comer, J.A., Stallknecht, D.E., Corn, J.L., and Nettles, V.F. (1991). *Lutzomyia shannoni* (Diptera: Psychodidae): a biological vector of the New Jersey serotype of vesicular stomatitis virus on Ossabaw Island, Georgia. Parassitologia *33 Suppl.*, 151–158.

Cotton, W.E. (1927). Vesicular stomatitis. Vet. Med. *22*, 169–175.

Cupp, E.W., Mare, C.J., Cupp, M.S., and Ramberg, F.B. (1992). Biological transmission of vesicular stomatitis virus (New Jersey) by *Simulium vittatum* (Diptera: Simuliidae). J. Med. Entomol. *29*, 137–140.

Dacheux, L., Berthet, N., Dissard, G., Holmes, E.C., Delmas, O., Larrous, F., Guigon, G., Dickinson, P., Faye, O., Sall, A.A., *et al.* (2010). Application of broad-spectrum resequencing microarray for genotyping rhabdoviruses. J. Virol. *84*, 9557–9574.

Davis, N.L., and Wertz, G.W. (1982). Synthesis of vesicular stomatitis virus negative-strand RNA in vitro: dependence on viral protein synthesis. J. Virol. *41*, 821–832.

Dietzgen, R.G., Calisher, C.H., Kurath, G., Kuzmin, I.V., Rodriguez, L.L., Stone, D.M., Tesh, R.B., Tordo, N., Walker, P.J., Wetzel, T., *et al.* (2011). Rhabdoviridae. In Virus Taxonomy: Classification and Nomenclature of Viruses, King, A.M.Q., Adams, M.J., Carstens, E.B., and Lefkowitz, E.J., eds (Elsevier, San Diego, CA), pp. 686–713.

Drolet, B.S., Stuart, M.A., and Derner, J.D. (2009). Infection of Melanoplus sanguinipes grasshoppers following ingestion of rangeland plant species harboring vesicular stomatitis virus. Appl. Environ. Microbiol. *75*, 3029–3033.

Emmenegger, E.J., and Kurath, G. (2008). DNA vaccine protects ornamental koi (Cyprinus carpio koi) against North American spring viremia of carp virus. Vaccine *26*, 6415–6421.

Federer, K.E., Burrows, R., and Brooksby, J.B. (1967). Vesicular stomatitis virus – the relationship between some strains of the Indiana serotype. Res. Vet. Sci. *8*, 103–117.

Francy, D.B., Moore, C.G., Smith, G.C., Jakob, W.L., Taylor, S.A., and Calisher, C.H. (1988). Epizootic vesicular stomatitis in Colorado, 1982: isolation of virus from insects collected along the northern Colorado Rocky Mountain Front Range. J. Med. Entomol. *25*, 343–347.

Gaidamovich, S., Altukhova, L.M., Obukhova, V.R., Ponirovskii, E.N., and Sadykov, V.G. (1980). Isolation of the Isfahan virus in Turkmenia. [In Russian.] Vopr. Virusol. *4*, 618–620.

Ge, P., Tsao, J., Schein, S., Green, T.J., Luo, M., and Zhou, Z.H. (2010). Cryo-EM model of the bullet-shaped vesicular stomatitis virus. Science *327*, 689–693.

Hanson, R.P. (1952). The Natural History of Vesicular Stomatitis. Bacteriol. Rev. *16*, 179–204.

Hassler, D., Braun, R., and Schwarz, T.F. (2003). Deadly encephalitis epidemic in India caused by the Chandipura virus. [In German] Dtsch. Med. Wochenschr. *128*, 1817.

House, J.A., House, C., Dubourget, P., and Lombard, M. (2003). Protective immunity in cattle vaccinated with a commercial scale, inactivated, bivalent vesicular stomatitis vaccine. Vaccine *21*, 1932–1937.

Howerth, E.W., Mead, D.G., Mueller, P.O., Duncan, L., Murphy, M.D., and Stallknecht, D.E. (2006). Experimental vesicular stomatitis virus infection in horses: effect of route of inoculation and virus serotype. Vet. Pathol. *43*, 943–955.

Hurd, H.S., McCluskey, B.J., and Mumford, E.L. (1999). Management factors affecting the risk for vesicular stomatitis in livestock operations in the western United States. J. Am. Vet. Med. Assoc. *215*, 1263–1268.

Jadi, R.S., Sudeep, A.B., Barde, P.V., Arankalle, V.A., and Mishra, A.C. (2011). Development of an inactivated candidate vaccine against Chandipura virus (Rhabdoviridae: Vesiculovirus). Vaccine *29*, 4613–4617.

Jenney, E.W., Mott, L.O., and Traub, E. (1958). Serological studies with the virus of vesicular stomatitis. I. Typing of vesicular stomatitis viruses by complement fixation. Am. J. Vet. Res. *19*, 993–998.

John, T.J. (2004). Chandipura virus, encephalitis, and epidemic brain attack in India. Lancet *364*, 2175; author reply 2175–2176.

Johnson, K.M., Vogel, J.E., and Peralta, P.H. (1966). Clinical and serological response to laboratory-acquired human infection by Indiana type vesicular stomatitis virus (VSV). Am. J. Trop. Med. Hyg. *15*, 244–246.

Kang, C.Y., and Prevec, L. (1970). Proteins of vesicular stomatitis virus. II. Immunological comparisons of viral antigens. J. Virol. *6*, 20–27.

Kretzschmar, E., Peluso, R., Schnell, M.J., Whitt, M.A., and Rose, J.K. (1996). Normal replication of vesicular stomatitis virus without C proteins. Virology *216*, 309–316.

Limesand, K.H., Higgs, S., Pearson, L.D., and Beaty, B.J. (2000). Potentiation of vesicular stomatitis New Jersey virus infection in mice by mosquito saliva. Parasite Immunol. *22*, 461–467.

Limesand, K.H., Higgs, S., Pearson, L.D., and Beaty, B.J. (2003). Effect of mosquito salivary gland treatment on vesicular stomatitis New Jersey virus replication and interferon alpha/beta expression in vitro. J. Med. Entomol. *40*, 199–205.

Liu, I.K., and Zee, Y.C. (1976). The pathogenesis of vesicular stomatitis virus, serotype Indiana, in *Aedes* aegypti mosquitoes. I. Intrathoracic injection. Am. J. Trop. Med. Hyg. *25*, 177–185.

McClellan, G.B. (1864). Report of Major Gen. George B. McClellan on the organization of the Army of the Potomac and of its campaigns in Virginia and Maryland, 26 July 1861 to 7 November 1862. In The Times Steam Printing House (Chicago, IL).

Mackett, M., Yilma, T., Rose, J.K., and Moss, B. (1985). Vaccinia virus recombinants: expression of VSV genes and protective immunization of mice and cattle. Science *227*, 433–435.

Martinez, I., and Wertz, G.W. (2005). Biological differences between vesicular stomatitis virus Indiana and New Jersey serotype glycoproteins: identification of amino acid residues modulating pH-dependent infectivity. J. Virol. *79*, 3578–3585.

Martinez, I., Rodriguez, L.L., Jimenez, C., Pauszek, S.J., and Wertz, G.W. (2003). Vesicular stomatitis virus glycoprotein is a determinant of pathogenesis in swine, a natural host. J. Virol. *77*, 8039–8047.

Martinez, I., Barrera, J.C., Rodriguez, L.L., and Wertz, G.W. (2004). Recombinant vesicular stomatitis (Indiana) virus expressing New Jersey and Indiana glycoproteins induces neutralizing antibodies to each serotype in swine, a natural host. Vaccine *22*, 4035–4043.

Mead, D.G., Ramberg, F.B., Besselsen, D.G., and Mare, C.J. (2000). Transmission of vesicular stomatitis virus from infected to noninfected black flies co-feeding on nonviremic deer mice. Science *287*, 485–487.

Mead, D.G., Gray, E.W., Noblet, R., Murphy, M.D., Howerth, E.W., and Stallknecht, D.E. (2004). Biological transmission of vesicular stomatitis virus (New Jersey serotype) by *Simulium vittatum* (Diptera: Simuliidae) to domestic swine (*Sus scrofa*). J. Med. Entomol. *41*, 78–82.

Mead, D.G., Lovett, K.R., Murphy, M.D., Pauszek, S.J., Smoliga, G., Gray, E.W., Noblet, R., Overmyer, J., and Rodriguez, L.L. (2009). Experimental transmission of vesicular stomatitis New Jersey virus from *Simulium vittatum* to cattle: clinical outcome is influenced by site of insect feeding. J. Med. Entomol. *46*, 866–872.

Narasimha Rao, S., Wairagkar, N.S., Murali Mohan, V., Khetan, M., and Somarathi, S. (2008). BrainStem encephalitis associated with chandipura in Andhra Pradesh outbreak. J. Trop. Pediatr. *54*, 25–30.

Nunamaker, R.A., Lockwood, J.A., Stith, C.E., Campbell, C.L., Schell, S.P., Drolet, B.S., Wilson, W.C., White, D.M., and Letchworth, G.J. (2003). Grasshoppers (Orthoptera: Acrididae) could serve as reservoirs and vectors of vesicular stomatitis virus. J. Med. Entomol. *40*, 957–963.

Pauszek, S.J., Allende, R., and Rodriguez, L.L. (2008). Characterization of the full-length genomic sequences of vesicular stomatitis Cocal and Alagoas viruses. Arch. Virol. *153*, 1353–1357.

Pauszek, S.J., Barrera, J.D., Goldberg, T., Allende, R., and Rodriguez, L.L. (2011). Genetic and antigenic relationships of vesicular stomatitis viruses from South America. Arch. Virol. *156*, 1961–1966.

Perez, A.M., Pauszek, S.J., Jimenez, D., Kelley, W.N., Whedbee, Z., and Rodriguez, L.L. (2010). Spatial and phylogenetic analysis of vesicular stomatitis virus over-wintering in the United States. Prev. Vet. Med. *93*, 258–264.

Perez de Leon, A.A., and Tabachnick, W.J. (2006). Transmission of vesicular stomatitis New Jersey virus to cattle by the biting midge *Culicoides* sonorensis (Diptera: Ceratopogonidae). J. Med. Entomol. *43*, 323–329.

Quiroz, E., Moreno, N., Peralta, P.H., and Tesh, R.B. (1988). A human case of encephalitis associated with vesicular stomatitis virus (Indiana serotype) infection. Am. J. Trop. Med. Hyg. *39*, 312–314.

Rainwater-Lovett, K., Pauszek, S.J., Kelley, W.N., and Rodriguez, L.L. (2007). Molecular epidemiology of vesicular stomatitis New Jersey virus from the 2004–2005 US outbreak indicates a common origin with Mexican strains. J. Gen. Virol. *88*, 2042–2051.

Rao, B.L., Basu, A., Wairagkar, N.S., Gore, M.M., Arankalle, V.A., Thakare, J.P., Jadi, R.S., Rao, K.A., and Mishra, A.C. (2004). A large outbreak of acute encephalitis with high fatality rate in children in Andhra Pradesh, India, in 2003, associated with Chandipura virus. Lancet *364*, 869–874.

Rao, T.R., Singh, K.R., Dhanda, V., and Bhatt, P.N. (1967). Experimental transmission of Chandipura virus by mosquitoes. Indian J. Med. Res. *55*, 1306–1310.

Reif, J.S., Webb, P.A., Monath, T.P., Emerson, J.K., Poland, J.D., Kemp, G.E., and Cholas, G. (1987). Epizootic vesicular stomatitis in Colorado, 1982: infection in occupational risk groups. Am. J. Trop. Med. Hyg. *36*, 177–182.

Reis, J.L., Jr, Rodriguez, L.L., Mead, D.G., Smoliga, G., and Brown, C.C. (2011). Lesion Development and Replication Kinetics During Early Infection in Cattle Inoculated With Vesicular Stomatitis New Jersey Virus Via Scarification and Black Fly (*Simulium vittatum*) Bite. Vet. Pathol. *48*, 547–57.

Rodriguez, L.L. (2002). Emergence and re-emergence of vesicular stomatitis in the United States. Virus Res. *85*, 211–219.

Rodriguez, L.L., Vernon, S., Morales, A.I., and Letchworth, G.J. (1990). Serological monitoring of vesicular stomatitis New Jersey virus in enzootic regions of Costa Rica. Am. J. Trop. Med. Hyg. *42*, 272–281.

Rodriguez, L.L., Fitch, W.M., and Nichol, S.T. (1996). Ecological factors rather than temporal factors dominate the evolution of vesicular stomatitis virus. Proc. Natl. Acad. Sci. U.S.A. *93*, 13030–13035.

Rodriguez, L.L., Pauszek, S.J., Bunch, T.A., and Schumann, K.R. (2002). Full-length genome analysis of natural isolates of vesicular stomatitis virus (Indiana 1 serotype) from North, Central and South America. J. Gen. Virol. *83*, 2475–2483.

Rose, J.K., Welch, W.J., Sefton, B.M., Esch, F.S., and Ling, N.C. (1980). Vesicular stomatitis virus glycoprotein is anchored in the viral membrane by a hydrophobic domain near the COOH terminus. Proc. Natl. Acad. Sci. U.S.A. *77*, 3884–3888.

Saidi, S., Tesh, R., Javadian, E., Sahabi, Z., and Nadim, A. (1977). Studies on the epidemiology of sandfly fever in Iran. II. The prevalence of human and animal infection with five phlebotomus fever virus serotypes in Isfahan province. Am. J. Trop. Med. Hyg. *26*, 288–293.

Scherer, C.F., O'Donnell, V., Golde, W.T., Gregg, D., Estes, D.M., and Rodriguez, L.L. (2007). Vesicular stomatitis New Jersey virus (VSNJV) infects keratinocytes and is restricted to lesion sites and local lymph nodes in the bovine, a natural host. Vet. Res. *38*, 375–390.

Schoeler, G.B., and Wikel, S.K. (2001). Modulation of host immunity by haematophagous arthropods. Ann. Trop. Med. Parasitol. *95*, 755–771.

Smith, P.F., Howerth, E.W., Carter, D., Gray, E.W., Noblet, R., and Mead, D.G. (2009). Mechanical transmission of vesicular stomatitis New Jersey virus by *Simulium vittatum* (Diptera: Simuliidae) to domestic swine (*Sus scrofa*). J. Med. Entomol. *46*, 1537–1540.

Smith, P.F., Howerth, E.W., Carter, D., Gray, E.W., Noblet, R., Smoliga, G., Rodriguez, L.L., and Mead, D.G. (2011). Domestic cattle as a non-conventional amplifying host of vesicular stomatitis New Jersey virus. Med. Vet. Entomol. *25*, 184–191.

Spiropoulou, C.F., and Nichol, S.T. (1993). A small highly basic protein is encoded in overlapping frame within the P gene of vesicular stomatitis virus. J. Virol. *67*, 3103–3110.

Stallknecht, D.E., Greer, J.B., Murphy, M.D., Mead, D.G., and Howerth, E.W. (2004). Effect of strain and serotype of vesicular stomatitis virus on viral shedding, vesicular lesion development, and contact transmission in pigs. Am. J. Vet. Res. *65*, 1233–1239.

Sudia, W.D., Fields, B.N., and Calisher, C.H. (1967). The isolation of vesiculay stomatitis virus (Indiana strain) and other viruses from mosquitoes in New Mexico, 1965. Am. J. Epidemiol. *86*, 598–602.

Tandale, B.V., Tikute, S.S., Arankalle, V.A., Sathe, P.S., Joshi, M.V., Ranadive, S.N., Kanojia, P.C., Eshwarachary, D., Kumarswamy, M., and Mishra, A.C. (2008). Chandipura virus: a major cause of acute encephalitis in children in North Telangana, Andhra Pradesh, India. J. Med. Virol. *80*, 118–124.

Tesh, R., Saidi, S., Javadian, E., Loh, P., and Nadim, A. (1977). Isfahan virus, a new *Vesiculovirus* infecting humans, gerbils, and sandflies in Iran. Am. J. Trop. Med. Hyg. *26*, 299–306.

Tesh, R.B., Peralta, P.H., and Johnson, K.M. (1969). Ecologic studies of vesicular stomatitis virus. I. Prevalence of infection among animals and humans living in an area of endemic VSV activity. Am. J. Epidemiol. *90*, 255–261.

Tesh, R.B., Peralta, P.H., and Johnson, K.M. (1970). Ecologic studies of vesicular stomatitis virus. II. Results of experimental infection in Panamanian wild animals. Am. J. Epidemiol. *91*, 216–224.

Tesh, R.B., Chaniotis, B.N., and Johnson, K.M. (1971). Vesicular stomatitis virus, Indiana serotype: multiplication in and transmission by experimentally infected phlebotomine sandflies (*Lutzomyia trapidoi*). Am. J. Epidemiol. 93, 491–495.

Tesh, R.B., Boshell, J., Modi, G.B., Morales, A., Young, D.G., Corredor, A., Ferro de Carrasquilla, C., de Rodriguez, C., Walters, L.L., and Gaitan, M.O. (1987). Natural infection of humans, animals, and phlebotomine sand flies with the Alagoas serotype of vesicular stomatitis virus in Colombia. Am. J. Trop. Med. Hyg. *36*, 653–661.

Vanleeuwen, J.A., Rodriguez, L.L., and Waltner-Toews, D. (1995). Cow, farm, and ecologic risk factors of clinical vesicular stomatitis on Costa Rican dairy farms. Am. J. Trop. Med. Hyg. *53*, 342–350.

Venkateswarlu, C.H., and Arankalle, V.A. (2009). Recombinant glycoprotein based vaccine for Chandipura virus infection. Vaccine *27*, 2845–2850.

Venkateswarlu, C.H., and Arankalle, V.A. (2010). Evaluation of the immunogenicity of a recombinant glycoprotein-based Chandipura vaccine in combination with commercially available DPT vaccine. Vaccine *28*, 1463–1467.

Walker, P.J., and Winton, J.R. (2010). Emerging viral diseases of fish and shrimp. Vet. Res. *41*, 51.

Walton, T.E., Webb, P.A., Kramer, W.L., Smith, G.C., Davis, T., Holbrook, F.R., Moore, C.G., Schiefer, T.J., Jones, R.H., and Janney, G.C. (1987). Epizootic vesicular stomatitis in Colorado, 1982: epidemiologic and entomologic studies. Am. J. Trop. Med. Hyg. *36*, 166–176.

Wertz, G.W. (1983). Replication of vesicular stomatitis virus defective interfering particle RNA in vitro: transition from synthesis of defective interfering leader RNA to synthesis of full-length defective interfering RNA. J. Virol. *46*, 513–522.

Wilks, C.R., and House, J.A. (1984). Susceptibility of various animals to the *Vesiculovirus* Piry. J. Hyg. (Lond) *93*, 147–155.

5 Genus *Lyssavirus*

Ivan V. Kuzmin and Noel Tordo

Abstract

Viruses from the genus *Lyssavirus* cause rabies: an acute progressive encephalomyelitis, with the highest fatality rate among conventional infectious diseases. Despite millennia of recognition and centuries of control attempts, rabies still kills over 55,000 people in the world every year. In contrast to the majority of rhabdoviruses, lyssaviruses do not require an arthropod vector, but rather are transmitted between mammals via bites. Although the domestic dog is still the major rabies reservoir, causing over 90% of human cases, lyssaviruses of different species broadly circulate in wildlife. All these facets clearly indicate that rabies is a significant problem for public health, veterinary and wildlife services, from the standpoint of the 'one health' concept. The present chapter describes recent advances in *Lyssavirus* taxonomy, phylogeny, pathobiology, and epidemiology. Other important topics, such as advances, challenges and perspectives of rabies prevention and control, are covered in Chapters 13 and 14.

Introduction

The genus *Lyssavirus* comprises a group of antigenically and genetically related viruses with a unique pathobiology. They specifically target the mammalian nervous system and cause rabies: an acute progressive encephalitis. This disease has been known since ancient times (Theodorides, 1986; Baer *et al.*, 1996). The suggestion of a relation to dog bites was mentioned in the pre-Mosaic Eshnunna Code of Mesopotamia, circa the twenty-third century BC: 'If a dog is mad and the authorities have brought the fact of the knowledge of its owner; if the dog is not kept in, and it bites a man and causes his death, then the owner shall pay two-thirds of a mina of silver, less than one-third for a slave' (Theodorides, 1986). In *The Iliad* (700 BC), Hector is compared to a rabid dog. Chinese scholars warned of the danger of rabid dogs in 500 BC and proposed treatments for bites, and Aristotle (fourth century BC) correctly associated the disease with mammals. In Rome, dogs were presumed more effective at transmitting rabies during the July period, 'dies caniculares', owing to the bad influence of Sirius, the dog constellation. Cordamus guessed that a poison (i.e. a 'virus') was present in dog saliva. Another Roman, Celsus, described clinical aspects of rabies in the first century AD: 'The patient is tortured at the same time by thirst and by invincible repulsion towards water' (Lyles and Rupprecht, 2007). Despite clear evidence that a disease comparable to rabies has been known to humankind for ages, recent estimate based on coalescent analysis of viral genes demonstrated that the age of the contemporary rabies virus lineages is close to a millennium (Badrane and Tordo, 2001; Hughes *et al.*, 2005; Nadin-Davis and Real, 2011).

In the New World, the occurrence of rabies is less documented because of a dearth of records before European colonization (Vos *et al.*, 2011). However, we can suspect that it was present before Columbus's arrival in the fifteenth century. Not long after discovery of the Americas, the bishop Petrus Martyr-Anglerius wrote in his *De Rebus Oceanicus et de Orbi Novi Decades Octo*, 'In several places bats not much smaller than turtle doves used to fly at them [the Spanish sailors and soldiers] in the early evening with brutal fury and with their venomous bites brought those injured to madness ... [and] bats ... come in from the marshes on the river and attack our men with deadly bite' (Koprowski, 1959). This may be the first description of rabies transmission by vampire bats although the disease symptoms were not clearly described (Vos *et al.*, 2011). Nevertheless, the true diagnosis of rabies in livestock was first made by the identification of Negri bodies in the brain of cattle during the outbreak of a previously undiagnosed disease in Santa Catarina, in southern Brazil in 1911 (Carini, 1911). The disease was reproduced for identification purposes by inoculation of rabbits. Wild animals were suspected as the virus reservoir, because attempts to control rabies in dogs did not reduce the number of rabies cases in cattle. Since vampire bats were observed biting cows, their involvement in rabies transmission to cattle was proposed and later confirmed by Haupt and Rehaag (1921).

Although bats are recognized today as the major reservoir of lyssaviruses, rabies in non-haematophagous bats was not recognized until the 1950s in the Americas, Eurasia and Africa, and until the 1990s in Australia. As to carnivores, and particularly to dogs which constitute the primary vector of rabies virus in the Old World, they were not recognized as rabies reservoirs in the Americas before the seventeenth century. The first presumably rabid dogs were documented in Haiti in 1688 by Jean Baptiste Du Tertre in *Historia General de las Antillas* (Vos *et al.*, 2011). Rabid dogs, foxes and wolves were documented in Mexico in 1709 (Koprowski, 1995; Lyles and Rupprecht, 2007).

Despite the long history of recognition, investigations and control attempts, rabies still continues to kill over 55,000 people in the world every year (Knobel *et al.*, 2005). However, significant progress has been achieved in prophylaxis of the disease (Tordo *et al.*, 2005). The first human rabies vaccine, developed by Louis Pasteur in 1885, from animal nervous tissue, was less immunogenic and frequently produced adverse effects, including severe disease and disability. Further advances in vaccine development improved the quality of biologics, and today the nervous tissue rabies vaccines have been mainly replaced with highly potent and safe vaccines, produced in cell culture (Rupprecht *et al.*, 2009). In domestic animals and wildlife, classical vaccines used parenterally have been complemented by a panel of recombinant and attenuated vaccines administrable by the oral route. Parenteral vaccination has led to elimination of dog-mediated rabies in west European countries, the USA and Canada, and significantly reduced the burden from dog rabies in Latin America (Belotto *et al.*, 2005). Oral vaccination has eliminated fox rabies from Western Europe (Pastoret and Brochier, 1999), and it is now broadly used for several wild rabies reservoir species in North America (Rupprecht *et al.*, 2008). However, rabies biologics and programmes for rabies elimination are costly, and not affordable in developing countries, particularly in Africa and Asia, where the burden of the disease is maximum (Knobel *et al.*, 2005). Moreover, increased surveillance with the use of powerful modern tools of virus identification continuously leads to discoveries of new lyssaviruses. Currently available rabies biologics do not provide reliable protection against some of these divergent lyssaviruses (Bahloul *et al.*, 1998; Badrane *et al.*, 2001; Hanlon *et al.*, 2005; Kuzmin *et al.*, 2005). Such problems require the development of novel, more affordable biologics with a broader spectrum, implementation of non-traditional strategies for rabies control, and other measures, are described in Chapters 13 and 14 of this book. In the present chapter, we focus on basic characteristics of lyssaviruses, their phylogeny, pathobiology, and epidemiology.

Taxonomy

Lyssavirus taxonomy is complex due to a combination of historical precedents based on serological methods, current knowledge based on phylogenetic analyses, and ecology of these viruses. Until the 1950s, it was believed that rabies virus was unique. The discoveries of serologically related viruses in Nigeria, Lagos bat virus from a Pteropodid bat (Boulger and Porterfield, 1958) and Mokola virus from a shrew (Shope *et al.*, 1970) and then from a child (Familusi and Moore, 1972) demonstrated that the structure of this virus group was more complex, and the terms 'rabies-related viruses' and 'rabies serogroup' were introduced (Shope *et al.*, 1970). Another serologically related virus, Duvenhage virus, was isolated from a man, who died of rabies after a bite of an insectivorous bat in 1970 in South Africa (Meredith *et al.*, 1971), inaugurating a fourth serotype (Schneider *et al.*, 1973). As viruses regularly isolated from bats in Europe since the 1950s were related serologically to Duvenhage virus, they were initially included in the Duvenhage serotype (Schneider, 1982; Schneider *et al.*, 1985).

Later, application of monoclonal antibodies (MAbs) refined the classification of the 'rabies serogroup' (Wiktor and Koprowski, 1978): European bat lyssaviruses were not only distinguished from the African Duvenhage virus (Dietzschold *et al.*, 1988), but separated into two distinct serotypes (Bourhy *et al.*, 1992), temporally termed 'biotypes' (King *et al.*, 1990). This differentiation was later supported by gene sequencing and phylogenetic analysis (Bourhy *et al.*, 1993). Extensive phylogenetic studies on the diversity of rabies-related viruses led to the creation of the novel operational term 'genotype', which has been broadly used in the scientific literature for over a decade. New genotypes were discovered, and quantitative criteria for their demarcation were proposed (Bourhy *et al.*, 1993; Fraser *et al.*, 1996; Kuzmin *et al.*, 2005).

To accommodate the growing variety of 'rabies-related' viruses, the genus *Lyssavirus* was established under the auspices of the International Committee on the Taxonomy of Viruses (ICTV). The name of the genus originated from Greek mythology: Lyssa (Λυσσα) was a goddess, or spirit of rage, fury, raging madness, and frenzy (Theodorides, 1986). The existing 'genotypes' served as a basis for lyssavirus taxonomy but were refined to satisfy the official ICTV taxonomy rules, which operate with more complex entities such as viral species. The present taxonomy of lyssaviruses includes 12 species (Table 5.1). One representative, Shimoni bat virus (Kuzmin *et al.*, 2010), provisionally included in the ninth ICTV report (Dietzgen *et al.*, 2011), was recently ratified by ICTV as a species. As a further potential extension of the genus, a novel Bokeloh bat lyssavirus was recently isolated from an insectivorous bat in Germany (Freuling *et al.*, 2011). Another divergent lyssavirus, tentatively named Ikoma lyssavirus, was detected in an African civet (*Civettictis civetta*) from Tanzania (Marston *et al.*, 2012).

Previously, even two arthropod-borne rhabdoviruses isolated in Africa, Obodhiang and kotonkan viruses, were included in the rabies serogroup, based on their limited antigenic cross-reactivity with Mokola virus, and to a significantly lesser extent with rabies virus. It was speculated that the proposed evolutionary pathway for the rabies-related viruses included arthropod-borne Obodhiang and kotonkan viruses as progenitors, through Mokola virus, which had the ability to replicate in insect cell lines (Buckley, 1975); however, no insect-derived isolates have been obtained from nature, to other highly neurotropic mammalian viruses, such as Lagos bat, Duvenhage and rabies viruses (Shope, 1982). Nevertheless, further rigorous serological analysis (Calisher *et al.*, 1989) and gene sequencing (Kuzmin *et al.*, 2006) demonstrated that Obodhiang and kotonkan viruses instead belong to the ephemeroviruses. The third arthropod-borne rhabdovirus occasionally included in the genus during a certain period of time as a tentative species, was Rochambeau virus. However, based on gene sequences, it was also removed from the genus, as it did not demonstrate any phylogenetic relatedness to lyssaviruses (Kuzmin *et al.*, 2006).

In general, demarcation criteria for lyssavirus species include:

1. Genetic distances, with the threshold of 80–82% nucleotide identity for the complete N gene or 80–81% nucleotide identity for the concatenated coding regions of N + P + M + G + L genes. Globally, all isolates belonging to the same species have higher identity values than the threshold, except the viruses currently included into the *Lagos bat virus* species. However, as these viruses are segregated into a monophyletic cluster in the majority of phylogenetic reconstructions, and in the absence of other demarcation characters, there is currently no plan to subdivide *Lagos bat virus* into several species, as has been suggested by several authors (Delmas *et al.*, 2008; Markotter *et al.*, 2008a).
2. Topology and consistency of phylogenetic trees, obtained with various evolutionary models.

Table 5.1 Viruses currently included in the genus *Lyssavirus*

Recognised and proposed species	Natural host	Geographical range	Comments
Rabies virus (type species)	Bats (Chiroptera) of multiple species, terrestrial mammals (predominantly Carnivora)	Terrestrial mammals – worldwide (except Australia, Antarctica and several insular territories); bats – New World only	Responsible for the vast majority of human rabies cases in the world. All currently available human and veterinary vaccine strains originate from this species
Australian bat lyssavirus	Pteropodid bats (at least four species of *Pteropus* genus) and insectivorous bats (*Saccolaimus albiventris*)	Australia (perhaps, with several surrounding islands)	Given limited surveillance, host range among insectivorous bats may be greater. Two human cases documented
European bat lyssavirus, type 1	Insectivorous bats (predominantly *Eptesicus serotinus*)	The major part of Europe, from Spain to the Ukraine	Given the limited surveillance in eastern Europe and Asia, may be distributed more broadly, along the reservoir species range. Spill-over infections in wild and companion animals, as well as a human case have been documented
European bat lyssavirus, type 2	Insectivorous bats (predominantly *Myotis daubentonii* and *M. dasycneme*)	North-western Europe	Two human cases have been documented
Khujand virus	Insectivorous bat *Myotis mystacinus*	Central Asia	Known by a single isolate. Given the limited surveillance in eastern Europe and Asia, may be distributed more broadly
Aravan virus	Insectivorous bat *Myotis blythi*	Central Asia	Known by a single isolate. Given the limited surveillance in eastern Europe and Asia, may be distributed more broadly
Bokeloh bat lyssavirus	Insectivorous bat *Myotis nattereri*	Germany	Known by a single isolate; currently does not have species status, and is not listed in the current ICTV documents
Irkut virus	Insectivorous bat *Murina leucogaster*	Eastern Asia	Known by two isolates, from a bat and from a human
Duvenhage virus	Insectivorous bats (?)	Sub-Saharan Africa	Known by four isolates, three of which came from humans, bitten by bats, and one from a bat, of the presumably *Miniopterus* species
Lagos bat virus	Pteropodid bats of several genera (*E. helvum, R. aegyptiacus, Epomophorus* spp.)	Sub-Saharan Africa	Constitutes several lineages with long genetic distances. Potentially, in the future may be subdivided into 2–3 separate species. Spill-over infections reported in wild and companion animals. No human cases documented to date
Mokola virus	Unknown	Sub-Saharan Africa	Twice isolated from shrews, once from a rodent. The majority of other isolates were obtained from companion animals, such as cats, as the result of spillover infection. Two human cases have been reported
Shimoni bat virus	Insectivorous bat *Hipposideros commersoni*	Kenya	Known by a single isolate. Serological surveys suggest that *H. commersoni* is likely the reservoir
West Caucasian bat virus	Insectivorous bats from genus *Miniopterus*	South-eastern Europe	Known by a single isolate. However, serological surveys suggest that WCBV (or other serologically related virus) may be present in *Miniopterus* bats in Africa (Kenya)
Ikoma lyssavirus	African civet (*Civettictis civetta*)	Tanzania	Known by a single isolate. Reservoir host is questionable

3 Antigenic patterns in reactions with anti-nucleocapsid MAbs and serological cross-reactivity using polyclonal antisera.
4 Whenever available, additional characteristics, such as ecological properties, host and geographic range, and pathological features are considered (Dietzgen *et al.*, 2011).

Based on genetic distances and serological cross-reactivity, the *Lyssavirus* genus has been subdivided into two phylogroups (Badrane *et al.*, 2001). Phylogroup I includes the species *Rabies virus, Duvenhage virus, Australian bat lyssavirus, Aravan virus, Khujand virus* and *Irkut virus*. Phylogroup II includes *Lagos bat virus, Mokola virus* and *Shimoni bat virus*. The remaining species of the genus, *West Caucasian bat virus*, cannot be included into either of these phylogroups, and is suggested to be considered as a representative of an independent phylogroup III (Hanlon *et al.*, 2005; Kuzmin *et al.*, 2005).

Genome organization, transcription and replication

The lyssavirus genome is a non-segmented single-stranded RNA of negative polarity. It varies slightly in length between 11.9 and 12.3 kb (Kuzmin *et al.*, 2008c). The genome organization is consistent with that of other rhabdoviruses and includes the leader sequence of ~ 50 nucleotides, followed by five genes encoding structural proteins: the nucleoprotein (N), the phosphoprotein (P), the matrix protein (M), the glycoprotein (G), the 'large' RNA-dependent RNA polymerase (RdRp; L), and is terminated by the trailer sequence of ~ 70 nucleotides (see Chapter 2, Fig. 2.2). The 10 to 16 nucleotides at the leader and trailer extremities are strongly conserved and inversely complementary, which indicates the presence of important signals for genome replication and encapsidation (Tordo *et al.*, 2005; Kuzmin *et al.*, 2008c). The template used for transcription and replication is the RNA genome encapsidated by the N protein, which is only locally and temporally unencapsidated to allow access for the RdRp and its cofactor, P protein. The M protein is located under the viral envelope while the G protein is in a transmembrane position. Both the M and G proteins are involved during the budding process, which allows production of the viral progeny.

During the transcription phase, the RdRp recognizes conserved signals along the (–) RNA genome that dictate initiation-capping (3′-UUGURRnGA-5′) as well as termination-polyadenylation (3′-A/UCUUUUUUUG-5′) of each mRNA. The poly(A) tail results from a reiterative copying of the 7 'U' by the RdRp. These conserved transcription signals are bordering each cistron and are separated by untranscribed intergenic sequences of variable lengths (2 to 100 nucleotides) that are eluded by the RdRp. During the replication phase, however, the RdRp becomes ignorant of the transcription signals, and copies a full-length genome, including intergenic sequences. It is not clear which viral and/or cellular factors are responsible for switching the RdRp from a transcriptase to a replicase state.

The quasi-permanent encapsidation of the RNA genome template during transcription and replication leaves possibly less opportunity for the RdRp, either in transcriptase or replicase mode, to freely jump from one template to another. This may explain why lyssaviruses, like other negative-strand RNA viruses that share a similar encapsidation process of their genomes, extremely rarely exhibit recombination (Sibold *et al.*, 1999), in contract to positive-strand RNA viruses (picornaviruses, coronaviruses, etc.) for which recombination is a powerful driver of evolution. For a non-segmented, negative-strand RNA genome, like that of the lyssaviruses, a predictable point for recombination would be at or near the intergenic regions where the *Vesiculovirus* RdRp was shown to pause during transcription to complete polyadenylation of the proximal mRNA and capping the distal one (Iverson and Rose, 1981). However, no recombination event has been reported in *Lyssavirus* genomes. Rather, they evolve through point mutations or short segments of insertion/deletion (indel) as the result of the a lack of proofreading mechanism of their RdRp. Polyadenylation signals are preferential places for indel occurrence, probably because they functionally perturb the fidelity of the RdRp for the necessary slippage along the poly-U template to generate the poly-A. Indeed, several lyssaviruses show polyadenylation signals with more than seven U residues (Tordo *et al.*, 1986; Kuzmin *et al.*, 2008c).

To optimize the coding capacity, the untranslated regions of *Lyssavirus* mRNAs are usually short. The first AUG codon of the mRNA, preferentially used to initiate translation, never occurs more than 30 nucleotides after the 5′ cap. The 3′ untranslated regions can be more variable, particularly in the G mRNA, which measures between 400–700 nucleotides in length. This observation was first made during the early stages of rabies virus genome sequencing, before PCR had been invented, forcing scientists to study vaccine strains well adapted to cell culture, as it was easier to produce such viruses in sufficient amounts to facilitate cDNA cloning (Tordo *et al.*, 1986). In some rabies virus vaccine strains, the long G mRNA 3′ untranslated region was found to contain sequences similar to incomplete initiation and termination transcription signals. This region was thus considered as a remnant gene, or 'pseudogene' (Tordo *et al.*, 1986). This hypothesis was further supported by the observation that fish novirhabdoviruses, like infectious haematopoietic necrosis virus (IHNV; Kurath *et al.*, 1985), bovine ephemeroviruses, like bovine ephemeral fever virus (BEFV; McWilliam *et al.*, 1997) and Adelaide river virus (ARV; Wang *et al.*, 1994) encode additional proteins at the same genome position. However, almost no trace of transcription initiation or termination signals were documented in the G mRNA 3′ untranslated region of the wild-type lyssaviruses that have been sequenced to date (Delmas *et al.*, 2008; Kuzmin *et al.*, 2008c). Nevertheless, in the West Caucasian bat virus (WCBV) genome, this region is not only significantly longer than in other lyssaviruses (~700 nucleotides; about 40%), but contains a stretch of 180 nucleotides without stop-codons (Kuzmin *et al.*, 2008c). Although the corresponding polypeptide has not been detected in WCBV-infected cells, this may be another case where a lyssavirus ancestor had a functional gene in this region. Otherwise, the occasional presence of such long nucleotide sequence without stop codons is hard to explain. Moreover, it would be surprising if all circulating contemporary lyssaviruses would have conserved a long untranslated region by chance. Reasons for its maintenance remain to be elucidated in the context of the regulation of genome transcription/replication or host–virus interactions.

Phylogeny

A representative phylogenetic tree of lyssaviruses is presented in Fig. 5.1. Similar results were obtained via comparisons of partial or complete N genes (Bourhy *et al.*, 1993, Kuzmin *et al.*, 2005), G genes (Badrane and Tordo, 2001), P genes (Nadin-Davis and Real, 2011) and L genes (Bourhy *et al.*, 2005). Nevertheless, as was demonstrated previously, the best quantitative separation of genotypes and species was achieved when complete N genes or concatenated sequences of all five structural genes were compared (Kuzmin *et al.*, 2005; Delmas *et al.*, 2008; Kuzmin *et al.*, 2010). Within Phylogroup I, Australian bat lyssavirus is most similar to the 'classical' rabies virus, whereas European bat lyssaviruses type 1 and type 2 constitute a monophyletic group with Duvenhage, Aravan, Khujand, Irkut and Bokeloh bat viruses. Within this cluster, the different lyssavirus species may group in a slightly different way, depending on the gene used for phylogeny (Kuzmin *et al.*, 2003).

Genetic distances between lyssavirus species are globally shorter than between species in other rhabdovirus genera (Fig. 5.2). At a comparative genetic scale, this may support a classification of all lyssaviruses in a single species, potentially named 'rabies virus' since they all cause the same disease, rabies. However, such an appreciation does not correspond to lyssavirus ecology, which has built the current taxonomy progressively. For example, rabies virus and Australian bat lyssavirus are relatively similar, phylogenetically and antigenically. However, the first is distributed globally, in carnivores and bats, including numerous documented host shifts from bats to carnivores (Daoust *et al.*, 1996; Leslie *et al.*, 2006, Aréchiga-Ceballos *et al.*, 2010), whereas the second virus was found to date only in Australia (Moore *et al.*, 2010) and, perhaps, some surrounding islands, based on serological suggestions (Arguin *et al.*, 2002), but, except for humans, no spill-over infections were documented in terrestrial mammals to date.

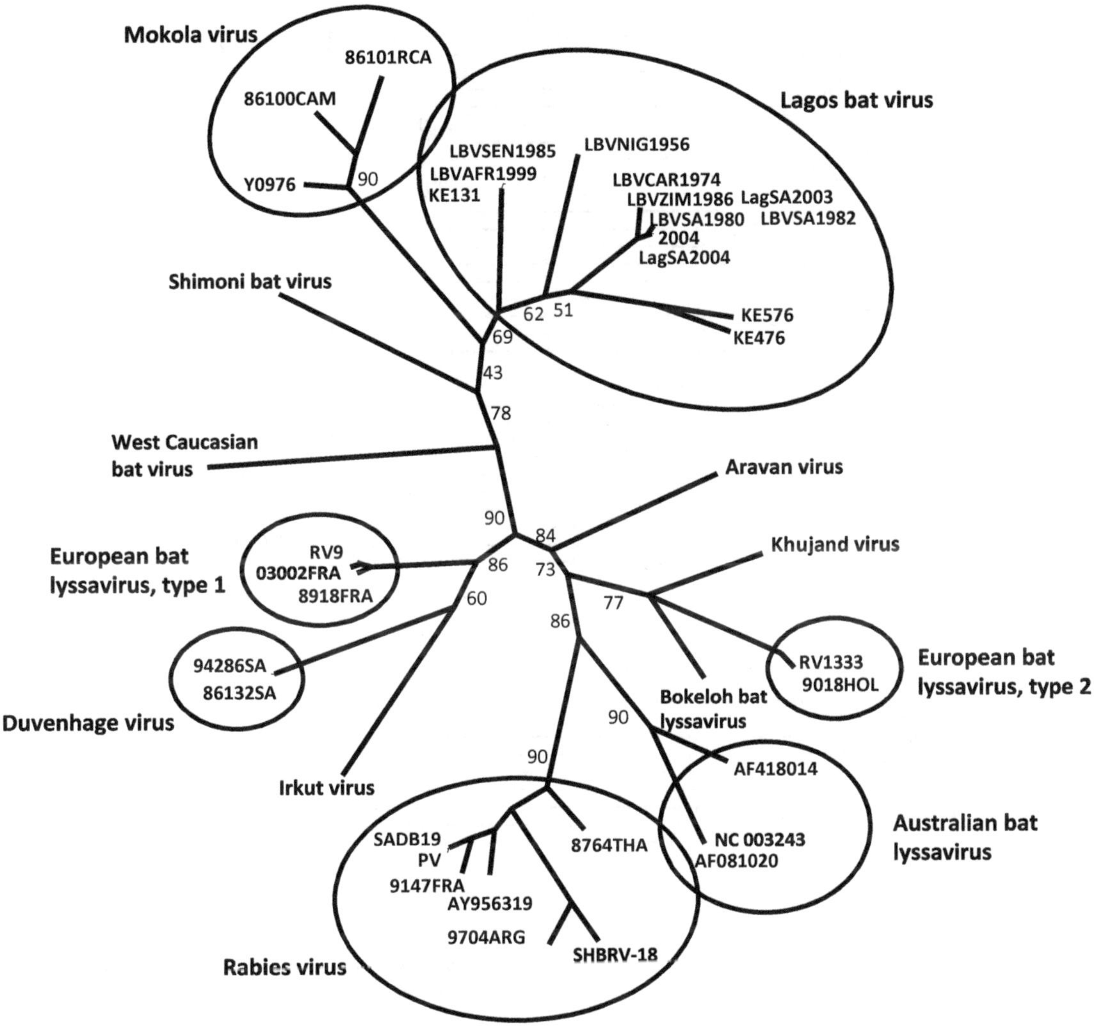

Figure 5.1 Maximum likelihood phylogenetic tree of lyssaviruses, based on the alignment of concatenated N+P+M+G+L gene sequences. Bootstrap values for 500 replicates are shown for the key nodes.

Moreover, experimental infection of carnivores with Australian bat lyssavirus did not result in frank disease development (McColl *et al.*, 2007). Similarly, European bat lyssaviruses type 1 and 2 have different ecological niches, with different principal host species (i.e. *Eptesicus serotinus* for the type 1 virus and *Myotis daubentonii* for the type 2 virus) (Fooks *et al.*, 2003). Very few spill-over infections into terrestrial mammals were documented for the type 1 virus only (Ronsholt, 2002; Tjørnehøj *et al.*, 2006; Dacheux *et al.*, 2009). We provide these observations only for the territories with a developed surveillance system, to demonstrate that taxonomic distinctions significantly correlate with ecological variables. By inference, we expect similar modalities for other lyssavirus species, circulating in the areas where no comprehensive surveillance system exists. Most likely, the limited genetic variability of lyssaviruses compared to other rhabdoviruses resulted from evolutionary constraints (i.e. purifying selection), applied by their unique pathobiology and host adaptation (Kuzmin *et al.*, 2006, 2009; Streicker *et al.*, 2010). This observation supports the ICTV standpoint that genetic distances alone are not sufficient for delineation of viral species. In addition, comparative lyssavirus phylogeny implies that we are certainly missing an evolutionary

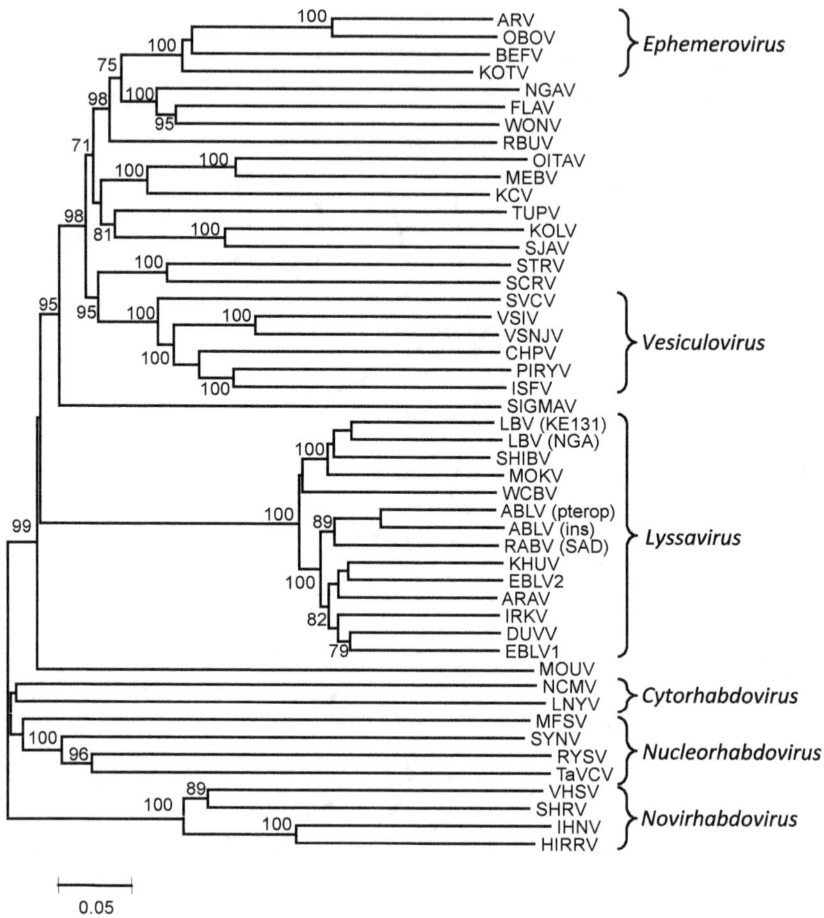

Figure 5.2 Neighbour-joining phylogenetic tree of *Rhabdoviridae* (*N* gene sequences), demonstrating the relative position of *Lyssavirus* genus within the family. Bootstrap values for 1000 replicates are shown for the key nodes.

link between lyssaviruses and other members of the *Rhabdoviridae* family. Recently described fossil rhabdoviral gene sequences, integrated into arthropod genomes (Fort *et al.*, 2012), may force us to refresh our minds and revisit the evolution of the whole family, in which arthropods likely played a key role.

Host range and distribution

Rabies virus (RABV) is distributed globally, with the exception of several insular territories, Australia and Antarctica (World Health Organization, 2004). In the New World, RABV circulates in bats and terrestrial carnivores. The number of phylogenetic lineages of RABV documented in bats corresponds roughly to the number of bat species tested so far, which suggests a long history of co-adaptation of each viral variant and its reservoir host (Kuzmin and Rupprecht, 2007; Streicker *et al.*, 2010). Moreover, some broadly distributed bat species, such as *Eptesicus fuscus* and *Tadarida brasiliensis*, harbour several lineages of rabies virus (Velasco-Villa *et al.*, 2006; Nadin-Davis *et al.*, 2010).

At least several well-established lineages of RABV, circulating in North American carnivores (i.e. the raccoon lineage, south-central skunk lineage, and other skunk lineages from Mexico) are linked phylogenetically to bat RABVs (Badrane and Tordo, 2001; Nadin-Davis and Real, 2011). The same is true for the RABV lineages identified in white-nosed coati (*Nasua narica*) in Mexico (Aréchiga-Ceballos *et al.*, 2010) and in marmoset monkeys (*Callithrix jacchus*) in Brazil (Favoretto *et al.*, 2001), although the two latter examples are not supported by a sufficient number of observations for distinguishing short-term local outbreaks from sustained virus circulation.

Among wild carnivores, skunks of different species appear very permissive to various RABV lineages. In addition to the south-central and Mexican skunk RABVs which are associated phylogenetically with bat viruses, there are several other lineages (i.e. north-central skunk, California skunk, and other skunk lineages from Mexico) which belong to a 'cosmopolitan' RABV lineage. The 'cosmopolitan' RABV has spread presumably to many parts of the world by human activity during the period of European colonization (Smith *et al.*, 1992; Badrane and Tordo, 2001; Velasco-Villa *et al.*, 2008; Nadin-Davis and Real, 2011). It likely originated from dogs, as 'cosmopolitan' viruses are linked phylogenetically to older vaccine strains and to the fox RABV currently circulating in Eurasia. Skunks in eastern Canada have acquired the Arctic RABV variant (Smith *et al.*, 1992; Nadin-Davis and Real, 2011), and skunks along the east coast of North America are frequently infected by the raccoon RABV (Blanton *et al.*, 2009, 2010).

In contrast, the North American raccoon (*Procyon lotor*) seems well adapted and constrained only to one specific RABV variant, which is believed to have emerged during the 1940s in Florida (Badrane and Tordo, 2001; Nadin-Davis and Real, 2011), and since then has rapidly spread along the east coast of North America (Biek *et al.*, 2007). It is not clear whether the raccoon RABV emerged from a bat RABV or from the south-central skunk RABV. Moreover, neither raccoon nor south-central skunk lineages can be associated with any specific bat RABV lineage, even though they are monophyletic with the whole cluster of American bat RABVs. Therefore, these viruses either evolved very fast (which is not supported by evolutionary reconstructions based on the available gene sequences), or the relevant host shift from bats occurred a long time ago, which allowed many years of independent evolution. Intensive oral rabies vaccination along the border of the enzootic area currently prevents the spread of raccoon rabies to the western parts of the continent (Slate *et al.*, 2009; see Chapter 13).

Other American lineages of RABV, circulating in dogs, coyotes (*Canis latrans*) and several species of foxes, belong to the 'cosmopolitan' group. Rabies control in dogs resulted in elimination of dog-mediated rabies in the USA and Canada, although it still occurs in the Latin America. Mongooses in several Caribbean islands acquired the 'cosmopolitan' RABV from dogs, and maintain its circulation (Nadin-Davis *et al.*, 2006b, 2008). Oral rabies vaccination of wild canids significantly reduced RABV circulation in their populations in North America. In addition to the representatives of 'cosmopolitan' RABV lineage, Arctic RABV circulates among Arctic foxes (*Vulpes lagopus*), red foxes (*V. vulpes*) and striped skunks (*Mephitis mephitis*) in northern territories of the continent and in eastern Canada (Nadin-Davis *et al.*, 2006a; Kuzmin *et al.*, 2008a).

In the Old World, RABV has been documented in carnivores only [with the exception of poorly understood epizootics in kudu antelopes (*Tragelaphus strepsiceros*) in southern Africa (Mansfield *et al.*, 2006)], predominantly in canids and herpestides. The 'cosmopolitan' lineage of RABV circulates in dogs and wild canids in vast territories of Eurasia and Africa (Smith *et al.*, 1992; Kissi *et al.*, 1995; Badrane and Tordo, 2001; Nadin-Davis and Real, 2011). Besides the 'cosmopolitan' lineage, introduced from Europe, distinct 'autochthonous' lineages of RABV circulate in dogs in several areas of Africa and south-east Asia. Although dog-mediated rabies has been eliminated in the developed countries of Europe via vaccination programmes and population control, dogs are still the major source of human cases in other parts of the Old World, particularly in the tropics, causing over 90% of human rabies. Arctic RABV circulates circumpolarly among Arctic and red foxes. Phylogenetic inferences suggest that the Arctic RABV lineage originated from so-called Arctic-like lineages, currently present in the Indian subcontinent, central and eastern Asia (Nadin-Davis *et al.*, 2007; Kuzmin *et al.*, 2008a). Raccoon dogs (*Nyctereutes procyonoides*), besides maintaining the circulation of Arctic-like RABV in their native habitats in eastern Asia, readily participate in circulation of 'cosmopolitan' RABV in the areas, enzootic for fox rabies, where this species was successfully introduced. The wolf (*Canis lupus*) was historically the major wildlife rabies reservoir in Eurasia. However, with population reductions, the significance of this species in circulation of rabies virus has diminished (Bourhy *et al.*, 1999; Kuzmin *et al.*, 2004). Following the rational use of oral vaccination programmes, rabies has been eliminated from the red fox and raccoon dog populations in Western Europe by the end of the twentieth century (Pastoret and Brochier, 1999). Nevertheless, rabies still remains present in red foxes, corsac foxes, jackals and raccoon dogs in Eastern Europe and Asia, which offer a risk of re-introduction of the disease to the countries that are currently considered 'rabies free' (Bourhy *et al.*, 1999; Kuzmin *et al.*, 2004, Singer *et al.*, 2009; Yang *et al.*, 2011).

Mustelids were almost never reported as important rabies reservoirs, although the European badger (*Meles meles*) might be relatively frequently infected in the areas affected by fox rabies, but these represent spill-over infections rather than sustained circulation (Kuzmin *et al.*, 2004). The only exception may be the Chinese ferret-badger (*Melogale moschata*), which was suggested as a reservoir of a specific RABV lineage, originating from the dog RABV (Liu *et al.*, 2010). Mongooses (*Herpestidae*) maintain circulation of dog-related RABV in southern Asia, but in Africa they harbour a specific RABV variant (Nel *et al.*, 2005; Davis *et al.*, 2007; Van Zyl *et al.*, 2010). Domestic and wild species of cats may be infected by other mammals and serve as a source of human exposures, but have never been documented as capable reservoirs that can maintain sustained circulation of RABV.

The presence of RABV in Old World bats was suggested several times but never corroborated (reviewed by Kuzmin and Rupprecht, 2007). In contrast, Old World bats harbour circulation of other lyssavirus species. In Europe, these include European bat lyssaviruses type 1 and 2, and West Caucasian bat virus, as well as the unclassified Bokeloh bat lyssavirus (Freuling *et al.*, 2011). In Asia, these include Aravan, Khujand and Irkut viruses (Kuzmin *et al.*, 2003, 2005). In Africa bats harbour Lagos bat, Duvenhage and Shimoni bat viruses (Markotter *et al.*, 2008b; Kuzmin *et al.*, 2010). Moreover, a recent survey suggested the presence of a virus, serologically related to the West Caucasian bat virus in African bats (Kuzmin *et al.*, 2008b). In the 'rabies free' Australia, lyssavirus presence to date is limited to the Australian bat lyssavirus (Fraser *et al.*, 1996; Guyatt *et al.*, 2003).

The only lyssavirus species that has never been documented in bats to date is Mokola virus. Initially isolated from shrews in Nigeria (Shope *et al.*, 1970), it was in several occasions isolated from companion animals and a rodent (Nel *et al.*, 2000; Sabeta *et al.*, 2007), but the natural reservoir host for Mokola virus remains unknown. Also, a novel lyssavirus, tentatively named Ikoma lyssavirus, was recently detected in an African civet from Tanzania (Marston *et al*, 2012).

Based on lyssavirus diversity, the Old World (likely Africa) was proposed as the area of the *Lyssavirus* genus origins (Shope, 1982). Host range and phylogenetic analysis suggest that bats are the principal hosts that drive lyssavirus evolution. At least two host switches from bats to terrestrial mammals were inferred from lyssavirus phylogeny and suggested for their natural history: the first switch occurred on the inter-species (that time termed 'inter-genotype') level, and the second one on the intra-species (intra-genotype) level (Badrane and Tordo, 2001). However, only the second switch is relatively well supported in North America by the phylogenetic relatedness of the raccoon and several skunk RABV lineages to the bat RABVs. The first switch remains indeterminate, as we do not know which lyssavirus species, which bat species, and which terrestrial species might be involved in this historical event. We still cannot answer the questions why RABV is not present in Old World bats, and why and how it exclusively colonized bats in the New World. Similarly, why other lyssavirus species, circulating in Old World bats, are not present in New World bats, is not understood. In addition, American bat RABV lineages, although related phylogenetically to the raccoon and several skunk RABV lineages, are significantly different from the 'terrestrial' Old World RABV lineages. How and where these originated is unknown.

The possibility exists that in some areas of the Old World with insufficient surveillance one can still find RABV in bats, and perhaps this virus will be more similar to the 'terrestrial' Old World RABV lineages than to the American bat RABV lineages. For example, no isolates of bat lyssaviruses are available from southern and south-eastern Asia, although anecdotal reports of rabid bats exist in the literature (Smith *et al.*, 1968; Pal *et al.*, 1980). This is supported by more recent serological surveys, that demonstrated the presence of lyssaviruses in bats from Bangladesh, Thailand, Philippines and Cambodia (reviewed by Kuzmin and Rupprecht, 2007).

Pathobiology

As mentioned, lyssaviruses cause an acute fatal meningo-encephalitis (rabies) in mammals, with almost a 100% case fatality rate. Lyssaviruses are quintessential neurotropic pathogens. The virus is generally delivered into a wound via a bite, or very rarely via contamination of mucous surfaces, via highly concentrated aerosols that were proposed to be a possible route of contamination for bats under the humid atmosphere in caves (Constantine *et al.*, 1972). The virus can replicate at the inoculation site, as was shown for skeletal muscle fibres (Charlton *et al.*, 1997). Virus typically penetrates neurons through the neuromuscular junction (Tordo *et al.*, 2005). The attachment to the host cell receptors is mediated by trimers of glycoproteins that form spikes protruding from the virion envelope. There are several types of putative receptors suggested for RABV attachment, including the nicotinic acetylcholine receptor, the neuronal cell adhesion molecule, and carbohydrate moieties, phospholipids and gangliosides (Tordo *et al.*, 2005). Once bound, the virus enters the neuron by endocytosis, and the endosomal vesicle is transported by retrograde axonal flow dependent on the microtubule network, using dynein motors. The pH decrease within the endosome provokes conformational changes of the glycoprotein, which mediates fusion between the endosomal and viral membranes, and the (–) strand RNA genome encapsidated by the N protein is liberated into the cytoplasm (Tordo *et al.*, 2005; Albertini, A., *et al.*, 2011). Transcription and replication occur at the neuron perikaryon.

The N-RNA template is transcribed by the viral polymerase (RdRp) and its cofactor, the P protein (see Chapter 2, Fig. 2.3). This generates in cascades the monocistronic (+) strand RNAs, corresponding to the leader RNA (Le$^+$), and the five mRNAs (Albertini, A.A., *et al.*, 2011). All but one of the monocistronic mRNAs produce a single protein from a single ORF initiated at the first AUG codon. The P mRNA produces three or four proteins, initiated from downstream in-frame AUG codons (Chenik *et al.*, 1995). Because the RdRp is pausing at each termination signal to polyadenylate the proximal mRNA, it eludes the intergenic region and only reinitiates at the distal initiation/capping signal, and the size of the untranscribed intergene regions influences the rate of re-initiation. Consequently, there is a progressive decrease in mRNA amounts from the 3′ encoded to the 5′ encoded cistrons. The N and P proteins, encoded by the two 3′ proximal mRNAs, are produced in greater quantities. The P protein plays a dual role during transcription and replication: on the one hand, it establishes a bridge between the N-RNA and the RdRp, and thus facilitates access to the template for the polymerase; on the other hand, it acts as a chaperone during the synthesis of the N, by forming N°–P complexes that keep the N soluble, preventing it from binding to cellular RNA (Albertini, A.A., *et al.*, 2011). The N protein is then transferred from these N°–P complexes to the nascent viral RNA to ensure encapsidation. There is one specific cis-acting encapsidation signal at the 5′-terminal ends of both the (+)-strand antigenome and the (–)-strand genome that is specifically recognized by the N°. The encapsidation proceeds in a cooperative manner in the 5′ to 3′ direction (Tordo *et al.*, 2005; Wunner, 2007). At the beginning of transcription, when limited quantities of N° are only sufficient to encapsidate the Le$^+$ transcript, the RdRp recognizes the transcription signals, and monocistronic mRNAs are produced. Since mRNAs do not harbour the encapsidation signal, they are not encapsidated, but directed to the translation machinery. This increases the amount of N° up to a critical concentration, at which the RdRp ignores the transcription signals and turns into the replicase mode. By then, it is producing a full length (+)-strand antigenome which is encapsidated co-transcriptionally by the soluble N° and forms the ribonucleoprotein that protects the RNA from enzymatic degradation. The full-length (+)-strand antigenome copies in turn become the template for (–)-strand genome RNA production, incorporated in the progeny virions (Wunner, 2007; Albertini, A.A., *et al.*, 2011).

The viral progeny is released at the synaptic junction and propagates up to the central nervous system (CNS) by using motor or sensory neurons and following neuronal connections. This avoids or limits surveillance by the host immune system, resulting in the absence of an early antibody response. The duration of an asymptomatic incubation period is two months on average (Tordo *et al.*, 2005), but may be very variable, from a week to several months and even years (Smith *et al.*, 1991). Having been delivered to the CNS, the virus disseminates rapidly. Nearly all regions of CNS may be affected, but the *Medulla oblongata* is a principal area from which the virus is transported to the salivary glands along the innervation pathways which allows the transmission to other susceptible host via a bite (Hanlon *et al.*, 1999).

The clinical period is rapid and severe (usually 1–2 weeks). The prodromal period is lacking specific clinical signs and is limited to fever, malaise, sometimes flu-like symptoms. Frequently, human patients feel paraesthesia and tingling in the inoculation site. With the development of encephalitis, cerebral symptoms manifest. Two major clinical forms of rabies have been described in animals and humans: furious and paralytic (although 'mixed' forms are very common). When the disease is furious, animals or humans become agitated and aggressive. Insomnia, irritability, and anxiety are commonly observed. Other signs, such as pupillary dilatation, altered phonation, aimless wandering, drooling of saliva, and muscle tremors and seizures may be noted. Humans often develop hallucinations and delirium. Some symptoms, which have been considered as 'classic' but are only observed in 50% or less of patients, are hydrophobia (painful throat seizures at attempts to drink or even due to seeing or hearing running water), aerophobia, photophobia, and phonophobia (seizures in response to airflow, bright lights, and loud sounds, respectively). Paralytic rabies is characterized by a greater prevalence of pareses and paralysis from the beginning of disease manifestations, whereas agitation and anxiety are moderate or absent. Subsequently, progressive paresis and paralysis appear. Sick animals or humans become comatose and die, usually due to respiratory failure or cardiac arrest. Once symptoms appear, death occurs usually within 1–10 days, unless intensive care is implemented (Hanlon *et al.*, 2007; Jackson, 2007). Before the fatal outcome, the infected animal should transmit the infection to a certain number of susceptible individuals (usually via bite) to ensure sustained viral circulation. In addition to salivary glands, during the late stage of the disease virus spreads from the CNS to periphery via neuronal pathways, and can be detected in various organs and tissues (which is critical for organ transplantation issues). Detection of rabies virus in nerves, surrounding hair follicles of skin, is one of the essential tests for ante-mortem diagnosis in humans (Dacheux *et al.*, 2008).

Epidemiology

Dog rabies, which causes over 90% of human mortalities, is vastly present in the developing tropical countries, particularly in Africa and Asia (World Health Organization, 2004; Warrell, 2008). Efforts of the Pan American Health Organization, with the support of national governments, succeeded in significant reduction of dog rabies in Latin America (Schneider *et al.*, 2007). In the developed countries of North America and Western Europe, extensive vaccination campaigns, coupled with a strict policy for housing and importation of companion animals led to elimination of dog-mediated rabies (Blancou, 2008; Slate *et al.*, 2009). In other areas dog rabies still represents an enormous burden not only for human populations, but also to wildlife, as was described, for example, for the endangered Ethiopian wolf (*Canis simensis*) (Randall *et al.*, 2004) and the unique Serengeti ecosystem (Lembo *et al.*, 2007).

Wildlife rabies is also distributed widely, although its importance may be masked in some areas, significantly affected by dog rabies. After successful elimination of dog-mediated rabies in North America and Western Europe, significant efforts have been implemented for reduction or elimination of wildlife canine rabies via extensive programmes of oral rabies vaccination (Pastoret and Brochier, 1999). These efforts brought several European countries to the 'rabies-free' status from the World Organization for Animal Health (OiE) standpoint, and ultimately reduced the risk of disease for companion animals and humans (Blancou, 2008; Velasco-Villa *et al.*, 2008; Slate *et al.*, 2009). Yet, bat lyssaviruses are still present in these territories, and therefore they cannot be considered 'rabies-free' from the WHO standpoint (World Health Organization, 2004), particularly given the cases of spill-over of bat lyssaviruses into terrestrial mammals, including companion animals (Ronsholt, 2002; Dacheux *et al.*, 2009). In addition, at least seven human rabies cases, caused by European bat lyssaviruses type 1 and type 2, and by Irkut virus, or by unidentified lyssaviruses of bat origin, were documented in Eurasia (Kuzmin and Rupprecht, 2007; Leonova *et al.*, 2009).

As mentioned above, Arctic RABV circulates circumpolarly in the Arctic and sub-Arctic areas of Eurasia and North America. Previously the Arctic fox (*V. lagopus*) was the primary host of this virus lineage (Kuzmin *et al.*, 2008a). However, with the ongoing global warming, the red foxes (*V. vulpes*) rapidly extend their range in the Arctic, with active involvement into circulation of the Arctic RABV (Blanton *et al.*, 2009, 2010). Given the scarce human population in the north, public health significance of Arctic rabies is limited, compared to more populated areas. To date, only one human case, caused by this RABV variant, was well documented (Kuzmin, 1999).

In Latin America, along with dog rabies, one of the major public health and veterinary threats is vampire bat rabies. Among the three known hematophagous bat species, the common vampire bat (*Desmodus rotundus*) is most broadly distributed and ecologically flexible. The distribution area of *D. rotundus* spreads from northern Mexico to Argentina. Deforestation, increasing of livestock population (i.e. food supply), and other anthropogenic changes affect the extension of vampire bat populations, and bring these bats in closer contact with humans and domestic mammals. Economic losses due to paralytic rabies in livestock are tremendous. In the enzootic area there is an at-risk population of more than 70 million head of cattle. Vampire bats usually bite many animals in a herd. The proportion of animals bitten may vary from 6% to 52%, and bites may be multiple. Baer (1991) reported a herd of 42 cows in Oaxaca, Mexico, where each animal had fresh bites, and 18 had over five bites. As reviewed by Acha (1967) and Baer (1991), the estimated annual mortality of bovines due to vampire bat rabies in Latin America during the 1960s was more than 500,000 cases with an economic loss of about 50 million US$. Clearly, cases of mortality due to vampire bat rabies were underestimated as cases reported annually varied from 3% to 60% of the actual mortality. Information from rural localities was often missing. However, even with limitations of the existing surveillance system, significant outbreaks in humans are regularly documented in the Amazon region (Peru, Brazil) (Salmón-Mulanovich *et al.*, 2009; Schneider *et al.*, 2009). The situation is complicated by the inability to create a sustained supply and storage of rabies biologics for indigenous human populations.

Methods for vampire bat population reduction include the use of anticoagulants (such as warfarin), which are deadly for vampire bats. The warfarin jelly is applied on the backs of captured bats, and the animals are released. In the roost, during self- and allo-grooming, other vampire bats consume the jelly on those that are treated and up to 20 or more vampire bats may be killed for each one treated. Bats also smear the jelly on the roost walls, excrete warfarin with urine and faeces, and finally the whole roost becomes deadly for bats. Another approach includes application of warfarin to the fresh vampire bat bite wounds on cattle, because bats often return to bites made the previous night (Greenhall, 1993). Experiments demonstrated potential efficiency of oral rabies vaccination of vampire bats maintained in captivity (Aguilar Setien *et al.*, 1998; Almeida *et al.*, 2005). However, the implementation of such a vaccination approach in the field remains problematic.

Fossil records indicate the presence of vampire bats in the south-western US from west Texas to northern California, and in the south-eastern US from Florida to West Virginia, from 30,000 until

5000 years ago (Ray *et al.*, 1988). The current tendency of climate changes may cause an extension of vampire bat range northwards up to the territories of Texas and California in the coming ~ 75 years (Mistry, 2009). In highland areas of Costa Rica, due to the mean minimum temperature increase, these bats are already found in altitudes where they had not been documented earlier (La Val, 2004).

Each American bat species may harbour a specific RABV variant. Viral phylogeny in general correlates with the phylogeny of bats, and it appears that successful spill-over infections and host shifts occur more frequently in phylogenetically related bat species (Streicker *et al.*, 2010). As bats fly and many of them migrate, they deliver pathogens over long distances in short periods of time. Therefore, distribution areas of bat RABVs correspond to the distribution areas of host species. Moreover, trans-continental delivery of the virus occurs during seasonal migrations. For example, a RABV isolated from a bat *Lasiurus ego* in Paraguay was genetically ~99% identical to the RABV variant, circulating in bats *L. cinereus* in North America (Sheeler-Gordon and Smith, 2001). In moderate latitudes, the majority of rabies cases in bats are documented during summer, as during the cold seasons these mammals either hibernate or migrate.

In general, circulation patterns of lyssaviruses in carnivores and in bats are different. Lyssaviruses cause acute fatal disease, and must ensure transmission to another susceptible individual before death of the host. Behavioural changes, such as increased aggression, are present in significant proportion of rabid animals, facilitating virus transmission via bites. Carnivores are highly susceptible to their homologous RABV variants. Sero-prevalence to RABV in fox populations in Europe (before initiation of oral vaccination programmes) was limited to 3–5%, which indicated that most exposures to the virus caused rabies and death of the animals, rather than development of adaptive immunity (Pastoret and Brochier, 1999).

Most bat species are gregarious. They frequently form large colonies, with high rates of physical contacts and cross-exposures. If bats were as susceptible to homologous lyssaviruses as carnivores, whole colonies and populations would die off during rabies outbreaks. Millennia of co-evolution made bats less susceptible to lyssaviruses, and the majority of exposures lead to abortive peripheral infection with the development of immune responses. Only a limited proportion of bats (perhaps, weak and immunocompromised individuals) develop productive CNS infection (rabies) and die. Therefore, sero-prevalence to lyssaviruses in bat population is as high as 30–70%, whereas the presence of lyssaviruses in CNS may be detected in 0–0.5% of apparently healthy bats (reviewed by Kuzmin and Rupprecht, 2007).

After elimination of dog-to-dog rabies transmission, and as the result of educational campaigns among public health providers and the general public, canine-mediated human rabies cases became progressively rare in the United States (US) and Canada from the 1950s. Instead, insectivorous bats became the most prominent source of human infections (Messenger *et al.*, 2002; De Serres *et al.*, 2008). For example, among 59 human rabies cases documented in the US during 1991–2009, only 13 were caused by dog exposure. Of these, two cases in the early 1990s were indigenous, whereas the other 11 were imported (i.e. people were exposed to dogs in developing countries, and did not follow the recommended post-exposure prophylaxis). Only one case of human infection was documented with the raccoon RABV, although circumstances of that exposure remain unknown (Krebs *et al.*, 2004). In contrast, bat RABVs caused 45 human deaths. Twenty of these cases were 'cryptic' (i.e. patients did not notice or did not recall any contact with bats), and four were associated with transplantation of organs and vessels from a rabies-infected donor (Srinivasan *et al.*, 2005; Blanton *et al.*, 2009, 2010).

The susceptibility of humans to various bat RABV variants appears different. Over 60% of human cases are caused by two phylogenetically similar RABV variants, associated with the silver-haired bat (*Lasionycteris noctivagans*) and the tri-coloured bat (*Perimyotis subflavus*). These bats are rarely found around humans or human dwellings. Investigations implied that these RABV variants may have an enhanced pathogenicity for humans. For instance, in one *in vitro* study the silver-haired bat RABV was better adapted to fibroblasts and epithelial cells compared to a coyote RABV (Morimoto *et al.*, 1996). This trait appeared to correlate with an ability to be more effective in replicating in the dermis at the inoculation site. By contrast, the big brown bats (*Eptesicus fuscus*), which routinely make maternity colonies in the attics of human dwellings across North America, maintain circulation of at least four lineages of RABV, but these very rarely were the source of human rabies.

The epidemiology of bat lyssaviruses in Europe is different of that in the Americas. European bat lyssaviruses, type 1 (EBLV-1) and 2 (EBLV-2), seem to be more conservative than RABV. The principal reservoir host of EBLV-1 is the serotine bat (*Eptesicus serotinus*), whereas the reservoir hosts of EBLV-2 are mouse-eared bats from the *Myotis* genus (Fooks *et al.*, 2003). European bats of other genera and families were reported positive for lyssaviruses only occasionally, most likely as the result of spill-over infections. The reasons of such limitations are unclear. They likely go deep to the historical roots of adaptation between bats and lyssaviruses along their evolutionary pathways and spread over the world. Given the high profile of rabies surveillance in developed European countries, the isolation of a novel Bokeloh bat lyssavirus in Germany in 2010 (Freuling *et al.*, 2011) was surprising. Besides the single isolate, no other data are available about ecology of this virus. The same in general is true for another bat lyssavirus, documented in Europe: the West Caucasian bat virus (WCBV), isolated on one occasion from the common long-fingered bat (*Miniopterus schreibersii*) in the Caucasus area (Kuzmin *et al.*, 2005). More recent studies demonstrated seroprevalence to WCBV in several species of *Miniopterus* bats from Kenya (Kuzmin *et al.*, 2008b), which may suggest broader circulation of this (or serologically related) lyssavirus, as *Miniopterus* bats are broadly distributed in tropics and subtropics of the Old World.

The epidemiology of African bat lyssaviruses is not well resolved (Markotter *et al.*, 2008b). Lagos bat virus circulates broadly in sub-Saharan Africa among Pteropodid bats, including *Eidolon helvum, Rousettus aegyptiacus, Epomophorus* spp. and others. On several occasions it was also isolated from spill-over infections in companion animals, and a mongoose, which demonstrated typical signs of rabies. The public health significance of Lagos bat virus is unknown, which is not surprising in developing countries, where the majority of cases are based on clinical signs, without laboratory confirmation and, moreover, without identification of viral isolates. Mokola virus may be broadly isolated as well, because it was documented in western and southern regions of Africa. However, except for two first isolations from shrews, all other isolates were obtained from companion animals, which were considered as spill-over infections from an unknown reservoir (Familusi and Moore, 1972; Sabeta *et al.*, 2007; Markotter *et al.*, 2008b). The epidemiology of Shimoni bat virus, known from a single isolation (Kuzmin *et al.*, 2010) remains to be resolved as well, although serological investigations suggests that insectivorous bats *Hipposideros commersoni* may be reservoir hosts of this virus (Kuzmin *et al.*, 2011). Duvenhage virus is associated with insectivorous bats, but the exact host range is unknown. Of the four available isolates, three were obtained from humans, who died of rabies after exposures to insectivorous bats, and on one occasion Duvenhage virus was isolated from a bat, presumably *Miniopterus* sp. (Markotter *et al.*, 2008b; Van Thiel *et al.*, 2009).

A significant public health and veterinary concern is that commercially available rabies biologics do not provide protection against Phylogroup II lyssaviruses and WCBV. Historical and modern approaches to overcome this limitation and create lyssavirus biologics with broader reactivity are described in Chapters 13 and 14 of this book.

Prior to 1996, Australia had been considered free of rabies. An outbreak of rabies involving several dogs occurred in the island state of Tasmania in 1867, but was quickly eradicated. Since then, only a few imported rabies cases were registered. The situation changed in 1996, when the Australian bat lyssavirus (ABLV) was isolated from a flying fox (Fraser *et al.*, 1996). There are four flying fox species in continental Australia: *P. alecto, P. poliocephalus, P. scapulatus* and *P. conspicillatus*, and the virus was recognized in each of them, along east coast of the continent (Guyatt *et al.*, 2003). Several samples of ABLV were obtained from the insectivorous bat *Saccolaimus flaviventris*.

Their nucleotide sequences formed another monophyletic clade, clearly distinguishable from the pteropodid clade. However, when compared with other lyssavirus genomes, these two lineages were still segregated confidently into one cluster, demonstrating that their differentiation from each other occurred later than differentiation of ABLV from other lyssavirus species (Gould *et al.*, 2002). It should be additionally clarified whether other species of Australian insectivorous bats participate in ABLV circulation, and if so, do they maintain the same virus strain as *S. flaviventris* bats, or does additional variability of the virus occur. Little is known about ABLV circulation patterns. The virus was detected in 6% of sick, injured or orphan flying foxes submitted to diagnostic laboratories. In one study, 9.4% of flying foxes submitted to diagnostic laboratories because they had bitten or scratched humans, or where testing was considered to be in the interest of public health, were ABLV-positive (Warrilow *et al.*, 2003). Serological surveys of a mixture of sick and apparently healthy bats demonstrated that 16% of the bats were seropositive to ABLV (Hooper *et al.*, 1997). Two human cases of rabies caused by ABLV have been described to date (Hanna *et al.*, 2000; Warrilow *et al.*, 2002).

Currently there are no efficient ways to control rabies in frugivorous or insectivorous bats. Reduction of the bat population is unacceptable. The obvious recommendation is to exclude bats from human proximity, making houses, attics, and other dwellings inaccessible, and use screens on the windows during summer months. Given the number of 'cryptic' cases, post-exposure prophylaxis is recommended for every case of physical contact with a bat, if a bite cannot be reasonable ruled out. Moreover, since 1996 the Advisory Committee on Immunization Practices recommends to consider rabies prophylaxis to persons, who awake and find a bat in the bedroom, if the bat cannot be captured and submitted for rabies testing (Advisory Committee on Immunization Practices, 2008).

Disclaimer

The findings and conclusions in this report are those of the authors and do not necessarily represent the views of the CDC.

References

Acha, P.N. (1967). Epidemiology of paralytic bovine rabies and bat rabies. Bull. Off. Int. Epizoot. *67*, 343–382.

Advisory Committee on Immunization Practices. (2008). Human rabies prevention – United States, 2008. Recommendations of the Advisory Committee on Immunization Practices. MMWR Morb. Mortal. Wkly Rep. 57/No. RR-3, 1–27.

Aguilar Setien, A., Brochier, B., Tordo, N., De Paz,O., Desmettre, P., Peharpre, D., and Pastoret, P.-P. (1998). Experimental rabies infection and oral vaccination in vampire bats (*Desmodus rotundus*). Vaccine *16*, 1122–1126.

Albertini, A., Bressanelli, S., Lepault, J., and Gaudin, Y. (2011). Structure and working of viral fusion machinery. Curr. Top. Membr. *68*, 49–80.

Albertini, A.A., Ruigrok, R.W., and Blondel, D. (2011). Rabies virus transcription and replication. Adv. Virus Res. *79*, 1–22.

Almeida, M.F., Martorelli, L.F., Aires, C.C., Sallum, P.C., and Massad, E. (2005). Indirect oral immunization of captive vampires, *Desmodus rotundus*. Virus Res. *111*, 77–82.

Aréchiga-Ceballos, N., Velasco-Villa, A., Shi, M., Flores-Chávez, S., Barrón, B., Cuevas-Domínguez, E., González-Origel, A., and Aguilar-Setién, A. (2010). New rabies virus variant found during an epizootic in white-nosed coatis from the Yucatan Peninsula. Epidemiol Infect. *138*, 1586–1589.

Arguin, P.M., Murray-Lillibridge, K., Miranda, M.E., Smith, J.S., Calaor, A.B., and Rupprecht, C.E. (2002). Serologic evidence of *Lyssavirus* infections among bats, the Philippines. Emerg. Infect. Dis. *8*, 258–262.

Badrane, H., and Tordo, N. (2001). Host switching in *Lyssavirus* history from the *Chiroptera* to the *Carnivora* orders. J. Virol. *75*, 8096–8104.

Badrane, H., Bahloul, C., Perrin, P., and Tordo, N. (2001). Evidence of two *Lyssavirus* phylogroups with distinct pathogenicity and immunogenicity. J. Virol. *75*, 3268–3276.

Baer, G.M. (1991). Vampire bat and bovine paralytic rabies. In The Natural History of Rabies, 2nd edn, Baer, G.M., ed. (CRC Press, Boca Raton, FL), pp. 390–406.

Baer, G.M., Neville, J., and Turner, G.S. (1996). Rabbis and Rabies: A Pictorial History through the Ages (Laboratorios Baer, Mexico City, Mexico).

Bahloul, C., Jacob, Y., Tordo, N., and Perrin, P. (1998). DNA-based immunisation for exploring the enlargement of immunological cross-reactivity against the lyssaviruses. Vaccine *16*, 417–425.

Belotto, A., Leanes, L.F., Schneider, M.C., Tamayo, H., and Correa, E. (2005). Overview of rabies in the Americas. Virus Res. *111*, 5–12.

Biek, R., Henderson, J.C., Waller, L.A., Rupprecht, C.E., and Real, L.A. (2007). A high-resolution genetic signature of demographic and spatial expansion in epizootic rabies virus. Proc. Natl. Acad. Sci. U.S.A. *104*, 7993–7998.

Blancou, J. (2008). The control of rabies in Eurasia: overview, history and background. Dev. Biol. (Basel) *131*, 3–15.

Blanton, J.D., Robertson, K., Palmer, D., and Rupprecht, C.E. (2009). Rabies surveillance in the United States during 2008. J. Am. Vet. Med. Assoc. *235*, 676–989.

Blanton, J.D., Palmer, D., and Rupprecht, C.E. (2010). Rabies surveillance in the United States during 2009. J. Am. Vet. Med. Assoc. *237*, 646–657.

Boulger, L.R., and Porterfield, J.S. (1958). Isolation of a virus from Nigerian fruit bats. Trans. Roy. Soc. Trop. Med. Hyg. *52*, 421–424.

Bourhy, H., Kissi, B., Lafon, M., Sacramento, D., and Tordo, N. (1992). Antigenic and molecular characterization of bat rabies virus in Europe. J. Clin. Microbiol. *30*, 2419–2426.

Bourhy, H., Kissi, B., and Tordo, N. (1993). Molecular diversity of the *Lyssavirus* genus. Virology *194*, 70–81.

Bourhy, H., Kissi, B., Audry, L., Smreczak, M., Sadkowska-Todys, M., Kulonen, K., Tordo, N., Zmudzinski, J.F., and Holmes, E.C. (1999). Ecology and evolution of rabies virus in Europe. J. Gen. Virol. *80*, 2545–2557.

Bourhy, H., Cowley, J.A., Larrous, F., Holmes, E.C., and Walker, P.J. (2005). Phylogenetic relationships among rhabdoviruses inferred using the L polymerase gene. J. Gen. Virol. *86*, 2849–2858.

Buckley, S.M. (1975). Arbovirus infection of vertebrate and insect cell cultures, with special emphasis on Mokola, Obodhiang, and kotonkan viruses of the rabies serogroup. Ann. N.Y. Acad. Sci. *266*, 241–250.

Calisher, C.H., Karabatsos, N., Zeller, H., Digoutte, J.P., Tesh, R.B., Shope, R.E., Travassos da Rosa, A.P., and St George, T.D. (1989). Antigenic relationships among rhabdoviruses from vertebrates and hematophagous arthropods. Intervirol. *30*, 241–257.

Carini, A. (1911). Sur une grande epizootie de rage. [In French.] Ann. Inst. Pasteur. Actualites *25*, 843–846.

Charlton, K.M., Nadin-Davis, S., Casey, G.A., and Wandeler, A.I. (1997). The long incubation period in rabies: delayed progression of infection in muscle at the site of exposure. Acta Neuropathol. *94*, 73–77.

Chenik, M., Chebli, K., and Blondel, D. (1995). Translation initiation at alternate in-frame AUG codons in the rabies virus phosphoprotein mRNA is mediated by a ribosomal leaky scanning mechanism. J. Virol. *69*, 707–712.

Constantine, D.G., Emmons, R.W., and Woodie, J.D. (1972). Rabies virus in nasal mucosa of naturally infected bats. Science *175*, 1255–1256.

Dacheux, L., Larrous, F., Mailles, A., Boisseleau, D., Delmas, O., Biron, C., Bouchier, C., Capek, I., Muller, M., Ilari, F., *et al.* (2008). European bat *Lyssavirus* transmission among cats, Europe. Emerg. Infect. Dis. *15*, 280–284.

Dacheux, L., Reynes, J.M., Buchy, P., Sivuth, O., Diop, B.M., Rousset, D., Rathat, C., Jolly, N., Dufourcq, J.B., Nareth, C., *et al.* (2009). A reliable diagnosis of human rabies based on analysis of skin biopsy specimens. Clin. Infect. Dis. *47*, 1410–1417.

Daoust, P.Y., Wandeler, A.I., and Casey, G.A. (1996). Cluster of rabies cases of probable bat origin among red foxes in Prince Edward Island, Canada. J. Wildl. Dis. *32*, 403–406.

Davis, P.L., Rambaut, A., Bourhy, H., and Holmes, E.C. (2007). The evolutionary dynamics of canid and mongoose rabies virus in Southern Africa. Arch. Virol. *152*, 1251–1258.

De Serres, G., Dallaire, F., Côte, M., and Skowronski, D.M. (2008). Bat rabies in the United States and Canada from 1950 through 2007: human cases with and without bat contact. Clin. Infect. Dis. *46*, 1329–1337.

Delmas, O., Holmes, E.C., Talbi, C., Larrous, F., Dacheux, L., Bouchier, C., and Bourhy, H. (2008). Genomic diversity and evolution of the lyssaviruses. PloS One *3*, e2057.

Dietzschold, B., Rupprecht, C.E., Tollis, M., Lafon, M., Mattei, J., Wiktor, T.J., and Koprowski, H. (1988). Antigenic diversity of the glycoprotein and nucleocapsid proteins of rabies and rabies-related viruses: implications for epidemiology and control of rabies. Rev. Infect. Dis. *10*, S4, 785–798.

Dietzgen, R.G., Calisher, C.H., Kurath, G., Kuzmin, I.V., Rodriguez, L.L., Stone, D.M., Tesh, R.B., Tordo, N., Walker, P.J., Wetzel, T., *et al.* (2011). Family Rhabdoviridae. In Virus Taxonomy: Ninth Report of the International Committee on Taxonomy of Viruses, King, A.M.Q., Adams, M.J., Carstens, E.B., and Lefkowitz, E.J., eds (Elsevier, Oxford, UK) pp. 686–714.

Familusi, J.B., and Moore, D.L. (1972). Isolation of a rabies related virus from the cerebrospinal fluid of a child with 'aseptic meningitis'. Afr. J. Med. Sci. *3*, 93–96.

Favoretto, S.R., de Mattos, C.C., Morais, N.B., Alves Araújo, F.A., and de Mattos, C.A. (2001). Rabies in marmosets (*Callithrix jacchus*), Ceará, Brazil. Emerg. Infect. Dis. *7*, 1062–1065.

Fooks, A.R., Brookes, S.M., Johnson, N., McElhinney, L.M., and Hutson, A.M. (2003). European bat lyssaviruses: an emerging zoonosis. Epidem. Infect. *131*, 1029–1039.

Fort, P., Albertini, A., Van-Hua, A., Berthomieu, A., Roche, S., Delsuc, F., Pasteur, N., Capy, P., Gaudin, Y., and Weill, M. (2012). Fossil rhabdoviral sequences integrated into arthropod genomes: ontogeny, evolution, and potential functionality. Mol. Biol. Evol. *29*, 381–390.

Fraser, G.C., Hooper, P.T., Lunt, R.A., Gould, A.R., Gleeson, L.J., Hyatt, A.D., Russell, G.M., and Kattenbelt, J.A. (1996). Encephalitis caused by a *Lyssavirus* in fruit bats in Australia. Emerg. Infect. Dis. *2*, 327–331.

Freuling, C., Beer, M., Conraths, F.J., Finke, S., Hoffmann, B., Keller, B., Kliemt, J., Mettenleiter, T.C., Mühlbach, E., Teifke, J.P., *et al.* (2011). Novel *Lyssavirus* in a Natterer's bat (*Myotis nattereri*), Germany. Emerg. Infect. Dis. *17*, 1519–1522.

Gould, A.R., Kattenbelt, J.A., Gumley, S.G., and Lunt, R.A. (2002). Characterisation of an Australian bat *Lyssavirus* variant isolated from an insectivorous bat. Virus Res. *89*, 1–28.

Greenhall, A.M. (1993). Ecology and bionomics of vampire bats in Latin America. In Bats and Rabies,

Greenhall, A.M., Artois, M., and Fekadu, M., eds (Foundation Marcel Merieux, Lyon, France), pp. 3–57.

Guyatt, K.J., Twin, J., Davis, P., Holmes, E.C., Smith, G.A., Smith, I.L., Mackenzie, J.S., and Young, P.L. (2003). A molecular epidemiological study of Australian bat *Lyssavirus*. J. Gen. Virol. *84*, 485–496.

Hanlon, C.A., Smith, J.S., and Anderson, G.R. (1999). Recommendations of a national working group on prevention and control of rabies in the United States. Article II: Laboratory diagnosis of rabies. The National Working Group on Rabies Prevention and Control. J. Am. Vet. Med. Assoc. *215*, 1444–1446.

Hanlon, C.A., Kuzmin, I.V., Blanton, J.D., Weldon, W.C., Manangan, J.S., and Rupprecht, C.E. (2005). Efficacy of rabies biologics against new lyssaviruses from Eurasia. Virus Res. *111*, 44–54.

Hanlon, C.A., Niezgoda, M., and Rupprecht, C.E. (2007). Rabies in terrestrial animals. In Rabies, Jackson, A., and Wunner, W., eds (Elsevier, London, UK), pp. 201–258.

Hanna, J.N., Carney, I.K., Smith, G.A., Tannenberg, A.E., Deverill, J.E., Botha, J.A., Serafin, I.L., Harrower, B.J., Fitzpatrick, P.F., and Searle, J.W. (2000). Australian bat *Lyssavirus* infection: a second human case, with a long incubation period. Med. J. Austral. *172*, 597–599.

Haupt, A., and Rehaag, H. (1921). Durch Fledermäuse verbreitete seuchenhafte Tollwut unter Viehbeständen in Santa Catharinia (Süd Brasilien). Zeitsch. Infektionsk. Haustiere 22, 104–127.

Hooper, P.T., Lunt, R.A., Gould, A.R., Samaratunga, H., Hyatt, A.D., Gleeson, L.J., Rodwell, B.J., Rupprecht, C.E., Smith, J.S., and Murray, P.K. (1997). A new *Lyssavirus* – the first endemic rabies-related virus recognized in Australia. Bull. Inst. Pasteur 95, 209–218.

Hughes, G.J., Orciari, L.A., and Rupprecht, C.E. (2005). Evolutionary timescale of rabies virus adaptation to North American bats inferred from the substitution rate of the nucleoprotein gene. J. Gen. Virol. *86*, 1467–1474.

Iverson, L.E., and Rose, J.K. (1981). Localized attenuation and discontinuous synthesis during vesicular stomatitis virus transcription. Cell 23, 477–484.

Jackson, A.C. (2007). Human disease. In Rabies, Jackson, A., and Wunner, W., eds (Elsevier, London, UK), pp. 309–340.

King, A., Davis, P., and Lawrie, A. (1990). The rabies viruses of bats. Vet. Microbiol. 23, 165–174.

Kissi, B., Tordo, N., and Bourhy, H. (1995). Genetic polymorphism in the rabies virus nucleoprotein gene. Virology *209*, 526–537.

Knobel, D., Cleaveland, S., Coleman, P.G., Fèvre, E.M., Meltzer, M.I., Miranda, M.E.G., Shaw, A., Zinsstag, J., and Meslin, F.X. (2005). Re-evaluating the burden of rabies in Africa and Asia. Bull. World Health Organ. *83*, 360–368.

Koprowski, H. (1959). Viruses. Trans. N.Y. Acad. Sci. *22*, 176–190.

Koprowski, H. (1995). Visit to an ancient curse. Sci. Med. *2*, 48–57.

Krebs, J.W., Mandel, E.J., Swerdlow, D.L., and Rupprecht, C.E. (2004). Rabies surveillance in the United States during 2003. J. Am. Vet. Med. Assoc. *225*, 1837–1849.

Kurath, G., Ahern, K.G., Pearson, G.D., and Leong, J.C. (1985). Molecular cloning of the six mRNA species of infectious hematopoietic necrosis virus, a fish rhabdovirus, and gene order determination by R-loop mapping. J. Virol. *53*, 469–476.

Kuzmin, I.V. (1999). An arctic fox rabies virus strain as the cause of human rabies in Russian Siberia. Arch. Virol. *144*, 627–629.

Kuzmin, I.V., and Rupprecht, C.E. (2007). Bat rabies. In Rabies, Jackson, A., and Wunner, W., eds (Elsevier, London, UK), pp. 259–307.

Kuzmin, I.V., Orciari, L.A., Arai, Y.T., Smith, J.S., Hanlon, C.A., Kameoka, Y., and Rupprecht, C.E. (2003). Bat lyssaviruses (Aravan and Khujand) from Central Asia: phylogenetic relationships according to N, P and G gene sequences. Virus Res. *97*, 65–79.

Kuzmin, I.V., Botvinkin, A.D., McElhinney, L.M., Smith, J.S., Orciari, L.A., Hughes, G.J., Fooks, A.R., and Rupprecht, C.E. (2004). Molecular epidemiology of terrestrial rabies in the former Soviet Union. J. Wildl. Dis. *40*, 617–631.

Kuzmin, I.V., Hughes, G.J., Botvinkin, A.D., Orciari, L.A., and Rupprecht, C.E. (2005). Phylogenetic relationships of Irkut and West Caucasian bat viruses within the *Lyssavirus* genus and suggested quantitative criteria based on the N gene sequence for *Lyssavirus* genotype definition. Virus Res. *111*, 28–43.

Kuzmin, I.V., Hughes, G.J., and Rupprecht, C.E. (2006). Phylogenetic relationships of seven previously unclassified viruses within the family Rhabdoviridae using partial nucleoprotein gene sequences. J. Gen. Virol. *87*, 2323–2331.

Kuzmin, I.V., Hughes, G.J., Botvinkin, A.D., Gribencha, S.G., and Rupprecht, C.E. (2008a). Arctic and Arctic-like rabies viruses: distribution, phylogeny and evolutionary history. Epidemiol. Infect. *136*, 509–519.

Kuzmin, I.V., Niezgoda, M., Franka, R., Agwanda, B., Markotter, W., Beagley, J.C., Urazova, O.Y., Breiman, R.F., and Rupprecht, C.E. (2008b). Possible emergence of West Caucasian bat virus in Africa. Emerg. Infect. Dis. *14*, 1887–1889.

Kuzmin, I.V., Wu, X., Tordo, N., and Rupprecht, C.E. (2008c). Complete genomes of Aravan, Khujand, Irkut and West Caucasian bat viruses, with special attention to the polymerase gene and non-coding regions. Virus Res. *136*, 81–90.

Kuzmin, I.V., Novella, I.S., Dietzgen, R.G., Padhi, A., and Rupprecht, C.E. (2009). The rhabdoviruses: biodiversity, phylogenetics, and evolution. Infect. Genet. Evol. *9*, 541–553.

Kuzmin, I.V., Mayer, A.E., Niezgoda, M., Markotter, W., Agwanda, B., Breiman, R.F., and Rupprecht, C.E. (2010). Shimoni bat virus, a new representative of the *Lyssavirus* genus. Virus Res. *149*, 197–210.

Kuzmin, I.V., Turmelle, A.S., Agwanda, B., Markotter, W., Niezgoda, M., Breiman, R.F., and Rupprecht, C.E. (2011). Commerson's leaf-nosed bat (*Hipposideros commersoni*) is the likely reservoir of Shimoni bat virus. Vector Borne Zoonotic Dis. *11*, 1465–1470.

La Val, R.K. (2004). Impact of global warming and locally changing climate on tropical cloud forest bats. J. Mammal. *85*, 237–244.

Lembo, T., Haydon, D.T., Velasco-Villa, A., Rupprecht, C.E., Packer, C., Brandão, P.E., Kuzmin, I.V., Fooks, A.R., Barrat, J., and Cleaveland, S. (2007). Molecular epidemiology identifies only a single rabies virus variant circulating in complex carnivore communities of the Serengeti. Proc. Biol. Sci. *274*, 2123–2130.

Leonova, G.N., Belikov, S.I., Kondratov, I.G., Krylova, N.V., Pavlenko, E.V., Romanova, E.V., Chentsova, I.V., and Petukhova, S.A. (2009). A fatal case of bat *Lyssavirus* infection in Primorye Territory of the Russian Far East. Rabies Bull. Europe. *4*, 5–8.

Leslie, M.J., Messenger, S., Rohde, R.E., Smith, J., Cheshier, R., Hanlon, C., and Rupprecht, C.E. (2006). Bat-associated rabies virus in Skunks. Emerg. Infect. Dis. *12*, 1274–1277.

Liu, Y., Zhang, S., Wu, X., Zhao, J., Hou, Y., Zhang, F., Velasco-Villa, A., Rupprecht, C.E., and Hu, R. (2010). Ferret badger rabies origin and its revisited importance as potential source of rabies transmission in Southeast China. BMC Infect. Dis. *10*, 234.

Lyles, D.S., and Rupprecht, C.E. (2007). Rhabdoviridae. In Fields Virology, Fifth edn, Knipe, D.M., Howley, P.M., Griffin, D.E., Lamb, R.A., Martin, M.A., Roizman, B., and Straus, S.E., eds (Wolters Kluwer/ Lippincott Williams & Wilkins, Philadelphia, PA), Vol. 1, pp. 1363–1408.

McColl, K.A., Chamberlain, T., Lunt, R.A., Newberry, K.M., and Westbury, H.A. (2007). Susceptibility of domestic dogs and cats to Australian bat *Lyssavirus* (ABLV). Vet. Microbiol. *123*, 15–25.

McWilliam, S.M., Kongsuwan, K., Cowley, J.A., Byrne, K.A., and Walker, P.J. (1997). Genome organization and transcription strategy in the complex GNS-L intergenic region of bovine ephemeral fever rhabdovirus. J. Gen. Virol. *78*, 1309–1317.

Mansfield, K., McElhinney, L., Hübschle, O., Mettler, F., Sabeta, C., Nel, L.H., and Fooks, A.R. (2006). A molecular epidemiological study of rabies epizootics in kudu (*Tragelaphus strepsiceros*) in Namibia. BMC Vet. Res. *2*, 2.

Markotter, W., Kuzmin, I., Rupprecht, C.E., and Nel, L.H. (2008a). Phylogeny of Lagos bat virus: challenges for *Lyssavirus* taxonomy. Virus Res. *135*, 10–21.

Markotter, W., Van Eeden, C., Kuzmin, I.V., Rupprecht, C.E., Paweska, J.T., Swanepoel, R., Fooks, A.R., Sabeta, C.T., Cliquet, F., and Nel, L.H. (2008b). Epidemiology and pathogenicity of African bat lyssaviruses. Dev. Biol. (Basel) *131*, 317–325.

Marston, D.A., Horton, D.L., Ngeleja, C., Hampson, K., McElhinney, L.M., Banyard, A.C., Haydon, D., Cleaveland, S., Rupprecht, C.E., Bigambo, M., *et al.* (2012). Ikoma lyssavirus: identification of a highly divergent novel *Lyssavirus* in an African civet (*Civettictis civetta*). Emerg. Infect. Dis. *18*, 664–667.

Meredith, C.D., Rossouw, A.P., and van Praag Koch, H. (1971). An unusual case of human rabies thought to be of chiropteran origin. S. Afr. Med. J. *45*, 767–769.

Messenger, S.L., Smith, J.S., and Rupprecht, C.E. (2002). Emerging epidemiology of bat-associated cryptic cases of rabies in humans in the United States. Clin. Infect. Dis. *35*, 738–747.

Mistry, S. (2009). Climate change, vampire bats, and rabies: Modeling range shifts on the US-Mexico border. In 94th Annual Meeting of the Ecological Society of America, August 2–7, Albuquerque, New Mexico. Abstract book. COS 85-2.

Moore, P.R., Jansen, C.C., Graham, G.C., Smith, I.L., and Craig, S.B. (2010). Emerging tropical diseases in Australia. Part 3. Australian bat *Lyssavirus*. Ann. Trop. Med. Parasitol. *104*, 613–621.

Morimoto, K., Patel, M., Corisdeo, S., Hooper, D.C., Fu, Z.F., Rupprecht, C.E., Koprowski, H., and Dietzschold, B. (1996). Characterization of a unique variant of bat rabies virus responsible for newly emerging human cases in North America. Proc. Natl. Acad. Sci. U.S.A. *93*, 5653–5658.

Nadin-Davis, S.A., and Real, L.A. (2011). Molecular phylogenetics of the lyssaviruses-insights from a coalescent approach. Adv. Virus Res. *79*, 203–238.

Nadin-Davis, S.A., Muldoon, F., and Wandeler, A.I. (2006a). Persistence of genetic variants of the arctic fox strain of Rabies virus in southern Ontario. Can. J. Vet. Res. *70*, 11–19.

Nadin-Davis, S.A., Torres, G., Ribas Mde, L., Guzman, M., De La Paz, R.C., Morales, M., and Wandeler, A.I. (2006b). A molecular epidemiological study of rabies in Cuba. Epidemiol. Infect. *134*, 1313–1324.

Nadin-Davis, S.A., Turner, G., Paul, J.P., Madhusudana, S.N., and Wandeler, A.I. (2007). Emergence of Arctic-like rabies lineage in India. Emerg. Infect. Dis. *13*, 111–116.

Nadin-Davis, S.A., Velez, J., Malaga, C., and Wandeler, A.I. (2008). A molecular epidemiological study of rabies in Puerto Rico. Virus Res. *131*, 8–15.

Nadin-Davis, S.A., Feng, Y., Mousse, D., Wandeler, A.I., and Aris-Brosou, S. (2010). Spatial and temporal dynamics of rabies virus variants in big brown bat populations across Canada: footprints of an emerging zoonosis. Mol. Ecol. *19*, 2120–2136.

Nel, L., Jacobs, J., Jafta, J., von Teichman, B., and Bingham, J. (2000). New cases of Mokola virus infection in South Africa: a genotypic comparison of Southern African virus isolates. Virus Genes *20*, 103–106.

Nel, L.H., Sabeta, C.T., von Teichman, B., Jaftha, J.B., Rupprecht, C.E., and Bingham, J. (2005). Mongoose rabies in southern Africa: a re-evaluation based on molecular epidemiology. Virus Res. *109*, 165–173.

Pal, S.R., Arora, B.M., Chuttani, P.N., Broer, S., Choudhury, S., and Joshi, R.M. (1980). Rabies virus infection of a flying fox bat, *Pteropus poliocephalus*, in Chandigarh, Northern India. Trop. Geogr. Med. *32*, 265–267.

Pastoret, P.P., and Brochier, B. (1999). Epidemiology and control of fox rabies in Europe. Vaccine *17*, 1750–1754.

Randall, D.A., Williams, S.D., Kuzmin, I.V., Rupprecht, C.E., Tallents, L.A., Tefera, Z., Argaw, K., Shiferaw, F., Knobel, D.L., Sillero-Zubiri, C., *et al.* (2004). Rabies in endangered Ethiopian wolves. Emerg. Infect. Dis. *10*, 2214–2217.

Ray, C.E., Linares, O.J., and Morgan, G.S. (1988). Palaeontology. In Natural History of Vampire Bats, Greenhall, A.M., and Schmidt, U., eds. (CRC Press, Boca Raton, FL), pp. 19–30.

Ronsholt, L.A. (2002). New case of European bat *Lyssavirus* (EBL) infection in Danish sheep. Rabies Bull. Europe. *26*, 15.

Rupprecht, C.E., Barrett, J., Briggs, D., Cliquet, F., Fooks, A.R., Lumlertdacha, B., Meslin, F.X., Müler, T., Nel, L.H., Schneider, C., *et al.* (2008). Can rabies be eradicated? Dev. Biol. (Basel) *131*, 95–121.

Rupprecht, C.E., Briggs, D., Brown, C.M., Franka, R., Katz, S.L., Kerr, H.D., Lett, S., Levis, R., Meltzer, M.I., Schaffner, W., *et al.* (2009). Evidence for a 4-dose vaccine schedule for human rabies post-exposure prophylaxis in previously non-vaccinated individuals. Vaccine *27*, 7141–7148.

Sabeta, C.T., Markotter, W., Mohale, D.K., Shumba, W., Wandeler, A.I., and Nel, L.H. (2007). Mokola virus in domestic mammals, South Africa. Emerg. Infect. Dis. *13*, 1371–1373.

Salmón-Mulanovich, G., Vásquez, A., Albújar, C., Guevara, C., Laguna-Torres, V.A., Salazar, M., Zamalloa, H., Cáceres, M., Gómez-Benavides, J., Pacheco, V., *et al.* (2009). Human rabies and rabies in vampire and nonvampire bat species, Southeastern Peru, 2007. Emerg. Infect. Dis. *15*, 1308–1310.

Sheeler-Gordon, L.L., and Smith, J.S. (2001). Survey of bat populations from Mexico and Paraguay for rabies. J. Wildl. Dis. *37*, 582–593.

Schneider, L.G., Dietzschold, B., Dierks, R.E., Matthaeus, W., Enzmann, P.J., and Strohmaier, K. (1973). Rabies group-specific ribonucleoprotein antigen and a test system for grouping and typing of rhabdoviruses. J. Virol. *11*, 748–755.

Schneider, L.G. (1982). Antigenic variants of rabies virus. Comp. Immunol. Microbiol. Infect. Dis. *5*, 101–107.

Schneider, L.G., Barnard, B.J.H., and Schneider, H.P. (1985). Application of monoclonal antibodies for epidemiological investigations and oral vaccination studies: I. African viruses. In Rabies in the tropics, Kuwert, E., Merieux, C., Koprowski, H., and Bogel, K., eds (Springer-Verlag, Berlin, Germany), pp. 49–53.

Schneider, M.C., Belotto, A., Adé, M.P., Hendrickx, S., Leanes, L.F., Rodrigues, M.J., Medina, G., and Correa, E. (2007). Current status of human rabies transmitted by dogs in Latin America. Cad. Saude. Publica. *23*, 2049–2063.

Schneider, M.C., Romijn, P.C., Uieda, W., Tamayo, H., da Silva, D.F., Belotto, A., da Silva, J.B., and Leanes, L.F. (2009). Rabies transmitted by vampire bats to humans: an emerging zoonotic disease in Latin America? Rev. Panam. Salud. Publica. *25*, 260–269.

Shope, R.E. (1982). Rabies-related viruses. Yale J. Biol. Med. *55*, 271–275.

Shope, R.E., Murphy, F.A., Harrison, A.K., Causey, O.R., Kemp, G.E., Simpson, D.I., and Moore, D.L. (1970). Two African viruses serologically and morphologically related to rabies virus. J. Virol. *6*, 690–692.

Sibold, C., Meisel, H., Krüger, D.H., Labuda, M., Lysy, J., Kozuch, O., Pejcoch, M., Vaheri, A., and Plyusnin, A. (1999). Recombination in Tula hantavirus evolution: analysis of genetic lineages from Slovakia. J. Virol. *73*, 667–675.

Singer, A., Kauhala, K., Holmala, K., and Smith, G.C. (2009). Rabies in northeastern Europe – the threat from invasive raccoon dogs. J. Wildl. Dis. *45*, 1121–1137.

Slate, D., Algeo, T.P., Nelson, K.M., Chipman, R.B., Donovan, D., Blanton, J.D., Niezgoda, M., and Rupprecht, C.E. (2009). Oral rabies vaccination in North America: opportunities, complexities, and challenges. PLoS Negl. Trop. Dis. *3*, e549.

Smith, J.S., Fishbein, D.B., Rupprecht, C.E., and Clark, K. (1991). Unexplained rabies in three immigrants in the United States. A virologic investigation. N. Engl. J. Med. *324*, 205–211.

Smith, J.S., Orciari, L.A., Yager, P.A., Seidel, H.D., and Warner, C.K. (1992). Epidemiologic and historical relationships among 87 rabies virus isolates as determined by limited sequence analysis. J. Infect. Dis. *166*, 296–307.

Smith, P.C., Lawhaswasdi, K., Vick, W.E., and Stanton, J.S. (1968). Isolation of rabies virus from fruit bats in Thailand. Nature *216*, 384.

Srinivasan, A., Burton, E.C., Kuehnert, M.J., Rupprecht, C., Sutker, W.L., Ksiazek, T.G., Paddock, C.D., Guarner, J., Shieh, W.J., Goldsmith, C., *et al.* (2005). Transmission of rabies virus from an organ donor to four transplant recipients. N. Engl. J. Med. *352*, 1103–1111.

Streicker, D.G., Turmelle, A.S., Vonhof, M.J., Kuzmin, I.V., McCracken, G.F., and Rupprecht, C.E. (2010). Host phylogeny constrains cross-species emergence and establishment of rabies virus in bats. Science. *329*, 676–679.

Theodorides, J. (1986). Histoire de la Rage 'Cave Canem' (Masson, Paris), p. 289.

Tjørnehøj, K., Fooks, A.R., Agerholm, J.S., and Rønsholt, L. (2006). Natural and experimental infection of sheep with European bat *Lyssavirus* type-1 of Danish bat origin. J. Comp. Pathol. *134*, 190–201.

Tordo, N., Poch, O., Ermine, A., Keith, G., and Rougeon, F. (1986). Walking along the rabies genome: is the large G-L intergenic region a remnant gene? Proc. Natl. Acad. Sci. U.S.A. *83*, 3914–3918.

Tordo, N., Ceccaldi, P.E., Gaudin, Y., and Wandeler, A. (2005). Rhabdovirus: Rabies. In Topley & Wilson's Microbiology and Microbial Infections, 10th edn, Collier, L., and Mahy, B., eds (Hodder Arnold, London, UK), pp. 1104–1136.

Van Thiel, P.P., de Bie, R.M., Eftimov, F., Tepaske, R., Zaaijer, H.L., van Doornum, G.J., Schutten, M., Osterhaus, A.D., Majoie, C.B., Aronica, E., *et al.* (2009). Fatal human rabies due to Duvenhage virus from a bat in Kenya: failure of treatment with coma-induction, ketamine, and antiviral drugs. PLoS Negl. Trop. Dis. *3*, e428.

Van Zyl, N., Markotter, W., and Nel, L.H. (2010). Evolutionary history of African mongoose rabies. Virus Res. *150*, 93–102.

Velasco-Villa, A., Orciari, L.A., Juárez-Islas, V., Gómez-Sierra, M., Padilla-Medina, I., Flisser, A., Souza, V., Castillo, A., Franka, R., Escalante-Mañe, M., *et al.* (2006). Molecular diversity of rabies viruses associated with bats in Mexico and other countries of the Americas. J. Clin. Microbiol. *44*, 1697–1710.

Velasco-Villa, A., Reeder, S.A., Orciari, L.A., Yager, P.A., Franka, R., Blanton, J.D., Zuckero, L., Hunt, P., Oertli,

E.H., Robinson, L.E., *et al.* (2008). Enzootic rabies elimination from dogs and reemergence in wild terrestrial carnivores, United States. Emerg. Infect. Dis. *14*, 1849–1854.

Vos, A., Nunan, C., Bolles, D., Müller, T., Fooks, A.R., Tordo, N., and Baer, G.M. (2011). The occurrence of rabies in pre-Columbian Central America: an historical search. Epidemiol. Infect. *139*, 1445–1452.

Wang, Y., McWilliam, S.M., Cowley, J.A., and Walker, P.J. (1994). Complex genome organization in the GNS-L intergenic region of Adelaide River rhabdovirus. Virology *203*, 63–72.

Warrell, M.J. (2008). Emerging aspects of rabies infection: with a special emphasis on children. Curr. Opin. Infect. Dis. *21*, 251–257.

Warrilow, D., Smith, I.L., Harrower, B., and Smith, G.A. (2002). Sequence analysis of an isolate from a fatal human infection of Australian bat *Lyssavirus*. Virology *297*, 109–119.

Warrilow, D., Harrower, B., Smith, I.L., Field, H., Taylor, R., Walker, C., and Smith, G.A. (2003). Public health surveillance for Australian bat *Lyssavirus* in Queensland, Australia, 2000–2001. Emerg. Infect. Dis. *9*, 262–264.

Wiktor, T.J., and Koprowski, H. (1978). Monoclonal antibodies against rabies virus produced by somatic cell hybridization: detection of antigenic variants. Proc. Natl. Acad. Sci. U.S.A. *75*, 3938–3942.

World Health Organization (2004). WHO Expert Consultation on Rabies: First Report (World Health Organization, Geneva, Switzerland).

Wunner, W.H. (2007). Rabies virus. In Rabies, Jackson, A., and Wunner, W., eds (Elsevier, London, UK), pp. 23–68.

Yang, D.K., Park, Y.N., Hong, G.S., Kang, H.K., Oh, Y.I., Cho, S.D., and Song, J.Y. (2011). Molecular characterization of Korean rabies virus isolates. J. Vet. Sci. *12*, 57–63.

Ephemeroviruses: Arthropod-borne Rhabdoviruses of Ruminants, with Large, Complex Genomes

6

Peter J. Walker, Kim R. Blasdell and D. Albert Joubert

Abstract

The genus *Ephemerovirus* comprises arthropod-borne rhabdoviruses that infect primarily ruminants. They include bovine ephemeral fever virus (BEFV) which is one of the most important vector-borne pathogens of cattle, Berrimah virus (BRMV), Kimberley virus (KIMV) and Adelaide River virus (ARV) which were isolated in Australia, Malakal virus (MALV), Obodhiang virus (OBOV) and kotonkan virus (KOTV) from Africa, and Puchong virus (PUCV) from Malaysia. Ephemeroviruses have similar morphology to other animal rhabdoviruses but have large complex genomes encoding a non-structural glycoprotein (G_{NS}) and several small accessory proteins. BEFV is enzootic in tropical and subtropical regions of Africa, Asia and Australia, extending into temperate climes in sometimes extensive epizootics. Clinical signs include fever, nasal and ocular discharge, stiffness, anorexia, rumenal stasis, joint swelling and limb paralysis. BEFV has been isolated from mosquitoes and biting midges (*Culicoides* spp.) but the specific vectors responsible for transmission are not defined. Disease outbreaks are usually associated with heavy rainfall or conditions that favour the proliferation of insect populations. KOTV also causes an ephemeral fever-like illness but other ephemeroviruses are not known to cause disease. Live-attenuated and inactivated BEFV vaccines are available but require multiple doses and the adoption rate, particularly in extensive beef industries, is low.

Introduction

The ephemeroviruses are arthropod-borne rhabdoviruses. They are enveloped (–) ssRNA viruses with the bullet- or cone-shaped morphology, and have the unusual characteristic of large and complex genomes. They appear to infect primarily cattle and other ruminants, and some are known to cause a debilitating and sometimes fatal disease. Bovine ephemeral fever virus (BEFV) is enzootic in ruminants in tropical regions of Africa, Asia and Australia, extending seasonally into subtropical and temperate zones to cause epizootics that can have serious economic impacts. Other ephemeroviruses occur commonly in cattle with a similar geographic distribution but have not been associated with disease, and some have been isolated only rarely from insects or cattle and their distribution and disease potential remains unknown. In this chapter, we review the molecular and biological characteristics of ephemeroviruses, focusing primarily on BEFV, the most intensively studied and economically most important of this somewhat neglected group of viruses.

The genus *Ephemerovirus*

The genus *Ephemerovirus* is one of six currently recognized genera in the *Rhabdoviridae* (Dietzgen *et al.*, 2011). It comprises the type species (*Bovine ephemeral fever virus*), three other recognized species (*Berrimah virus, Adelaide River virus* and *Kotonkan virus*) and one proposed new species (*Obodhiang virus*). Three other viruses (*Kimberley virus, Malakal virus* and *Puchong virus*) are closely related serologically and genetically to the recognized ephemeroviruses and are probable members of the genus (Walker *et al.*, 2011). Several other groups of related arthropod (dipteran)-borne rhabdoviruses infect vertebrates, including the vesiculoviruses and viruses in the Hart Park, Almpiwar, Le Dantec and Tibrogargan serological groups (Tesh *et al.*, 1983; Calisher *et al.*, 1989). Many of these viruses are presently unclassified but they share a common genetic lineage and have similar biological characteristics to the ephemeroviruses with which they cluster phylogenetically as the Dimarhabdovirus supergroup (Bourhy *et al.*, 2005).

History

BEFV was first recognized as an infectious agent in South Africa in the early twentieth century when it was passed experimentally in cattle by intravenous blood inoculation (Bevan, 1912). It was first isolated from cattle blood in 1967 by intracerebral passage in suckling mice (van der Westhuizen, 1967) and subsequently adapted to growth in cell culture (Inaba *et al.*, 1968). It has since been isolated on many occasions from cattle, as well as biting midges (*Culicoides* spp.) and mosquitoes. It occurs as a single cross-neutralizing serotype globally and is considered one of the most important vector-borne pathogens of livestock.

Berrimah virus (BRMV) is very closely related to BEFV but is distinguishable both serologically and genetically. It was isolated on only one occasion from a healthy sentinel steer near Darwin, Australia, in 1981 (Gard *et al.*, 1983). Adelaide River virus (ARV) was also first isolated in 1981 from a healthy sentinel steer at Tortilla Flat Research Farm, 90 km south-east of Darwin (Gard *et al.*, 1984). The isolation of these viruses was the consequence of the implementation of an extensive sentinel herd system in Australia and Papua New Guinea (St. George, 1980; Gard *et al.*, 1988). ARV is clearly distinguishable from BEFV and BRMV in serum neutralization tests and has distinct genetic characteristics.

Kotonkan virus (KOTV) was isolated from a mixed pool of biting midges collected at the University Agricultural Farm at Ibadan, Nigeria, in December 1967 (Kemp *et al.*, 1973a). It was originally considered to be a probable lyssavirus based on evidence of serological cross-reactions with rabies virus and rabies-related viruses, and rapid adaptation to growth in cultured murine neuroblastoma cells (Bauer and Murphy, 1975; Clark, 1980). However, more recent serological and genetic evidence, together with its morphology, arthropod vector, vertebrate host range and disease association, have clearly identified KOTV as an ephemerovirus. It is the only ephemerovirus other than BEFV that has been shown to cause disease in cattle (Tomori *et al.*, 1974).

Obodhiang virus (OBOV) was isolated on two occasions from mosquitoes collected near Malakal in southern Sudan in 1963 (Schmidt *et al.*, 1965; Shope, 1975). Little is known of the natural history of OBOV but it was identified as a rhabdovirus based on typical morphological characteristics and shown to be capable of replication in cultured insect cells (Buckley, 1973, 1975; Bauer and Murphy, 1975). Like KOTV, it was provisionally classified as a lyssavirus because of weak cross-reactions in group-specific serological tests and evidence of adaptation to murine neuroblastoma cells (Bauer and Murphy, 1975; Clark, 1980). However, it is now known that OBOV is more closely related genetically and serologically to the ephemeroviruses. It is most closely related to ARV with which it cross-reacts weakly in serum neutralization tests (Blasdell *et al.*, 2012).

Kimberley virus (KIMV) was first isolated in 1973 from a pool of mosquitoes collected in the Ord River Valley of Western Australia (Liehne *et al.*, 1981). It was subsequently isolated from healthy sentinel cattle at the Tortilla Flat Research Farm in the Northern Territory in 1980 (Cybinski and Zakrzewski, 1983) and at Peachester near Brisbane in South-East Queensland in 1982 (Cybinski and Muller, 1990), as well as from biting midges collected at Kairi in North Queensland in 1982 (Zakrzewski and Cybinski, 1984). KIMV is serologically distinct from most other ephemeroviruses but cross-reacts strongly in neutralization tests with Malakal virus (MALV) (K.R. Blasdell and P.J. Walker, unpublished dada). MALV was isolated on a single occasion from mosquitoes collected at Malakal in the Upper Nile Province of Sudan during the month prior to the first isolation of OBOV from the same location in 1963 (Schmidt *et al.*, 1965; Karabatsos, 1985). The potency of cross-reaction between MALV and KIMV in serum neutralization tests suggests they should be considered strains of the same virus.

Puchong virus (PUCV) was isolated on only one occasion from a pool of gravid female mosquitoes collected near Kuala Lumpur in Selangor, Malaysia in 1965 (Karabatsos, 1985). PUCV cross-reacts strongly in complement-fixation and indirect immunofluorescence tests with MALV (Karabatsos, 1985; Calisher *et al.*, 1989) but the viruses do not cross-react in serum-neutralization tests and are genetically quite distinct (P.J. Walker and K. Blasdell, unpublished data).

Analysis of the phylogenetic relationships amongst a wide range of animal rhabdoviruses using short segments of L or N protein amino acid sequences supports the clustering of BEFV, BRMV, KIMV, ARV, OBOV and KOTV in a single genetic lineage (Bourhy *et al.*, 2005; Kuzmin *et al.*, 2006). This is also evident from the phylogenies of complete sequences of the G proteins (Fig. 6.1) which indicate that all ephemeroviruses cluster as a distinct genetic lineage and are most closely related to Tibrogargan virus (TIBV) and Coastal Plains virus (CPV) in the proposed genus *Tibrovirus* (Gubala *et al.*, 2011; Walker *et al.*, 2011). TIBV has been isolated from biting midges (*Culicoides brevitarsis*) and both TIBV and CPV have been isolated from healthy sentinel cattle in Australia (Cybinski *et al.*, 1980; Cybinski and Gard, 1986). Serological surveys have indicated that, like the ephemeroviruses, they infect primarily cattle and water buffalo but CPV antibody has also been detected in dogs, horses and goats. TIBV is closely related serologically to Bivens Arm virus (BAV) which was isolated from biting midges (*Culicoides insignis*) in the USA and has been shown to infect cattle and water buffalo in Florida and the Caribbean Islands (Gibbs *et al.*, 1989; Tuekam *et al.*, 1991). The tibroviruses and ephemeroviruses appear therefore to share a similar ecology but they form distinct genetic lineages and, as discussed below, they differ significantly in genome organization.

Virion morphology, structure and morphogenesis

Ephemerovirus virions display the bullet-shaped morphology that is characteristic of animal rhabdoviruses but have often been described as more cone-like in appearance with the particle tapered inwards towards the apex (Ito *et al.*, 1969; Lecatsas *et al.*, 1969; Murphy *et al.*, 1972; Bauer and Murphy, 1975). For BEFV, estimations of the size of virions viewed by negative-contrast electron microscopy vary but the average central diameter is approximately 75 nm and the maximum length from early cell culture harvests is 185 nm. Interestingly, estimations of the length of OBOV (170 nm) and KOTV (182 nm) suggest variations that reflect the length of their respective genomes (Bauer and Murphy, 1975). Like other rhabdoviruses, BEFV, OBOV and KOTV virions display an axial channel intruding from the base and surface projections on the envelope that have been estimated to extend 8 nm from the lipid membrane (Ito *et al.*, 1969; Murphy *et al.*, 1972; Bauer and Murphy, 1975). Nucleocapsids are helical with a periodicity of 4.8 nm and an appearance that mimics the conical shape of virions, apparently with the thread-like nucleoprotein wrapped around a central core (Murphy *et al.*, 1972; Bauer and Murphy, 1975). Defective-interfering (DI) particles occur commonly and take the form of variously truncated or otherwise deformed virions (Murphy *et al.*, 1972; Bauer and Murphy, 1975). The presence of DI particles in negatively stained virus preparations is likely to have contributed to variations in the reported length of BEFV virions.

Like other rhabdoviruses, BEFV contains five major structural proteins that can be released progressively from purified virions by treatment with mild non-ionic detergent and salt (Walker *et al.*, 1991). Nucleocapsids comprise the (–) ssRNA genome, the nucleoprotein (N) and the large (L) and small (P) subunits of the RNA-dependent RNA polymerase. The matrix protein (M) remains associated with nucleocapsids following treatment of virions with detergent in low-salt conditions which remove the envelope containing a single class 1 transmembrane glycoprotein (G) that forms the surface projections (Walker *et al.*, 1991). The N, L, P, M and G proteins have been detected in purified virions

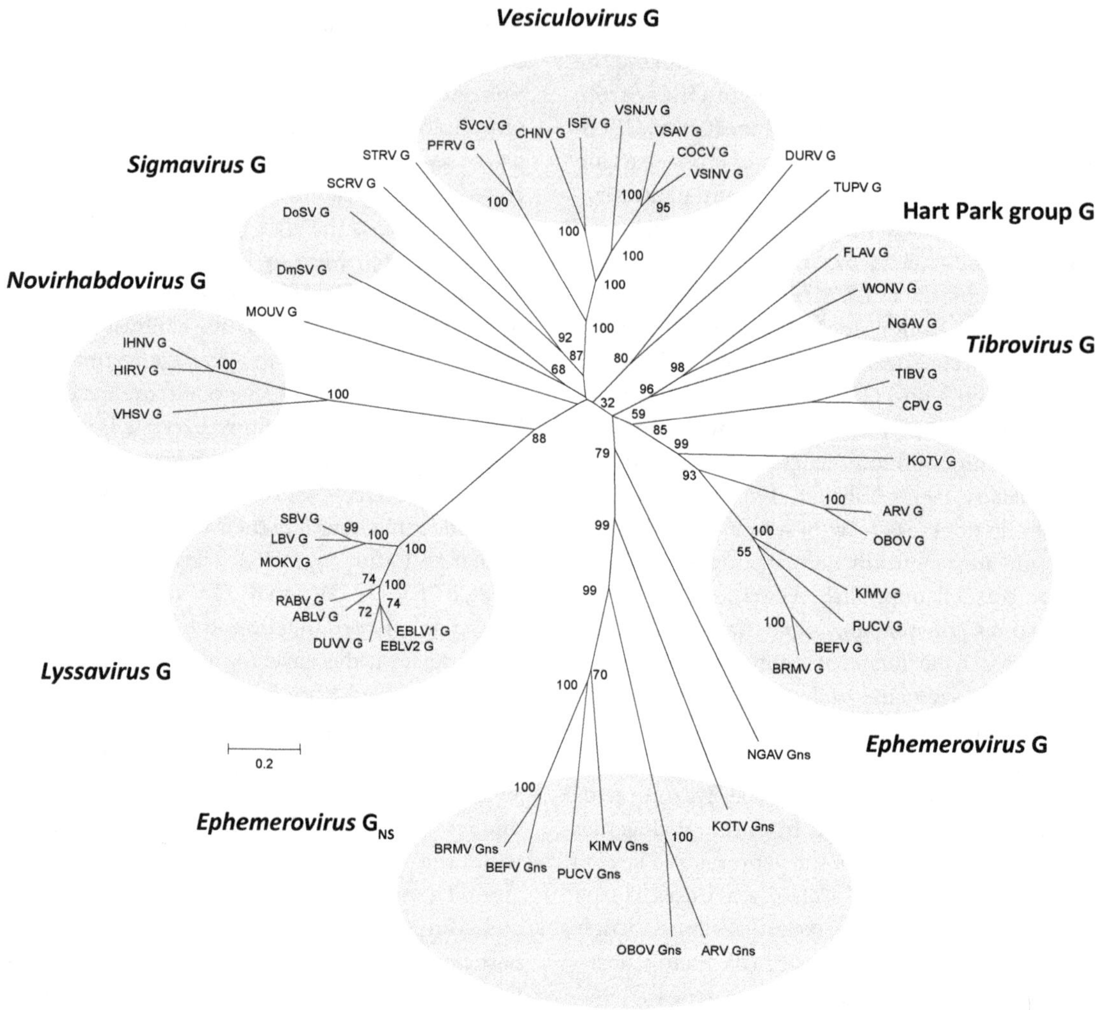

Figure 6.1 Phylogenetic analysis of full length sequences of animal rhabdovirus G and G_{NS} proteins. The minimum evolution consensus bootstrap tree (1000 replicates) was constructed in MEGA from a Clustal W sequence alignment using default parameters. Rhabdoviruses included in the analysis include (clockwise from top) pike fry rhabdovirus (PFRV), spring viraemia of carp virus (SVCV), Chandipura virus (CHNV), Isfahan virus (ISFV), vesicular stomatitis New Jersey virus (VSNJV), vesicular stomatitis Alagoas virus (VSAV), Cocal virus (COCV), vesicular stomatitis Indiana virus (VSIV), Durham virus (DURV), tupia virus (TUPV), Flanders virus (FLAV), Wongabel virus (WONV), Ngaingan virus (NGAV), Tibrogargan virus (TIBV), Coastal Plains virus (CPV), kotonkan virus (KOTV), Adelaide River virus (ARV), Obodhiang virus (OBOV), Kimberley virus (KIMV), Puchong virus (PUCV), bovine ephemeral fever virus (BEFV), Berrimah virus (BRMV), European bat *Lyssavirus* 1, European bat *Lyssavirus* 2, Duvenhage virus (DUVV), Australian bat *Lyssavirus* (ABLV), rabies virus (RABV), Mokola virus (MOKV), Lagos bat virus (LBV), Shimoni bat virus (SBV), vesicular haemorrhagic septicaemia virus (VHSV), Hirame rhabdovirus (HIRV), infectious haematopoietic necrosis virus (IHNV), Moussa virus (MOUV), *Drosophila melanogaster* sigmavirus (DmSV), *Drosophila obscura* sigmavirus (DoSV), *Simiperca chuatsi* rhabdovirus (SCRV) and sea trout rhabdovirus (STRV). Clusters of viruses representing assigned genera and other recognized groups are shaded.

of other ephemeroviruses (Blasdell *et al.*, 2012; K. Blasdell and P.J. Walker, unpublished data). There is also evidence that several smaller proteins (<20kDa) encoded in ephemerovirus accessory genes may be present in virions (see below).

Ephemeroviruses replicate in the cytoplasm of infected cells. Budding of mature virions takes place at cytoplasmic membranes or into intracytoplasmic vesicles, with virions accumulating in vesicles and in intracellular spaces (Ito *et al.*, 1969; Bauer and Murphy, 1975). Cytoplasmic inclusions similar to the massed nucleocapsid inclusions of rabies virus have been observed in association with the sites of budding of BEFV, OBOV and KOTV (Murphy *et al.*, 1972; Bauer and Murphy, 1975). Infected cells often display hypertrophied membranes of the endoplasmic reticulum and Golgi complex and may become highly vacuolated late in infection (Murphy *et al.*, 1972). Similar patterns of morphogenesis and cytopathology have been reported in mammalian cell cultures and in the neurons of infected mice (Ito *et al.*, 1969; Lecatsas *et al.*, 1969; Holmes and Doherty, 1970; Murphy *et al.*, 1972; Bauer and Murphy, 1975). Little is yet known of the processes of infection and virion maturation in insect cells.

Genome organisation and transcription strategy

The non-segmented (–) ssRNA genomes of the ephemeroviruses are amongst the largest and most complex known in rhabdoviruses, ranging in size from approximately 14.6–15.9 kb and containing up to 12 long open reading frames (ORFs) (Table 6.1 and Fig. 6.2). The five structural protein genes are arranged in the canonical order (3′-N–P–M–G–L-5′) and are bounded by partially complementary 3′ and 5′ untranslated regions that contain promoter elements for RNA-dependent RNA polymerase binding during replication and transcription. Conserved transcription initiation (TI) and transcription termination/polyadenylation (TTP) sequences flank each gene to direct transcription of capped and polyadenylated mRNAs (Table 6.2). However, the unique and defining feature of the ephemeroviruses is the presence of an additional complex coding region between the G and L genes containing a set of accessory genes. The structure of this region is similar in all ephemeroviruses but varies in viruses of different species.

In all ephemeroviruses, the G-L intergenic region commences with a gene encoding a large non-structural glycoprotein (G_{NS}) that is related in structure and amino acid sequence to the G protein. As discussed below, ephemerovirus G_{NS} genes share a common ancestor with rhabdovirus G genes and appear to have arisen by an ancient gene duplication or recombination event (Wang *et al.*, 1994; Walker *et al.*, 2011). The G_{NS} gene is followed invariably by two genes (α and β) which are each bounded by TI and TTP sequences. The α gene contains two consecutive ORFs (α1 and α2); α1 encodes a protein that has the characteristics of a viroporin and α2 encodes a protein of unknown function. The α2 ORF is conserved in all ephemeroviruses so appears to be of functional significance. In most cases, the α2 initiation codon is in close proximity to the α1 termination codon (i.e. within 27 nucleotides) and so expression may involve translation re-initiation as has been described for several other RNA viruses including the mononegavirus, respiratory syncytial virus (Ahmadian *et al.*, 2000; Powell, 2010). In BEFV, a third small ORF (α3) occurs within α2 in a different reading frame. The BEFV α3 protein does not occur in other ephemeroviruses and it is not known if it is expressed. Like the G_{NS} gene, the α and β genes occur in all known ephemeroviruses but the region downstream is more variable. In BEFV, BRMV, KIMV and KOTV, the β gene is followed by the γ gene; and in KOTV, the γ gene is followed by the δ gene. The functions of proteins encoded in the β, γ and δ genes are also presently unknown. Several ephemeroviruses also contain an alternative ORF in the P gene encoding a small basic protein (P′). The function of P′ proteins, if they are indeed expressed, may be analogous to the small basic proteins (C and C′) encoded in the P genes of vesiculoviruses (Spiropoulou and Nichol, 1993; Peluso *et al.*, 1996).

With the exception of the α and P genes, in which there are alternative or consecutive ORFs, TI and TTP sequences flanking each gene usually result in transcription of monocistronic mRNAs. However, in some cell-culture-adapted strains, corruptions in the stretch of seven uridine residues in TTP sequences result in expression of bicistronic or polycistronic mRNAs. In BEFV, the cell-culture-adapted BB7721 strain, a β-γ bicistronic mRNA is transcribed owing to abbreviation of the TTP sequence following the β gene ($GUAC[U]_6$) (McWilliam *et al.*, 1997). This strain also has a

Table 6.1 Characteristics of ephemeroviruses

Virus	Abbreviation	Countries isolated	Source of isolation	Disease/antibody reported historically[a]	Genome size (nt)	Accessory genes	GenBank accession
Bovine ephemeral fever virus	BEFV	South Africa, Zimbabwe, Nigeria, Kenya, Israel, Turkey, India, China, Taiwan, Japan, Australia	Cattle, biting midges, mosquitoes	South Africa, Lesotho, Botswana, Swaziland, Madagascar, Zimbabwe, Zambia, Angola, Malawi, Tanzania, Burundi, Rwanda, Kenya, Ethiopia, Uganda, Congo, Central African Republic, Nigeria, Guinea-Bissau, Mauritania, Chad, Egypt, Yemen, Kuwait, Jordan, Israel, Syria, Turkey, Iraq, Iran, Pakistan, India, Nepal, Sri Lanka, Laos, China, Japan, Taiwan, Malaysia, Singapore, Indonesia, Papua New Guinea, Australia,	14900	G_{NS}, α1, α2, α3, β, γ	AF234533
Berrimah virus	BRMV	Australia	Cattle	Australia, Indonesia	15024	G_{NS}, α1, α2, β, γ	HM461974
Kimberley virus	KIMV	Australia	Cattle, biting midges, mosquitoes	Australia, Indonesia, China	15430+	G_{NS}, α1, α2, β, γ	NA
Malakal virus	MALV	Sudan	Mosquitoes	NA	15388+	G_{NS}, α1, α2, β, γ	NA
Puchong virus	PUCV	Malaysia	Mosquitoes	NA	14812+	G_{NS}, α1, α2, β, γ	NA
Kotonkan virus	KOTV	Nigeria	Biting midges	Nigeria	15870	G_{NS}, α1, α2, β, γ, δ	HM474855
Adelaide River virus	ARV	Australia	Cattle	Australia, Indonesia, China	14614+	G_{NS}, α1, α2, β	U05987
Obodhiang virus	OBOV	Sudan	Mosquitoes	NA	14717	G_{NS}, α1, α2, β	HM856902

NA, not available.
[a]Burgess (1971) and other sources.

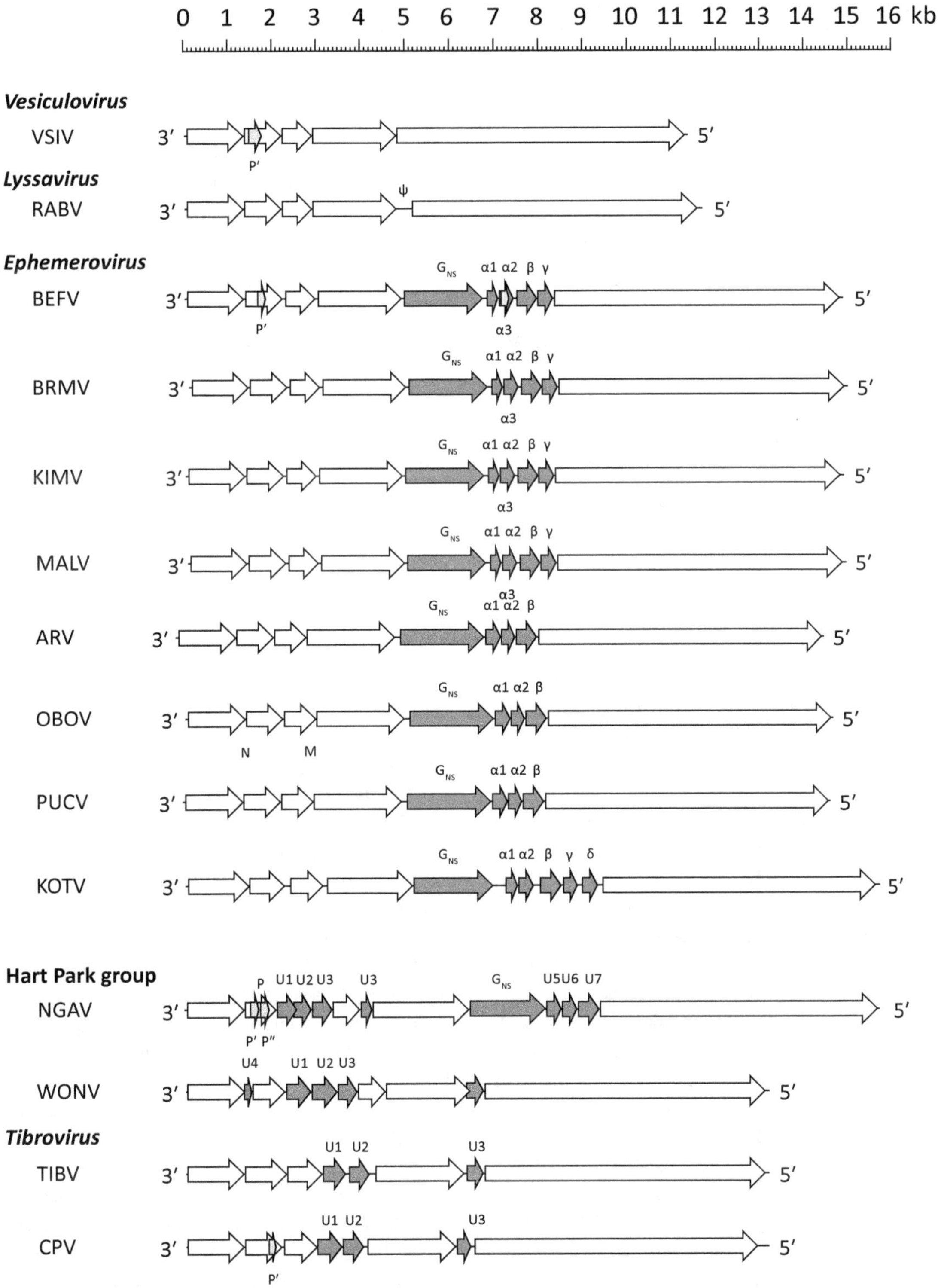

Figure 6.2 Genome organization of viruses representing the genus *Vesiculovirus*, genus *Lyssavirus*, genus *Ephemerovirus*, the Hart Park group and the proposed genus *Tibrovirus*. The positions of ORFs are indicated and those encoding accessory proteins are shaded. Abbreviations of virus names are as in Fig. 6.1.

Table 6.2 Relatively conserved transcriptional control sequences in ephemeroviruses

Virus	Transcriptional control sequences	
	TI[a]	TTP[b]
BEFV	**UUGUC**C	G(U/A)A**CU**$_7$
BRMV	**UUGUC**C	G(U/A)A**CU**$_7$
KIMV	**UUGUC**C	(G/U)UA**CU**$_7$
MALV	**UUGUC**C	(G/U/A)UA**CU**$_7$
PUCV	**UUGUC**(C/A/U)	G/U)(U/A)A**CU**$_7$
ARV	**UUGUC**(C/G/U)	G(U/C)A**CU**$_7$
OBOV	**UUGUC**(C/G/U)	G(U/A) A**CU**$_7$
KOTV	**UUGUC**(C/G/U)	(G/U)(U/A/)(U/A)**CU**$_7$
Consensus	UUGUCC	GUACU$_7$

Sequences in bold are universally conserved.
BEFV, bovine ephemeral fever virus; BRMV, Berrimah virus; KIMV, Kimberley virus; MALV, Malakal virus; PUCV, Puchong virus; KOTV, Kotonkan virus; ARV, Adelaide River virus; OBOV, Obodhiang virus.
[a]Transcription initiation.
[b]Transcription termination.

lethal mutation which severely truncates the β ORF, suggesting that normal expression of the β and γ genes is unnecessary for efficient growth in cell culture. In the cell-culture-attenuated vaccine strain (919) derived from the same BEFV isolate, the TTP sequence following the G_{NS} gene is similarly corrupted (L. Trinidad and P.J. Walker, unpublished data), and in a cell-culture-adapted strain of ARV, TTP sequences following the G and G_{NS} genes are corrupted, resulting in a polycistronic G-G_{NS}-α mRNA (Wang *et al.*, 1994). These corruptions do not occur in ephemeroviruses that have been passaged only in cattle, or at low passage in cell culture, suggesting that the accessory genes are likely to have important functions *in vivo*.

Tibroviruses share a similar ecology with ephemeroviruses and are the most closely related viruses phylogenetically. However, tibrovirus genomes are smaller and less complex (Gubala *et al.*, 2011). The 13.3 kb TIBV genome contains two small ORFs (U1 and U2) between the M and G genes, and one small ORF (U3) between the G and L genes, each bounded by TI and TTP sequences. The 13.2 kb CPV genome has a similar organization but there is a fourth small ORF (U4) overlapping the G ORF. In each virus, ORF U3 encodes a protein that appears to be the orthologue of the ephemerovirus α1 protein with the structural characteristics of a viroporin. However, there is no G_{NS} gene in tibroviruses and the other accessory genes appear to be unrelated to those of ephemeroviruses (Gubala *et al.*, 2011; Walker *et al.*, 2011).

Evolution of genome complexity

The increasing availability of animal and plant rhabdovirus genome sequences has revealed extraordinary diversity and varying degrees of complexity that are not evident in the relatively simple genomes of vesiculoviruses and lyssaviruses. Rhabdoviruses are now known commonly to encode accessory proteins in various locations throughout the genome as independent genes or as alternative overlapping or consecutive ORFs. Indeed, in various rhabdoviruses, accessory genes have been identified in each of the junctions between the N, P, M, G and L genes and within all genes except the L gene (Walker *et al.*, 2011). This, of course, raises the interesting if somewhat speculative question as to the origin of these accessory genes and whether the evolution of rhabdoviruses genomes has favoured increasing or decreasing genetic complexity.

It may appear superficially logical to argue that purifying selection would drive rhabdovirus evolution towards small efficient genomes and the frequent generation of DI genomes during rhabdovirus replication seems to provide a suitable mechanism for elimination of redundant sequences (Yang and Lazzarini, 1983). However, this simple proposition invokes the assumption that rhabdoviruses first emerged with large complex genomes, perhaps derived by regressive evolution from endosymbiotic bacteria. As discussed by Holmes (2009), there are several reasons to suggest that this is rather improbable. The most compelling of these is the fitness cost associated with the high mutation frequency of RNA polymerase which restricts the size of RNA virus genomes. In contrast, there are convincing arguments that evolution has generally favoured increasing complexity, with RNA viruses originating in a primitive RNA world that preceded DNA-based replication (Holmes, 2009). The evolution of increasing complexity in rhabdovirus genomes is evident in the P gene of VSV which encodes proteins with regions of sufficient structural flexibility to allow the high rate of nucleotide substitution necessary for the evolution of alternative ORFs (Bilsel *et al.*, 1990). The P gene of rabies virus (RABV) and M gene of VSV also encode multiple proteins translated from alternative initiation codons (Chenik *et al.*, 1995; Jayakar and Whitt, 2002). The high level of amino acid sequence similarity and conserved structural characteristics of ephemerovirus G and G_{NS} proteins suggest the evolution of increasing complexity may also occur through gene duplication or homologous recombination (Walker *et al.*, 1992; Wang and Walker, 1993). Indeed, evidence of nucleotide or amino acid sequence similarity in other adjacent genes of ephemeroviruses and other rhabdoviruses suggest that the evolution of increasing genome complexity by these mechanisms may be a common theme (McWilliam *et al.*, 1997; Walker *et al.*, 2011). Gene duplication could occur through a simple process of polymerase jumping similar to that proposed for the generation of DI genomes (Yang and Lazzarini, 1983) but involving upstream (rather than downstream) relocation of the polymerase and nascent strand during replication (Wang and Walker, 1993). However, it could be argued that a perfectly duplicated gene may offer no selective advantage and the larger genome would be rapidly eliminated by purifying selection before a new or differentiated function could evolve. Alternatively, homologous recombination could provide immediate selective advantage by various means including increased host range, cell tropism, replication efficiency or by more effective engagement of host defensive responses (Walker *et al.*, 2011). Although recombination is rare in negative-sense RNA viruses, there is increasing evidence that it can occur in evolutionary timeframes (Sibold *et al.*, 1999; Chare *et al.*, 2003; Lukashev, 2005; Han *et al.*, 2008; Qin *et al.*, 2008) and so it must be considered an equally plausible mechanism for increasing genome complexity.

Until recently, the G_{NS} gene was considered to be a defining characteristic of the ephemeroviruses but it is now known that Ngaingan virus (NGAV), an arthropod-borne rhabdovirus that was isolated from a mixed pool of biting midges (*Culicoides* spp.) in Australia and infects primarily marsupials, also contains a G_{NS} gene (Doherty *et al.*, 1973; Gubala *et al.*, 2010). Phylogenetic analysis indicates that the ephemerovirus and NGAV G_{NS} genes form a monophyletic clade, suggesting they share a common ancestor (Fig. 6.1). However NGAV has a very different genome organization from the ephemeroviruses and the G, N and L proteins are phylogenetically distinct, clustering with Flanders virus (FLAV) and Wongabel virus (WONV) to form the Hart Park group (Figs 6.1 and 6.2). As FLAV and WONV lack G_{NS}, it appears that NGAV G_{NS} may have been obtained by homologous recombination with an ancestral ephemerovirus. Alternatively, the Hart Park group and the ephemeroviruses may share a common ancestor with a G_{NS} gene which has subsequently been lost during the evolution of FLAV and WONV. Either way, the presence of the G_{NS} gene in NGAV does present a taxonomic conundrum that may only be resolved when the genome sequences of further arthropod-borne rhabdoviruses have been determined. On balance, it appears most likely that the evolution of genome complexity in ephemeroviruses and other rhabdoviruses is not unidirectional but rather is a dynamic process through which the new genes may emerge and subsequently be lost through the forces of natural selection.

Major structural proteins

The five major structural proteins of ephemeroviruses appear to be functionally homologous to the corresponding proteins of other rhabdoviruses, at least with respect to their primary functions.

N proteins

Ephemerovirus nucleoproteins (N) are acidic molecules that share in excess of 46% amino acid sequence identity and range in size from 47.9 to 49.7 kDa. The BEFV N protein has been shown to bind tightly to the RNA genome following treatment of virions with salt and detergent (Walker *et al.*, 1991). Sequence alignments with the nucleoproteins of other rhabdoviruses suggest the ephemerovirus N proteins adopt a very similar folded structure to those of VSV and RABV when complexed with the RNA, with N-terminal and C-terminal lobes comprising mainly a series of α-helices forming upper and lower jaws around a cavity that accommodates the RNA (Albertini *et al.*, 2006; Green *et al.*, 2006). They also share a highly conserved region near the cavity between the two lobes with the common sequence GxSxK-SPYSS that has important roles in encapsidation, replication and transcription (Wang *et al.*, 1995; Nayak *et al.*, 2009). Interestingly, the C-terminal loop structure between helices α12 and α13 that is involved in interactions between adjacent N molecules in the VSV ribonucleoprotein is extended in all ephemeroviruses by an additional highly acidic domain containing runs of multiple aspartate and glutamate residues (Green *et al.*, 2006). This extension also occurs as a disordered region in RABV N protein but does not have the same highly acidic characteristic (Albertini *et al.*, 2006). In VSV, mutations in this loop stimulate replication, possibly by facilitating polymerase access to the RNA template (Harouaka and Wertz, 2009).

P proteins

Ephemerovirus P proteins range in size from 29.9 to 34.3 kDa. They are the least conserved in amino acid sequence of the major structural proteins but share several common structural characteristics including a highly acidic N-terminal domain, a relatively hydrophobic central region and a C-terminal domain containing two short highly basic regions. In VSV and RABV, the P proteins have a modular structure in which functional domains are interspersed with disordered regions. The N-terminal domain interacts with the L protein and soluble N protein (N^0), the central domain is responsible for P protein dimerization, and the C-terminal domain binds with the N–RNA complex (Gerard *et al.*, 2009; Delmas *et al.*, 2010). The RABV P protein is expressed as five isoforms from alternative initiation codons (Chenik *et al.*, 1995). The smaller three of these forms (P3–P5) are transported to the nucleus owing to the presence of a nuclear localization signal (NLS) located in the C-terminal region of the protein, whereas the cytoplasmic distribution of the longer forms (P and P2) is due to a nuclear export signal (NES) in the N-terminal region of the protein (Pasdeloup *et al.*, 2005). The charge distribution and secondary structure predictions for ephemerovirus P proteins are similar to those of vesiculoviruses and lyssaviruses, suggesting that they are likely to adopt a similar modular conformation.

M proteins

The M proteins of ephemeroviruses range in predicted size from 24.3 to 26.3 kDa. Although relatively variable in sequence, particularly between the more distantly related species (e.g. ARV and KOTV), they share in excess of 27% amino acid sequence identity. They are also clearly related in sequence to the M proteins of VSV and RABV, sharing several conserved regions including a highly basic domain near the N-terminus and a hydrophobic sequence in the central domain which is preceded by the common sequence motif $Dx_{1-2}YxDxGx_{2-4}(K/R)$. This motif includes the sequence that appears to be associated with the induction of apoptosis by lyssavirus M proteins (Larrous *et al.*, 2010) and the formation of nucleocapsid–M complexes in VSV (Dancho *et al.*, 2009). Ephemerovirus M proteins also contain the late budding domain (L-domain) motifs in the N-terminal region that occur in many enveloped viruses and are associated with the recruitment of cellular proteins involved in the budding process. The motif PPxY interacts with Nedd4-related E3 ubiquitin ligases and the motif

P(S/T)AP recruits the protein TSG101, each involved in the formation of multivesicular bodies (Freed, 2002; Gottwein *et al.*, 2003). Both of these motifs occur in VSV M but only PPxY has been shown to be involved in efficient budding, and in RABV M, only the PPxY motif is present (Harty *et al.*, 1999; Jayakar *et al.*, 2000; Irie *et al.*, 2004; Raux *et al.*, 2010). Amongst ephemeroviruses, BEFV, BRMV, KIMV, MALV and PUCV also each contain the PPxY motif but in ARV, OBOV and KOTV only the P(S/T)AP motif is present. Interestingly, ARV, OBOV and KOTV are also more similar to VSV in overall sequence of the M protein, whereas the other ephemeroviruses are more similar to RABV.

G proteins

As in other rhabdoviruses, ephemerovirus G proteins are class I transmembrane glycoproteins that protrude from the surface of the viral envelope (Walker *et al.*, 1991). The encoded proteins range in size from 72.3 to 77.2 kDa prior to signal peptide cleavage and glycosylation. Each contains four or five potential N-glycosylation sites that vary in location and a set of 18 conserved cysteine residues in the globular head of the ectodomain (Walker *et al.*, 1992; Wang and Walker, 1993). Six of the cysteine residues appear to form a set of three disulfide bridges that are common to all animal rhabdoviruses (Walker and Kongsuwan, 1999; Walker, 2008) (Fig. 6.3). The 81 kDa mature BEFV G protein is the target of neutralizing antibodies and induces protective immunity in cattle (Cybinski *et al.*, 1990; Uren *et al.*, 1994). Four independent neutralizing antigenic sites (G1-G4) have been identified by competition ELISA using monoclonal antibodies and three of these have been mapped by sequence analysis of neutralization escape mutants (Cybinski *et al.*, 1990, 1992; Kongsuwan *et al.*, 1998) (Fig. 6.4). Site G1 spans amino acids Y^{487}-K^{503} in the stalk region of the G protein, comprising two overlapping domains that appear to have some local conformation. The G proteins of the ephemeroviruses and the related arthropod-borne tibroviruses are the largest of known rhabdoviruses due primarily to a longer stalk region in which the G1 antigenic site is located (Walker and Kongsuwan, 1999; Gubala *et al.*, 2011). G2 is a conformational site spanning conserved residues C^{172} and C^{182} that appear to form the C_{VI}-C_{VII} disulfide bridge in the mature BEFV G protein. In the loop between these cysteine residues, there is a potential N-glycosylation site in BEFV but it does not occur in other ephemeroviruses. G3 is the major conformational neutralization site comprising two partially overlapping elements (G3a and G3b) and encompassing three different regions of the G protein (Q^{49}-D^{57}; K^{215}-E^{229}; and Q^{265}) that appear to be aligned in the folded structure (Fig. 6.4). In the VSV G protein this region (designated domain III) comprises four β-stranded sheets and two helices and has the fold of a pleckstrin homology (PH) domain (Roche *et al.*, 2006). The major neutralization sites of other rhabdovirus G proteins are also located in this PH domain and the lateral domain (designated domain I) that together form a cleft which may serve as the receptor-binding site (Roche *et al.*, 2007). The BEFV G protein has also been shown to be fusogenic at low pH and all ephemerovirus share with VSV a poly-proline helix and critical aromatic residues at the tips of the cd and Pe fusion loops of the G protein (Johal *et al.*, 2008). These residues ($W^{92}Y^{93}$ and $F^{135}W^{136}$ in the BEFV unprocessed G protein sequence) are juxtaposed by a conserved disulfide bridge (C_{III}-C_V; Fig. 6.4) and fully exposed in the post-fusion form of the G protein to allow membrane penetration (Roche *et al.*, 2006, 2007).

L proteins

The L protein is the large multifunctional RNA-dependent RNA polymerase that occurs in all non-segmented negative-sense RNA viruses (mononegaviruses). Ephemerovirus L proteins range in size from 245 to 250 kDa (2128–2144 amino acids). They are the most conserved of the structural proteins, sharing in excess of 49% amino acid sequence identity. They contain all six conserved functional regions (CRs) that are common to mononegavirus L proteins (Poch *et al.*, 1990), including those associated with P protein binding (CRI), RNA template binding (CRII), RNA-dependent RNA polymerase (CRIII), polyribonucleotide transferase (CRV)

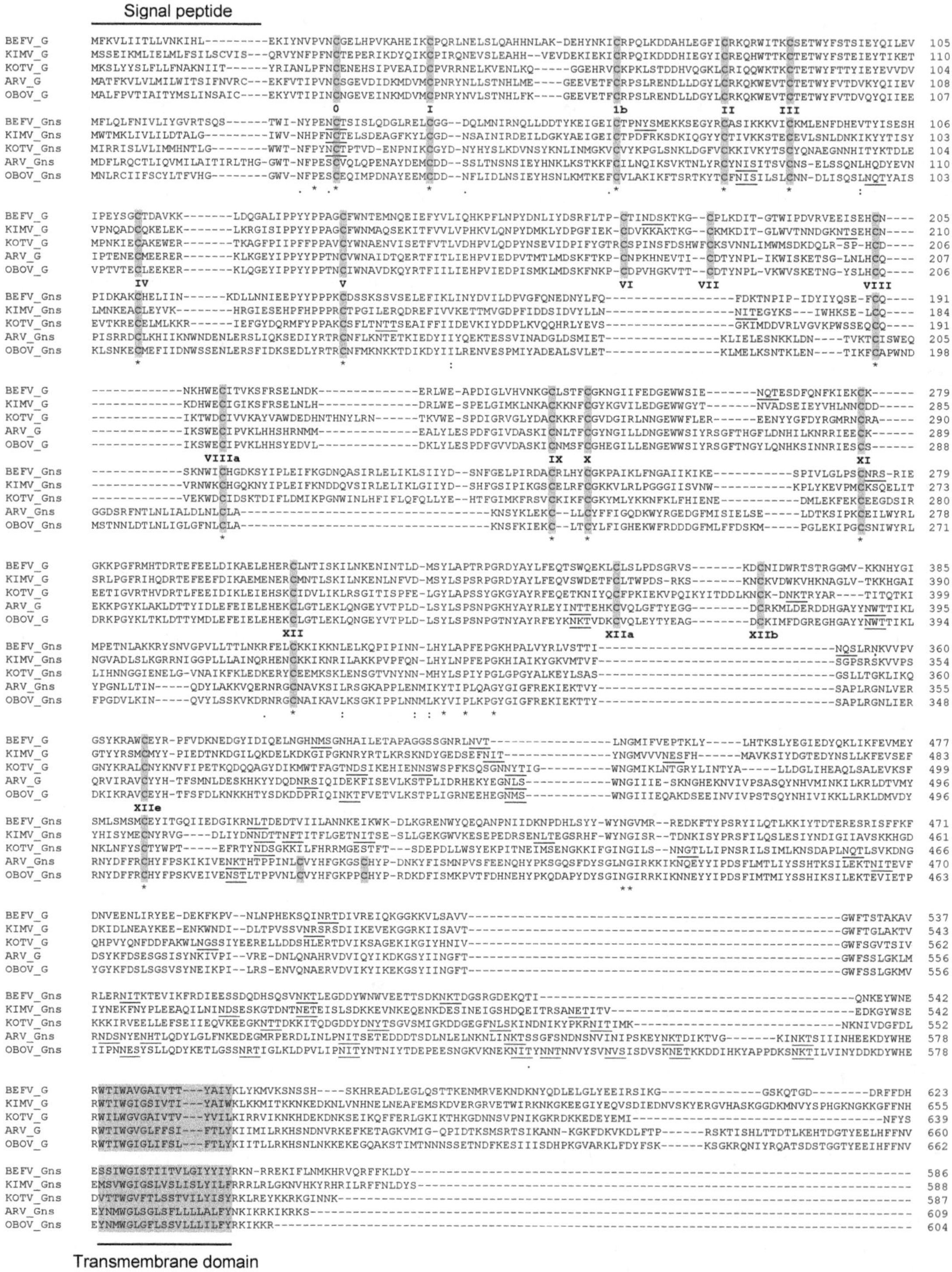

Figure 6.3 An alignment of ephemerovirus G proteins and G_{NS} proteins illustrating the conservation of cysteine residues. The alignment was generated in Clustal X using default parameters and then adjusted manually. Cysteine residues are shaded and numbered according to the system described in Walker and Kongsuwan (1999). Predicted N-glycosylation sites are underlined. Predicted signal peptides and transmembrane domains are also indicated. Other fully conserved (*), strongly conserved (:) or weakly conserved (.) amino acids are indicated below the alignment.

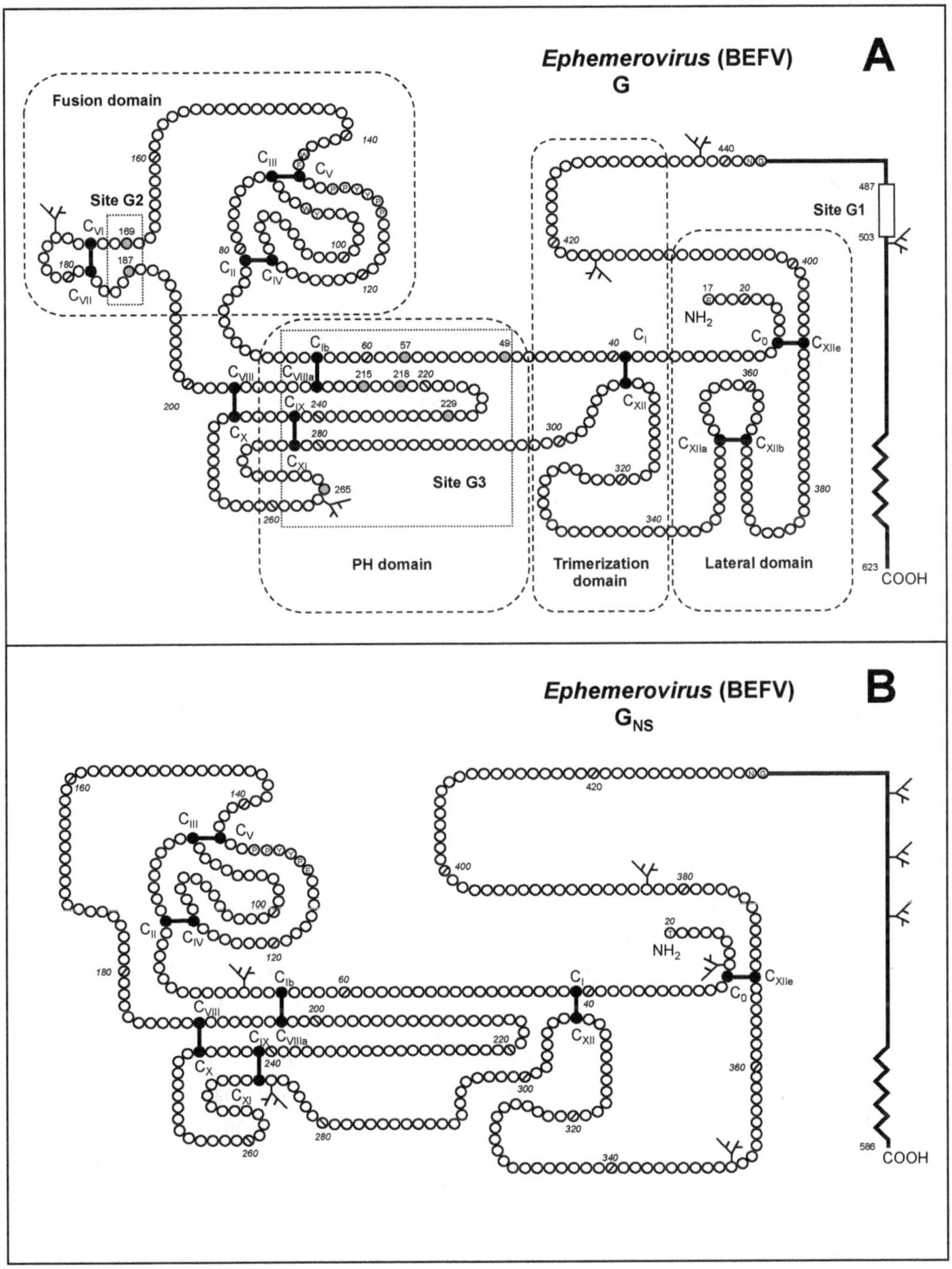

Figure 6.4 Schematic illustration of the likely similar folded structures of (A) the BEFV G protein and (B) the BEFV G_{NS} protein. The locations of disulfide bridges are based on the resolved crystal structures of the low-pH and prefusion forms of the VSV G protein (Roche *et al*., 2006, 2007) and the predicted structural conservation of animal rhabdovirus G proteins (Walker and Kongsuwan, 1999). Regions of the BEFV G protein aligning appropriately with the lateral domain (DI), trimerization domain (DII), pleckstrin homology (PH) domain (DIII), and fusion domain (DIV) of the VSV G protein are boxed in dashed lines. BEFV neutralizing antigenic sites G1, G2 and G3 are also indicated with amino acids known to be located in these sites shown as shaded circles. The positions of predicted N-glycosylation sites in the G and G_{NS} proteins are underlined. Amino acids are numbered from the N-terminus of the unprocessed G and G_{NS} proteins as the predicted signal peptidase cleavage sites have not been confirmed experimentally.

and 2′-O methyltransferase (CRVI) activities (Rahmeh *et al.*, 2009, 2010).

Accessory proteins

By analogy with other RNA viruses with relatively complex genomes (e.g. coronaviruses, paramyxoviruses, lentiviruses), the proteins encoded in the long G-L intergenic region of ephemeroviruses have been termed accessory proteins (Walker *et al.*, 2011). The functions of these proteins are presently poorly understood and most are unrelated in sequence to any other known proteins. However, their expression may be blocked selectively during adaptation to cell culture by mutations that inactivate transcription signalling or truncate translation (Wang and Walker, 1993; McWilliam *et al.*, 1997). This implies that the proteins are not essential for virus replication *in vitro* and are likely to have accessory roles associated with the enhancement of viral replication efficiency *in vivo* or engagement of the host to block innate immune responses, regulate apoptosis or increase cytopathology (Walker *et al.*, 2011).

G_{NS} proteins

The ephemerovirus G_{NS} proteins are class I transmembrane glycoproteins that are similar in size and have many of the structural characteristics of rhabdovirus G proteins from which they appear to have evolved (Walker *et al.*, 1992; Wang and Walker, 1993). The unprocessed G_{NS} proteins range in size from 67.1–70.9 kDa. The mature proteins appear to be more highly glycosylated than G proteins with 7 to 11 N-glycosylation sites located primarily in the C-terminal third of the protein just upstream of the transmembrane domain. An alignment of ephemerovirus G_{NS} proteins indicates that they also share 14 conserved cysteine residues that are likely to form seven disulfide bridges to stabilize the structure of the globular head domain (Fig. 6.3). These align with 14 of the 18 conserved cysteine residues of ephemerovirus G proteins, including the six cysteine residues that are common to all animal rhabdovirus G proteins and six others that show genus-specific variations (Walker and Kongsuwan, 1999). These structural similarities and the moderate level of sequence conservation suggest that G_{NS} proteins adopt a similar folded structure to G proteins and possibly could have a similar function (Fig. 6.4). However, although transported to the cell surface, BEFV G_{NS} is not present in virions and, unlike the G protein, does not induce cell fusion when expressed at low pH in insect cells (Walker *et al.*, 1991; Johal *et al.*, 2008). Lack of fusion activity is consistent with the absence of critical aromatic residues at the tips of the predicted fusogenic loops (Johal *et al.*, 2008). Furthermore, BEFV G_{NS} does not cross-react antigenically with the BEFV G protein, sharing none of the three major antigenic sites and, unlike the G protein, it does not protect against homologous challenge in cattle (Hertig *et al.*, 1996; Johal *et al.*, 2008). It has also been shown that polyclonal rabbit antibody to G_{NS} fails to neutralize BEFV produced in either mammalian or insect cells (P.J. Walker and S.S. Davis, unpublished data). Therefore, although clearly related in structure, G_{NS} does not appear to share the principal role of the G protein in cell attachment and entry and its function is presently unknown.

α Proteins

The ephemerovirus *α* gene encodes two proteins in consecutive ORFs. The *α*1 proteins range in size from approximately 10.5 to 14.5 kDa. Although the level of sequence identity is quite variable between more distantly related ephemeroviruses (e.g. 12.8% identity between KOTV and PUCV), all *α*1 proteins have a similar structure with an N-terminal domain of 36–40 amino acids containing clusters of aromatic residues followed by a transmembrane domain and a highly basic C-terminal domain of more variable length (Wang *et al.*, 1994; McWilliam *et al.*, 1997). These features suggest that the *α*1 proteins are likely to function as viroporins, generating pores in the plasma membrane to increase permeability and facilitate virus budding (Gonzalez and Carrasco, 2003). Similar viroporin-like proteins are also encoded in the genomes of other arthropod-borne rhabdoviruses including the U5 ORF of Wongabel virus, the SH ORF of Flanders virus and the U3 ORFs of the tibroviruses (Gubala *et al.*, 2008, 2011; Walker *et al.*, 2011).

The ephemeroviruses *α*2 proteins range in size

from approximately 10.7 kDa to 15.5 kDa. There is a relatively high level of amino acid sequence conservation between α2 proteins of the more closely related ephemerovirus (e.g. 58% identity for BEFV and BRMV; 89% identity for KIMV and MALV) but little sequence identity between other viruses (e.g. 5% identity for OBOV and KOTV). The α2 proteins have no distinguishing features that would suggest function.

β Proteins

The 16.6- to 18.5-kDa ephemerovirus β proteins are expressed in infected cells and may occur in virions (Walker *et al.*, 2011). They share high levels of sequence identity between closely related viruses (e.g. >55% identity amongst BEFV, BRMV, KIMV, MALV and PUCV) and less but identifiable homology between the more distantly related viruses, but there are no universally conserved sequence motifs. Although its function is unknown, the β protein is not essential for virus replication *in vitro* as, in some highly cell culture-adapted strains of BEFV, it is not expressed owing to a mutation that severely truncates the coding sequence. This suggests the β protein function is associated with pathogenesis or the host immune response to infection *in vivo* (Wang *et al.*, 1994; McWilliam *et al.*, 1997; Walker *et al.*, 2011).

γ Proteins

Ephemerovirus γ proteins range in size from 11.7 kDa to 13.7 kDa and occur in all known ephemeroviruses other than ARV and OBOV. They share >55% amino acid sequence identity, except for KOTV which is somewhat more distantly related (17.2–23.5% identity). In BEFV, expression of the γ proteins is blocked or attenuated in some cell-culture-adapted strains by mutations that inactivate the β gene transcription termination sequence, resulting in transcription of a bicistronic β-γ mRNA (McWilliam *et al.*, 1997; Walker *et al.*, 2011). This suggests that, like the β protein, the γ proteins are not essential for replication *in vitro*.

δ Protein

Only KOTV amongst known ephemeroviruses encodes a δ protein (Blasdell *et al.*, 2012). This 12.4 kDa protein shares significant amino acid sequence homology with proteins in the coactivator-associated arginine methyl transferase 1 (CARM1) superfamily. These multidomain enzymes (also known as protein arginine methyl transferase 4, PRMT4) are recruited by transcription factors to methylate histone H3 and interact with the p160 family of co-activators, leading to chromatin remodelling and gene activation (Troffer-Charlier *et al.*, 2007). CARM1-knockout mammalian cells display impaired NF-κB dependent expression of genes associated with immune and inflammatory responses (Covic *et al.*, 2005; Miao *et al.*, 2006; Hassa *et al.*, 2008) and CARM1 cooperates with PRMT1 to activate STAT5-mediated transcription upon stimulation with interleukin-4 (Kleinschmidt *et al.*, 2008). The N-terminal region of CARM1 contains a PH domain that is responsible for high-specificity binding to proline-rich regions of target proteins; the central region contains the catalytic domain; and the C-terminal region contains the S-adenosylmethionine binding domain (Troffer-Charlier *et al.*, 2007). Although the role of KOTV δ is presently unknown, it aligns specifically with the 112 amino-acid N-terminal PH domain of CARM1 and may function to mimic this region, blocking its interaction with its target proteins and inhibiting the activation of transcription (Blasdell *et al.*, 2012).

Other possible accessory proteins

As in VSV and RABV (Spiropoulou and Nichol, 1993; Nadin-Davis *et al.*, 2002), alternative ORFs that could potentially encode small basic proteins occur in the P genes of some ephemeroviruses. The largest of these occurs in ARV in which a 7.4 kDa (63 amino-acid) protein of pI 10.3 is encoded. However, this ORF is not preserved in closely related OBOV. The BEFV P gene contains an alternative ORF encoding a 5.8 kDa (48 amino acid) protein of pI 11.2 that is preserved in several strains but does not occur in closely related BRMV. In other ephemeroviruses, alternative ORFs in the P gene comprise less than 25 codons and it is not know if any are expressed. The BEFV α2 ORF also contains a small alternative ORF encoding a potential 5.7 kDa protein (α3) of pI 10.0. It is preserved in several BEFV strains but does not occur in closely related BRMV or other ephemeroviruses and its significance is presently unknown (Blasdell *et al.*, 2012; K.R. Blasdell and P.J. Walker, unpublished data).

Geographic distribution, host range and vectors

Bovine ephemeral fever (BEF) is enzootic or periodically epizootic in most tropical and subtropical to temperate regions of Africa, the Middle-East, Asia and Australia (Table 6.1). In the Eastern Hemisphere the range of BEFV infection can extend from northern China (44° N), through Japan, Taiwan and South-East Asia to southern Australia (36° S) (Shiel *et al.*, 1989; Bai *et al.*, 1991; Bai, 1993). In the Western Hemisphere, it ranges from southern Africa (34° S) almost to the Caspian Sea (36° N) including Kenya, Saudi Arabia and countries bordering the southern and eastern Mediterranean (Burgess, 1971; Madbouly *et al.*, 2006; Yeruham *et al.*, 2007; Roya, 2008). There have also been reports of BEF in countries across south central Asia and the Indian subcontinent (Burgess, 1971). BEFV antibody has been detected in cattle in European Turkey (Karaoglu *et al.*, 2007) but it has not been reported elsewhere in Europe or in the Americas. BEF is non-contagious, even when mucosal surfaces are exposed to nasal or ocular discharges or to saliva (Mackerras *et al.*, 1940). Epizootics are seasonal and are commonly associated with recent rainfall, with widespread epizootics often following prolonged wet weather and the flooding of river systems (Murray, 1970; Davies *et al.*, 1975; St George and Standfast, 1988; Finlaison *et al.*, 2010). However, there are also reports of BEF outbreaks during dry summer periods (Davies *et al.*, 1990; Yeruham *et al.*, 2007).

Although certainly transmitted by haematophagous insects, the vectors are poorly defined. In Africa, BEFV has been isolated once from a mixed pool of biting midges (*Culicoides kingi, C. novosus, C. bedfordi, C. imicola* and *C. cornutus*) and once from *Culicoides coarctatus* (Davies and Walker, 1974; Blackburn *et al.*, 1985). The virus has been isolated in Australia, once from a mixed pool of four mosquito species (*Culex* sp., *Verrallina carmenti, Uranotaenia nivipes* and *Ur. albescens*) and twice from *Anopheles bancroftii* (Standfast *et al.*, 1976; Muller and Standfast, 1986). It has also been isolated from biting midges (*Culicoides brevitarsis*) in Australia (Cybinski and Muller, 1990). Experimental infections by artificial membrane feeding on BEFV-infected blood have provided some evidence of replication in one species of mosquito (*Culex annulirostris*) and two species of biting midge (*Culicoides brevitarsis* and *Culicoides marksi*) (Kay *et al.*, 1975; St George and Standfast, 1988). However, other workers have failed to recover virus from a wide range of South African livestock-associated *Culicoides* species 10 days after experimental feeding on BEFV-infected sheep or cattle blood (Venter *et al.*, 2003). Based on the observation that experimental infection of cattle requires direct injection of virus into the bloodstream (Burgess, 1971), disparities with the epizootiological spread of other midge-borne diseases of livestock (Kirkland, 1993), and evidence that the explosive dynamics of epizootics are consistent with the rapid emergence of large mosquito populations, it has been suggested that mosquitoes, rather than biting midges, are the principal vectors of BEFV (St. George, 1993; Murray, 1997). However, as KIMV has also been isolated from both biting midges (*C. brevitarsis*) and mosquitoes (*Cx. annulirostris*) (Liehne *et al.*, 1981; Zakrzewski and Cybinski, 1984) and other ephemeroviruses have been isolated either from biting midges (KOTV) or mosquitoes (MALV, OBOV and PUCV) (Schmidt *et al.*, 1965; Kemp *et al.*, 1973a; Karabatsos, 1985), ephemerovirus could potentially employ a diverse range of susceptible vectors for transmission and this may vary by location and season, depending on ambient temperatures and the availability and abundance of vectors.

BEF is primarily a disease of cattle and water buffalo but an ephemeral fever-like illness has been reported in camels in East Africa (Dirie and Abudurahman, 2003) and there is evidence of subclinical BEFV infection in red deer in Australia and a wide range of wild ruminants in Africa (Davies *et al.*, 1975; Cybinski and Zakrzewski, 1983; Barnard, 1997; Anderson and Rowe, 1998). The wide host range and apparent absence of clinical disease suggests that African wild ruminants may be natural reservoirs of BEFV infection and the ancestral origin of ephemeroviruses (Davies *et al.*, 1975). BEFV can also cause subclinical infection in sheep following intravenous experimental inoculation (Hall *et al.*, 1975) and neutralizing antibody was detected at high prevalence in goats and low prevalence in sheep during an epizootic in Taiwan in 1984 (Chiu and Lu, 1998). A serological survey of swine in Korea has also indicated a high prevalence of BEFV infection (Lim *et al.*, 2007). It has been suggested that birds, particularly those such as cattle egret (*Bubulcus coromandus*) that feed in close association with cattle, may be involved in the epizootiology. Surveys of birds to date have not been sufficiently extensive to draw any firm conclusions (St George and Standfast, 1988) but, as discussed below, recent molecular epizootiological data suggests the role of birds in the ecology of BEFV should be revisited.

KIMV has not been associated with disease but appears to have a similar geographic distribution and ecology to BEFV. The virus has been isolated from both biting midges (*Culicoides brevitarsis*) and mosquitoes (*Culex annulirostris*) and antibodies to KIMV have been detected in cattle across much of northern and eastern Australia and southern and eastern China, and in Indonesia (Cybinski and Zakrzewski, 1983; Jiang and Yan, 1989; Soleha *et al.*, 1993a,b). Cattle and water buffalo appear to be the principal vertebrate hosts but there is also serological evidence of infection in goats and horses (Cybinski and Zakrzewski, 1983; Soleha *et al.*, 1993a). As serological and genetic studies indicate that KIMV and MALV are strains of the same virus (K.R. Blasdell and P.J. Walker, unpublished data), it can be reasonably interpreted that the distribution extends at least to East Africa and may span across much of the known distribution or BEFV. MALV was isolated on three occasions from mosquitoes (*Mansonia uniformis*) but the host range and geographic distribution beyond East Africa have not yet been determined.

Of the other ephemeroviruses, BRMV antibody has been reported in cattle in Australia, and cattle, water buffalo, sheep and goats in Indonesia (Soleha *et al.*, 1993a; Daniels *et al.*, 1995). However, the close relationship to BEFV and absence of a readily accessible type-specific serological test has prevented any useful assessment to date of the host range or geographic distribution of BRMV in Australia or elsewhere. Like BRMV, ARV has never been isolated from insects but its distribution in Australia suggests *Culicoides* and/or mosquitoes are likely to be involved in transmission (Muller and Standfast, 1986). ARV antibody has been detected in cattle, water buffalo and pigs (Gard *et al.*, 1984). Its geographic distribution includes Australia, New Guinea, Indonesia and China but serological surveys have not yet been conducted elsewhere across the known range of ephemeroviruses (Bai, 1993; Soleha *et al.*, 1993a; Daniels *et al.*, 1995). KOTV has been isolated from biting midges and a serological survey conducted in Nigeria indicated a high prevalence of KOTV antibody in cattle, hedgehog (*Atelerix albiventris*) and the giant rat (*Cricetomys gambianus*), and some evidence of infection in sheep, horses and possibly humans (Kemp *et al.*, 1973a). The distribution of KOTV in other parts of the world is currently unknown.

Nothing is yet known of the host range or wider geographic distribution of OBOV or PUCV. Like MALV, each of these viruses was isolated from *Mansonia uniformis* mosquitoes. This species breeds in permanent or semi-permanent swamps and water holes, has a preference for feeding on cattle and has a global distribution that largely reflects the limits of known distribution of ephemeroviruses (Lee *et al.*, 1988). It has also been the source of isolation of several other arthropod-borne viruses of medical and veterinary importance including Rift Valley fever virus, Japanese encephalitis virus and Ross River virus (Peiris *et al.*, 1994; Ryan *et al.*, 2000; Sang *et al.*, 2010). The role of *M. uniformis* in the ecology and epizootiology of BEFV and other ephemeroviruses is worthy of further study.

Pathology and pathogenesis

Bovine ephemeral fever in cattle (also called three-day sickness, bovine enzootic fever, bovine influenza and stiffsiekte) is a disabling disease that often is associated with high rates of morbidity during epizootics. Case fatality rates are generally low but can vary in different epizootics and were reported to be as high as 8.6% in the Jordan Valley of Israel in 1999 (Yeruham *et al.*, 2002,

2010). During an epizootic in Taiwan in 1996, a case fatality rate of 11.3% resulted from disease and culling (Liao *et al.*, 1998). BEFV infection of cattle can be subclinical (St. George *et al.*, 1977; St. George, 1981). In mild cases, clinical signs can include loss of appetite, fever, nasal and ocular discharge, stiffness, muscle fasciculation and temporary lameness. In more severe cases, cattle may show signs of depression, anorexia, profuse salivation, rumenal stasis, joint swelling and limb paralysis. Disease signs may last for only 1–3 days and although recovery is generally rapid, there are some reports of prolonged paralysis and ataxia following recovery from the acute effects of infection (Mackerras *et al.*, 1940; Basson *et al.*, 1970; Hill and Schultz, 1977). Severe disease and mortalities occur primarily in heavy or lactating animals. In dairy cattle, there is a reduction in milk production that may persist post-recovery until the next lactation period (Theodoridis *et al.*, 1973b; Davis *et al.*, 1984). There is also evidence of temporary infertility in bulls and delayed oestrus and mid-term abortions in cows (Theodoridis *et al.*, 1973b; Burgess and Chenoweth, 1975; Uren *et al.*, 1987). Clinical ephemeral fever has also been reported in water buffalo with similar characteristics to the disease in cattle (Malviya and Prasad, 1977; Bai *et al.*, 1989). Disablement of draught animals at the time of harvest can have significant economic impacts.

The clinical signs of ephemeral fever appear to be due primarily to vascular permeability and an associated inflammatory response (Young and Chung, 1986; St. George, 1993). Viraemia is brief and virus is detected primarily in leucocytes and blood plasma. Peak viraemia generally precedes the onset of fever which is accompanied by a marked neutrophilic leucocytosis, lymphopaenia, eosinopenia, elevated fibrinogen, hypocalcaemia and elevated levels of cytokines. However, there is no evidence of extensive tissue damage other than a vasculitis affecting the epithelium of small vessel synovial membranes, tendon sheaths, muscles facia and skin (Mackerras *et al.*, 1940; Basson *et al.*, 1970; St George *et al.*, 1984; St. George, 1993). Severe Wallerian degeneration associated with prolonged paralysis and ataxia is considered to be due to pressure on the spinal cord (Hill and Schultz, 1977). Anti-inflammatory drugs are effective in treating acute clinical signs which have been attributed to an immunopathological response similar to milk fever (Uren *et al.*, 1989; St. George, 1993; St George *et al.*, 1995). Cytokines induced following experimental BEFV infection include INF-α, IL-I and TNF which are potent mediators of inflammation. Peak cytokine levels precede or coincide with the onset of fever (Uren and Zakrzewski, 1989).

At the cellular level, BEFV infection has been shown to induce caspase-dependent apoptosis in several continuous cell lines, including Vero, BHK and MDBK bovine kidney cells (Chang *et al.*, 2004). Apoptosis requires BEFV uncoating and viral gene expression, and involves signalling through the Fas death receptor pathway, resulting in caspase 8 activation, the release of cytochrome c from mitochondria, upregulation of apoptosis protease activating factor 1 (Apaf-1) and subsequent activation of caspase 9 and caspase 3 that is responsible for many of the cytophysiological characteristics of apoptosis (Lin *et al.*, 2009). It has also been shown that BEFV-induced apoptosis involves Src-dependent c-Jun N-terminal kinase (JNK) signalling (Chen *et al.*, 2010) which is known to have a critical role in both the extrinsic and intrinsic apoptotic pathways (Elmore, 2007). There is also evidence that BEFV infection induces phosphorylation of protein kinase B (PKB or Akt) which, for several other negative-sense RNA viruses, has been shown to be important for efficient replication through phosphorylation of the viral P protein (Sun *et al.*, 2008; Ji *et al.*, 2011). However, unexpectedly, BEFV replication was also shown to be enhanced by inhibition of the upstream kinase PI3K and the downstream target of activated Akt, mammalian target of rifampicin (mTOR), in the signalling cascade. This suggests that BEFV may have the ability to reverse Akt dephosphorylation, which may be by an alternative pathway, and that BEFV may be susceptible to the downstream effectors of the PI3K/Akt/mTOR pathway in mammalian cells (Ji *et al.*, 2011).

Natural disease associated with KOTV has only been reported in Swiss Brown heifers imported to Nigeria from Germany in 1969 in which seroconversion to KOTV neutralizing antibody followed an acute illness of short duration and some mortalities (Kemp *et al.*, 1973a). Subsequent experimental infections conducted in white Fulvani calves using an inoculum derived from third mouse-brain passage of the KOTV insect isolate produced a mild ephemeral fever-like illness characterized by nasal discharge, elevated temperature, anorexia and lethargy, and associated leucocytosis and relative neutrophilia (Tomori *et al.*, 1974). There is widespread neutralizing antibody to KOTV in cattle in northern Nigeria (Kemp *et al.*, 1973a) but little is known of its wider distribution or importance as a pathogen in Africa or elsewhere in the world.

Epizootiology

BEFV exists as a single serotype worldwide (Inaba *et al.*, 1969a; Snowdon, 1970; Kemp *et al.*, 1973b; Tian *et al.*, 1987; Kato *et al.*, 2009). However, although detailed and current epizootiological information is available from few countries, the available evidence suggests that the virus circulates in at least four distinct epizootiological systems. These are defined largely by geography and weather patterns, lying north and south of the Equator in the Eastern and Western Hemispheres.

In Africa, BEFV appears to be enzootic or periodically epizootic across much of the continent. The first recorded epizootic in southern Africa occurred in 1906–1907, although it was not regarded as a new disease (Bevan, 1907; Freer, 1910). It moved in a wave from north to south from north-western Zimbabwe (then Rhodesia) through Gauteng (then Transvaal) and Kwazulu-Natal and eventually reached the Eastern Cape (Freer, 1910; St George, 1981). More recent reports from South Africa indicate that the disease occurs regularly in most parts of the country, with annual occurrence of localized disease outbreaks and more severe epizootics associated with weather conditions favouring the emergence of large vector populations (Coetzer, 1993). The enzootic status of BEFV in southern Africa is also indicated by reports of high seroprevalence in various wild ruminants which may serve as a reservoir of infection (Anderson and Rowe, 1998). A similar pattern has been observed in equatorial Kenya with sporadic outbreaks generally following local rains and more widespread epizootics associated with years of greater than average rainfall (Davies *et al.*, 1975, 1990). Seroconversions are observed in cattle and wild ruminants during periods between outbreaks when clinical disease is absent (Davies *et al.*, 1975). A similar association of BEF with periods of high rainfall has been observed in Nigeria (Kemp *et al.*, 1973b). An epizootic in Egypt in 1909 was reported to have moved northwards along the Nile Valley during the dry summer period but may have been associated with the flooding of vector breeding sites by water released from the Aswan Low Dam, which had been completed several years previously (Rabagliati, 1909). More recently, BEFV appears to have become endemic in the Nile Delta (Yeruham *et al.*, 2010). Epizootics in eastern Saudi Arabia and countries bordering the eastern Mediterranean have been reported at irregular and sometimes long intervals, occurring during dry summer weather and have been attributed to the displacement of vectors from the Nile Delta, Saudi Arabia or East Africa on prevailing southerly or south-westerly winds (Abu Elzein *et al.*, 1997; Yeruham *et al.*, 2002, 2005).

In Asia, BEF epizootics in subtropical and temperate regions of China, Taiwan and Japan are reported to have progressed during the summer and autumn in a northerly direction out of enzootic areas in the south of China. The disease has been known in China since 1934 with epizootics subsequently recorded in 1944 and 1949 (Bai, 1993). Since the mid-1950s, outbreaks have occurred regularly across much of the country although less frequently in the more northern regions (Bai, 1993; Shirakawa *et al.*, 1994). Epizootics in Taiwan, Japan and the Korean Peninsula have occurred less regularly (Shirakawa *et al.*, 1994; Hsieh *et al.*, 2005). Outbreaks in Korea and Japan have been linked chronologically to epizootics in nearby provinces in China and the direction and speed of a low-level jet stream which may have allowed long-distance displacement of infected vectors (Shirakawa *et al.*, 1994). Molecular epizootiological studies of Taiwanese and Japanese isolates using nucleotide and deduced amino acid sequences of the ectodomain region of the G gene have identified three lineages of isolates that cluster chronologically, implicating the same genotypes as the cause of simultaneous epizootics in each country (Hsieh *et al.*, 2005; Kato *et al.*, 2009). The clusters comprised isolates from Taiwan in 1984 and Japan in 1988–1989 (genotype I), Taiwan in 1996–2004 and Japan in 2001–2004 (genotype II) and the 1966 Japanese (Yamaguchi) vaccine strain (genotype III). The Asian lineages were also shown to cluster separately from viruses isolated in Australia (Walker, 2005; Kato *et al.*, 2009) (Fig. 6.5). Although there is significant cross-neutralization of isolates from Japan, Taiwan and Australia, there are multiple amino acid substitutions in neutralizing antigenic sites G1 and G3 of the Asian isolates, including two common mutations that introduce potential N-glycosylation sites (Kato *et al.*, 2009).

BEF has been known in Australia since the late summer of 1936 when a major epizootic swept out of the far north of the Northern Territory, progressing westwards to the Kimberley region and on a southward path through much of eastern Australia. The epizootic ultimately reached Victoria, some 3000 km to the south, in late autumn of 1937 (Seddon, 1938). BEFV does not appear to have been present in Australia prior to this epizootic, perhaps owing to the small cattle population at that time in the remote north, and appears have been introduced in vectors displaced across the Timor Sea from the Indonesian Archipelago (Seddon, 1938). Similar sweeping epizootics occurred in Australia in 1955–56, 1967–68 and then, in rapid succession, in 1970–71, 1972–74 and 1974–76 (St. George *et al.*, 1977). Epizootics have generally coincided with periods of extreme weather and heavy rainfall associated with the La Niña phase of the Southern Oscillation in the Western Pacific (http://www.bom.gov.au/climate/enso/lnlist/). In the 1972–74 and 1974–76 epizootics, the southward sweep was interrupted by the winter months, resurging again in the spring and early summer. During the periods between these sweeping epizootics, BEF was enzootic in the far north of Australia and sporadic outbreaks occurred in most years in the Northern Territory, Western Australia or Queensland (St. George *et al.*, 1977). However, following the 1974–1976 epizootic, BEF became enzootic as far south as the Central Coast of New South Wales, with disease outbreaks interspersed with sporadic cases accompanying the summer rains (Uren *et al.*, 1983, 1987). This may have been due to adaptation of the virus to vector(s) with a capacity for overwintering in the southern temperate region. The return of La Niña conditions after a record period of drought saw a return of epizootic BEF in eastern Australia in 2007–2008 when a progressive front of disease in the late summer was associated with the rapid southward movement of an intense low pressure cell (Finlaison *et al.*, 2010). Similar epizootic conditions prevailed in subsequent years with disease recorded in Victoria in 2010 for the first time in 20 years. The rapid spread of BEF across the Australian continent during sweeping epizootics has been attributed to wind-borne movement of insect vectors associated with the summer monsoons (Murray, 1970; Newton and Wheatley, 1970).

Molecular epizootiological studies using G protein ectodomain sequences have been undertaken on a large set of BEFV isolates collected from various locations in Australia during the period 1956–2010 (P.J. Walker *et al.*, unpublished data) (Fig. 6.5). Surprisingly, the analysis indicates that BEFV exists in Australia as a single major lineage that is distinct from Asian lineages and has evolved progressively across the continent. Although a new subclade has emerged since 1984, there has been no indication of a south-eastern endemic focus in which sublineages should have had the opportunity to evolve. The reason for this apparent discrepancy with gross epizootiological observations is not yet clear but the data suggests BEFV moves rapidly across the continent, even during inter-epizootic periods to evolve as a single population. Indeed, the phylogenetic pattern is similar to that of Murray Valley encephalitis, Kunjin and sindbis viruses in Australia for which water bird migrations are believed to have an important role in the epidemiology (Coelen and Mackenzie, 1988; Flynn *et al.*, 1989; Sammels *et al.*, 1999; Poidinger *et al.*, 2000). The phylogenetic analysis has also identified mutations in two specific amino acids that appear to be of epizootiological significance, each located in the major conformational antigenic site G3. The first occurs at amino acid 218 in which an R>K mutation has resulted in a loss of reactivity with site G3a monoclonal antibody 8B6 (Cybinski and Davis,

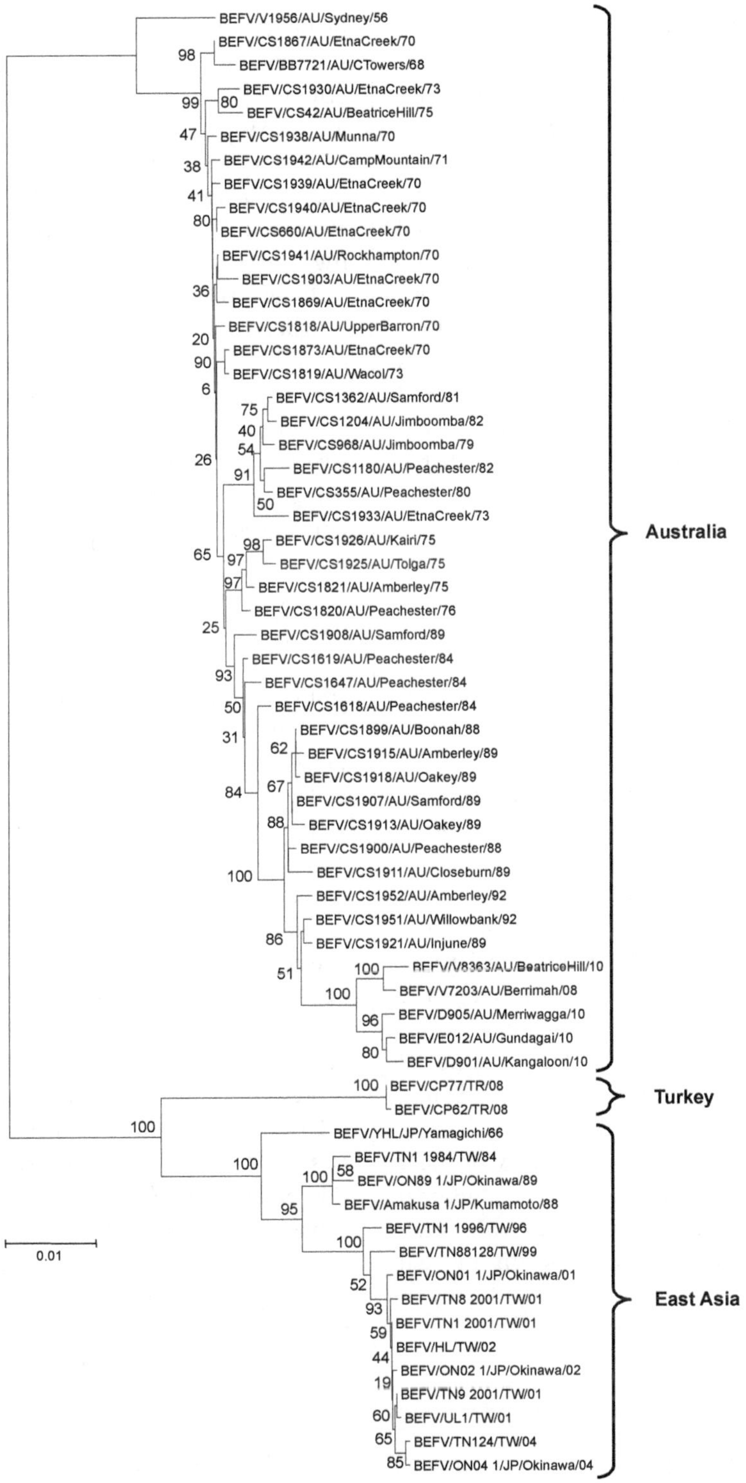

Figure 6.5 Unrooted minimum evolution tree of nucleotide sequences of the ectodomain region of the G protein BEFV isolates collected between 1956 and 2010 from Australia, Turkey, Taiwan and Japan. The tree was generated in MEGA from a Clustal W alignment using default parameters. Bootstrap values (1000 repetitions shown as a percentage) are shown for the major nodes. Branch lengths are proportional to the distance between sequences with the bar indicating a distance of 1%.

1993; Walker, 2005). Almost all viruses isolated since 1975 share this mutation which happens to coincide with the emergence of enzootic BEF in south-eastern Australia. The second critical site is amino acid 215 which is under strong positive selection as evidenced by various mutations (K>N/E/T) in multiple isolates collected from different locations at different times. These mutations result in loss of reactivity with antigenic site G3b monoclonal antibody 8D3 which targets the only BEFV epitope that cross-reacts with KIMV (Cybinski *et al.*, 1990; Cybinski and Davis, 1993; Walker, 2005). BEFV and KIMV have long been known to have a similar distribution in Australia (Cybinski and Zakrzewski, 1983). KIMV appears to cause only subclinical infections in cattle and there is evidence that cattle with pre-existing KIMV neutralizing antibody can subsequently be infected with BEFV and develop clinical ephemeral fever (Cybinski, 1987). It appears that KIMV neutralizing antibody may be the source of the strong positive selection on amino acid 215.

Very little is known of the epizootiology of BEF throughout much of Central, South and South-East Asia. Analysis of G protein ectodomain sequences of two BEFV isolates from the 2008 outbreak in Turkey indicates they are far more closely related to the East Asian than the Australian clade (P.J. Walker, unpublished data). The East Asian and Middle-East/North-East Africa episystems may be linked by transcontinental movement of hosts and/or vectors across South, Central and West Asia. There is also a scarcity of information about the epizootiology of other ephemeroviruses. Sequence analysis of additional isolates of BEFV and related viruses from other sites across their natural geographic range would reveal much about origins and epizootiology of this important disease of livestock.

Control

Vaccination of cattle against bovine ephemeral fever is practiced in several countries including South Africa, Egypt, China, Japan, South Korea, Taiwan and Australia. Live-attenuated vaccines have been developed by serial passage of BEFV in mice and/or cell culture. BEFV has been reported to lose both pathogenicity and immunogenicity when passaged in cell culture (Inaba *et al.*, 1969b). Live-attenuated vaccines are usually delivered in Freund's incomplete adjuvant or saponin (Quil A) (Theodoridis *et al.*, 1973a; Vanselow *et al.*, 1985; Erasmus, 1986; Vanselow *et al.*, 1995). Killed vaccines comprising cell culture-adapted, formalin-inactivated virus adsorbed to aluminium phosphate gel have also been used, either in a 2-dose regimen (Inaba *et al.*, 1973; Chiu and Lu, 1987) or as a boosting dose following administration of live-attenuated virus (Inaba *et al.*, 1974). Although more effective under experimental conditions, protection using a formalin-inactivated aluminium gel vaccine in the field has been reported to be of short duration, requiring three or four annual doses. A stronger and more durable immune response has been demonstrated using binary ethyleneimine (BEI) inactivation and a double oil emulsion adjuvant (Hsieh *et al.*, 2006).

Natural immunity to BEFV in cattle appears to be solid and durable (Mackerras *et al.*, 1940) and it has been established that colostral antibody will protect calves against experimental BEFV challenge. There is good evidence that protection is mediated by neutralizing antibody and experimental vaccines using G protein split by non-ionic detergent treatment of purified virions or crude pelleted virus have been shown to be very effective (Bai *et al.*, 1993; Uren *et al.*, 1994). Protection has also been achieved using BEFV G protein expressed from a recombinant vaccinia virus (Hertig *et al.*, 1996) but incomplete protection was reported for G protein delivered in a lumpy skin disease virus (capripox) vector (Wallace and Viljoen, 2005). Nevertheless, single-dose vaccines using recombinant G protein, administered either in new generation slow-release adjuvants, as plasmid DNA or in a live virus vector, would be very attractive for extensive beef industries for which there is poor uptake of existing multiple-dose products.

As viraemia is brief and there is no evidence of long-term persistence of virus in infected animals, the risk of spread of BEFV through international trade in livestock is very low if suitable quarantine procedures are applied (Kirkland, 2002). Long-distance spread of the virus is far more likely to occur by wind-borne movement of infected vectors and there is epidemiological evidence that this has occurred periodically in the Middle East and Asia (see above). Nevertheless, movement of cattle from an enzootic area appears to have precipitated an unseasonal BEF outbreak in southern Australia in 1988 (Shiel *et al.*, 1989), suggesting that rapid transport of cattle from infected to transmission-receptive virus-free zones should be avoided.

Conclusion

Despite the significance of BEFV as a pathogen of cattle and the sometimes serious economic impacts of its periodic re-emergence, there is much that remains unknown about the ephemeroviruses. What is the relative importance of mosquitoes and biting midges as vectors of BEFV and what are the major vector species that maintain it in enzootic cycles and during epizootics in various parts of the world? What is the prevalence and distribution of BEFV in Central Africa and Central, South and South-East Asia and how do those regions intersect with and potentially influence the epizootiological patterns in other parts of the world? The molecular epidemiological picture in Australia indicates that BEFV is a single evolving lineage. How are the northern and southern limits of the episystem so intimately connected? Are birds involved in BEFV transmission, allowing rapid transcontinental mixing of the virus population between the sweeping epizootics? How will climate change and associated extreme weather events impact on BEF epizootiology? BEFV has been shown to be present in European Turkey. Could climate change or other factors drive BEFV in the footsteps of bluetongue virus to become endemic or periodically epizootic across much of Europe? Do other ephemeroviruses contribute to the epizootiology of ephemeral fever? Despite evidence of life-long immunity, there are anecdotal reports of multiple attacks of ephemeral fever and disease outbreaks have been observed in sentinel herds in Australia without virus isolation or seroconversion. What is the global distribution of KOTV and does it or other ephemeroviruses contribute to the aetiology of ephemeral fever?

There is also a need for a better understanding of the pathogenesis of BEF. There are indications that it is an immunopathological condition. If so, how and why does BEFV subvert innate immune responses to promote rather than prevent disease? Are the viral accessory proteins involved in pathogenesis? What are the sites of BEFV replication? Why are high morbidity rates often associated with BEFV infection while other closely related ephemeroviruses do not cause disease? Clearly, the availability of a BEFV infectious clone would greatly facilitate studies of pathogenesis and other aspects of infection and immunity in ruminants and insects.

At the molecular level, the ephemeroviruses also present some intriguing questions. The ephemerovirus genome is far more complex than those of many other rhabdoviruses. How did this complexity arise and what might it tell us about the evolutionary origins of rhabdoviruses? What are the functions of G_{NS} and the small accessory proteins? G_{NS} has the essential structural characteristics of a rhabdovirus G protein. Why is G_{NS} not packaged into virions and why has the G protein superfamily structure been preserved? Are the functions of the smaller accessory proteins related to replication and transmission processes in insects? For other viruses, accessory proteins are commonly involved in blocking host innate immune responses but little is known of the antiviral defences of insects. Defining the functions of ephemerovirus accessory proteins may help inform us how insects respond to viral infection and lead to a better understanding of vector competence, an important aspect of the epidemiology of vector-borne diseases. These and many other unresolved questions mark the ephemeroviruses as a rich field of study for future research.

References

Abu Elzein, E.M., Gameel, A.A., al Afaleq, A.I., al Gundi, O., and Bukhari, A. (1997). Bovine ephemeral fever in Saudi Arabia. Vet. Record *140*, 630–631.

Ahmadian, G., Randhawa, J.S., and Easton, A.J. (2000). Expression of the ORF-2 protein of the human respiratory syncytial virus M2 gene is initiated by a ribosomal termination-dependent reinitiation mechanism. EMBO J. *19*, 2681–2689.

Albertini, A.A.V., Wernimont, A.K., Muziol, T., Ravelli, R.B.G., Clapier, C.R., Schoehn, G., Weissenhorn, W., and Ruigrok, R.W.H. (2006). Crystal structure of the rabies virus nucleoprotein–RNA complex. Science *313*, 360–363.

Anderson, E.C., and Rowe, L.W. (1998). The prevalence of antibody to the viruses of bovine virus diarrhoea, bovine herpes virus 1, rift valley fever, ephemeral fever and bluetongue and to *Leptospira* sp in free-ranging wildlife in Zimbabwe. Epidemiol. Infect. *121*, 441–449.

Bai, W. (1993). Epidemiology and control of bovine ephemeral fever in China. In Bovine Ephemeral Fever and Related Rhabdoviruses, St. George, T.D., Uren, M.F., Young, P.L., and Hoffman, D., eds (ACIAR, Canberra, Australia), pp. 20–22.

Bai, W., Zhang, Z.G., Jiang, C.L., Uren, M.F., and Davis, S.S. (1989). Clinical and pathological responses of water buffalo (*Bubalus bubalis*) to experimental infection with bovine ephemeral fever virus – Chinese strain. Aust. Vet. J. *66*, 373–374.

Bai, W., Jiang, C.L., and Davis, S.S. (1991). Preliminary observations on the epidemiology of bovine ephemeral fever in China. Trop. Anim. Health Prod. *23*, 22–26.

Bai, W., Yan, J., Zhang, Z., Jiang, C., and Lin, X. (1993). Studies on a vaccine against ephemeral fever. In Bovine Ephemeral Fever and Related Rhabdoviruses, St. George, T.D., Uren, M.F., Young, P.L., and Hoffman, D., eds (ACIAR, Canberra, Australia), pp. 111–114.

Barnard, B.J. (1997). Antibodies against some viruses of domestic animals in southern African wild animals. Onderstepoort J. Vet. Res. *64*, 95–110.

Basson, P.A., Pienaar, J.G., and van der Westhuizen, B. (1970). The pathology of ephemeral fever: a study of the experimental disease in cattle. J. S. Afr. Vet. Med. Assoc. *40*, 385–397.

Bauer, S.P., and Murphy, F.A. (1975). Relationship of two arthropod-borne rhabdoviruses (kotonkan and Obodhiang) to the rabies serogroup. Infect. Immun. *12*, 1157–1172.

Bevan, L.E.W. (1907). Preliminary report on the so-called stiff-sickness or three-day-sickness of cattle. J. Comp. Pathol. Ther. *20*, 104–113.

Bevan, L.E.W. (1912). Ephemeral fever, or three day sickness in cattle. Vet. J. *68*, 458–461.

Bilsel, P.A., Rowe, J.E., Fitch, W.M., and Nichol, S.T. (1990). Phosphoprotein and nucleocapsid protein evolution of vesicular stomatitis virus New Jersey. J. Virol. *64*, 2498–2504.

Blackburn, N.K., Searle, L., and Phleps, R.J. (1985). Viruses isolated from *Culicoides* (Diptera: *Ceratopogonidae*) caught at a veterinary research farm, Mazowe, Zimbabwe J. Entomol. Soc. South Africa *48*, 331–336.

Blasdell, K.R., Voysey, R., Bulach, D., Joubert, D.A., Tesh, R.B., Boyle, D.B., and Walker, P.J. (2012). Kotonkan and Obodhiang viruses: African ephemeroviruses with large and complex genomes. Virology *425*, 143–153.

Bourhy, H., Cowley, J.A., Larrous, F., Holmes, E.C., and Walker, P.J. (2005). Phylogenetic relationships among rhabdoviruses inferred using the L polymerase gene. J. Gen. Virol. *86*, 2849–2858.

Buckley, S.M. (1973). Singh's *Aedes albopictus* cell cultures as helper cells for the adaptation of Obodhiang and kotonkan viruses of the rabies serogroup to some vertebrat cell cultures. Appl. Microbiol. *25*, 695–696.

Buckley, S.M. (1975). Arbovirus infection of vertebrate and insect cell cultures, with special emphasis on Mokola, Obodhiang, and kotonkan viruses of the rabies serogroup. Ann. N.Y. Acad. Sci. *266*, 241–250.

Burgess, G.W. (1971). Bovine ephemeral fever. A review. Vet. Bull. *41*, 887–895.

Burgess, G.W., and Chenoweth, P.J. (1975). Mid-piece abnormalities in bovine semen following experimental and natural cases of bovine ephemeral fever. Brit. Vet. J. *131*, 536–544.

Calisher, C.H., Karabatsos, N., Zeller, H., Digoutte, J.-P., Tesh, R.B., Shope, R.E., Travassos da Rosa, A.P.A., and St. George, T.D. (1989). Antigenic relationships among rhabdoviruses from vertebrates and hematophagous arthropods. Intervirol. *30*, 241–257.

Chang, C.J., Shih, W.L., Yu, F.L., Liao, M.H., and Liu, H.J. (2004). Apoptosis induced by bovine ephemeral fever virus. J. Virol. Meth. *122*, 165–170.

Chare, E.R., Gould, E.A., and Holmes, E.C. (2003). Phylogenetic analysis reveals a low rate of homologous recombination in negative-sense RNA viruses. J. Gen. Virol. *84*, 2691–2703.

Chen, C.Y., Chang, C.Y., Liu, H.J., Liao, M.H., Chang, C.I., Hsu, J.L., and Shih, W.L. (2010). Apoptosis induction in BEFV-infected Vero and MDBK cells through Src-dependent JNK activation regulates caspase-3 and mitochondria pathways. Vet. Res. *41*, e15.

Chenik, M., Chebli, K., and Blondel, D. (1995). Translation initiation at alternate in-frame AUG codons in the rabies virus phosphoprotein mRNA is mediated by a ribosomal leaky scanning mechanism. J. Virol. *69*, 707–712.

Chiu, S.Y., and Lu, Y.S. (1987). The prophylaxis of bovine ephemeral fever in Taiwan. J. Chin. Soc. Vet. Sci. *13*, 198–195.

Chiu, S.Y., and Lu, Y.S. (1998). Epidemiology of bovine ephemeral fever infection in Taiwan. Microbiol. Res. *153*, 289–295.

Clark, H.F. (1980). Rabies serogroup viruses in neuroblastoma cells: propagation, 'autointerference,' and apparently random back-mutation of attenuated viruses to the virulent state. Infect. Immun. *27*, 1012–1022.

Coelen, R.J., and Mackenzie, J.S. (1988). Genetic variation of Murray Valley encephalitis virus. J. Gen. Virol. *69*, 1903–1912.

Coetzer, J.A.W. (1993). The diagnosis and differential diagnosis of bovine ephemeral fever in southern Africa. In Bovine Ephemeral Fever and Related Rhabdoviruses, St. George, T.D., Uren, M.F., Young, P.L., and Hoffman, D., eds (ACIAR, Canberra, Australia), pp. 70–73.

Covic, M., Hassa, P.O., Saccani, S., Buerki, C., Meier, N.I., Lombardi, C., Imhof, R., Bedford, M.T., Natoli, G., and Hottiger, M.O. (2005). Arginine methyltransferase CARM1 is a promoter-specific regulator of NF-κB-dependent gene expression. EMBO J. *24*, 85–96.

Cybinski, D.H. (1987). Homologous and heterologous antibody reactions in sera from cattle naturally infected with bovine ephemeral fever group viruses. Vet. Microbiol. *13*, 1–9.

Cybinski, D.H., and Davis, S.S. (1993). Antigenic variation in the glycoprotein of bovine ephemeral fever virus. In Arbovirus Research in Australia: Proceedings of the Sixth Symposium, Uren, M.F., and Kay, B.H., eds (CSIRO/QIMR, Brisbane, Australia), pp. 166–173.

Cybinski, D.H., and Gard, G.P. (1986). Isolation of a new rhabdovirus in Australia related to Tibrogargan virus. Aust. J. Biol. Sci. *39*, 225–232.

Cybinski, D.H., and Muller, M.J. (1990). Isolation of arboviruses from cattle and insects at two sentinel sites in Queensland, Australia, 1979–85. Aust. J. Zool. *38*, 25–32.

Cybinski, D.H., and Zakrzewski, H. (1983). The isolation and preliminary characterization of a rhabdovirus in Australia related to bovine ephemeral fever virus. Vet. Microbiol. *8*, 221–235.

Cybinski, D.H., St. George, T.D., Standfast, H.A., and McGregor, A. (1980). Isolation of Tibrogargan virus, a new Australian rhabdovirus, from *Culicoides brevitarsis*. Vet. Microbiol. *5*, 301–308.

Cybinski, D.H., Walker, P.J., Byrne, K.A., and Zakrzewski, H. (1990). Mapping of antigenic sites on the bovine ephemeral fever virus glycoprotein using monoclonal antibodies. J. Gen. Virol. *71*, 2065–2072.

Cybinski, D.H., Davis, S.S., and Zakrzewski, H. (1992). Antigenic variation of the bovine ephemeral fever virus glycoprotein. Arch. Virol. *124*, 211–224.

Dancho, B., McKenzie, M.O., Connor, J.H., and Lyles, D.S. (2009). Vesicular stomatitis virus matrix protein mutations that affect association with host membranes and viral nucleocapsids. J. Biol. Chem. *284*, 4500–4509.

Daniels, P.W., Sendow, I., Soleha, E., Hunt, S.N.T., and Bahri, S. (1995). Australian–Indonesian collaboration in veterinary arbovirology – a review. Vet. Microbiol. *46*, 151–174.

Davies, F.G., and Walker, A.R. (1974). The isolation of ephemeral fever virus from cattle and *Culicoides* midges in Kenya. Vet. Record *95*, 63–64.

Davies, F.G., Ochieng, P., and Walker, A.R. (1990). The occurrence of ephemeral fever in Kenya, 1968–1988. Vet. Microbiol. *22*, 129–136.

Davies, F.G., Shaw, T., and Ochieng, P. (1975). Observations on the epidemiology of ephemeral fever in Kenya. J. Hyg. *75*, 231–235.

Davis, S.S., Gibson, D.S., and Clark, R. (1984). The effect of bovine ephemeral fever on milk production. Aust. Vet. J. *61*, 128.

Delmas, O., Assenberg, R., Grimes, J.M., and Bourhy, H. (2010). The structure of the nucleoprotein binding domain of *Lyssavirus* phosphoprotein reveals a structural relationship between the N-RNA binding domains of Rhabdoviridae and *Paramyxoviridae*. RNA Biol. *7*, 322–327.

Dietzgen, R.G., Calisher, C.H., Kurath, G., Kuzman, I.V., Rodriguez, L.L., Stone, D.M., Tesh, R.B., Tordo, N., Walker, P.J., Wetzel, T., *et al.* (2011). Rhabdoviridae. In *Virus Taxonomy*, Ninth Report of the International Committee on Taxonomy of Viruses, King, A.M.Q., Adams, M.J., Carstens, E.B., and Lefkowitz, E.J., eds (Elsevier, San Diego, CA), pp. 654–681.

Dirie, M.F., and Abudurahman, O. (2003). Observations on little known diseases of camels (*Camelus dromedarius*) in the Horn of Africa. Rev. Sci. Tech. OIE *22*, 1043–1049.

Doherty, R.L., Carley, J.G., Standfast, H.A., Dyce, A.L., Kay, B.H., and Snowdon, W.A. (1973). Isolation of arboviruses from mosquitoes, biting midges, sandflies and vertebrates collected in Queensland, 1969 and 1970. Trans. R. Soc. Trop. Med. Hyg. *67*, 536–543.

Elmore, S. (2007). Apoptosis: a review of programmed cell death. Toxicol. Pathol. *35*, 495–516.

Erasmus, B.J. (1986). The use of live ephemeral fever vaccine in South Africa. In Arbovirus Research in Australia: Proceedings of the Fourth Symposium, St. George, T.D., Kay, B.H., and Blok, J., eds (CSIRO/QIMR, Brisbane, Australia), pp. 318–319.

Finlaison, D.S., Read, A.J., and Kirkland, P.D. (2010). An epizootic of bovine ephemeral fever in New South Wales in 2008 associated with long-distance dispersal of vectors. Aust. Vet. J. *88*, 301–306.

Flynn, L.M., Coelen, R.J., and Mackenzie, J.S. (1989). Kunjin virus isolates of Australia are genetically homogeneous. J. Gen. Virol. *70*, 2819–2824.

Freed, E.O. (2002). Viral late domains. J. Virol. *76*, 4679–4687.

Freer, G.W. (1910). Ephemeral fever or three-day sickness of cattle. Vet. J. *66*, 19–22.

Gard, G.P., Cybinski, D.H., and St George, T.D. (1983). The isolation in Australia of a new virus related to bovine ephemeral fever virus. Aust. Vet. J. *60*, 89–90.

Gard, G.P., Cybinski, D.H., and Zakrzewski, H. (1984). The isolation of a fourth bovine ephemeral fever group virus. Aust. Vet. J. *61*, 332.

Gard, G.P., Shorthose, J.E., Weir, R.P., Walsh, S.J., and Melville, L.F. (1988). Arboviruses recovered from sentinel livestock in northern Australia. Vet. Microbiol. *18*, 109–118.

Gerard, F.C.A., Ribeiro, E.d.A., Leyrat, C., Ivanov, I., Blondel, D., Longhi, S., Ruigrok, R.W.H., and Jamin, M. (2009). Modular organization of rabies virus phosphoprotein. J. Mol. Biol. *388*, 978–996.

Gibbs, E.P.J., Calisher, C.H., Tesh, R.B., Lazuick, J.S., Bowen, R., and Greiner, E.C. (1989). Bivens Arm virus: a new rhabdovirus isolated from *Culicoides insignis* in Florida and related to Tibrogargan virus of Australia. Vet. Microbiol. *19*, 141–150.

Gonzalez, M.E., and Carrasco, L. (2003). Viroporins. FEBS Lett. *552*, 28–34.

Gottwein, E., Bodem, J., Muller, B., Schmechel, A., Zentgraf, H., and Krausslich, H.-G. (2003). The Mason–Pfizer monkey virus PPPY and PSAP motifs both contribute to virus release. J. Virol. *77*, 9474–9485.

Green, T.J., Zhang, X., Wertz, G.W., and Luo, M. (2006). Structure of the vesicular stomatitis virus nucleoprotein–RNA complex. Science *313*, 357–360.

Gubala, A., Davis, S., Weir, R., Melville, L., Cowled, C., Walker, P., and Boyle, D. (2010). Ngaingan virus, a macropod-associated rhabdovirus, contains a second glycoprotein gene and seven novel open reading frames. Virology *399*, 98–108.

Gubala, A., Davis, S., Weir, R., Melville, L., Cowled, C., and Boyle, D. (2011). Tibrogargan and Coastal Plains rhabdoviruses: genomic characterisation, evolution of novel genes and seroprevalence in Australian livestock. J. Gen. Virol. *92*, 2160–2170.

Gubala, A.J., Proll, D.F., Barnard, R.T., Cowled, C.J., Crameri, S.G., Hyatt, A.D., and Boyle, D.B. (2008). Genomic characterisation of Wongabel virus reveals novel genes within the Rhabdoviridae. Virology *376*, 13–23.

Hall, W.T., Daddow, K.N., Dimmock, C.K., George, T.D., and Standfast, H.A. (1975). The infection of merino sheep with bovine ephemeral fever virus. Aust. Vet. J. *51*, 344–346.

Han, G.-Z., He, C.-Q., Ding, N.-Z., and Ma, L.-Y. (2008). Identification of a natural multi-recombinant of Newcastle disease virus. Virology *371*, 54–60.

Harouaka, D., and Wertz, G.W. (2009). Mutations in the C-terminal loop of the nucleocapsid protein affect vesicular stomatitis virus RNA replication and transcription differentially. J. Virol. *83*, 11429–11439.

Harty, R.N., Paragas, J., Sudol, M., and Palese, P. (1999). A proline-rich motif within the matrix protein of vesicular stomatitis virus and rabies virus interacts with WW domains of cellular proteins: implications for viral budding. J. Virol. *73*, 2921–2929.

Hassa, P.O., Covic, M., Bedford, M.T., and Hottiger, M.O. (2008). Protein arginine methyltransferase 1 coactivates NF-κB-dependent gene expression synergistically with CARM1 and PARP1. J. Mol. Biol. *377*, 668–678.

Hertig, C., Pye, A.D., Hyatt, A.D., Davis, S.S., McWilliam, S.M., Heine, H.G., Walker, P.J., and Boyle, D.B. (1996). Vaccinia virus-expressed bovine ephemeral fever virus G but not G(NS) glycoprotein induces neutralizing antibodies and protects against experimental infection. J. Gen. Virol. *77*, 631–640.

Hill, M.W.M., and Schultz, K. (1977). Ataxia and paralysis associated with bovine ephemeral fever infection. Aust. Vet. J. *53*, 217–221.

Holmes, E.C. (2009). The Evolution and Emergence of RNA Viruses (Oxford University Press, Oxford).

Holmes, I.H., and Doherty, R.L. (1970). Morphology and development of bovine ephemeral fever virus. J. Virol. *5*, 91–96.

Hsieh, Y.-C., Chen, S.-H., Chou, C.C., Ting, L.-J., Itakaru, C., and Wang, F.-I. (2005). Bovine ephemeral fever in Taiwan (2001–2002). J. Vet. Med. Sci. *67*, 411–416.

Hsieh, Y.-C., Wang, S.-Y., Lee, Y.-F., Chen, S.-H., Mak, P.O.T., and Chu, C.-Y. (2006). DNA sequence analysis of glycoprotein G gene of bovine ephemeral fever virus and development of a double oil emulsion vaccine against bovine ephemeral fever. J. Vet. Med. Sci. *68*, 543–548.

Inaba, Y., Tanaka, Y., Sato, K., Ito, H., and Omori, T. (1968). Propagation in laboratory animals and cell cultures of a virus from cattle with bovine epizootic fever. Jpn J. Microbiol. *12*, 253–255.

Inaba, Y., Sato, K., Tanaka, Y., Ito, H., and Omori, T. (1969a). Serological identification of bovine epizootic fever virus as ephemeral fever virus. Jpn J. Microbiol. *13*, 388–389.

Inaba, Y., Tanaka, Y., Sato, K., Ito, H., and Omori, T. (1969b). Bovine epizootic fever. 3. Loss of virus pathogenicity and immunogenicity for the calf during serial passage in various host systems. Jpn J. Microbiol. *13*, 181–186.

Inaba, Y., Kurogi, H., Sato, K., Goto, Y., Omori, T., and Matumoto, M. (1973). Formalin-inactivated, aluminium phosphate gel-adsorbed vaccine of bovine ephemeral fever virus. Arch. Virol. *42*, 42–53.

Inaba, Y., Kurogi, H., Takahash, A., Sato, K., Omori, T., Goto, Y., Hanaki, T., Yamamoto, M., Kishi, S., Kodama, K., *et al.* (1974). Vaccination of cattle against bovine ephemeral fever with live attenuated virus followed by killed virus. Arch. Virol. *44*, 121–132.

Irie, T., Licata, J.M., McGettigan, J.P., Schnell, M.J., and Harty, R.N. (2004). Budding of PPxY-containing rhabdoviruses is not dependent on host proteins TGS101 and VPS4A. J. Virol. *78*, 2657–2665.

Ito, Y., Tanaka, Y., Inaba, Y., and Omori, T. (1969). Electron microscopic observations of bovine epizootic fever virus. Natl. I. Anim. Health Q. *9*, 35–44.

Jayakar, H.R., and Whitt, M.A. (2002). Identification of two additional translation products from the matrix (M) gene that contribute to vesicular stomatitis virus cytopathology. J. Virol. *76*, 8011–8018.

Jayakar, H.R., Murti, K.G., and Whitt, M.A. (2000). Mutations in the PPPY motif of vesicular stomatitis virus matrix protein reduce virus budding by inhibiting a late step in virion release. J. Virol. *74*, 9818–9827.

Ji, W.T., Wang, Y.C., Lin, F.L., Liao, M.H., Shih, W.L., and Liu, H.J. (2011). Inhibitors of phosphatidylinositol 3-kinase and mTOR but not Akt enhance replication of bovine ephemeral fever virus. Vet. J. *187*, 119–123.

Jiang, C.L., and Yan, J.D. (1989). Evidence of Kimberley virus infection of cattle in China. Trop. Anim. Health Prod. *21*, 85–86.

Johal, J., Gresty, K., Kongsuwan, K., and Walker, P.J. (2008). Antigenic characterization of bovine ephemeral fever rhabdovirus G and GNS glycoproteins expressed from recombinant baculoviruses. Arch. Virol. *153*, 1657–1665.

Karabatsos, N., ed. (1985). International Catalogue of Arboviruses Including Certain other Viruses of Vertebrates, 3rd edn (American Society for Tropical Medicine and Hygiene, San Antonio, TX).

Karaoglu, T., Ozgunluk, I., Demir, B., Ozkul, A., and Bergu, I. (2007). Seroprevalence of culicoides-borne disease in cattle in European Turkey. Vet. J. Ankara Univ. *54*, 121–125.

Kato, T., Aizawa, M., Takayoshi, K., Kokuba, T., Yanase, T., Shirafuji, H., Tsuda, T., and Yamakawa, M. (2009). Phylogenetic relationships of the G gene sequence of bovine ephemeral fever virus isolated in Japan, Taiwan and Australia. Vet. Microbiol. *137*, 217–123.

Kay, B.H., Carley, J.G., and Filippich, C. (1975). The multiplication of Queensland and New Guinean arboviruses in *Culex annulirostris* Skuse and *Aedes vigilax* (Skuse) (Diptera: Culicidae). J. Med. Entomol. *12*, 279–283.

Kemp, G.E., Lee, V.H., Moore, D.L., Shope, R.E., Causey, O.R., and Murphy, F.A. (1973a). Kotonkan, a new rhabdovirus related to Mokola virus of the rabies serogroup. Amer. J. Epidemiol. *98*, 43–49.

Kemp, G.E., Mann, E.D., Tomori, O., Fabiyi, A., and O'Connor, E. (1973b). Isolation of bovine ephemeral fever virus in Nigeria. Vet. Record 93, 107–108.

Kirkland, P.D. (1993). The epidemiology of bovine ephemeral fever in south-eastern Australia: evidence for a mosquito vector. In Bovine Ephemeral Fever and Related Rhabdoviruses, St. George, T.D., Uren, M.F., Young, P.L., and Hoffman, D., eds (ACIAR, Canberra, Australia), pp. 33–37.

Kirkland, P.D. (2002). Akabane and bovine ephemeral fever virus infections. Vet. Clin. Food Anim. Pract. *18*, 501–514.

Kleinschmidt, M.A., Streubel, G., Samans, B., Krause, M., and Bauer, U.-M. (2008). The protein arginine methyltransferases CARM1 and PRMT1 cooperate in gene regulation. Nucleic Acids Res. *36*, 3202–3213.

Kongsuwan, K., Cybinski, D.H., Cooper, J., and Walker, P.J. (1998). Location of neutralizing epitopes on the G protein of bovine ephemeral fever rhabdovirus. J. Gen. Virol. *79*, 2573–2581.

Kuzmin, I.V., Hughes, G.J., and Rupprecht, C.E. (2006). Phylogenetic relationships of seven previously unclassified viruses within the family Rhabdoviridae using partial nucleoprotein gene sequences. J. Gen. Virol. *87*, 2323–2331.

Larrous, F., Gholami, A., Mouhamad, S., Estaquier, J., and Bourhy, H. (2010). Two overlapping domains of a *Lyssavirus* matrix protein that acts on different cell death pathways. J. Virol. *84*, 9897–9906.

Lecatsas, G., Theodoridis, A., and Erasmus, B.J. (1969). Electron microscopic studies on bovine ephemeral fever virus. Arch. Virol. *28*, 390–398.

Lee, D.J., Hicks, M.M., Debenham, M.L., Griffiths, M., Bryan, J.H., and Marks, E.N. (1988). The Culicidae of the Australasian Region. In Entomology Monograph No. 2 (The University of Queensland and The University of Sydney, Brisbane, Australia), pp. 145–160.

Liao, Y.K., Inaba, Y., Li, N.J., Chain, C.Y., Lee, S.L., and Liou, P.P. (1998). Epidemiology of bovine ephemeral fever virus in Taiwan. Microbiol. Res. *153*, 289–295.

Liehne, P.F.S., Anderson, S., Stanley, N.F., Liehne, C.G., Wright, A.E., Chan, K.H., Leivers, S., Britten, D.K., and Hamilton, N.P. (1981). Isolation of Murray Valley encephalitis virus and other arboviruses in the Ord River Valley 1972–1976. Aust. J. Exp. Biol. Med. Sci. *59*, 347–356.

Lim, S.I., Kweon, C.H., Tark, D.S., Kim, S.H., and Yang, D.K. (2007). Sero-survey on Aino, Akabane, Chuzan, bovine ephemeral fever and Japanese encephalitis virus of cattle and swine in Korea. J. Vet. Sci. *8*, 45–49.

Lin, C.H., Shih, W.L., Lin, F.L., Hsieh, Y.C., Kuo, Y.R., Liao, M.H., and Liu, H.J. (2009). Bovine ephemeral fever virus-induced apoptosis requires virus gene expression and activation of Fas and mitochondrial signaling pathway. Apoptosis *14*, 864–877.

Lukashev, A.N. (2005). Evidence for recombination in Crimean-Congo hemorrhagic fever virus. J. Gen. Virol. *86*, 2333–2338.

Mackerras, I.M., Mackerras, M.J., and Burnet, F.M. (1940). Experimental studies of ephemeral fever in Australian cattle. CSIRO Bull. *136*, 1–116.

McWilliam, S.M., Kongsuwan, K., Cowley, J.A., Byrne, K.A., and Walker, P.J. (1997). Genome organization and transcription strategy in the complex GNS-L intergenic region of bovine ephemeral fever rhabdovirus. J. Gen. Virol. *78*, 1309–1317.

Madbouly, H.M., Hegazi, A.G., El Shabrawy, M., Tamam, S.M., and Gamil, G.S. (2006). Isolation and identification of bovine ephemeral fever virus from cattle in middle Egypt. Vet. Med. J. Giza *54*, 877–886.

Malviya, H.K., and Prasad, J. (1977). Ephemeral fever – a clinical and epidemiological study in cross bred cows and buffaloes. Indian Vet. J. *54*, 440–444.

Miao, F., Li, S., Chavez, V., Lanting, L., and Natarajan, R. (2006). Coactivator-associated arginine methyltransferase-1 enhances nuclear factor-κB-mediated gene transcription through methylation of histone H3 at arginine 17. Mol. Endocrinol. *20*, 1562–1573.

Muller, M.J., and Standfast, H.A. (1986). Vectors of ephemeral fever group viruses. In Arbovirus Research in Australia: Proceedings of the Fourth Symposium, St George, T.D., Kay, B.H., and Blok, J., eds (CSIRO/QIMR, Brisbane, Australia), pp. 295–300.

Murphy, F.A., Taylor, W.P., Mims, C.A., and Whitfield, S.G. (1972). Bovine ephemeral fever virus in cell culture and mice. Arch. Virol. *38*, 234–249.

Murray, M.D. (1970). The spread of ephemeral fever of cattle during the 1967–68 epizootic in Australia. Aust. Vet. J. *46*, 77–82.

Murray, M.D. (1997). Possible vectors of bovine ephemeral fever in the 1967/68 epizootic in northern Victoria. Aust. Vet. J. *75*, 220–220.

Nadin-Davis, S.A., Abdel-Malik, M., Armstrong, J., and Wandeler, A.I. (2002). *Lyssavirus* P gene characterisation provides insights into the phylogeny of the genus and identifies structural similarities and diversity within the encoded phosphoprotein. Virology *298*, 286–305.

Nayak, D., Panda, D., Das, S.C., Luo, M., and Pattnaik, A.K. (2009). Single amino acid alterations in a highly conserved central region of vesicular stomatitis virus N protein differentially affect the viral nucleocapsid template functions. J. Virol. *83*, 5525–5534.

Newton, L.G., and Wheatley, C.H. (1970). The occurrence and spread of ephemeral fever of cattle in Queensland. Aust. Vet. J. *46*, 561–568.

Pasdeloup, D., Poisson, N., Raux, H., Gaudin, Y., Ruigrok, R.W., and Blondel, D. (2005). Nucleocytoplasmic shuttling of the rabies virus P protein requires a nuclear localization signal and a CRM1-dependent nuclear export signal. Virology *334*, 284–293.

Peiris, J.S.M., Amerasinghe, P.H., Amerasinghe, F.P., Calisher, C.H., Perera, L.P., Arunagiri, C.K., Munasingha, N.B., and Karunaratne, S.H. (1994). Viruses isolated from mosquitoes collected in Sri Lanka. Amer. J. Trop. Med. Hyg. *51*, 154–161.

Peluso, R.W., Richardson, J.C., Talon, J., and Lock, M. (1996). Identification of a set of proteins (C′ and C) encoded by the bicistronic P gene of the Indiana serotype of vesicular stomatitis virus and analysis of their effect on transcription by the viral RNA polymerase. Virology *218*, 335–342.

Poch, O., Blumberg, B.M., Bougueleret, L., and Tordo, N. (1990). Sequence comparison of five polymerases (L proteins) of unsegmented negative-strand RNA viruses: theoretical assignment of functional domains. J. Gen. Virol. *71*, 1153–1162.

Poidinger, M., Hall, R.A., Lindsay, M.D., Broom, A.K., and Mackenzie, J.S. (2000). The molecular epidemiology of Kokobera virus. Virus Res. *68*, 7–13.

Powell, M.L. (2010). Translational termination-reinitiation in RNA viruses. Biochem. Soc. Transact. *38*, 1558–1564.

Qin, Z., Sun, L., Ma, B., Cui, Z., Zhu, Y., Kitamura, Y., and Liu, W. (2008). F gene recombination between genotype II and VII Newcastle disease virus. Virus Res. *131*, 299–303.

Rabagliati, D.S. (1909). Three days fever or stiff sickness in cattle. Vet. Record *4*, 503–505.

Rahmeh, A.A., Li, J., Kranzusch, P.J., and Whelan, S.P.J. (2009). Ribose 2′-O methylation of the vesicular stomatitis virus mRNA cap precedes and facilitates subsequent guanine-N-7 methylation by the large polymerase protein. J. Virol. *83*, 11043–11050.

Rahmeh, A.A., Schenk, A.D., Danek, E.I., Kranzusch, P.J., Liang, B., Walz, T., and Whelan, S.P. (2010). Molecular architecture of the vesicular stomatitis virus RNA polymerase. Proc. Natl. Acad. Sci. U.S.A. *107*, 20075–20080.

Raux, H., Obiang, L., Richard, N., Harper, F., Blondel, D., and Gaudin, Y. (2010). The matrix protein of vesicular stomatitis virus binds dynamin for efficient viral assembly. J. Virol. *84*, 12609–12618.

Roche, S., Bressanelli, S., Rey, F.A., and Gaudin, Y. (2006). Crystal structure of the low-pH form of the vesicular stomatitis virus glycoprotein G. Science *313*, 187–191.

Roche, S., Rey, F.A., Gaudin, Y., and Bressanelli, S. (2007). Structure of the prefusion form of the vesicular stomatitis virus glycoprotein G. Science *315*, 843–848.

Roya, S. (2008). Survey on serological diagnosis of bovine ephemeral fever (BEF) by IR-BK and Vero cell line in southern provinces of Iran by *in vitro* method. Paper presented at: FAVA – OIE Joint Symposium on Emerging Diseases (Bangkok, TVMA).

Ryan, P.A., Do, K.A., and Kay, B.H. (2000). Definition of Ross River virus vectors at Maroochy Shire, Australia. J. Med. Entomol. *37*, 146–152.

Sammels, L.M., Lindsay, M.D., Poidinger, M., Coelen, R.J., and Mackenzie, J.S. (1999). Geographic distribution and evolution of Sindbis virus in Australia. J. Gen. Virol. *80*, 739–748.

Sang, R., Kioko, E., Lutomiah, J., Warigia, M., Ochieng, C., O'Guinn, M., Lee, J.S., Koka, H., Godsey, M., Hoel, D., *et al.* (2010). Rift Valley fever virus epidemic in Kenya, 2006/2007: the entomologic investigations. Amer. J. Trop. Med. Hyg. *83*, 28–37.

Schmidt, J.R., Williams, M.C., Lulu, M., Mivule, A., and Mujombe, E. (1965). Viruses isolated from mosquitoes collected in the Southern Sudan and Western Ethiopia. East Afr. Virus Res. Inst. Rep. *15*, 24.

Seddon, H.R. (1938). The spread of ephemeral fever (three-day sickness) in Australia 1936–37. Aust. Vet. J. *14*, 90–101.

Shiel, M.J., Campbell, J., Davis, K.E., and Davis, G.P. (1989). An outbreak of bovine ephemeral fever in northeast Victoria. Aust. Vet. J. *66*, 222–223.

Shirakawa, H., Ishibashi, K., and Ogawa, T. (1994). A comparison of the epidemiology of bovine ephemeral fever in South Korea and south-western Japan. Aust. Vet. J. *71*, 50–52.

Shope, R.E. (1975). Rabies virus antigenic relationships. In The Natural History of Rabies, Baer, G.M., ed. (Academic Press, New York), pp. 141–152.

Sibold, C., Meisel, H., Kruger, D.H., Labuda, M., Lysy, J., Kozuch, O., Pejcoch, M., Vaheri, A., and Plyusnin, A. (1999). Recombination in Tula hantavirus evolution: analysis of genetic lineages from Slovakia. J. Virol. *73*, 667–675.

Snowdon, W.A. (1970). Bovine ephemeral fever: the reaction of cattle to different strains of ephemeral fever virus and the antigenic comparison of two strains of virus. Aust. Vet. J. *46*, 258–266.

Soleha, E., Daniels, P.W., Sendow, I., and Sukarsih (1993a). Seroepidemiological studies of bovine ephemeral fever group viral infections in Indonesia. Paper presented at: 6th Symposium on Arbovirus Research in Australia (CSIRO/QIMR, Brisbane, Australia).

Soleha, E., Daniels, P.W., Sukarsih, and Sendow, I. (1993b). A study of bovine ephemeral fever group rhabdoviral infections in West Java, Indonesia. In Bovine Ephemeral Fever and Related Rhabdoviruses, St. George, T.D., Uren, M.F., Young, P.L., and Hoffman, D., eds (ACIAR, Canberra, Australia), pp. 45–50.

Spiropoulou, C.F., and Nichol, S.T. (1993). A small highly basic protein is encoded in overlapping frame within the P gene of vesicular stomatitis virus. J. Virol. *67*, 3103–3110.

St. George, T.D. (1980). A sentinel herd system for the study of arbovirus infections in Australia and Papua New Guinea. Vet. Sci. Comm. *4*, 39–51.

St. George, T.D. (1981). Ephemeral fever. In Virus Diseases of Food Animals, Gibbs, E.P.J., ed. (Academic Press, London), pp. 541–564.

St. George, T.D. (1993). The natural history of ephemeral fever of cattle. In Bovine Ephemeral Fever and Related Rhabdoviruses, St. George, T.D., Uren, M.F., Young, P.L., and Hoffman, D., eds (ACIAR, Canberra, Australia), pp. 13–19.

St George, T.D., and Standfast, H.A. (1988). Bovine ephemeral fever. In The Arboviruses: Epidemiology and Ecology, Monath, T.P., ed. (CRC Press, Boca Raton, FL), pp. 71–86.

St George, T.D., Standfast, H.A., Christie, D.G., Knott, S.G., and Morgan, I.R. (1977). The epizootiology of bovine ephemeral fever in Australia and Papua New Guinea. Aust. Vet. J. *53*, 17–28.

St George, T.D., Cybinski, D.H., Murphy, G.M., and Dimmock, C.K. (1984). Serological and biochemical factors in bovine ephemeral fever. Aust. J. Biol. Sci. *37*, 341–349.

St George, T.D., Murphy, G.M., Burren, B., and Uren, M.F. (1995). Studies on the pathogenesis of bovine ephemeral fever. IV: A comparison with the inflammatory events in milk fever of cattle. Vet. Microbiol. *46*, 131–142.

Standfast, H.A., St George, T.D., and Dyce, A.L. (1976). The isolation of ephemeral fever virus from mosquitoes in Australia. Aust. Vet. J. *52*, 242.

Sun, M., Fuentes, S.M., Timani, K., Sun, D., Murphy, C., Lin, Y., August, A., Teng, M.N., and He, B. (2008). Akt plays a critical role in replication of nonsegmented negative-stranded RNA viruses. J. Virol. *82*, 105–114.

Tesh, R.B., Travassos da Rosa, A.P.A., and Travassos da Rosa, J.S. (1983). Antigenic relationship among rhabdoviruses infecting terrestrial vertebrates. J. Gen. Virol. *64*, 169–176.

Theodoridis, A., Boshoff, S.E., and Botha, M.J. (1973a). Studies on the development of a vaccine against bovine ephemeral fever. Onderstepoort J. Vet. Res. *40*, 77–82.

Theodoridis, A., Giesecke, W.H., and Du Toit, I.J. (1973b). Effects of ephemeral fever on milk production and reproduction of dairy cattle. Onderstepoort J. Vet. Res. *40*, 83–91.

Tian, F.G., Jiang, C.L., Zakrzewski, H., and Davis, S.S. (1987). A comparison of a Chinese and an Australian strain of bovine ephemeral fever virus. Aust. Vet. J. *64*, 159.

Tomori, O., Fagbami, A., and Kemp, G. (1974). Kotonkan virus: experimental infection of white Fulani calves. Bull. Epizoot. Dis. Afr. *22*, 195–200.

Troffer-Charlier, N., Cura, V., Hassenboehler, P., Moras, D., and Cavarelli, J. (2007). Functional insights from structures of coactivator-associated arginine methyltransferase 1 domains. EMBO J. *26*, 4391–4401.

Tuekam, T., Greiner, E.C., and Gibbs, E.P.J. (1991). Seroepidemiology of Bivens Arm virus infections of cattle in Florida, St Croix and Puerto Rico. Vet. Microbiol. *28*, 121–127.

Uren, M.F., and Zakrzewski, H. (1989). Mechanisms of immunity to BEF. In Arbovirus Research in Australia: Proceedings of the Fifth Symposium, Uren, M.F., Blok, J., and Manderson, L.H., eds (CSIRO/QIMR, Brisbane, Australia), pp. 274–276.

Uren, M.F., St George, T.D., and Stranger, R.S. (1983). Epidemiology of ephemeral fever of cattle in Australia 1975–1981. Aust. J. Biol. Sci. *36*, 91–100.

Uren, M.F., St George, T.D., Kirkland, P.D., Stranger, R.S., and Murray, M.D. (1987). Epidemiology of bovine ephemeral fever in Australia 1981–1985. Aust. J. Biol. Sci. *40*, 125–136.

Uren, M.F., St George, T.D., and Zakrzewski, H. (1989). The effect of anti-inflammatory agents on the clinical expression of bovine ephemeral fever. Vet. Microbiol. *19*, 99–111.

Uren, M.F., Walker, P.J., Zakrzewski, H., St George, T.D., and Byrne, K.A. (1994). Effective vaccination of cattle using the virion G protein of bovine ephemeral fever virus as an antigen. Vaccine *12*, 845–850.

van der Westhuizen, B. (1967). Studies on bovine ephemeral fever. I. Isolation and preliminary characterization of a virus from natural and experimentally produced cases of bovine ephemeral fever. Onderstepoort J. Vet. Res. *34*, 29–40.

Vanselow, B.A., Abetz, I., and Trenfield, K. (1985). A bovine ephemeral fever vaccine incorporating adjuvant Quil A: a comparative study using adjuvants Quil A, aluminium hydroxide gel and dextran sulphate. Vet. Record *117*, 37–43.

Vanselow, B.A., Walthall, J.C., and Abetz, I. (1995). Field trials of ephemeral fever vaccines. Vet. Microbiol. *46*, 117–130.

Venter, G.J., Hamblin, C., and Paweska, J.T. (2003). Determination of the oral susceptibility of South African livestock-associated biting midges, *Culicoides* species, to bovine ephemeral fever virus. Med. Vet. Entomol. *17*, 133–137.

Walker, P.J. (2005). Bovine ephemeral fever in Australia and the world. Curr. Topics Microbiol. Immunol. *292*, 57–80.

Walker, P.J. (2008). Bovine ephemeral fever virus. In Encyclopedia of Virology, Mahy, B.W.J., and van Regenmortel, M.H.V., eds (Academic Press, Oxford, UK), pp. 354–362.

Walker, P.J., and Kongsuwan, K. (1999). Deduced structural model for animal rhabdovirus glycoproteins. J. Gen. Virol. *80*, 1211–1220.

Walker, P.J., Byrne, K.A., Cybinski, D.H., Doolan, D.L., and Wang, Y. (1991). Proteins of bovine ephemeral fever virus. J. Gen. Virol. *72*, 67–74.

Walker, P.J., Byrne, K.A., Riding, G.A., Cowley, J.A., Wang, Y., and McWilliam, S. (1992). The genome of bovine ephemeral fever rhabdovirus contains two related glycoprotein genes. Virology *191*, 49–61.

Walker, P.J., Dietzgen, R.G., Joubert, D.A., and Blasdell, K.R. (2011). Rhabdovirus accessory genes. Virus Res. *162*, 110–125.

Wallace, D.B., and Viljoen, G.J. (2005). Immune responses to recombinants of the South African vaccine strain of lumpy skin disease virus generated by using thymidine kinase gene insertion. Vaccine *23*, 3061–3067.

Wang, Y., and Walker, P.J. (1993). Adelaide river rhabdovirus expresses consecutive glycoprotein genes as polycistronic mRNAs: new evidence of gene duplication as an evolutionary process. Virology *195*, 719–731.

Wang, Y., McWilliam, S.M., Cowley, J.A., and Walker, P.J. (1994). Complex genome organization in the GNS-L intergenic region of Adelaide River rhabdovirus. Virology *203*, 63–72.

Wang, Y., Cowley, J.A., and Walker, P.J. (1995). Adelaide River virus nucleoprotein gene: analysis of phylogenetic relationships of ephemeroviruses and other rhabdoviruses. J. Gen. Virol. *76*, 995–999.

Yang, F., and Lazzarini, R.A. (1983). Analysis of the recombination event generating a vesicular stomatitis virus deletion defective interfering particle. J. Virol. *45*, 766–772.

Yeruham, I., Braverman, Y., Yadin, H., Van Ham, M., Chai, D., Tiomkin, D., and Frank, D. (2002). Epidemiological investigations of outbreaks of bovine ephemeral fever in Israel. Vet. Record *151*, 117–121.

Yeruham, I., Yadin, H., Abramovitch, I., and Shama, Z. (2005). Bovine ephemeral fever in a dairy cattle herd in the Jordan Valley. Vet. Record *156*, 284–286.

Yeruham, I., Gur, Y., and Braverman, Y. (2007). Retrospective epidemiological investigation of an outbreak of bovine ephemeral fever in 1991 affecting dairy cattle herds on the Mediterranean coastal plain. Vet. J. *173*, 190–193.

Yeruham, I., Van Ham, M., Stram, Y., Friedgut, O., Yadin, H., Mumcuoglu, K.Y., and Braverman, Y. (2010). Epidemiological investigation of bovine ephemeral fever outbreaks in Israel. Vet. Med. Int. *2010*, e290541.

Young, P.L., and Chung, Y.S. (1986). The pathology of experimentally produced bovine ephemeral fever. In Proceedings of the Fourth Symposium, T.D. St George, B.H. Kay, and J. Blok, eds. (Brisbane, CSIRO/QIMR), pp. 301–302.

Zakrzewski, H., and Cybinski, D.H. (1984). Isolation of Kimberley virus, a rhabdovirus, from *Culicoides brevitarsis*. Aust. J. Exp. Biol. Med. Sci. *62*, 779–780.

Fish Novirhabdoviruses

7

Gael Kurath

Abstract

The genus *Novirhabdovirus* contains several of the important rhabdoviruses that infect fish hosts. There are four established virus species: *Infectious haematopoietic necrosis virus* (IHNV), *Viral haemorrhagic septicaemia virus* (VHSV), *Hirame rhabdovirus* (HIRRV), and *Snakehead rhabdovirus* (SHRV). Viruses of these species vary in host and geographic range, and they have all been studied at the molecular and genomic level. As globally significant pathogens of cultured fish, IHNV and VHSV have been particularly well studied in terms of molecular epidemiology and evolution. Phylogenic analyses of hundreds of field isolates have defined five major genogroups of IHNV and four major genotypes of VHSV worldwide. These phylogenies are informed by the known histories of IHNV and VHSV, each involving a series of viral emergence events that are sometimes associated with host switches, most often into cultured rainbow trout. In general, IHNV has relatively low genetic diversity and a narrow host range, and has been spread from its endemic source in North American to Europe and Asia due to aquaculture activities. In contrast, VHSV has broad host range and high genetic diversity, and the source of emergence events is virus in widespread marine fish reservoirs in the northern Atlantic and Pacific Oceans. Common mechanisms of emergence and host switch events include use of raw feed, proximity to wild fish reservoirs of virus, and geographic translocations of virus or naive fish hosts associated with aquaculture.

Introduction to novirhabdoviruses and fish rhabdoviruses in general

Rhabdoviruses have been very successful at evolving and diversifying within fish hosts (Kurath and Winton, 2008; Walker and Winton, 2010). Fish are the most diverse class of vertebrate animals, with over 25,000 species. They are also of special interest because jawed fishes are the earliest animals on the evolutionary tree of life to have developed adaptive immunity. Thus, fish have innate, cellular adaptive, and humoral adaptive immune systems comparable to those of other vertebrates. Due to the cold-blooded nature of fish hosts, fish viruses have evolved with low temperature optima for replication, ranging from 12°C to 25°C. In general fish viruses are inactivated at temperatures above 20–30°C, and they are not considered a threat for zoonosis in humans. Within fish rhabdoviruses the first taxonomic genus to be established officially was the *Novirhabdovirus* genus, approved by the International Committee on Taxonomy of Viruses in 1998 (Walker *et al.*, 2000). This was based on decades of research motivated by the widespread and economically significant diseases caused by infectious haematopoietic necrosis virus (IHNV), viral haemorrhagic septicaemia virus (VHSV), and Hirame rhabdovirus (HIRRV). In the 1980s characterization of the genome of IHNV revealed five genes that were analogous to those already described in vesiculovirus and lyssavirus genomes, but in addition there was a sixth gene, designated non-virion, or 'NV', between the G and L genes (Kurath and Leong, 1985). This NV gene is present in all novirhabdovirus genomes (Kurath *et al.*, 1997; Schutze *et al.*, 1999; Johnson *et al.*, 2000), and it is not found in any other rhabdovirus, making it the molecular hallmark of the *Novirhabdovirus* genus.

To date, all known members of the *Novirhabdovirus* genus have been isolated from fish hosts. However, it is important to note that the genus *Novirhabdovirus* does not contain all rhabdoviruses of fish. There are many other fish rhabdoviruses, including some that are well known fish pathogens such as Spring viraemia of carp virus (SVCV), that clearly represent distinct evolutionary lineages in fish hosts (Ahne *et al.*, 2002). The genomes of SVCV and related viruses do not have an NV gene, and phylogenetic analyses of other genes do not indicate a close relationship with the *Novirhabdovirus* genus. Based on phylogenetic analyses of recent sequence data for several fish viruses, two new genera of fish rhabdoviruses are currently being considered by the ICTV *Rhabdoviridae* study group (D.M. Stone, pers. comm.). These are tentatively designated the *Sprivivirus* genus, with type species SVCV, and the *Perhavirus* genus, with type species *Perch rhabdovirus*. Fig. 7.1 illustrates the position of the *Novirhabdovirus* genus within the family *Rhabdoviridae*, and its separation from these two newly proposed genera. The existence of these distinct lineages of fish rhabdoviruses suggests that during the evolution of the *Rhabdoviridae* there likely have been multiple events in which ancestral rhabdoviruses crossed the barrier between aquatic and terrestrial hosts.

With the exception of SVCV, the viruses in the two newly proposed fish rhabdovirus

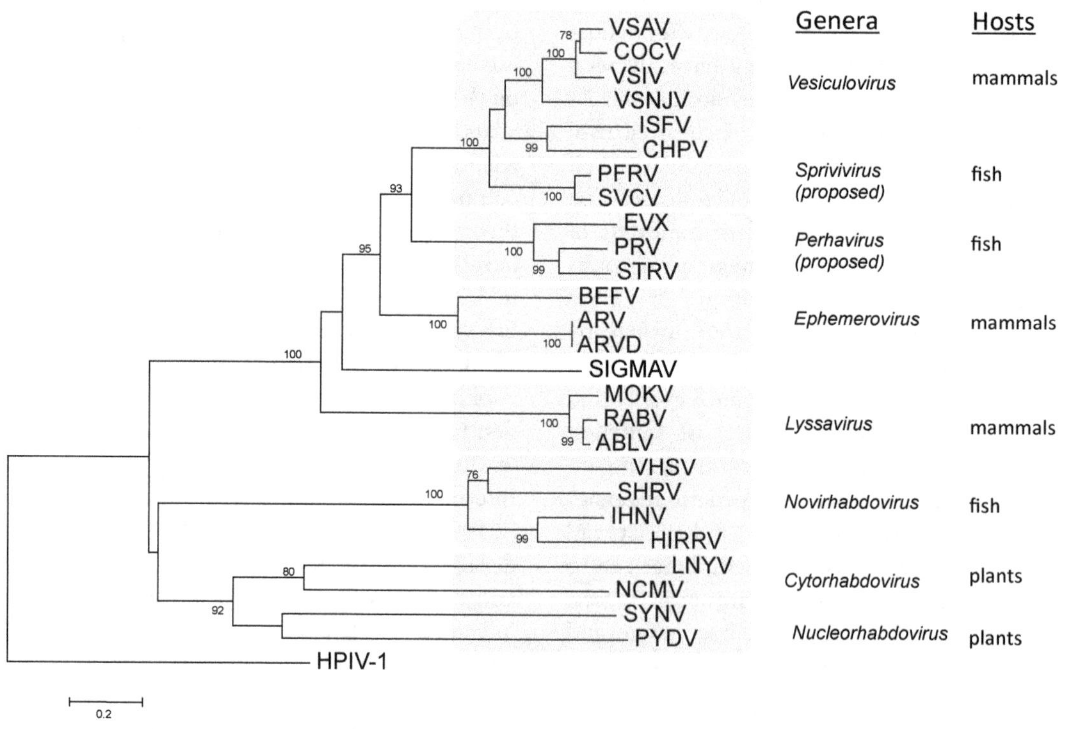

Figure 7.1 Phylogenetic tree of the *Rhabdoviridae* illustrating relationship of the genus *Novirhabdovirus* with other established rhabdovirus genera and two proposed new genera that also contain rhabdoviruses of fish. The tree is based on complete N protein sequences. Fish virus abbreviations are: PFR, pike fry rhabdovirus; SVCV, Spring viraemia of carp virus; EVX, Eel virus European X; PRV, perch rhabdovirus; STRV, sea trout rhabdovirus; VHSV, viral haemorrhagic septicaemia virus; SHRV, snakehead rhabdovirus; IHNV, infectious haematopoietic necrosis virus; HIRRV, hirame rhabdovirus. The paramyxovirus HPIV-1, Human parainfluenza virus-1, is included as an outgroup. Neighbour-joining tree, courtesy of Dr David Stone, Centre for Environment, Fisheries, and Aquaculture Sciences, Weymouth, UK.

genera are not as thoroughly characterized as the novirhabdoviruses. Therefore, although it is important to recognize that there are several other fish rhabdoviruses, this chapter will focus on novirhabdoviruses in the genomics era.

Virus species within the *Novirhabdovirus* genus

The ninth report of the ICTV recognizes four species within the genus *Novirhabdovirus* (Table 7.1) (Dietzgen *et al.*, 2011). IHNV is the type species of the genus. Both IHNV and VHSV are globally important fish pathogens that are found in wild fish and also cause significant disease burdens on fish reared in various aquaculture settings. IHNV and VHSV are among the seven finfish virus species listed as 'notifiable' by the 2011 Aquatic Animal Health Code of the World Organization for Animal Health (World Organization for Animal Health, 2011), indicating their recognition as serious pathogen threats to global animal production systems. They have been studied intensively since the 1950s and are among the best characterized fish viruses (reviewed in Bootland and Leong, 1999; Smail, 1999). For both IHNV and VHSV several hundred field isolates have been analysed by genetic typing in molecular epidemiology studies, and complete G gene sequences for over 100 isolates are available in GenBank. Hirame rhabdovirus (HIRRV) is also an important pathogen of cultured fish, but with a more limited geographic range, and few isolates have been described in the literature. The fourth virus, snakehead rhabdovirus (SHRV), was isolated from warm-water fish with ulcerative disease in Asia but subsequent studies showed that the disease was caused by a fungal agent rather than the rhabdovirus. Genetic diversity of SHRV field isolates has not been reported. Complete genome sequences are available for selected strains of all four *Novirhabdovirus* species.

As shown in Fig. 7.1, IHNV and HIRRV are most closely related within the *Novirhabdovirus* genus. For example, the complete G gene sequences of IHNV and HIRRV share 69% nucleotide (nt) and 74% amino acid (aa) sequence identity, but they each share only 48–49% nt and 38–40% aa identity with the G sequences of representatives from the other two *Novirhabdovirus* species. The G genes of VHSV and SHRV have 54% nt identity, with 48% aa identity in the predicted G proteins. Additional fish rhabdoviruses that are considered possible members of the *Novirhabdovirus* genus include eel viruses EEV-B12 and EEV-C26, but these require further genetic characterization to define their taxonomic status.

The viruses in the four *Novirhabdovirus* species differ in several important aspects, including host and geographic range. IHNV has a relatively narrow host range restricted to cold-water salmon and trout fish, collectively referred to as 'salmonids', in the families *Oncorhynchus* and *Salmo*. IHNV is endemic to western North America and it has been inadvertently spread to Europe

Table 7.1 Virus species in the genus *Novirhabdovirus*, and status of genomics information as of 2011

Species name (abbreviation)	Number of isolates characterized genetically[a]	Number of full G gene sequences in Genbank	Number of full genome sequences in Genbank	Reverse genetics system(s)	DNA vaccines developed	Molecular analysis of host response
Infectious haematopoietic necrosis virus (IHNV)	Several hundred	115	3	Yes	Yes	Yes
Viral haemorrhagic septicaemia virus (VHSV)	Several hundred	209	13	Yes	Yes	Yes
Hirame rhabdovirus (HIRRV)	3	3	3	No	Yes	Yes
Snakehead rhabdovirus (SHRV)	1	1	1	Yes	Yes[b]	No

[a]Virus isolates characterized by genetic typing using various genomic regions.
[b]A DNA vaccine for SHRV has been used to demonstrate non-specific protection against IHNV challenge, but it has not been tested for efficacy against SHRV.

and Asia by aquaculture related activities. It has also been detected in north-western and eastern Russia since the year 2000. In contrast, VHSV has an extremely broad host range including over 80 marine and freshwater fish species from diverse taxonomic families in Europe, North America, and Asia. Although the major burden of VHSV in aquaculture is disease in rainbow trout (*Oncorhynchus mykiss*), it also causes disease epidemics in wild marine fish and it has recently emerged to cause large-scale epidemics in wild fish of the North American Great Lakes region. Hirame rhabdovirus causes mass mortalities and is widely distributed in farmed Japanese flounder (*Paralichthys olivaceous*) and other marine fish since the 1980s, but it has not been isolated outside of Asia. SHRV has been reported only from snakehead fish (*Ophicephalus striatus*) in South-East Asia.

Novirhabdoviruses can, under some circumstances, cause rapid mortality to high levels in infected fish populations (Wolf, 1988a; Kurath and Winton, 2008). As with any viral pathogen, incidence and severity of disease are influenced by a combination of viral, host, and environmental factors (LaPatra, 1998). Relative to their mammalian counterparts, environmental temperature is a much greater factor in novirhabdoviral disease, as temperature variations have a strong impact on both virus replication and host immune system function. IHNV, VHSV, and HIRRV replicate optimally at 12–15°C, and they are inactivated at temperatures above 18–20°C. In contrast, SHRV is a warm water fish virus that replicates optimally at temperatures from 15–25°C and can replicate for some time at 30°C, but it is inactivated by incubation at 30–37°C.

The diseases caused by IHNV, VHSV, and HIRRV have been described in their natural hosts in the field, and disease pathogenesis has been studied in various experimental models. In general they cause acute haemorrhagic septicemia that affects multiple organs and causes necrosis of haematopoietic tissues. Common external clinical signs include skin darkening, exophthalmia, haemorrhage, and ascites, although these are often difficult to observe in the field. Disease outbreaks are most common in juvenile fish, with the development of resistance as fish increase in size and age, as documented for some important fish hosts.

Genomics and molecular research on novirhabdoviruses

With regard to the application of modern genomics and molecular science to these viruses, well established *in vivo* laboratory infection models and *ex vivo* leucocyte systems have been used extensively to study molecular aspects of viral replication and gene expression, and also the host innate and adaptive immune response to viral infection. For VHSV and IHNV the majority of studies have been conducted in rainbow trout, and HIRRV studies have been in Japanese flounder. Infections and transfections in various fish cell lines have also been used for numerous studies of viral gene and protein function as well as host response. Since 2000, reverse genetics systems have been developed for SHRV, two strains of IHNV (European and North American), and three strains of VHSV (European, North American and Korean). These have been used to investigate viral gene function, virulence and host specificity determinants, and potential applications as RNA viral vectors or live attenuated vaccines (Biacchesi *et al.*, 2010; Biacchesi, 2011). The viability of numerous recombinant viruses with deletions, foreign gene insertions, or heterologous gene exchanges demonstrated great flexibility of *Novirhabdovirus* genomes, and suggested that these systems will be highly informative as they are applied creatively in the future.

DNA vaccines carrying the viral glycoprotein genes have been generated for all four *Novirhabdovirus* species and they have proven to be highly efficacious in experimental trials for protection against IHNV, VHSV, and HIRRV. Due to the exceptionally high level of protection elicited by these vaccines they have also been used as basic research models for DNA vaccines, and for characterization of the protective host immune response in the absence of viral infection. The host transcriptional responses to both viral infection and G protein vaccination have been characterized extensively for IHNV and VHSV in rainbow trout, and for HIRRV in Japanese flounder. Studies using quantitative reverse transcriptase polymerase chain reaction assays (qRT-PCR), microarrays, and deep sequencing technologies have defined a characteristic host transcriptional response pattern dominated by interferon stimulated genes, including some that are conserved throughout vertebrates, and some that are unique to fish hosts. Excellent review articles have been published recently on *Novirhabdovirus* reverse genetics (Biacchesi, 2011), DNA vaccines (Lorenzen and LaPatra, 2005; Kurath *et al.*, 2007; Kurath, 2008), and the host immune response to viral infection or DNA vaccination (Aoki *et al.*, 2011; Purcell *et al.*, 2012; Verrier *et al.*, 2011). Therefore this chapter will describe general *Novirhabdovirus* genomics, and then focus on aspects relevant to *Novirhabdovirus* molecular epidemiology and evolution, with information from various molecular approaches included where appropriate. Finally there will be brief descriptions of a novel *in vivo* viral fitness and competition system that has been developed for IHNV, and databases available online for additional information on novirhabdoviruses.

Novirhabdovirus genomes and protein function

The genomes of novirhabdoviruses contain the five common rhabdovirus genes, with the sixth NV gene inserted between the G and L genes (Fig. 7.2). Thus, the gene order is 3′-N-P-M-G-NV-L-5′, where N is nucleoprotein, P is phosphoprotein, M is matrix protein, G is glycoprotein, NV is non-virion protein, and L is polymerase. The 3′ leader RNAs upstream of the N gene are approximately 50–60 nt in length, and the 5′ trailer downstream of the L gene is approximately 80–100 nt. Individual genes have upstream and downstream untranslated regions of variable length that generally range from 30–100 nt, but the NV gene has no downstream untranslated region in some genomes. Putative transcription start [5′-GGCAC-3′] and stop [5′-AGAYAG$(A)_7$–3′] signals are highly conserved between genes and

Gene Junction		Transcription!Signals Stop	Intergenic	Start
N5P	IHNV	AGATAG$(A)_7$	T	GGCAC
	HIRRV	AGATAG$(A)_7$	T	GGCAC
	VHSV	AGATAG$(A)_7$	C	GGCAC
	SHRV	AGATAG$(A)_7$	C	GGCAC
P5M	IHNV	AGATAG$(A)_7$	T	GGCAC
	HIRRV	AGATAG$(A)_7$	T	GGCAC
	VHSV	AGATAG$(A)_7$	C	GGCAC
	SHRV	AGATAG$(A)_7$	T	GGCAC
M5G	IHNV	AGATAG$(A)_7$	T	GGCAC
	HIRRV	AGATAG$(A)_7$	T	GGCAC
	VHSV	AGATAG$(A)_7$	T	GGCAC
	SHRV	AGATAG$(A)_7$	T	GGCAC
G5NV	IHNV	AGACAG$(A)_7$	C	GGCAC
	HIRRV	AGATAG$(A)_7$	C	GGCAC
	VHSV	AGATAG$(A)_7$	T	GGCAC
	SHRV	AGACAG$(A)_7$	C	GGCAC
NV5L	IHNV	AGATAG$(A)_7$	T	GGCAC
	HIRRV	AGATAG$(A)_7$	T	GGCAC
	VHSV	AGATAG$(A)_7$	T	GGCTC
	SHRV	AGATAG$(A)_7$	C	GGCAC

Figure 7.2 Generalized *Novirhabdovirus* genome showing organization of the six viral genes, and conserved transcription signals and intergenic regions at each gene junction for the four viral species. Variant nucleotides are highlighted in grey.

between novirhabdoviruses of different species, and they differ from consensus signals of viruses in other genera (Fig. 7.2). Intergenic regions are single nucleotides, either T or C (Bjorklund *et al.*, 1996; Dietzgen *et al.*, 2011).

In general the functions of novirhabdovirus N, P, M, G, and L proteins have been assumed to be the same as demonstrated for the homologous proteins in the well studied mammalian virus model systems, such as vesicular stomatitis and rabies viruses. Although this is likely valid in many cases due to the similarity of virion and genome structure, it is important to note whether a function has been proven specifically for novirhabdoviruses, and to recognize where functions may differ. The G protein of novirhabdoviruses has been demonstrated to be present as trimers on the virion surface, and the VHSV G protein disulfide bonds and glycan structures have been assessed empirically, revealing an overall folding structure that is conserved with other rhabdovirus G proteins despite low sequence similarity (Einer-Jensen *et al.*, 1998). The novirhabdovirus G protein facilitates virus entry into host cells by binding to a cellular receptor, which has been identified as fibronectin for VHSV (Bearzotti *et al.*, 1999). The G protein is also the major antigenic protein of novirhabdoviruses, eliciting strong innate and adaptive host immune responses. Studies with subunit and DNA vaccines have clearly indicated that the G protein alone is necessary and sufficient to stimulate protective immunity for VHSV, IHNV, and HIRRV (Engelking and Leong, 1989; Anderson *et al.*, 1996; Lorenzen *et al.*, 1998; Takano *et al.*, 2004), and it has been suggested that novirhabdovirus G proteins may function as pathogen-associated molecular patterns (PAMPs) (Kurath *et al.*, 2007). Monoclonal antibodies raised against IHNV and VHSV G proteins have been used to differentiate between viral strains, and several antigenic epitopes have been defined (Huang *et al.*, 1996; Fernandez-Alonso *et al.*, 1998; Lorenzen *et al.*, 2000). Among the other viral proteins, the novirhabdovirus M protein has been shown to induce host cell shut-off and stimulate apoptosis when transfected into a fish cell line (Chiou *et al.*, 2000). This indicates a role for the IHNV M protein in immune evasion strategies similar to that of the VSV M protein. Transfection of the IHNV P protein did not induce apoptosis in fish cells, demonstrating a difference from the rabies virus immune evasion mechanism. The structural and enzymatic functions of the novirhabdovirus N, P, and L proteins are assumed to be comparable to their roles in mammalian rhabdoviruses, although this has not been specifically proven in most cases.

The NV gene and protein

Ever since the first report of the NV gene and protein in 1985 (Kurath and Leong, 1985), an NV gene has been found with a functional open reading frame in the genome of every isolate of a novirhabdovirus that has been characterized. The genes range in size from approximately 370–425 nt in length, and encode proteins of 111–122 amino acids that are predicted to be 12.7–13.8 kDa in molecular mass (Kurath *et al.*, 1997; Johnson *et al.*, 2000). Maintenance of an open reading frame suggests an extant function, but until recently the role of NV was only speculative. Although the NV gene is present in all novirhabdoviruses, there is little sequence conservation between species. For example, the NV genes of IHNV and VHSV share only 39% nt and 14% aa identity, compared with 48% nt and 38% aa identities in the G genes. Among the representatives of the four *Novirhabdovirus* species, the NV sequences of IHNV and HIRRV share 61% nt and 51% aa identities, and all other pairs share 39–43% nt and 14–17% aa identity. In general, the NV gene is less conserved than the other viral genes among isolates within a species, or between novirhabdoviruses from different species.

The gene was originally named the 'non-virion' gene because it encodes a small protein that was not found in purified virus particles. Although the NV protein is difficult to detect in infected cells, polyclonal antibodies raised against prokaryotically expressed VHSV and IHNV proteins were used to demonstrate that NV protein is present in trace quantities in both VHSV and IHNV infected fish cells (Schuetze *et al.*, 1996). More recently studies with an IHNV NV protein tagged with green fluorescent protein (NV-GFP) revealed that the NV protein is actively transported into the nucleus of fish cells (Choi *et al.*, 2011). This nuclear transport is directed by a unique nuclear localization signal (NLS) of four sequential amino acids in the NV, ^{32}EGDL^{35}, and is an inherent property of the NV protein that does not require any other viral proteins or viral infection. This NLS signal is not present in the VHSV NV protein, and nuclear localization has not been examined for other novirhabdovirus NV proteins.

The function of the NV protein has long been speculated to involve evasion of the fish innate immune response, but proof has been elusive until the development of reverse genetics systems for novirhabdoviruses. Reverse genetics systems for SHRV, IHNV and VHSV have all been used to assess the role of NV by comparing full length recombinant viruses with viruses in which the NV gene has been deleted. All three systems demonstrated that the NV gene was not essential for production and replication of viable virus in cell lines, but results were not consistent with regard to phenotype. For SHRV, NV deletion mutants replicated in fish cells with the same efficiency as wild type recombinant viruses, and the NV deletion mutant was equally pathogenic in zebrafish infections (Johnson *et al.*, 2000; Alonso *et al.*, 2004). In contrast, IHNV and VHSV NV deletion mutants had much lower replication than wild type recombinants in fish cells, and they were highly attenuated in the course of *in vivo* infections, suggesting that for these viruses the NV is required for efficient replication and pathogenicity (Biacchesi *et al.*, 2000; Thoulouze *et al.*, 2004; Ammayappan *et al.*, 2011; Biacchesi *et al.*, 2011).

The same researchers who demonstrated nuclear localization of the IHNV NV protein conducted further studies using a series of recombinant IHNV viruses including a mutant without the NV nuclear localization signal in the NV gene. Their results demonstrated that NV, and specifically nuclear localization of NV, is required for optimal virus growth in cells, and several lines of evidence indicated that NV nuclear localization correlates with inhibition of the host cell interferon response (Choi *et al.*, 2011). This provided the first direct evidence for a role of NV in supporting efficient viral replication by interfering with the host interferon system. At the same time other researchers using recombinant VHSV viruses reported a role for the NV protein in suppressing early apoptosis caused by VHSV infection in cell culture (Ammayappan and Vakharia, 2011). They demonstrated that recombinant VHSV with the NV gene deleted, or mutants with stop codons at the beginning of the NV gene, induced apoptosis faster than wild type recombinant viruses in cell culture. They also found that this effect was reversed by providing the IHNV NV protein in trans, suggesting that both VHSV and IHNV NV proteins are capable of this function despite the absence of any significant sequence similarity. Viral suppression of host apoptosis early in infection is a mechanism used by some viruses to prolong virus replication, but it has not been reported for other rhabdoviruses. It is balanced by the induction of apoptosis later in infection to aid in viral release, as shown for VSV and RABV. Thus, for novirhabdoviruses the early anti-apoptotic role of the NV protein and the induction of apoptosis by the viral M protein may comprise careful regulation of host apoptosis to the benefit of the viruses (Ammayappan and Vakharia, 2011). The demonstration of an early anti-apoptotic function for the VHSV NV, and nuclear localization and inhibition of the host interferon response by the IHNV NV, indicate roles in two different mechanisms of immune evasion.

Genetic diversity and global molecular epidemiology of novirhabdoviruses

Due to the large number of field isolates characterized for IHNV and VHSV, their genetic diversity is quite well known and major phylogenetic groups have been identified. This has led to detailed understanding of the molecular epidemiology of these viruses and insights into their evolution in diverse geographic areas, at both local and global scales. Because several fundamental features of these two virus species differ, they will be described separately.

IHNV genetic diversity: five genogroups

Phylogenetic analyses of complete or partial G gene sequences of IHNV field isolates from North America, Asia, and Europe have defined five major genetic groups, referred to in the literature as 'genogroups' (Kurath *et al.*, 2003; Enzmann *et al.*, 2005; Nishizawa *et al.*, 2006b). Major features of the genogroups are presented in Table 7.2 and their global geographic distribution is shown in Fig. 7.3A. A phylogenetic tree of complete G gene open reading frame (ORF) sequence data from various research groups is shown in Fig. 7.4. The U, M, and L genogroups contain virus isolates from the endemic range of IHNV in western

Table 7.2 Five major global phylogenetic genogroups of IHNV

Genogroup (subgroups)	Geographic range	Predominant host(s)[a]	Year of first detection[b]
U	North America (west coast)	Sockeye and Chinook salmon	1950s
	North America (west coast)	Atlantic salmon	1992
	Japan	Sockeye, kokanee, and masu salmon, rainbow trout	1971
	Russia (northwest)	Rainbow trout	2000
	Russia (east coast)	Sockeye salmon	2001
M (A-F)	Idaho, Oregon, Washington, USA	Rainbow and steelhead trout	1978
L (I and II)	California and southern Oregon coast, USA	Chinook salmon	1966
J	Japan	Rainbow trout	1980
	Korea	Rainbow trout	1991
E (A-G)	Continental Europe	Rainbow trout	1987

[a]Host species are as follows: sockeye and kokanee salmon, *Oncorhynchus nerka*; Chinook salmon, *O. tshawytscha*; Atlantic salmon, *Salmo salar*; masu salmon, *O. masou*; rainbow and steelhead trout, *O. mykiss*.
[b]The year of first detection for a genogroup is often a disease event caused by a virus that was later determined to be in the genogroup.

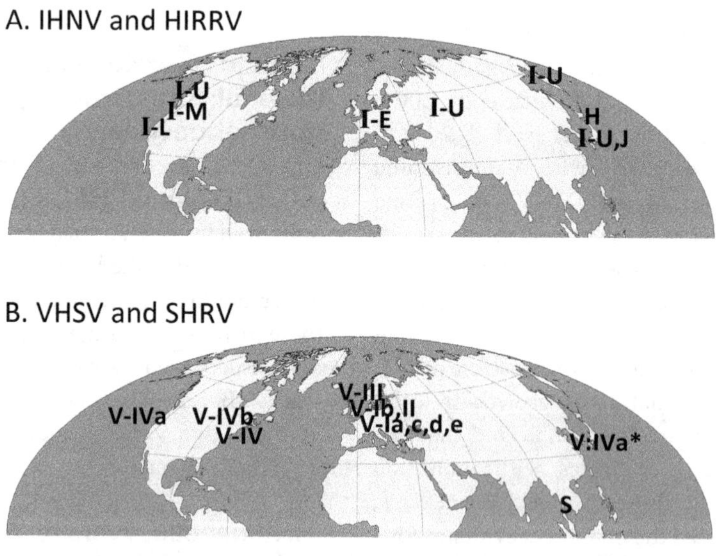

Figure 7.3 Global geographic distribution of novirhabdoviruses. (A) Five IHNV genogroups indicated as I-U, M, L, J, E, and HIRRV as H. (B) Four VHSV genotypes as V-I, II, III, IV, with genotype I subgroups indicated as a-e, and SHRV as S. The asterisk for VHSV in Asia is to note that although VHSV-IVa is predominant, VHSV-Ia has been detected once. Maps show only the northern hemisphere because novirhabdoviruses have not been reported from the southern hemisphere.

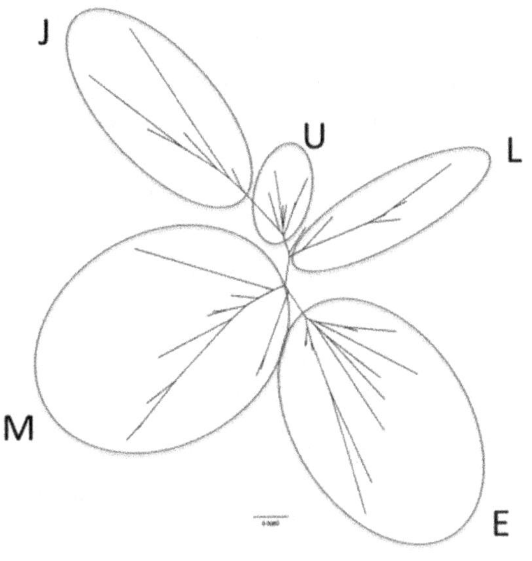

Figure 7.4 Phylogenetic tree showing five major subgroups within the species *Infectious haematopoietic necrosis virus*, based on complete IHNV G gene open reading frame sequences. Genogroups U, M, and L originated in North American Pacific salmonids, and U and M have been introduced to Asia and Europe, respectively, because of the movement of fish eggs for aquaculture purposes. The J and E genogroups occur in Japan and Europe, respectively, in farmed rainbow trout. Bayesian tree courtesy of Dr Rachel Life, University of Washington and USGS Western Fisheries Research Center, Seattle, USA.

North America, where they occur in the upper, middle, and lower portions of the virus geographic range, respectively (Kurath *et al.*, 2003). These genogroups also differ in general host specificity, with U group viruses occurring mostly in sockeye and kokanee salmon (both *Oncorhynchus nerka*), M in rainbow and steelhead trout (both *O. mykiss*), and L in Chinook salmon (*O. tshawytscha*). Among these North American genogroups U is the most homogeneous, with a maximum of 1.6% nt (1.8% aa) diversity in the G gene, despite having the largest geographic range and including a large proportion of the characterized virus isolates. Within the U genogroup a combination of modern phylogenetic and compartmentalization analyses have indicated two geographically separated subgroups, but these have low genetic divergence. The U genogroup also includes older virus isolates from Japan and recent isolates from Russia, but these do not expand the genetic diversity of the group. In contrast, the more localized M genogroup is the most diverse, with up to 5.3% nt (6.7% aa) diversity in the G gene, and it contains several well supported phylogenetic subgroups designated MA-MF. These subgroups have been shown to co-circulate stably in intensive trout culture in Southern Idaho. The L genogroup also has relatively low diversity with a maximum of 1.8% nt (3.0% aa) divergence in the available G gene sequences. It contains two defined subgroups, designated LI and LII, that are separated spatially within the L genogroup range in California, and appear to differ in estimated rates of evolution.

Outside of North America a fourth genogroup, designated J, contains viruses isolated since 1980 from rainbow trout farms in Japan and Korea (Nishizawa *et al.*, 2006; Kim *et al.*, 2007). Phylogenetically the J genogroup is derived from the U genogroup, and it is thought to have arisen by adaptation to higher virulence after a host switch into rainbow trout, as will be described below. Within the J genogroup there is up to 5% nt (5.7% aa) diversity in the viral G genes. The fifth genogroup, designated E, contains numerous isolates from continental Europe, mostly from France, Italy, and Germany (Enzmann *et al.*, 2010). The G gene sequences of E isolates have up to 4.6% nt (5.9% aa) diversity. This genogroup is most closely related to the North American M group, supporting its hypothesized origin as an introduction of North American M virus into Europe, also described below. The segregation of IHNV into these five genogroups is strongly supported by various phylogenetic methods, but the overall global genetic diversity of IHNV is low, with a maximum of 6.6% nt and 8.1% aa divergence in the known G gene sequences. Among the five IHNV genogroups the greatest genetic difference is between isolates in the J and E genogroups.

IHNV molecular epidemiology and evolution in North America

Historical records clearly indicate that IHNV originated in Pacific salmonids of North America, and its current global distribution is a result of inadvertent translocations associated with aquaculture. In North America the molecular epidemiology of IHNV is complex due to biological and anthropogenic aspects of Pacific salmonids. The anadromous lifestyle of Pacific salmonids means that they hatch and rear for some period of time in freshwater rivers and lakes, and then migrate to the ocean where various stocks co-mingle for two to four years before returning to their natal freshwater sites to spawn as sexually mature adults. The length and timing of this migratory pattern varies among fish species and stocks, but some salmon migrate hundreds of miles in both freshwater and marine environments, making the epidemiological analyses of their pathogens complicated. Most wild Pacific salmon populations in North America have IHNV prevalence at some level. As wild salmon decline they are increasingly cultured not only for food, but also for recreation, natural resource supplementation, and restoration programmes. The wide variety of fish culture systems including fish farms, hatcheries, spawning channels, and freshwater or marine netpens, adds additional complexities, with high potential for anthropogenic influence on virus epidemiology and evolution.

The written literature on IHNV describes a series of events that have been used to inform and interpret the results of molecular epidemiological studies (Wolf, 1988a; Bootland and Leong, 1999). IHNV was first reported as the cause of severe epidemics among juvenile sockeye salmon (*O. nerka*) in Washington State hatcheries in the early 1950s, and it has been postulated as the likely cause of epidemics reported in California Chinook salmon (*O. tshawytscha*) hatcheries as early as the 1940s. Surveys of salmonid populations in the 1970s revealed that IHNV was endemic in Alaskan sockeye, but it was limited to specific localized areas in Washington, Oregon, and California. From 1978–1980, IHNV emerged as a major pathogen in the intensive rainbow trout farming industry in southern Idaho, and it has now been both endemic and frequently epidemic in this region for over 30 years. From 1980–1983, IHNV also emerged in the lower Columbia River, which forms the border between Washington and Oregon, and it continues to be the most significant viral pathogen in the extensive salmonid hatchery system in the Columbia River basin today. More recently, IHNV emerged in marine netpens of Atlantic salmon (*Salmo salar*) off the coast of British Columbia, causing two waves of epidemics from 1992 to 1996 and from 2001 to 2003 (Saksida, 2006). Finally, since 2007, IHNV has been emerging in steelhead trout (ocean-run *O. mykiss*) on the coast of Washington. The molecular epidemiology associated with these events will be presented below. The current geographic range of IHNV in North America extends from Alaska to California, and inland to Idaho. Within this range the virus has been found in nearly all watersheds that have Pacific salmon.

Genetic typing systems for IHNV field isolates were first developed in the 1990s when several studies were done using RNase T1 fingerprinting and RNase protection analyses. These methods were relatively quickly supplanted by a typing system based on the sequencing of a 303 nt variable region in the middle of the G gene, referred to as the 'midG'. A series of regional studies using midG typing revealed low genetic diversity of IHNV in Alaska, British Columbia, and coastal Washington, and significantly higher genetic diversity in a localized trout farming region in southern Idaho (reviewed in Kurath *et al.*, 2003). In 2003 the midG typing data from several regional studies was combined to generate a phylogenetic analysis of 323 isolates collected throughout the IHNV range in North America. The resulting tree defined the three distinct genogroups designated U, M, and L, with an overall maximum of 8.6% nt diversity observed in the variable midG typing region (Kurath *et al.*, 2003). These genogroups were also consistently resolved in subsets of isolates using a different sequencing window of 412 nt at the 5′ end of the N gene, and they have been observed in various analyses using complete G or N gene sequences. A continuing system for midG typing of North American IHNV isolates has been established as an aid for fish health management agencies, and the typing data are available for public use in an on-line database (http://gis.nacse.org/ihnv). As of 2011, midG typing of over 1600 IHNV field isolates confirms the resolution of North American IHNV into genogroups U, M, and L, with a maximum overall nt diversity still at 8.6% (R. Life and G. Kurath, unpublished). Within each of these three genogroups the maximum nucleotide divergence in the midG region is 4.3% for U, 8.6%

for M, and 3.3% for L. Importantly, the typing of additional isolates in recent years has not resulted in an increase in the overall genetic diversity of the genogroups as determined in 2003, suggesting that the current extent of genetic diversity of IHNV in North America is known.

Among the IHNV genogroups in North America, the U group has the lowest genetic diversity and is hypothesized to represent the ancestral IHNV type that is highly fit and co-adapted due to a long-term, stable association with the sockeye salmon host throughout the northern range of the virus (Emmenegger *et al.*, 2000; Emmenegger and Kurath, 2002). In contrast, the M genogroup exhibits 3–4-fold higher genetic diversity than the U group. Estimates of evolutionary divergence rates indicate that the M genogroup is evolving 6-fold faster than the U, with rates of 1.2×10^{-3} and 2×10^{-4} substitutions per site per year reported for the M and U genogroups, respectively (Kurath *et al.*, 2003). M group IHNV occurs as multiple phylogenetically distinct subgroups that co-circulate within and between trout farms in Southern Idaho. Several lines of evidence are consistent with the hypothesis that the M genogroup arose from a sockeye-specific U group ancestral virus by a host switch and adaptation to rainbow trout, and also adaptation to the higher water temperature in the Idaho trout farming region. The subsequent development of subgroups and higher genetic diversity in the M genogroup are likely due to adaptation to the relatively recent host–pathogen association, influenced by numerous anthropogenic factors such as high rearing densities, continuous introduction of naive susceptible hosts, fragmented host populations, and general lack of temporal or spatial bottlenecks in intensive fish farming (Troyer *et al.*, 2000; Troyer and Kurath, 2003).

The Idaho trout farming region where M virus is endemic is within the extensive Columbia River watershed, which drains 670,000 km^2 of land in Washington, Oregon and Idaho, and has numerous salmonid hatcheries and fish farms. Within this watershed, IHNV in the U group has been isolated consistently for several decades as a stable background of widely dispersed endemic virus, and specific M group types have been associated with two emergence events that each persisted for several years (Garver *et al.*, 2003). Until recently, M group IHNV did not occur outside of the Columbia River basin, but since 2007 viruses in subgroup MD have emerged to cause several epidemics in steelhead trout (the anadromous form of rainbow trout) in coastal Washington rivers. Molecular epidemiology of virus isolates associated with the MD IHNV emergence events revealed new virus transmission routes between steelhead in the Columbia River basin and coastal rivers.

On the Pacific coast of Canada, IHNV in the U group has been endemic in most stocks of wild and hatchery sockeye salmon for several decades. In 1992–1996, and again in 2001–2003, epidemics of IHN occurred in seapen-reared Atlantic salmon in farms off the coast of British Columbia (Saksida, 2006). Molecular epidemiology of the isolates associated with these epidemics revealed they were U genogroup viruses very similar to those known to be endemic in local Pacific salmonids. This was useful in defining the source of the virus that caused the epidemics, and ruled out the possibility that a new virus might have been introduced by the Atlantic salmon. In general, monitoring of the genetic types of IHNV field isolates has been a useful tool for understanding and responding to viral emergence and epidemic events.

The L genogroup of IHNV is dominant without exception in California watersheds and does not overlap the geographic range of the U or M genogroups, occurring only in fish stocks south of a known bio-geographical transition zone on the southern coast of Oregon (Kurath *et al.*, 2003; Kelley *et al.*, 2007). Within the L group the LI subgroup occurs in coastal rivers of northern California and southern Oregon and has a low divergence rate indicating genetic stasis similar to the U genogroup. In contrast, the LII subgroup is found in inland valley watersheds of the Sacramento and San Joaquin Rivers, and it appears to be evolving more rapidly, similar to the M genogroup. Phylogenetically the L genogroup is closer to the U than the M genogroup, but the evolutionary relationship of the L and U groups is unclear, such that it is not known at present which is ancestral.

IHNV molecular epidemiology and evolution in Asia, Europe and Russia

IHNV was first detected in Japan in 1971, in juvenile kokanee (land-locked sockeye salmon) at a hatchery that had received Alaskan sockeye salmon eggs. Thus, the original source of introduction of IHNV into Japan was speculated to be contaminated sockeye salmon eggs from Alaska (Kimura and Yoshimizu, 1991). The virus subsequently spread to other hatcheries on both Hokaiddo and Honshu Islands, causing significant losses in juvenile kokanee, masu salmon (*O. masou*), and rainbow trout. Although the disease burden was significantly reduced by egg disinfection and other biosecurity policies, IHNV re-emerged in large rainbow trout in Japan during the 1990s (Nishizawa *et al.*, 2006), and it was first detected in Korea in rainbow trout in 1991 (Kim *et al.*, 2007). This history correlates well with phylogenetic relationships of Asian IHNV isolates based on complete G gene ORF sequences (Nishizawa *et al.*, 2006). Older Japanese IHNV isolates from kokanee, chum salmon (*O. keta*), and rainbow trout in 1971–1982 fall clearly within the U genogroup, with short branch lengths similar to those of North American U group viruses. This is consistent with the speculated source of IHNV introduction into Japan because all Alaskan IHNV examined to date have been in the U genogroup. Subsequently, more recent Japanese isolates from 1980–1996 constitute the new J genogroup that is derived from the U genogroup as a monophyletic subgroup with long branches (Fig. 7.4), indicating higher genetic diversification. Appearance of the J genogroup correlated with the re-emergence of IHNV in large Japanese rainbow trout in the 1980s (Nishizawa *et al.*, 2006). The lack of U group isolates from Japan after 1982 suggests that the J may have displaced U group viruses, but this hypothesis is based on a limited number of virus isolates. IHNV was first detected in Korea in 1991, and four isolates collected during 1991–2001 fell into the J genogroup. This strongly suggests that the appearance of IHNV in Korea was due to direct introduction of virus from Japan, since importation of fish eggs from Japan to Korea is common, and Japan is the only known location where J group IHNV has been documented. In Korea IHNV is now established in cultured rainbow trout and free-ranging chum salmon.

In Europe IHNV was first reported in cultured rainbow trout in both France and Italy in 1987, and in Germany in 1992 (reviewed in Enzmann *et al.*, 2010). The virus subsequently became established, posing a significant threat to the European trout farming industry that continues to today. It was originally speculated that the source of IHNV in Europe was contaminated trout eggs imported from North America. This is supported by phylogenetic analyses of complete G gene sequences that show that European IHNV is most closely related to M genogroup IHNV, which is dominant in North American cultured rainbow trout (Enzmann *et al.*, 2005; Johansson *et al.*, 2009). Subsequently, analyses of over 70 European isolates of IHNV collected between 1987–2007 showed that they form a monophyletic group that has diverged significantly from the American M genogroup viruses (Enzmann *et al.*, 2010). This genogroup, designated E, has relatively high genetic diversity similar to that of the M (see branch lengths in Fig. 7.4) with seven subgroups designated A–G. Analysis of the E subgroups in light of existing records of fish movements led to several important insights into the molecular epidemiology of IHNV in Europe. First, the majority of disease outbreaks were caused by transmission due to trade in infected fish, so the major risk factor was introduction of infected fish into a farm. Second, the phylogeny showed clear evidence of 'viral tourism', in which IHNV crosses national borders in Europe apparently without restriction. Finally, the tree provided clear evidence that an entire subgroup could be eliminated by strict sanitation practices. In this case subgroup F was established with high prevalence in a localized area from 1995–1997 but it was targeted by strict sanitation policies, and thus far (through 2010) there have been no subsequent isolations of viruses in this subgroup. Each of these observations was soundly supported by the phylogenetic analysis, providing an excellent example of the value of genetic typing for understanding and management of fish viral diseases.

The first detection of IHNV in Russia was in a trout farm northwest of Moscow in 2000. Since that time the virus has been isolated from trout farms in the same region, with occasional outbreaks of IHN disease (Shchelkunov, 2009). Genetic typing of these IHNV isolates indicated they are in the U genogroup, on a branch that is similar in length to those of isolates from North America. No potential sources of the virus have been identified and there is no known epidemiological link with North American fish or fish eggs. Despite the genetic similarity with North American U group IHNV, these isolates are unusual from a host specificity perspective because U group viruses typically do not infect rainbow trout efficiently or cause disease. This observation suggests that there may be exceptions to the generalized host specificity of the IHNV U genogroup, and merits further investigation.

On the Pacific coast of Russia IHNV was first detected in 2001 in adult sockeye salmon returning to Kamchatka Peninsula rivers to spawn (Rudakova *et al.*, 2007). The first IHNV epidemics in Kamchatka occurred in juvenile sockeye salmon at two Kamchatka hatcheries in 2001 and 2004, and epidemics have now been documented also in wild sockeye (Rudakova and Bochkova, 2009). Extensive surveillance for virus has detected IHNV in several hatchery and wild sockeye salmon stocks, but not in any other salmonid species. Genetic typing using both midG and 5′N gene sequences indicated that IHNV from Kamchatka is clearly within the U genogroup, closely related and in some cases identical to North American U group IHNV. Again, no epidemiological link involving import of fish or fish eggs from North America is known. It was hypothesized that Russian and North American sockeye populations could exchange virus by transmission during their ocean migrations, where they are known to overlap in geographic range. Although this is true, other factors may also be involved to explain the timing of the appearance of IHNV in Kamchatka, since surveillance for virus was negative for some years prior to 2001. Continuing virus surveillance and genetic typing are being conducted to monitor and respond to the emergence of IHNV in Kamchatka.

In summary, genetic typing and phylogenetics support the following hypothetical history of IHNV: it was originally endemic in sockeye salmon in Alaska and possibly Chinook salmon in California, and it diverged into three genogroups U, M, and L, and spread to its current range throughout Pacific watersheds of the United States during the 1950s to 1980s. It is uncertain whether U or L viruses were originally ancestral, but evidence suggests the M genogroup arose from a U group ancestor following a host switch into rainbow trout. In North America IHNV continues to be endemic and sometimes epidemic in Pacific salmon, with a series of local or regional emergence events that indicate continuing virus spread and evolution. IHNV was inadvertently introduced to both Japan and Europe through aquaculture related transport of contaminated fish or fish eggs, and its subsequent establishment in these regions has led to evolution of two additional distinct lineages that are now separate genogroups, J and E.

VHSV genetic diversity: four genotypes

Genetic typing of VHSV field isolates has defined four major phylogenetic groups, designated genotypes I–IV (Einer-Jensen *et al.*, 2004; Snow *et al.*, 2004) (Fig. 7.5 and Table 7.3). These groups were consistent in phylogenetic trees of partial N gene or complete G gene ORF sequences from over 70 VHSV isolates. The genotypes differ in geographic range, and in some cases also in pathogenicity for rainbow trout. All VHSV isolates from freshwater rainbow trout farms in western Europe are in genotype I, which has several subgroups, Ia-Ie. The majority of European trout farm isolates are in subgroup Ia, but there are also small numbers within subgroups Ic, Id, and Ie. Subgroup Ib contains isolates from wild marine fish in the Baltic sea and waters of northern Europe, and these isolates are not virulent for rainbow trout. Genotypes II, III, and IV also contain marine fish isolates, with genotype II found in a localized area of the Baltic sea, genotype III in the North Sea and coastal waters around the United Kingdom and Ireland, and genotype IV in North America and Asia. In North America genotype IV occurs as two subgroups designated IVa on the Pacific coast in marine and anadromous fish, and IVb that emerged recently in various freshwater fish of the North American Great Lakes

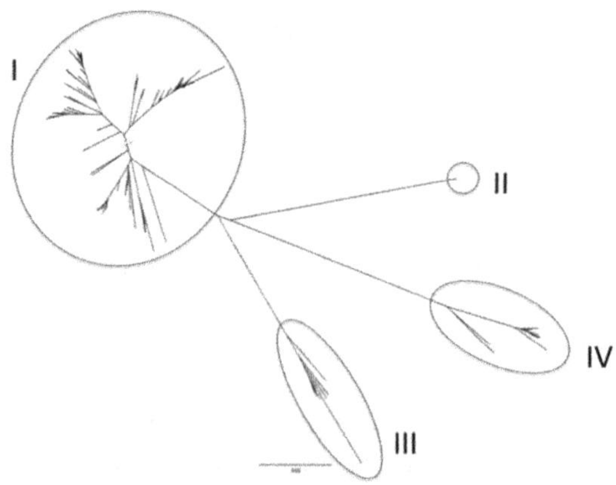

Figure 7.5 Phylogenetic tree showing four major subgroups within the species *Viral haemorrhagic septicaemia virus*, based on complete VHSV G gene open reading frame sequences. Genotypes I, II, and III occur in wild fish in the European marine environment, and a large subgroup Ia within genotype I contains the majority of all isolates from European freshwater rainbow trout farms. Genotype IV occurs in wild marine and freshwater fish in North America, and in wild and cultured marine fish in Asia. Bayesian tree courtesy of Dr Katja Einer-Jensen, Danish Technical University, Arhus, Denmark.

region. There have been only a small number of VHSV isolates described from fish on the Atlantic coast of North America, and they are also in genotype IV (Gagne *et al.*, 2007). The marine isolates in genotypes Ib, II, III, and IV represent an extensive reservoir of VHSV in a wide range of marine fish host species in the northern Atlantic and Pacific Oceans. There is no evidence for differences in host specificity between genotypes Ib, II, III, or IV, but they have low virulence for rainbow trout. Historically VHSV isolates have been referred to using the terms 'freshwater' or 'marine', where 'freshwater' was synonymous with being pathogenic for rainbow trout in freshwater farms. However, in VHSV genotype IV there are now many isolates that are not virulent for rainbow trout, but they occur in freshwater fish. Therefore the term 'freshwater VHSV' will not be used in this chapter: instead the genotype I subgroups from European rainbow trout farms will be referred to specifically as 'European freshwater VHSV'.

Overall the genetic diversity of VHSV is significantly higher than that of IHNV, with full G ORF sequences showing up to 18% nt diversity. Within each of the four VHSV genotypes the G genes have up to a maximum of 1–6% nt diversity, and genotypes I, II, and III differ from each other by 9–13%. Genotype IV is more divergent, differing from the other three genotypes by 13–16% at the nt level.

Table 7.3 Four major global phylogenetic genotypes of VHSV

Genotype (subgroups)	Geographic range	Predominant host[a]	Year of first detection[b]
I (a)	Continental Europe freshwater	Rainbow trout (cultured)	1938
I (b-e)	Marine waters of Europe	Various marine fish (e.g. Atlantic herring, cod)	1979
II	Baltic Sea	Marine fish (e.g. Atlantic herring, sprat)	1996
III	France, North Sea, marine waters of UK and Ireland	Marine fish (e.g. turbot), eel	1987
IV (a)	North America, Pacific coast	Various marine fish (e.g. Pacific herring)	1988
	North America, Atlantic coast	Various marine fish (e.g. halibut)	1994
	Japan and Korea	Olive flounder	1999
IV (b)	North America, Great Lakes	Various freshwater fish (e.g. muskellunge, yellow perch)	2003

[a]Host species are as follows: rainbow trout, *O. mykiss*; Atlantic herring, *Clupea harengus*; cod, *Gadus morhua*; sprat, *Sprattus sprattus*; turbot, *Scophthalmus maximus*; eel, *Anguilla anguilla*; Pacific herring, *Clupea pallasi*; Pacific sardine, *Sardinops sagax*; Olive flounder, *Paralichthys olivaceus*; muskellunge, *Esox masquinongy*; yellow perch, *Perca flavescens*.
[b]The year of first detection for a genotype is often a disease event caused by a virus that was later determined to be in that genotype.

VHSV history and molecular epidemiology in Europe

The relationships in the global VHSV phylogeny have been interpreted with respect to the known history and features of VHSV isolates as described in literature (Wolf, 1988b; Smail, 1999). The disease viral haemorrhagic septicaemia (VHS) was first described in Danish rainbow trout farms in 1938, and the VHSV was first isolated and reported as the causative agent of the disease in 1963. For the next 25 years VHSV was understood as an important rainbow trout farm pathogen that was limited to continental Europe, causing outbreaks in hundreds of freshwater farms and occasionally in trout reared in marine netpens. Recent typing of these viruses showed that they all were within genotype I, mostly within subgroup Ia (Einer-Jensen *et al.*, 2004). VHSV was the subject of extensive research in Europe, and it was also the target of an eradication programme that was largely successful in Denmark (Olesen, 1998). During this period a few reports of occasional VHSV isolations from marine fish were considered anomalies. However, the detection of VHSV in North America in 1988 led to the discovery of a reservoir of VHSV in marine fish off the North American Pacific coast, as described below. Subsequently, several large scale oceanographic surveys were conducted in European waters during the 1990s, in which over 40,000 wild marine fish were assayed for the virus (King *et al.*, 2001b; Brudeseth and Evensen, 2002; Skall *et al.*, 2005). The combined results of these surveys defined an extensive reservoir of VHSV in diverse species of marine fish in waters surrounding Europe and the United Kingdom. Most virus isolates obtained were from apparently healthy fish. This thorough analysis of the host and geographic range of VHSV in wild marine fish constitutes an exceptionally well-characterized example of a wildlife reservoir for a virus of veterinary importance. The discovery of this European marine reservoir for VHSV was complimented by controlled laboratory challenge studies demonstrating that the marine VHSV isolates were uniformly of low or no pathogenicity for rainbow trout (reviewed in Skall *et al.*, 2005).

Genetic typing of the European marine VHSV isolates identified VHSV genotypes II and III and subtype Ib. The greater genetic diversity of VHSV in European marine fish and the finding that European genotype I trout farm isolates are nested within this broader phylogeny led to the conclusion that viruses in the extensive marine reservoir for VHSV were ancestral to genotype I lineages of trout farm isolates, which most likely diverged from a marine type Ib-like ancestor by host switches and adaptation into rainbow trout (Einer-Jensen *et al.*, 2004). Careful analysis of the phylogeny provided evidence for at least three or four independent host switch events into rainbow trout that occurred most likely during the 1950s, leading to segregation of subgroups Ia and Ic, and to other isolates related to the most recent common ancestor of genotype I. With the discovery of the marine reservoirs of VHSV, the general understanding of the virus was modified to being a wide-spread virus of marine fish that had historically adapted to rainbow trout, leading to the evolution of genotype I lineages that continue to cause a significant disease burden on European rainbow trout farms.

In addition to disease in farmed rainbow trout, outbreaks of VHS were reported in turbot (*Scophthalmus maximus*) farms during the 1990s in Germany, Scotland and Ireland. In these cases full G gene molecular typing demonstrated that the turbot farm isolates were in genotype III, very similar to VHSV isolates from wild fish in the adjacent North Sea (Stone *et al.*, 1997). This was the first strong epidemiological link between VHSV in wild marine fish and cultured fish. Virus isolates from wild fish in the local marine reservoir were pathogenic to turbot (Snow and Smail, 1999; King *et al.*, 2001a;), suggesting that this transmission of virus from wild to cultured fish may not require a host switch, and therefore constitutes a greater threat to cultured fish species.

More recently, between 1998 and 2007, outbreaks of VHS in rainbow trout reared in marine netpens were reported in Sweden, Finland, and Norway (Nordblom and Norell, 2000; Raja-Halli *et al.*, 2006; Dale *et al.*, 2009). Genetic typing revealed that these outbreaks were caused by VHSV isolates in subgroups Ib and Id and genotype III, respectively, indicating they originated not from the Ia viruses currently circulating in freshwater trout farms, but from viruses known to exist previously in wild marine reservoir fish. The molecular data clearly suggested three independent host switch events in which a virus from the marine reservoir gained access to cultured rainbow trout, resulting in overt disease.

Molecular clock analyses of the VHSV G gene sequences indicated that the rates of substitution were not equal throughout the phylogenetic tree, but that a molecular clock hypothesis could be accepted for two subsets of the data representing either trout farm subgroup Ia isolates, or marine fish isolates in a group comprised subgroup Ib and genotypes II, III, and IV (Einer-Jensen *et al.*, 2004). The substitution rates were 1.74×10^{-3} and 7.06×10^{-4} per codon per year, for the trout farm and marine isolate groups, respectively. Thus, the substitution rate for VHSV in the major trout farm subtype is 2.5 times higher than the rate in European marine VHSV, again suggesting novel selection pressures in the trout farm environment.

VHSV molecular epidemiology and evolution in western North America and Asia

The first detections of VHSV in North America occurred in 1988 when the virus was isolated from adult salmonids returning to two different hatcheries on the Pacific coast of Washington state (reviewed in Meyers and Winton, 1995). Based on the appreciation at that time of VHSV as a devastating pathogen in European trout farms, it was originally assumed that the virus was introduced from Europe. Control measures based on federal regulations and regional policies required destruction of all potentially infected fish and complete disinfection of facilities where the virus was detected. These measures were implemented at several fish culture facilities, as the virus was detected in several additional Pacific Northwest hatcheries during 1989–1994. Characterization of the VHSV isolates from these Pacific watersheds showed that they were indistinguishable from European strains by serology, with complete cross-neutralization by polyclonal antibodies. However, molecular studies revealed genetic differences, such that the N genes of European and North American VHSV had 13% divergence at the nucleotide level (Bernard *et al.*, 1992). The detection of VHSV in Pacific watersheds increased the awareness of the virus in the region, and VHSV was subsequently isolated in Pacific cod and Pacific herring in coastal waters of Alaska and British Columbia. Combined results from several surveillance studies in marine fish determined that the virus was endemic among Pacific herring and Pacific sardine as the primary hosts and reservoir, with Pacific cod, Chinook and coho salmon, and several other marine species as minor reservoirs (Meyers and Winton, 1995; Kent *et al.*, 1998; Hedrick *et al.*, 2003). The marine reservoir for VHSV in Pacific waters of North America extends from Alaska to California, and the virus continues to be isolated occasionally in apparently healthy salmonids returning to hatcheries to spawn. Within this reservoir VHSV is also associated with regular natural epidemics among Pacific herring and sardines in British Columbia and Alaska (Hershberger *et al.*, 1999), where it is considered an opportunistic pathogen triggered by additional stressors such as unusual temperatures, high densities, or pollution. The scale of these epidemics has in some cases been massive. For example, mortality from VHS and other diseases represents a leading hypothesis accounting for the decline and failed recovery of Pacific herring populations in Prince William Sound, Alaska. These population-level impacts resulted in a trophic cascade, where interconnected fish, bird, and marine mammal populations were also impacted.

Controlled laboratory challenge studies with Pacific isolates of VHSV determined that they had low pathogenicity for rainbow trout and other salmonids. Although the potential for adaptation to cultured trout was recognized based on the experience with VHSV in Europe (Meyers and Winton, 1995), this has not occurred in North America to date. Genetic typing of Pacific coast North American VHSV provided the first evidence for genotype IV, which is the most divergent of the four VHSV genotypes. Molecular clock analyses estimated that the divergence of genotype IV from the other three genotypes found in Europe occurred approximately 500 years ago, well before the development of modern intensive aquaculture (Einer-Jensen *et al.*, 2004). Within North American Pacific VHSV isolates there is very low diversity, with a maximum of 1.5% divergence in a partial G gene region (669 nt) that was sequenced for eleven isolates obtained between 1988–2002, from sites ranging from Alaska to California (Hedrick *et al.*, 2003). These same isolates differed from representative European VHSV strains by a minimum of 14% in the same genetic region.

In the Asian Pacific, VHSV was first detected in Japan in an outbreak of VHS in marine farmed Japanese flounder (*Paralichthyes olivaceous*) in 1996 (Isshiki *et al.*, 2001). It was detected in additional flounder farms in Japan in 1998 and 2000, and in 1999 and 2000 VHSV was isolated from wild flounder during surveys of viruses in wild marine fish of Japan. Genetic typing of Japanese VHSV isolates from both cultured and wild flounder revealed that they were in subgroup IVa, most closely related to the VHSV found off the Pacific coast of North America (Nishizawa *et al.*, 2002). A single exception was an isolate from the 1996 farm outbreak that was in subgroup Ia. This suggested an introduction of virus from Europe, but the mechanism of introduction is unknown and that virus does not appear to have persisted. Further testing of additional isolates from both wild and farmed flounder found only viruses from subgroup IVa, suggesting that there is a widespread marine reservoir of this type of VHSV in coastal areas of Japan, and that this is a source of virus infection in farmed flounder. The Japanese isolates comprised a monophyletic group within subgroup IVa, suggesting that they may have evolved independently from North American Pacific VHSV for some period of time. In Korea the first detection of VHSV was in an outbreak in farmed flounder in 2001, and since then the virus has been isolated from both cultured and wild flounder (Kim *et al.*, 2009). Genetic typing demonstrated that the Korean isolates have over 99% identity with the types found in the wild marine flounder reservoir around Japan, indicating that this reservoir extends to Korean waters.

Emergence of VHSV in the North American Great Lakes

A major emergence event is currently ongoing in the Great Lakes region of North America, where VHSV was isolated for the first time in 2003 (Elsayed *et al.*, 2006). The Great Lakes are the largest group of freshwater lakes in the world by surface area, with over 20% of the world's surface freshwater. Within the Great Lakes basin fish populations constitute a very diverse component of the ecosystem and they are highly valued for recreational fishing. The emergence of VHSV began with two disease outbreaks that caused significant mortality in wild fish in 2005, and continued with several large-scale multispecies outbreaks in 2006–2007 (Groocock *et al.*, 2007; Lumsden *et al.*, 2007; Kane-Sutton *et al.*, 2010). The virus spread to all five major lakes and several inland lakes within the basin, and it has been detected in 28 diverse fish host species, several of which were not previously known to be hosts of VHSV. As a result, federal regulations have strictly limited fish transport in the region in an effort to prevent further spread of the virus. Numerous surveillance studies have been conducted to monitor the host and geographic range of VHSV in the Great Lakes, and research studies have revealed variation in susceptibility and persistence of the virus in different Great Lakes fish species (Kim and Faisal, 2010a,b).

Sequence analysis of the G gene of the original index virus isolate from 2003 indicated that it was most closely related to genotype IV VHSV from the Pacific coast of North America, but with sufficient divergence to be designated as a separate subgroup (Elsayed *et al.*, 2006). Thus, subgroup IVa now contains VHSV isolates from the North American Pacific coast and Asia, and subgroup IVb contains the Great Lakes VHSV. Subsequent typing of 108 VHSV isolates collected during 2003–2009 from diverse fish hosts throughout the Great Lakes revealed that they were extremely homogeneous genetically (Thompson *et al.*, 2011). Among these isolates there were a total of eleven different genetic types that differed on average by only 1 out of 669 nt (0.15%), with a maximum difference of 7 out of 669 nt (1.05%) in the partial G gene region analysed. In phylogenetic analyses all Great Lakes VHSV isolates grouped with the index isolate in subgroup IVb, as shown in Fig. 7.6. This illustrates the low genetic diversity expected for a virus that has been recently introduced and emerged in naive host populations, without sufficient time for genetic diversification. The emergence of VHSV in the Great Lakes represents an invasion of the largest freshwater ecosystem in the world, and it is the first time that VHSV has caused epidemic

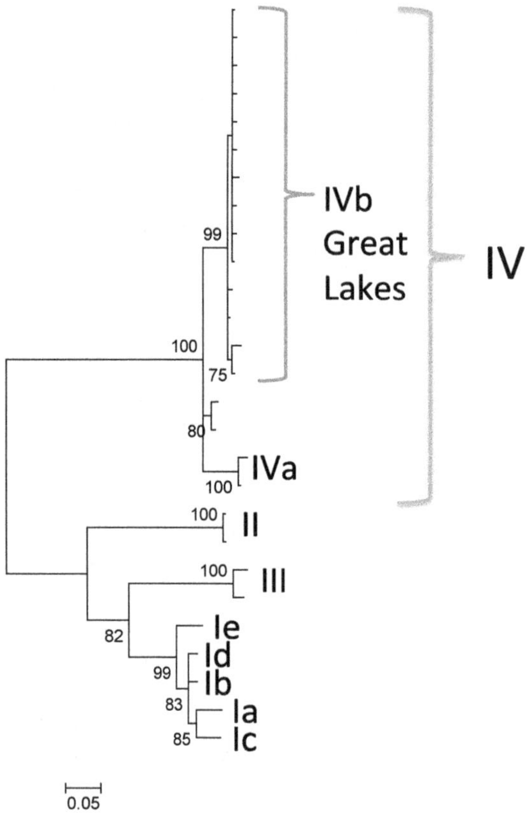

Figure 7.6 Phylogenetic tree showing viral genotypes associated with the emergence of viral haemorrhagic septicaemia virus (VHSV) subgroup IVb in the Great Lakes region of North America. Genetic typing was based on 669 nt partial G gene sequences of 108 VHVS isolates collected from Great Lakes fish during 2003–2009 (Thompson *et al*., 2011). These isolates were resolved into 11 sequence types that had extremely low genetic diversity and clustered together into a new subgroup designated IVb, within genotype IV. This subgroup is most closely related to VHSV type IV isolates from the east coast (closest to most recent common ancestral node for genotype IV) and west coast (subgroup IVa) of North America. The separation of Great Lakes type IVb isolates from genotypes I, II, and III clearly demonstrated that the virus was not introduced from Europe. Bayesian tree courtesy of Tarin Thompson, USGS Western Fisheries Research Center, Seattle, USA.

outbreaks in wild freshwater fish. It also represents a unique opportunity to observe and study the course of a large-scale viral emergence in a natural aquatic ecosystem as the virus adapts to new hosts and potentially evolves over time.

In summary, VHSV was first known in Europe as a devastating pathogen of farmed rainbow trout, and several decades later an extensive marine fish reservoir was discovered as the source of the virus that emerged in trout. The history of VHSV in North America is very different, as the virus was first detected in the marine reservoir, and to date it has not caused significant disease impacts in aquacultured fish, including trout. However, VHSV in North America has caused several documented epidemics among wild fish populations in both the Pacific marine environment and in the Great Lakes. In Asia VHSV has only relatively recently begun to cause disease in farmed flounder, emerging from a wild marine reservoir that also includes flounder.

Ecology of novirhabdoviruses: mechanisms of emergence and host switch events

For both VHSV and IHNV, disease occurs predominantly in juvenile fish and involves a combination of factors including viral genotype, overall fish health status, stress levels, environmental conditions, and presence of other pathogens (LaPatra, 1998). Survivors of IHNV or VHSV infections develop broad protective immunity to re-infection.

For NV the existence and/or biological significance of carrier fish is not well understood and may vary for different virus–host associations. In contrast, for VHSV the existence of carrier fish is well documented. The ecology of IHNV and VHSV differs significantly owing to distinctions in host range and viral reservoirs. IHNV has a narrow host range and is endemic in Pacific salmonids, and there are no known alternate vectors or reservoirs. In natural fish populations IHNV ecology is driven by host migration patterns, and persistence is hypothesized to be due in some cases to overlapping migration timing of different salmonid populations (Kelley *et al.*, 2007). Epidemics of IHNV have been documented in wild sockeye or kokanee salmon but not in any other Pacific salmonid species. In contrast, VHSV has an exceptionally broad host range and a widespread reservoir in diverse species of marine fish in both the North Atlantic and North Pacific Oceans. Little is known about the ecology of VHSV within this reservoir, but overt disease has only been reported in herring and sardine populations off the west coast of North America. HIRRV is also found in a marine reservoir in Asia, but disease has not been reported among wild fish. The ecology of IHNV, VHSV, and HIRRV in cultured fish is mainly driven by security of the water supply and anthropogenic impacts including physical aspects of culture facilities and husbandry practices. Since the development of high-quality pasteurized fish feeds in the 1960s and 1970s, the introduction of virus into fish culture facilities happens most often by import of contaminated fish or fish eggs, or by waterborne virus in the water supply. Transmission of virus within and between facilities is mitigated by strict biosecurity policies observed in most fish farms and hatcheries. Globally, IHNV has been spread from North America to other continents by aquaculture-related activities, while VHSV occurrence appears to be due to emergence from wild fish reservoirs that are widely distributed throughout the northern hemisphere.

The histories of IHNV and VHSV described earlier include series of viral emergence events associated with a relatively small number of hypothesized sources and mechanisms of transmission (Walker and Winton, 2010). Sources of virus include wild fish reservoirs, and fish or fish eggs associated with aquaculture. Mechanisms include the use of raw fish products in fish feed, proximity to infected wild reservoir fish, geographic translocations of infected fish or virus, or introduction of naive exotic fish species into a virus-endemic area. As described above, in many cases genetic typing of current or historic virus isolates has been used to identify probable sources and mechanisms of viral emergence.

Many early emergence events were associated with the historical use of raw unpasteurized wild fish products in feed for cultured fish. This was common in aquaculture worldwide through the 1960s, when it became clear that this practice was a potential source of disease. The cause of the original emergence of VHSV in farmed rainbow trout in Europe was hypothesized by various researchers to be due to the use of raw feed (Dixon, 1999; Einer-Jensen *et al.*, 2004). Similarly, researchers investigating the original emergence of IHNV in Washington sockeye hatcheries in the 1950s determined that the cause of the disease outbreaks was most likely the use of raw by-product from wild sockeye salmon canneries in the hatchery fish feed (reviewed in Kurath *et al.*, 2003). This practice may also account for the host switch of IHNV into rainbow trout in North America, as raw feed was used in trout farms during the 1960s and into the 1970s. Since then high quality pasteurized processed fish feeds have been developed and refined. They have replaced the use of raw feed in most aquaculture settings, although there are still some cases in which raw feed is used, especially for culture of novel fish species for which nutritional requirements have not been defined.

With the elimination of the use of raw feed, the next most probable mechanism of viral emergence involves proximity to infected fish, often in a wild marine reservoir. Transmission due to proximity could involve waterborne virus that is shed by infected wild fish into the water supply for land-based farms, or into water surrounding netpens. These mechanisms are likely responsible for the recent emergence of VHSV into marine netpen rainbow trout in Northern Europe, as described below. Examples of viral emergence due to geographic translocations of virus in infected fish or fish eggs for aquaculture include the introductions of IHNV of the U group into Japanese sockeye farms (Kimura and Yoshimizu, 1991), IHNV of the M group into European trout farms (Enzmann *et al.*, 2005), and IHNV of the J group into Korean trout farms (Kim *et al.*, 2007). In contrast, geographic translocation of a wild virus seems to be responsible for the emergence of VHSV in the Great Lakes, although the specific mechanism responsible for movement of the virus is unknown. An example illustrating emergence due to introduction of a naive exotic fish species into a location with endemic virus is Atlantic salmon reared in marine netpens in Western Canada, where raw feed is not used, but proximity to endemic IHNV and VHSV in wild fish has resulted in viral transmission and emergence (Saksida, 2006; K.A. Garver, personal communication).

Among the novirhabdovirus emergence events there are several that involved viral host switches, sometimes with evidence for viral adaptation

Table 7.4 *Novirhabdovirus* host jump events

Virus years	Location	Probable source	Virus genotype[a]	New host[b]	Environment	Hypothesized mechanism(s)	Notes
VHSV							
1950–1970s	Europe	Wild marine fish reservoir	Ib-like	Rainbow trout	Freshwater farm	Raw wild fish products in feed for farmed fish, adaptation to new host	Historical emergence leading to establishment of endemic virus in Europe, 3–4 independent events likely, origin of genotype I subgroups Ia, Ic, id, Ie, followed by expansion of subtype Ia
1998, 2000	Sweden	Wild marine fish reservoir	Ib	Rainbow trout	Marine netpen	Proximity to virus reservoir fish, adaptation to new host	Recent localized emergence
2000–2004	Finland	Wild marine fish reservoir	Id	Rainbow trout	Marine netpen	Raw feed, proximity to virus reservoir fish, adaptation to new host	Recent localized emergence
2007	Norway	Wild marine fish reservoir	III	Rainbow trout	Marine netpen	Proximity to virus reservoir fish (raw feed specifically ruled out), adaptation to new host	Recent localized emergence
1990s–present	Canada, west coast	Wild marine fish reservoir	IVa	Atlantic salmon	Marine netpen	Introduction of naive exotic host, proximity to endemic virus reservoir in wild fish, virus adaptation possibly not necessary	Possible indications of early adaptation to increased virulence
2003–present	US Great Lakes	Wild marine fish reservoir	IVb-like	Several freshwater fish species	Wild	Viral invasion of a new geographic range with naive susceptible hosts, virus adaptation possibly not necessary	Mechanism of invasion uncertain
IHNV							
1970s	US west coast	Wild sockeye salmon	U	Rainbow trout	Freshwater farm	Raw wild fish product from salmon canneries in fish feed, adaptation to new host and to higher temperature	Historical emergence leading to establishment of endemic virus in Idaho trout farms, origin of M genogroup, followed by diversification of subgroups
1970s	Japan	Farmed sockeye or masu salmon	U	Rainbow trout	Freshwater farm	Transmission from other fish species within fish farm, virus adaptation to new host	Host jump followed later in 1980s by virus adaptation to increased virulence in large trout, origin of J genogroup
1992–2003	Canada west coast	Wild sockeye salmon	U	Atlantic salmon	Marine netpen	Introduction of naive exotic host, proximity to virus reservoir fish, no evidence of virus adaptation	Likely three independent events, followed by eradication

[a] Virus genotypes Ia-Ie, II, IVa and IVb for VHSV, and U, M, and JRt for IHNV are as described in text.
[b] host species are as follows: rainbow trout, *Oncorhynchus mykiss*; sockeye salmon, *O. nerka*; masu salmon, *O. masou*; Atlantic salmon, *Salmo salar*.

to higher virulence (Table 7.4). A host switch is defined as a virus infecting a host species it had not previously infected and adaptation refers to subsequent evolution of the virus to allow more efficient infection and transmission in the new host, often leading to disease outbreaks (Parrish *et al.*, 2008). For novirhabdoviruses the great majority of these host switches are into cultured rainbow trout, possibly due to the long history and immense scale of intensive trout aquaculture world-wide. Evidence for viral adaptation is based on experimental challenges that demonstrate differences in virulence between an emerged virus and a virus that represents the presumed source of emergence. For example, the original emergence of VHSV in European rainbow trout is considered to involve host switches with adaptation because phylogenetic analyses indicate the most probable ancestor was VHSV type Ib, which has been shown to be of low virulence for trout (Skall *et al.*, 2005; Campbell *et al.*, 2009). Similarly, the emergence of IHNV in rainbow trout in North America appears to have involved a host switch with adaptation because the presumed ancestral U type IHNV was not virulent for trout (Rucker *et al.*, 1953; Garver *et al.*, 2006). As noted above, the mechanism for these early host switches was most likely the use of raw feed.

More recently, the three independent host switches into marine netpen reared rainbow trout in northern Europe between 1998–2007 were caused by VHSV types Ib, Id, and III, all of which have been shown to have low pathogenicity for rainbow trout (Skall *et al.*, 2005). Although one case may have involved use of raw fish feed, this mechanism was ruled out in the Norwegian outbreak, indicating the existence of other mechanisms capable of facilitating a host switch. Due to the location of the rainbow trout netpens in the marine environment within the range of the endemic marine fish reservoir for VHVS, probable mechanisms involve proximity, resulting in either transmission of waterborne virus or possibly co-mingling of the cultured fish with infected marine fish that may enter netpens. A more complex case of proximity resulting in a host switch is the origin of the J genogroup of IHNV, that appears to have occurred as a two step process in cultured Japanese rainbow trout. In this case the U group source virus was detected in rainbow trout soon after its introduction into cultured Japanese sockeye salmon, but the re-mergence of IHNV to cause severe disease in the 1990s involved a later additional adaptation to increased virulence in larger rainbow trout (Nishizawa *et al.*, 2006; Mochizuki *et al.*, 2009). In this example both steps in the origin of the J genogroup occurred within fish culture facilities, most likely due to proximity based mechanisms.

The IHNV M and J genogroups and VHSV subgroup Ia are all major phylogenetic groups that are hypothesized to have arisen following evolutionary host switches into cultured rainbow trout, with adaptation to higher virulence. In each case, phylogenies indicate that these groups have higher genetic diversity than their ancestral virus groups. Evolutionary rate estimates indicate that the IHNV M genogroup has evolved 6-fold faster than its ancestral U genogroup (Kurath *et al.*, 2003), and European freshwater subgroup Ia has evolved 2.5-fold faster than ancestral isolates from marine reservoir fish (Einer-Jensen *et al.*, 2004). The observation of these same phenomena in the phylogenies of both IHNV and VHSV supports the hypothesis that host switches for fish novirhabdoviruses into rainbow trout can be followed by a period of relatively rapid evolution and diversification, which likely involves adaptation to the new host and also to unique conditions within the trout farm environment.

It is interesting to note that VHSV has not adapted to rainbow trout in western North America. During over 20 years since the first detection of VHSV in North America there has been no evidence for adaptation to rainbow trout, despite detections of the virus in anadromous salmonids that could carry VHSV IVa into freshwater regions where trout are reared. However, during this time there has been no use of raw feed for trout, and no culture of rainbow trout in marine netpens, providing little opportunity for adult salmonids to co-mingle or be in close proximity to cultured trout. In the absence of these two mechanisms it may be that host switches into trout can be avoided. It is also interesting that there are no documented cases of virus host switches into wild fish, possibly due to a lack of positive selection pressures provided by high density rearing conditions in aquaculture fish populations (Kurath and Winton, 2011).

Molecular studies of host specificity and virulence

The ecology and geographic distribution of viruses are intimately associated with variation in the susceptibility of potential host species. For both IHNV and VHSV various experimental approaches have been used to investigate the basis of differential host specificity phenotypes associated with different virus isolates. Host specificity and virulence factors for VHSV are mostly associated with whether or not isolates are adapted to cultured rainbow trout. Full length or nearly full length genome sequences of several VHSV isolates have been analysed for possible genetic determinants associated with high virulence in rainbow trout. Two such studies identified either 4 or 10 amino acids, in various viral proteins, that differed consistently between isolates of high and low virulence, suggesting that a limited number of genetic changes may be involved (Betts and Stone, 2000; Campbell *et al.*, 2009). These putative VHSV virulence determinants have not been confirmed experimentally, but recently developed reverse genetics systems for VHSV provide the tools needed to do so. Another interesting study involved the use of trout gill epithelial cell cultures, demonstrating that high virulence isolates traversed a single layer of trout cells much more rapidly and with higher efficiency than low virulence isolates (Brudeseth *et al.*, 2008). This provides a tractable *in vitro* system for further analysis of the possible mechanism responsible for the different phenotypes. In a third approach, a comparison of rainbow trout infections with high and low virulence VHSV isolates was conducted using qRT-PCR assays to assess viral load and host innate immune response over time (Campbell *et al.*, 2011). Although both viruses replicated and stimulated a host innate immune response after injection delivery, the low virulence isolate was not detectable after immersion exposure and did not stimulate an innate immune response, suggesting that it was not able to establish an infection by the natural exposure route.

For IHNV in North America there is clear evidence of important ecological consequences based on differential virulence patterns in different Pacific salmon species. Isolates in the U, M, and L genogroups in North America are most commonly obtained in the field from sockeye salmon, rainbow trout, and Chinook salmon, respectively (Garver *et al.*, 2003; Kurath *et al.*, 2003). This host specificity is not absolute, so that viruses from each genogroup are found in each species occasionally, but they appear to have evolved differential virulence phenotypes that impact their prevalence, and thus ecology, in these host species. For U and M group IHNV several controlled laboratory challenge studies have confirmed that this field pattern is based on biological differences in host-specific virulence: U viruses are virulent in sockeye salmon but not in rainbow trout, and M viruses are virulent in rainbow trout but not sockeye salmon (Garver *et al.*, 2006). Comparative studies of infections with representative IHNV isolates from the U and M genogroups have revealed that both viruses were able to enter both sockeye salmon and rainbow trout and replicate to some degree, but higher virulence in each host (U in sockeye salmon, or M in rainbow trout) was associated with faster viral replication to significantly higher viral load levels, as assessed by qRT-PCR (Peñaranda *et al.*, 2009; Purcell *et al.*, 2009). Analyses of the host Mx gene induction also by qRT-PCR showed that in each host the more virulent isolate stimulated a significantly greater host innate immune response that generally correlated with the viral load over time. Thus, high virulence isolates did not suppress the host innate immune response, but they were able to persist and replicate in the presence of this response, while low virulence isolates were cleared from the host.

An *in vitro* model in which M virus replicates more rapidly than U virus in a rainbow trout cell line has been used to explore the mechanistic basis of host specificity at the cellular level (Park *et al.*, 2010, 2011). This revealed that the growth restriction of the U strain was due to reduced genome replication and transcription, leading to a decreased yield of infectious virus. Further studies with recombinant chimeric IHNV showed that the viral G and/or NV genes were not responsible for the difference in phenotype, and the host interferon response was not implicated as a major factor. However, a difference in response to a casein kinase II inhibitor was demonstrated, suggesting that U and M strain viruses may differ in requirement for host kinases. Thus, to date, several possible mechanisms or determinants of host specificity for IHNV and VHSV have been suggested by various methods. With the active application of reverse genetic for novirhabdoviruses our understanding of these mechanisms should be greatly improved in the near future.

In vivo viral fitness and competition assay system for IHNV

The concept of fitness is central to evolutionary theory. For viruses fitness is defined as the ability to produce infectious progeny in a given environment, and it is typically measured for vertebrate viruses as competitive fitness in co-infections of cultured cell lines (Domingo *et al.*, 1997). Experimental systems using IHNV have been developed to measure both absolute and relative fitness of viruses *in vivo*, using single and mixed infections of selected IHNV strains in juvenile rainbow trout (Troyer *et al.*, 2008; Wargo *et al.*, 2010). In standard assays groups of 20–30 trout were exposed by bath immersion to a 1:1 mixture of two virus strains of interest, and control groups were exposed to each virus alone. Fish were then separated into individual tanks to avoid cross-infection, and held for a three day period of viral growth and competition *in vivo*. Total RNA was extracted from individual fish, and the viral load of each strain was quantified using carefully validated

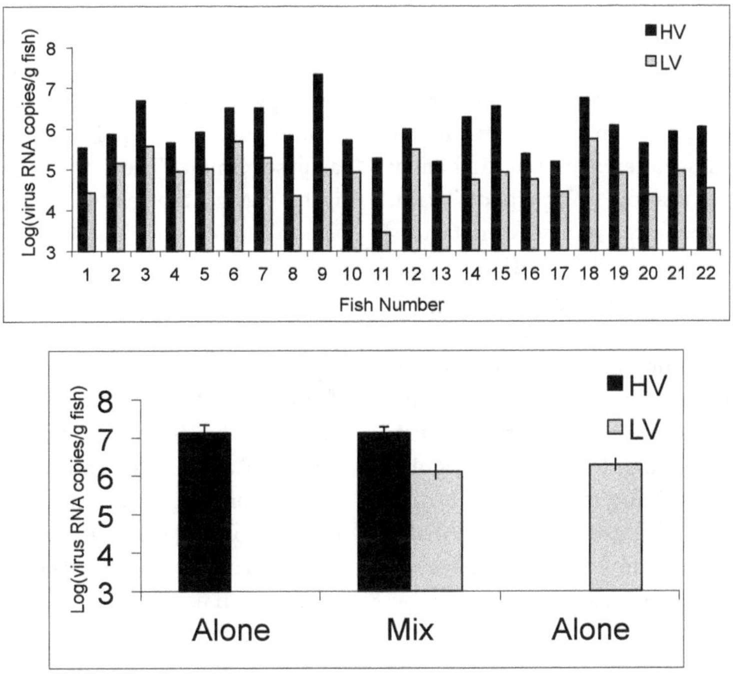

Figure 7.7 Example of an *in vivo* viral competitive fitness assay illustrating a significant difference in fitness that correlates with virulence for two strains of infectious haematopoietic necrosis virus (IHNV). The two viral strains had either high virulence (HV) or low virulence (LV) for rainbow trout. Groups of fish were exposed to a 1:1 mixture of the strains, or to each strain alone, as described in the text. The top panel shows individual fish viral loads for HV (black) and LV (grey) IHNV in 22 fish exposed to the mixture of the two strains. The bottom panel shows mean viral load for the groups, allowing us to conclude that HV is statistically more fit than LV under these conditions. For each virus there was no significant impact of being in a mixed infection versus a single infection, indicating an absence of direct competition. Graphs begin at log 3 on the *y*-axis because this is the detection limit of the genotype-specific qRT-PCR assays used to quantify each viral genotype. Figure provided by Dr Andrew Wargo, University of Washington and USGS Western Fisheries Research Center, Seattle, USA.

genotype-specific qRT-CPR assays. Fig. 7.7 shows typical results of an assay using two IHNV strains that differ in virulence in rainbow trout, designated HV and LV, for high and low virulence, respectively. The results demonstrated biological variation that is typical for individual hosts within an infected population. It also showed that all fish were co-infected, and that HV replicated to higher levels than LV in individual fish. Comparison of the mean viral loads in each group show that HV has significantly higher fitness than LV, and that there was no evidence of direct competition, since the viral loads of each strain in mixed infections did not differ from those in single infections.

This system has been used in several studies showing that virulence correlates with fitness of IHNV strains (Wargo *et al.*, 2010), strains of equal virulence have equal fitness (Troyer *et al.*, 2008), and that fitness is host specific for representative U and M IHNV strains tested in both rainbow trout and sockeye salmon (Peñaranda *et al.*, 2011). Modifications of the system have been used to examine individual steps of the viral replication cycle, showing that in-host replication is the major determinant of fitness differences, but viral entry and shedding also play a role (Wargo and Kurath, 2011). It has also been used to assess the factors that contribute to host-to-host variation in viral load, and to assess fitness throughout the course of viral infections by measuring virus shed over time from groups of individual fish held for 30 days. A major benefit of this system is the capacity to relatively easily handle groups of infected fish hosts that are large enough for statistical significance. This system could be modified for fitness studies of other novirhabdoviruses, potentially providing useful information on *in vivo* fitness of other vertebrate rhabdoviruses.

On-line databases for novirhabdoviruses

General information about important viral pathogens in the *Novirhabdovirus* genus is available at internet websites dedicated to aquatic animal diseases. These include the Aquatic Animal Pathogen and Quarantine Information System (AAQPIS) developed by the United Nations Food and Agriculture Organization, the Permanent Advisory Network for Diseases in Aquaculture (PANDA) funded by the European Union, and the International Database of Aquatic Animal Diseases (IDAAD) developed by the World Organization for Animal Health. In addition to these species-level descriptions, there are now also online databases that provide detailed epidemiological and genetic records for individual isolates of both IHNV and VHSV. Databases focused on North American isolates of IHNV and VHSV were released in 2006 and 2010 respectively, as Molecular Epidemiology of Aquatic Pathogens (MEAP)-IHNV (http://gis.nacse.org/ihnv) and MEAP-VHSV (http://gis.nacse.org/vhsv) (Emmenegger *et al.*, 2011). Databases focused on European and Asian isolates of VHSV and IHNV were published in 2009 and 2010 as Fish Pathogens-EU databases (http://www.fishpathogens.eu/vhsv/index.php and http://www.fishpathogens.eu/ihnv/index.php, respectively) (Jonstrup *et al.*, 2009, 2010). All four of these databases provide genetic sequence data as well as epidemiological features of the source isolates, with search, sort, and mapping capabilities to facilitate analyses online. The purpose of these databases is to engage the greater network of interested specialists by providing data that they can use to generate new insights into the epidemiology or evolution of these viruses.

Acknowledgements

Great thanks are due to David Stone, Katja Einer-Jensen, Rachel Life, Tarin Thompson, and Bill Batts for providing sequence comparison information and phylogenetic tree figures, and to Andrew Wargo for fitness Fig. 7.7. The use of trade, firm, or corporation names does not constitute an official endorsement or approval by the US Department of Interior or the US Geological Survey.

References

Ahne, W., Bjorklund, H., Essbauer, S., Fijan, N., Kurath, G., and Winton, J. (2002). Spring viremia of carp – review. Dis. Aquat. Organ. 52, 261–272.

Alonso, M., Kim, C., Johnson, M., Pressley, M., and Leong, J.-A. (2004). The NV gene of snakehead Rhabdovirus (SHRV) is not required for pathogenesis, and a heterologous glycoprotein can be incorporated into the SHRV envelope. J. Virol. 78, 5875–5882.

Ammayappan, A., and Vakharia, V.N. (2011). Nonvirion protein of *Novirhabdovirus* suppresses apoptosis at the early stage of virus infection. J. Virol. *85*, 8393–8402.

Ammayappan, A., Kurath, G., Thompson, T.M., and Vakharia, V.N. (2011). A reverse genetics system for the Great Lakes strain of viral hemorrhagic septicemia virus: the NV gene is required for pathogenicity. Mar. Biotechnol. (NY) *13*, 672–683.

Anderson, E.D., Mourich, D.V., Fahrenkrug, S.C., LaPatra, S., Shepherd, J., and Leong, J.A. (1996). Genetic immunization of rainbow trout (*Oncorhynchus mykiss*) against infectious hematopoietic necrosis virus. Molec. Marine Biol. Biotechnol. 5, 114–122.

Aoki, T., Hirono, I., Kondo, H., Hikama, J., and Jung, T. (2011). Microarray technology is an effective tool for identifying genes related to the aquacultural improvement of Japanese flounder, Paralychthys olivaceus. Comp. Biochem. Physiol. Part D Genomics Proteomics *6*, 39–43.

Bearzotti, M., Delmas, B., Lamoureux, A., Loustau, A.M., Chilmonczyk, S., and Bremont, M. (1999). Fish rhabdovirus cell entry is mediated by fibronectin. J. Virol. 73, 7703–7709.

Bernard, J., Bremont, M., and Winton, J. (1992). Nucleocapsid gene sequence of a North American isolate of viral haemorrhagic septicaemia virus, a fish rhabdovirus. J. Gen. Virol. 73, 1011–1014.

Betts, A.M., and Stone, D.M. (2000). Nucleotide sequence analysis of the entire coding regions of virulent and avirulent strains of viral haemorrhagic septicaemia virus. Virus Genes *20*, 259–262.

Biacchesi, S. (2011). The reverse genetics applied to fish RNA viruses. Vet. Res. *42*, 12.

Biacchesi, S., Thoulouze, M.I., Bearzotti, M., Yu, Y.X., and Bremont, M. (2000). Recovery of NV knockout infectious hematopoietic necrosis virus expressing foreign genes. J. Virol. *74*, 11247–11253.

Biacchesi, S., Lamoureux, A., Merour, E., Bernard, J., and Bremont, M. (2010). Limited interference at the early stage of infection between two recombinant novirhabdoviruses: viral hemorrhagic septicemia virus and infectious hematopoietic necrosis virus. J. Virol. *84*, 10038–10050.

Bjorklund, H.V., Higman, K.H., and Kurath, G. (1996). The glycoprotein genes and gene junctions of the fish rhabdoviruses spring viremia of carp virus and hirame rhabdovirus: Analysis of relaitonships with other rhabdoviruses. Virus Res. 42, 65–80.

Bootland, L.M., and Leong, J.C. (1999). Infectious hematopoietic necrosis virus. In Fish diseases and disorders, Woo, P.T.K., and Bruno, D.W., eds (CAB International, Wallngford, UK), pp. 57–112.

Brudeseth, B.E., and Evensen, O. (2002). Occurrence of viral haemorrhagic septicaemia virus (VHSV) in wild marine fish species in the coastal regions of Norway. Dis. Aquat. Organ. 52, 21–28.

Brudeseth, B.E., Skall, H.F., and Evensen, O. (2008). Differences in virulence of marine and freshwater isolates of viral hemorrhagic septicemia virus in vivo correlate with in vitro ability to infect gill epithelial cells and macrophages of rainbow trout (*Oncorhynchus mykiss*). J. Virol. *82*, 10359–10365.

Campbell, S., Collet, B., Einer-Jensen, K., Secombes, C.J., and Snow, M. (2009). Identifying potential virulence determinants in viral haemorrhagic septicaemia virus (VHSV) for rainbow trout. Dis. Aquat. Organ. *86*, 205–212.

Campbell, S., McBeath, A., Secombes, C., Snow, M., and Collet, B. (2011). Interferon response following infection with genetically similar isolates of viral haemorrhagic septicaemia virus (VHSV) exhibiting contrasting virulence in rainbow trout. Fish Shellfish Immunol. *30*, 287–294.

Chiou, P.P., Kim, C.H., Ormonde, P., and Leong, J.A. (2000). Infectious hematopoietic necrosis virus matrix protein inhibits host-directed gene expression and induces morphological changes of apoptosis in cell cultures. J. Virol. *74*, 7619–7627.

Choi, M.K., Moon, C.H., Ko, M.S., Lee, U.H., Cho, W.J., Cha, S.J., Do, J.W., Heo, G.J., Jeong, S.G., Hahm, Y.S., *et al.* (2011). A nuclear localization of the infectious haematopoietic necrosis virus NV protein is necessary for optimal viral growth. PLoS One *6*, e22362.

Dale, O.B., Orpetveit, I., Lyngstad, T.M., Kahns, S., Skall, H.F., Olesen, N.J., and Dannevig, B.H. (2009). Outbreak of viral haemorrhagic septicaemia (VHS) in seawater-farmed rainbow trout in Norway caused by VHS virus Genotype III. Dis. Aquat. Organ. *85*, 93–103.

Dietzgen, R., Calisher, C., Kurath, G., Kuzmin, I., Rodriguez, L., Stone, D., Tesh, R., Tordo, N., Walker, P., Wetzel, T., *et al.* (2011). Rhabdoviridae. In Virus Taxonomy: Classification and Nomenclature of Viruses, King, A., Adams, M., Carstens, E., and Lefkowitz, E., eds (Elsevier, Oxford, UK), pp. 686–713.

Dixon, P.F. (1999). VHSV came from the marine environment: clues from the literature, or just red herrings? Bull. Eur. Ass. Fish Pathol. *19*, 60–65.

Domingo, E., Menéndez-Arias, L., and Holland, J.J. (1997). RNA virus fitness. Rev. Med. Virol. 7, 87–96.

Einer-Jensen, K., Krogh, T.N., Roepstorff, P., and Lorenzen, N. (1998). Characterization of intramolecular disulfide bonds and secondary modifications of the glycoprotein from viral hemorrhagic septicemia virus, a fish rhabdovirus. J. Virol. 72, 10189–10196.

Einer-Jensen, K., Ahrens, P., Forsberg, R., and Lorenzen, N. (2004). Evolution of the fish rhabdovirus viral haemorrhagic septicaemia virus. J. Gen. Virol. *85*, 1167–1179.

Elsayed, E., Faisal, M., Thomas, M., Whelan, G., Batts, W.N., and Winton, J. (2006). Isolation of viral haemorrhagic septicaemia virus from muskellunge, Esox masquinongy (Mitchill), in Lake St. Clair, Michigan, USA reveals a new sublineage of the North American genotype. J. Fish Dis. *29*, 611–619.

Emmenegger, E.J., and Kurath, G. (2002). Genetic characterization of infectious hematopoietic necrosis virus of coastal salmonid stocks in Washington state. J Aquat Anim Health *14*, 25–34.

Emmenegger, E.J., Meyers, T.R., Burton, T.O., and Kurath, G. (2000). Genetic diversity and epidemiology of infectious hematopoietic necrosis virus in Alaska. Dis. Aquat. Organ. *40*, 163–176.

Emmenegger, E.J., Kentop, E., Thompson, T.M., Pittam, S., Ryan, A., Keon, D., Carlino, J.A., Ranson, J., Life, R.B., Troyer, R.M., *et al.* (2011). Development of an aquatic pathogen database (AquaPathogen X) and its utilization in tracking emerging fish virus pathogens in North America. J. Fish. Dis. *34*, 579–587.

Engelking, H.M., and Leong, J.C. (1989). Glycoprotein from infectious hematopoietic necrosis virus (IHNV) induces protective immunity against five IHNV types. J. Aquat. Anim. Health *1*, 291–300.

Enzmann, P.J., Kurath, G., Fichtner, D., and Bergmann, S. (2005). Infectious hematopoietic necrosis virus: monophyletic origin of European isolates from North American Genogroup M. Dis. Aquat. Organ. *66*, 187–195.

Enzmann, P.J., Castric, J., Bovo, G., Thiery, R., Fichtner, D., Schutze, H., and Wahli, T. (2010). Evolution of infectious hematopoietic necrosis virus (IHNV), a fish rhabdovirus, in Europe over 20 years: implications for control. Dis. Aquat. Organ. *89*, 9–15.

Fernandez-Alonso, M., Lorenzo, G., Perez, L., Bullido, R., Estepa, A., Lorenzen, N., and Coll, J.M. (1998). Mapping of linear antibody epitopes of the glycoprotein of VHSV, salmonid rhabdovirus. Dis. Aquat. Organ. *34*, 167–176.

Gagne, N., Mackinnon, A.M., Boston, L., Souter, B., Cook-Versloot, M., Griffiths, S., and Olivier, G. (2007). Isolation of viral haemorrhagic septicaemia virus from mummichog, stickleback, striped bass, and brown trout in eastern Canada. J. Fish Dis. *30*, 213–223.

Garver, K., Troyer, R., and Kurath, G. (2003). Two distinct phylogenetic clades of infectious hematopoietic necrosis virus overlap within the Columbia River Basin. Dis. Aquat. Organ. *55*, 187–203.

Garver, K., Batts, W., and Kurath, G. (2006). Virulence comparisons of infectious hematopoietic necrosis virus U and M genogroups in sockeye salmon and rainbow trout. J. Aquat. Anim. Health *18*, 232–243.

Groocock, G.H., Getchell, R.G., Wooster, G.A., Britt, K.L., Batts, W.N., Winton, J.R., Casey, R.N., Casey, J.W., and Bowser, P.R. (2007). Detection of viral hemorrhagic septicemia in round gobies in New York State (USA) waters of Lake Ontario and the St. Lawrence River. Dis. Aquat. Organ. 76, 187–192.

Hedrick, R.P., Batts, W.N., Yun, S., Traxler, G.S., Kaufman, J., and Winton, J.R. (2003). Host and geographic range extensions of the North American strain of viral hemorrhagic septicemia virus. Dis. Aquat. Organ. *55*, 211–220.

Hershberger, P.K., Kocan, R.M., Elder, N.E., Meyers, T.R., and Winton, J.R. (1999). Epizootiology of viral hemorrhagic septicemia virus in Pacific herring from the spawn-on-kelp fishery in Prince William Sound, Alaska, USA. Dis. Aquat. Organ. *37*, 23–31.

Huang, C., Chien, M.S., Landolt, M., Batts, W., and Winton, J. (1996). Mapping the neutralizing epitopes on the glycoprotein of infectious haematopoietic necrosis virus, a fish rhabdovirus. J. Gen. Virol. *77*, 3033–3040.

Isshiki, T., Nishizawa, T., Kobayashi, K., Nagano, T., and Miyazaki, T. (2001). An outbreak of VHSV (viral hemorrhagic septicemia virus) infection in farmed Japanese flounder *Paralichthys olivaceus* in Japan. Dis. Aquat. Organ. *47*, 87–99.

Johansson, T., Einer-Jensen, K., Batts, W., Ahrens, P., Bjorkblom, C., Kurath, G., Bjorklund, H., and Lorenzen, N. (2009). Genetic and serological typing of European infectious hematopoietic necrosis virus (IHNV) isolates. Dis. Aquat. Organ. *86*, 213–221.

Johnson, M.C., Simon, B.E., Kim, C.H., and Leong, J.A. (2000). Production of recombinant snakehead rhabdovirus: the NV protein is not required for viral replication. J. Virol. *74*, 2343–2350.

Jonstrup, S.P., Gray, T., Kahns, S., Skall, H.F., Snow, M., and Olesen, N.J. (2009). FishPathogens.eu/vhsv: a user-friendly viral haemorrhagic septicaemia virus isolate and sequence database. J. Fish Dis. *32*, 925–929.

Jonstrup, S.P., Schuetze, H., Kurath, G., Gray, T., Jensen, B.B., and Olesen, N.J. (2010). An isolate and sequence database of infectious haematopoietic necrosis virus (IHNV). J. Fish Dis. *33*, 469–471.

Kane-Sutton, M., Kinte, B., Dennis, P., and Koonce, J. (2010). Viral hemorrhagic septicemia virus infection in yellow perch, *Perca flavescens*, in Lake Erie. J. Great Lakes Res. *36*, 37–43.

Kelley, G.O., Bendorf, C.M., Yun, S., Kurath, G., and Hedrick, R.P. (2007). Genotypes and phylogeographical relationships of infectious hematopoietic necrosis virus in California, USA. Dis. Aquat. Organ. 77, 29–40.

Kent, M.L., Traxler, G.S., Kieser, D., Richard, J., Dawe, S.C., Shaw, R.W., Prosperi-Porta, G., Ketcheson, J., and Evelyn, T.P.T. (1998). Survey of salmonid pathogens in ocean-caught fishes in British Columbia, Canada. J. Aquat. Anim. Health *10*, 211–219.

Kim, R., and Faisal, M. (2010a). Comparative susceptibility of representative Great Lakes fish species to the North American viral hemorrhagic septicemia virus Sublineage IVb. Dis. Aquat. Organ. *91*, 23–34.

Kim, R., and Faisal, M. (2010b). Experimental studies confirm the wide host range of the Great Lakes viral haemorrhagic septicaemia virus genotype IVb. J. Fish Dis. *33*, 83–88.

Kim, W.-S., Oh, M.-J., Nishizawa, T., Park, J.W., Kurath, G., and Yoshimizu, M. (2007). Genotyping of Korean isolates of infectious hematopoietic necrosis virus (IHNV) based on the glycoprotein gene. Arch. Virol. *152*, 2119–2124.

Kim, W.-S., Kim, S.-R., Kim, D., Kim, J.-O., Park, M.-A., Kitamura, S.-I., Kim, H.-Y., Kim, D.-H., Han, H.-J., Jung, S.-J., *et al.* (2009). An outbreak of VHSV (viral hemorrhagic septicemia virus) infection in farmed olive flounder *Paralichthys olivaceus* in Korea. Aquaculture *296*, 165–168.

Kimura, T., and Yoshimizu, M. (1991). Viral diseases of fish in Japan. Annu. Rev. Fish Dis. *1*, 67–82.

King, J.A., Snow, M., Skall, H.F., and Raynard, R.S. (2001a). Experimental susceptibility of Atlantic salmon *Salmo salar* and turbot *Scophthalmus maximus* to European freshwater and marine isolates of viral haemorrhagic septicaemia virus. Dis. Aquat. Organ. *47*, 25–31.

King, J.A., Snow, M., Smail, D.A., and Raynard, R.S. (2001b). Distribution of viral haemorrhagic septicaemia virus in wild fish species of the North Sea, North East Atlantic Ocean and Irish Sea. Dis. Aquat. Organ. *47*, 81–86.

Kurath, G. (2008). Biotechnology and DNA vaccines for aquatic animals. Rev. Sci. Tech. *27*, 175–196.

Kurath, G., and Leong, J.C. (1985). Characterization of infectious hematopoietic necrosis virus mRNA species reveals a nonvirion rhabdovirus protein. J. Virol. *53*, 462–468.

Kurath, G., and Winton, J. (2008). Fish rhabdoviruses. In Encyclopedia of Virology, Mahy. B.W.J., and VanRegenmortel, M.H.V., eds (Elsevier, Oxford, UK), pp. 221–227.

Kurath, G., and Winton, J. (2011). Complex dynamics at the interface between wild and domestic viruses of finfish. Curr. Opin. Virol. *1*, 73–80.

Kurath, G., Higman, K.H., and Bjoerklund, H.V. (1997). Distribution and variation of NV genes in fish rhabdoviruses. J. Gen. Virol. *78*, 113–117.

Kurath, G., Garver, K.A., Troyer, R.M., Emmenegger, E.J., Einer-Jensen, K., and Anderson, E.D. (2003). Phylogeography of infectious hematopoietic necrosis virus in North America. J. Gen. Virol. *84*, 803–814.

Kurath, G., Purcell, M., and Garver, K. (2007). Fish rhabdovirus models for understanding host response to DNA vaccines. CAB Reviews: perspectives in agriculture, veterinary sciences, and natural resources *2*, 1–12.

LaPatra, S.E. (1998). Factors affecting pathogenicity of infectious hematopoietic necrosis virus (IHNV) for salmonid fish. J. Aquat. Anim. Health *10*, 121–131.

Lorenzen, N., and LaPatra, S.E. (2005). DNA vaccines for aquacultured fish. Rev. Sci. Tech. *24*, 201–213.

Lorenzen, N., Cupit, P.M., Secombes, C.J., and Cunningham, C. (2000). Three monoclonal antibodies to the VHS virus glycoprotein: comparison of reactivity in relation to differences in immunoglobulin variable domain gene sequences. Fish Shellfish Immunol. *10*, 129–142.

Lorenzen, N., Lorenzen, E., Einer-Jensen, K., Heppell, J., Wu, T., and Davis, H. (1998). Protective immunity to VHS in rainbow trout (*Oncorhynchus mykiss*, Walbaum) following DNA vaccination. Fish Shellfish Immunol. *8*, 261–270.

Lumsden, J.S., Morrison, B., Yason, C., Russell, S., Young, K., Yazdanpanah, A., Huber, P., Al-Hussinee, L., Stone, D., and Way, K. (2007). Mortality event in freshwater drum Aplodinotus grunniens from Lake Ontario, Canada, associated with viral haemorrhagic septicemia virus, Type IV. Dis. Aquat. Organ. *76*, 99–111.

Meyers, T.R., and Winton, J.R. (1995). Viral hemorrhagic septicemia virus in North America. Annu. Rev. Fish Dis. *5*, 3–24.

Mochizuki, M., Kim, H., Kasai, H., Nishizawa, T., and Yoshimizu, M. (2009). Virulence change of Infectious hematopoietic necrosis virus against rainbow trout *Oncorhynchus mykiss* with viral molecular evoluton. Fish Pathol. *44*, 159–165.

Nishizawa, T., Iida, H., Takano, R., Isshiki, T., Nakajima, K., and Muroga, K. (2002). Genetic relatedness among Japanese, American and European isolates of viral hemorrhagic septicemia virus (VHSV) based on partial G and P genes. Dis. Aquat. Organ. *48*, 143–148.

Nishizawa, T., Kinoshita, S., Kim, W.-S., Higashi, S., and Yoshimizu, M. (2006). Nucleotide diversity of Japanese isolates of infectious hematopoietic necrosis virus (IHNV) based on the glycoprotein gene. Dis. Aquat. Organ. *71*, 267–272.

Nordblom, B., and Norell, A. (2000). Report on an outbreak of VHS (Viral Haemorrhagic Septicaemia) in farmed fish in Sweden. Report for the Standing Veterinary Committee. Jonkoping: Swedish Boardfo Agriculture, Department of Animal Production and Health.

Olesen, N. (1998). Sanitation of viral haemorrhagic septicaemia (VHS). J. App. Ichthyol. *14*, 173–177.

Park, J.W., Moon, C.H., Wargo, A.R., Purcell, M.K., and Kurath, G. (2010). Differential growth of U and M type Infectious hematopoietic necrosis virus in a rainbow trout derived cell line, RTG-2. J. Fish Dis. *33*, 583–591.

Park, J.W., Moon, C.H., Harmache, A., Wargo, A.R., Purcell, M.K., Bremont, M., and Kurath, G. (2011). Restricted growth of U-type infectious haematopoietic necrosis virus (IHNV) in rainbow trout cells may be linked to casein kinase II activity. J. Fish Dis. *34*, 115–129.

Parrish, C.R., Holmes, E.C., Morens, D.M., Park, E.C., Burke, D.S., Calisher, C.H., Laughlin, C.A., Saif, L.J., and Daszak, P. (2008). Cross-species virus transmission and the emergence of new epidemic diseases. Microbiol. Mol. Biol. Rev. *72*, 457–470.

Peñaranda, M.M.D., Purcell, M.K., and Kurath, G. (2009). Differential virulence mechanisms of infectious hematopoietic necrosis virus (IHNV) in rainbow trout (*Oncorhynchus mykiss*) include host entry and virus replication kinetics. J. Gen. Virol. *90*, 2172–2182.

Peñaranda, M.M.D., Wargo, A.R., and Kurath, G. (2011). *In vivo* fitness correlates with host-specific virulence of Infectious hematopoietic necrosis virus (IHNV) in sockeye salmon and rainbow trout. Virology *417*, 312–319.

Purcell, M., Laing, K., and Winton, J. (2012). Immunity to fish rhabdoviruses. Viruses *4*, 140–166.

Purcell, M.K., Garver, K.A., Conway, C., Elliott, D.G., and Kurath, G. (2009). Infectious haematopoietic necrosis virus genogroup-specific virulence mechanisms in sockeye salmon, *Oncorhynchus nerka* (Walbaum), from Redfish Lake, Idaho. J. Fish Dis. 32, 619–631.

Raja-Halli, M., Vehmas, T.K., Rimaila-Paernaenen, E., Sainmaa, S., Skall, H.F., Olesen, N.J., and Tapiovaara, H. (2006). Viral haemorrhagic septicaemia (VHS) outbreaks in Finnish rainbow trout farms. Dis. Aquat. Organ. 72, 201–211.

Rucker, R.R., Whipple, W.J., Parvin, J.R., and Evans, C.A. (1953). A contagious disease of sockeye salmon possibly of virus origin. U.S. Fish Wild. Serv. Fish. Bull. *54*, 35–46.

Rudakova, S.L., Kurath, G., and Bochkova, E.V. (2007). Occurence and genetic typing of infectious hematopoietic necrosis virus in Kamchatka, Russia. Dis. Aquat. Organ. *75*, 1–11.

Rudakova, S., and Bochkova, E. (2009). Analysis of two natural epizootics of Infectious hematopoietic necrosis virus among sockeye salmon fry (*Oncorhynchus nerka*) in spawning Lake Nachikinskoe (Kamchatka). In Proceedings of the Third Bilateral Conference on Aquatic Animal Health: Russia and the United States (Shepherdstown, West Virginia, Khaled bin Sultan Living Oceans Foundation), pp. 130–137.

Saksida, S. (2006). Infectious haematopoietic necrosis epidemic (2001 to 2003) in farmed Atlantic salmon *Salmo salar* in British Columbia. Dis. Aquat. Organ. *72*, 213–223.

Schuetze, H., Enzmann, P.J., Mundt, E., and Mettenleiter, T.C. (1996). Identification of the non-virion (NV) protein of fish rhabdoviruses viral haemorrhagic septicaemia virus and infectious haematopoietic necrosis virus. J. Gen. Virol. *77*, 1259–1263.

Schutze, H., Mundt, E., and Mettenleiter, T.C. (1999). Complete genomic sequence of viral hemorrhagic septicemia virus, a fish rhabdovirus. Virus Genes *19*, 59–65.

Shchelkunov, I. (2009). Emergent fish diseases in Russia. In Proceedings of the Third Bilateral Conference on Aquatic Animal Health: Russia and the United States (Shepherdstown, West Virginia, Khaled bin Sultan Living Oceans Foundation), pp. 1–5.

Skall, H.F., Olesen, N.J., and Mellergaard, S. (2005). Viral haemorrhagic septicaemia virus in marine fish and its implications for fish farming – a review. J. Fish Dis. *28*, 509–529.

Smail, D. (1999). Viral haemorrhagic septicaemia. In Fish Diseases and Disorders, Woo, P., and Bruno, D., eds (CAB International, Wallingford, UK), pp. 123–147.

Snow, M., and Smail, D.A. (1999). Experimental susceptibility of turbot *Scophthalmus maximus* to viral haemorrhagic septicaemia virus isolated from cultivated turbot. Dis. Aquat. Organ. *38*, 163–168.

Snow, M., Bain, N., Black, J., Taupin, V., Cunningham, C.O., King, J.A., Skall, H.F., and Raynard, R.S. (2004). Genetic population structure of marine viral haemorrhagic septicaemia virus (VHSV). Dis. Aquat. Organ. *61*, 11–21.

Stone, D.M., Way, K., and Dixon, P.F. (1997). Nucelotide sequence of the glycoprotein gene of viral haemorrhagic septicaemia (VHS) viruses from different geographical areas: A link between VHS in farmed fish species and viruses isolated from North Sea cod (Gadus morhua L.). J. Gen. Virol. *78*, 1319–1326.

Takano, T., Iwahori, A., Hirono, I., and Aoki, T. (2004). Development of a DNA vaccine against hirame rhabdovirus and analysis of the expression of immune-related genes after vaccination. Fish Shellfish Immunol. *17*, 367–374.

Thompson, T.M., Batts, W.N., Faisal, M., Bowser, P., Casey, J.W., Phillips, K., Garver, K.A., Winton, J., and Kurath, G. (2011). Emergence of Viral hemorrhagic septicemia virus in the North American Great Lakes region is associated with low viral genetic diversity. Dis. Aquat. Organ. *96*, 29–43.

Thoulouze, M.-I., Bouguyon, E., Carpentier, C., and Bremont, M. (2004). Essential Role of the NV Protein of *Novirhabdovirus* for Pathogenicity in Rainbow Trout. J. Virol. *78*, 4098–4107.

Troyer, R.M., and Kurath, G. (2003). Molecular epidemiology of infectious hematopoietic necrosis virus reveals complex virus traffic and evolution within southern Idaho aquaculture. Dis. Aquat. Organ. *55*, 175–185.

Troyer, R.M., LaPatra, S.E., and Kurath, G. (2000). Genetic analyses reveal unusually high diversity of infectious haematopoietic necrosis virus in rainbow trout aquaculture. J. Gen. Virol. *81*, 2823–2832.

Troyer, R.M., Garver, K.A., Ranson, J.C., Wargo, A.R., and Kurath, G. (2008). In vivo virus growth competition assays demonstrate equal fitness of fish rhabdovirus strains that co-circulate in aquaculture. Virus Res. *137*, 179–188.

Verrier, E.R., Langevin, C., Benmansour, A., and Boudinot, P. (2011). Early antiviral response and virus-induced genes in fish. Dev. Comp. Immunol. *35*, 1204–1214.

Walker, P.J., and Winton, J.R. (2010). Emerging viral diseases of fish and shrimp. Vet. Res. *41*, 51–75.

Walker, P.J., Benmansour, A., Calisher, C.H., Dietzgen, R., Fang, R.X., Jackson, A.O., Kurath, G., Leong, J.C., Nadin-Davies, S., Tesh, R.B., *et al.* (2000). Family Rhabdoviridae. In The Seventh Report of the International Committee for Taxonomy of Viruses, van Regenmortel, M.H.V., Fauquet, C.M., Bishop, D.H.L., Carstens, E.B., Estes, M.K., Lemon, S.M., Maniloff, J., Mayo, M.A., McGeoch, D.J., Pringle, C.R., *et al.*, eds (Academic Press, San Diego, CA), pp.623–644.

Wargo, A.R., and Kurath, G. (2011). *In vivo* fitness associated with high virulence in a vertebrate virus is a complex trait regulated by host entry, replication, and shedding. J. Virol. *85*, 3959–3967.

Wargo, A.R., Garver, K.A., and Kurath, G. (2010). Virulence correlates with fitness in vivo for two M group genotypes of Infectious hematopoietic necrosis virus (IHNV). Virology *404*, 51–58.

Wolf, K. (1988a). Infectious hematopoietic necrosis virus. In Fish Viruses and Fish Viral Diseases (Cornell University Press, Ithaca, New York), pp. 83–114.

Wolf, K. (1988b). Viral hemorrhagic septicemia. In Fish Viruses and Fish Viral Diseases (Cornell University Press, Ithaca, New York), pp. 217–249.

World Organization for Animal Health. (2011). Aquatic animal health code. Available at: http://www.oie.int/en/international-standard-setting/aquatic-code/access-online/.

The Sigma Viruses of *Drosophila*

8

Ben Longdon, Lena Wilfert and Francis M. Jiggins

Abstract

The sigma virus of *Drosophila melanogaster* (DMelSV) was discovered accidentally over 70 years ago owing to it causing infected flies to become paralysed and die on exposure to CO_2. Recently, five other species of *Drosophila* and a species of Muscidae fly have also been found to be infected with sigma viruses, and together these viruses form a major clade of rhabdoviruses that is a sister clade to the dimarhabdoviruses. In those cases where the transmission of these viruses has been investigated, sigma viruses are transmitted purely vertically through infected eggs or sperm. In natural populations of insects the spread of these viruses can be very rapid, and has led to flies evolving resistance to sigma viruses. Two resistance genes have been identified in *D. melanogaster*, one of which is involved in autophagy.

Discovery of the sigma virus of *Drosophila melanogaster* (DMelSV)

The sigma virus of *Drosophila melanogaster* (DMelSV) was discovered by chance in 1937, when French researchers found that some flies became paralysed and died when exposed to CO_2 during routine anesthetization, rather than recovering as normal (L'Heritier and Teissier, 1937). They found that the trait was transmitted vertically from parent to offspring, but was independent of the host chromosomes. It could also be transferred by injecting haemolymph from flies that were sensitive to CO_2 into non-sensitive flies, suggesting that the trait was caused by an infectious agent. The trait was later attributed to the presence of a virus-sized particle, which they named sigma (L'Heritier, 1948). The bullet-shaped morphology, antigenic profile and partial genome sequences subsequently identified DMelSV as a rhabdovirus (Berkalof *et al.*, 1965; Teninges, 1968; Calisher *et al.*, 1989; Teninges *et al.*, 1993).

The sigma virus clade

D. melanogaster is not the only species to be infected with a sigma virus. By screening species for CO_2 sensitivity, six additional species of flies were found to be infected with sigma viruses (Longdon *et al.*, 2010, 2011c). These viruses infect five species of *Drosophila* – *D. affinis*, *D. obscura*, *D. tristis*, *D. immigrans* and *D. ananassae* – and one member of the Muscidae, *Muscina stabulans*. These viruses have been tentatively named as DAffSV, DObsSV, DTriSV, DImmSV, DAnaSV and MStaSV, respectively (using the standard *Drosophila* species name abbreviation and SV for sigma virus). As these new viruses have been only recently discovered we know very little about their biology, so the majority of this chapter describes the sigma virus of *D. melanogaster* (DMelSV).

Together the sigma viruses form a clade of viruses that infect *Diptera* (flies), that are closely related to the dimarhabdoviruses (Bourhy *et al.*, 2005); a clade containing the *Ephemerovirus* and *Vesiculovirus* clades of rhabdoviruses (Fig. 8.1 and Fig. 3.2) (Longdon *et al.*, 2010, 2011c). The sigma virus clade contains a greater amount of genetic divergence between its members than seen in four out of the six accepted rhabdovirus genera, suggesting that they are a diverse and major group of

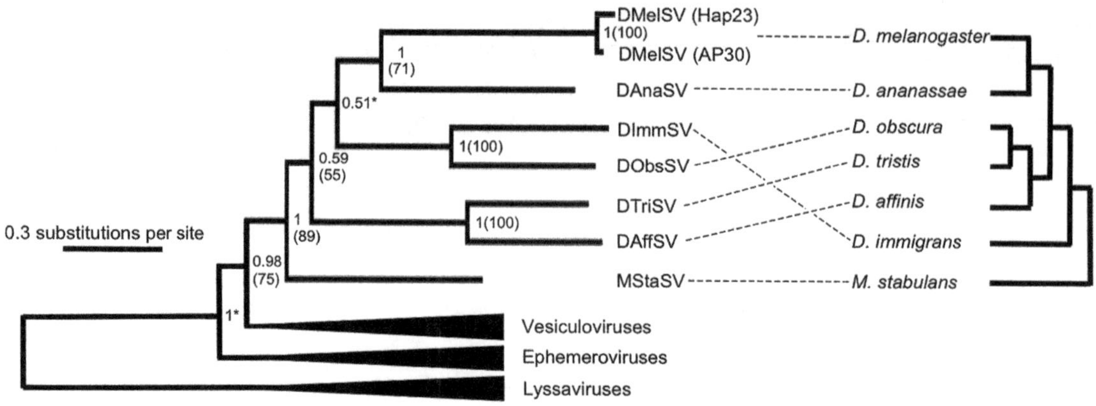

Figure 8.1 Phylogeny of the sigma viruses (left) and their hosts (right), with incongruence between host and virus phylogenies visualized. Node labels represent Bayesian posterior supports with maximum likelihood bootstrap support in brackets. The tree is rooted with the *Lyssavirus* clade. Non-sigma virus clades are collapsed. Nodes marked with an asterisk had bootstrap support of <50%. Figure is reproduced from Longdon *et al.* (2011c).

rhabdoviruses (Longdon *et al.*, 2010). The three viruses (DMelSV, DAffSV and DObsSV) whose transmission mode has been investigated have been shown to be vertically transmitted (Brun and Plus, 1980; Longdon *et al.*, 2011b), suggesting that it is likely that the other sigma viruses are also vertically transmitted.

It is likely that sigma viruses infect a wide range of dipterans if not insects as a whole. Populations of thirteen other species of *Drosophila* (Brun and Plus, 1980) and *Culex quinquefasciatus* mosquitoes (Shroyer and Rosen, 1983) have all been found to contain individuals that are paralysed or killed by CO_2. Furthermore, in *Drosophila athabasca* and *C. quinquefasciatus*, the CO_2 sensitivity is inherited independently of the host chromosomes in a biparental manner similar to that of DMelSV (Williamson, 1959, 1961; Shroyer and Rosen, 1983). As CO_2 sensitivity is a common trait of many insect rhabdoviruses (Bussereau and Contamine, 1980; Rosen, 1980; Shroyer and Rosen, 1983; Sylvester and Richardson, 1992), it seems likely rhabdoviruses, and perhaps sigma viruses, infect a range of different dipterans. Additionally, rhabdovirus sequences have inserted into the genomes of various arthropod species (Katzourakis and Gifford, 2010) and rhabdovirus-like particles have been found in firebug testes (Afzelius *et al.*, 1989). Recently a new rhabdovirus has been found in *Culex tritaeniorhynchus*, which has been suggested to cluster within the sigma virus clade (Kuwata *et al.*, 2011). However, this may be an artefact of the method used to infer the phylogeny (B. Longdon, unpublished data) and the virus actually represents a distinct lineage that sits as an out-group to the sigma and dimarhabdovirus clades. Nevertheless, this and the previous results highlight that there are likely many more sigma-like viruses, and other rhabdoviruses, to be discovered in invertebrates.

The discovery of a sigma virus outside the genus *Drosophila* (in *M. stabulans*) is of particular interest as the closely related dimarhabdoviruses are vector-borne pathogens of vertebrates, some of which are vectored by other dipterans (Dacheux *et al.*, 2010). It is currently impossible to distinguish whether the ancestor of the sigma viruses and dimarhabdoviruses evolved from vertebrate-infecting viruses that switched to being purely entomopathogenic in the sigma virus clade, or whether the ancestors of the dimarhabdoviruses were viruses of arthropods that came to infect vertebrates.

Genetics

The genome organization of DMelSV is similar to that of other rhabdoviruses, with 3′-N–P–M–G–L-5′ genes, but with an additional gene of unknown function (the X gene) located between the P and M genes (Teninges *et al.*, 1993; Contamine and Gaumer, 2008; Longdon *et al.*, 2010). The X gene (sometimes referred to as gene 3) shares structural similarities to viral RNA polymerases. It is predicted to have an N-terminal signal peptide and has three predicted N-glycosylation sites suggesting it may be a secreted glycoprotein (Landesdevauchelle *et al.*, 1995; Longdon *et al.*, 2010; Walker *et al.*, 2011). Some other rhabdoviruses also contain one or more accessory genes (Walker *et al.*, 2011); the plant-arthropod infecting rhabdoviruses also have an additional gene located between the P and the M genes (Ammar *et al.*, 2009) and some dimarhabdoviruses also contain additional genes (Gubala *et al.*, 2008, 2010; Zhu *et al.*, 2011). In total the DMelSV genome is 12,625 nucleotides (nt) long. Following each gene are motifs with seven U residues (3′-GUACUUUUUUU-5′), which are thought to be transcription termination sequences triggering polyadenylation of the mRNA (Teninges *et al.*, 1993). In DMelSV the putative transcription initiation site has been identified as 3′-GUUGUNG-5′ for all genes except the N gene, where it is 3′UUGUUG-5′ (Teninges *et al.*, 1993). The only other sigma virus with a near complete genome sequence is DObsSV which has the same genome organization as DMelSV, and is 12,676 nt long (excluding the 5′ trailer) (Longdon *et al.*, 2010). The full length L gene has been sequenced for DAffSV, and other sigma viruses have partial L gene sequences available (Longdon *et al.*, 2010; Longdon *et al.*, 2011c). The viral substitution rate has been estimated (Carpenter *et al.*, 2007) by sequencing a DMelSV strain that had been split and maintained in two different *Drosophila* strains for 10–20 years. It was estimated that there were 4.6×10^{-5} substitutions/site/year, which is equivalent to approximately 0.6 substitutions per genome per year. The laboratory-derived mutation rate was not found to be significantly different from that of isolates collected in the field (Carpenter *et al.*, 2007).

Recombination is very rare in negative-sense RNA viruses *per se* (Chare *et al.*, 2003) but is perhaps even less likely in sigma viruses where the chance of multiple infections is low as this would require both parent's viral type to infect the offspring (see below). Despite early suggestions that DMelSV might recombine (Ohanessian-Guillemain, 1959), statistical analyses of sigma virus RNA sequences (DMelSV and DObsSV) from the field have found that recombination is very rare or absent (Carpenter *et al.*, 2007; Longdon *et al.*, 2011b; L. Wilfert, unpublished data).

Mode of natural transmission

DMelSV is transmitted vertically through both sperm and eggs with horizontal transmission being rare or absent (L'Heritier 1957; Brun and Plus, 1980; Fleuriet, 1988). This is a rather unusual mode of transmission, as although numerous vertically transmitted pathogens are known from insects, most of these are only transmitted through eggs, or they are also transmitted horizontally (Buchner, 1965; Engelstadter and Hurst, 2009). How males transmit the virus is unclear, but viral particles have been observed in sperm cells (Teninges, 1968). This biparental mode of transmission is essential for the virus to spread through the host population, as a purely maternally transmitted parasite will go extinct, even if transmitted perfectly, if it is costly to its host. It has been shown that by transmission through both males and females, parasites can persist and spread through host populations even if infected hosts suffer a fitness cost (L'Heritier, 1970; Fine, 1975).

The rate of DMelSV transmission differs between the sexes (Fig. 8.2) (L'Heritier, 1948, 1957, 1970; Fleuriet, 1988). Firstly, females transmit the virus at a higher rate than males. Secondly, the transmission rate of a fly is reduced when it is infected by its father rather than its mother. Females inheriting the virus from their mother are termed 'stabilized', as their rate of transmission is close to 100%. However, if a female is infected by her father (termed 'unstabilized'), her average transmission drops to a much lower rate of about 80% (Fleuriet, 1988). Similarly, while a 'stabilized' male infected by his mother can transmit the virus to his offspring (albeit at a lower rate than females), a male that is infected by his father does not transmit the virus at all. Therefore, it seems the virus cannot be transmitted through males for two successive generations.

The lower rates of transmission in flies infected

from their father rather than their mother seems to have a physiological cause, as sperm transfer a lower viral titre to the developing embryo (Plus, 1955; L'Heritier, 1957, 1970). This is thought to lead to a failure to infect the early-stage germ line, and the gametes may never become infected later in development, even if the viral titres recover to higher levels in other tissues later in life (Plus, 1955; L'Heritier, 1957, 1970). Further evidence for the difference in viral titres transmitted in sperm and eggs comes from crosses when both parents are infected, which results in the maternal viral type being inherited in the vast majority of cases (Ohanessian-Guillemain, 1959). This is consistent with the observation that few wild caught flies show evidence of multiple infections (Longdon *et al.*, 2011b).

Other factors can also affect the rate of transmission. Females with high ovarian viral titres have high rates of transmission (Bregliano, 1970; Brun and Plus, 1980). After a female has mated with an infected male, the proportion of infected offspring declines over time, which could suggest infected sperm have reduced longevity or the virus is lost from sperm over time (Plus, 1955; L'Heritier, 1957). Transmission of most DMelSV viral strains can also be prevented by placing infected flies at 30°C, and transmission will resume if the parents are returned to lower temperatures (20°C).

Interestingly, viral titres can be about 3–5 times greater in flies infected by their fathers rather than mothers (L'Heritier, 1957, 1970) despite these flies transmitting the virus at a lower rate (Plus, 1955; L'Heritier, 1957). A similar effect is seen after artificial inoculation of the virus, which results in high viral titres but low transmission rates (see Fig. 1 in L'Heritier, 1970). It has been suggested that this may be a viral response to failure to invade the zygote's pole cells (L'Heritier, 1970), causing over-replication in the host in an 'attempt' to get into the germline (L'Heritier, 1970). As evidence for this, in stabilized flies about one half of the total viral load is found in the ovarian cysts (clusters of germ line cells) of egg laying females (L'Heritier, 1970), whereas in non-stabilized females only a few cysts harbour similar viral titres. This means that the majority of the viral load in non-stabilized flies is in somatic tissues, which may represent a failure to invade and replicate to high levels in germ line cells.

Similar patterns of transmission have been observed for DAffSV and DObsSV. While in *D. affinis* males have a lower rate of transmission than females (45% compared with 98%), this is not the case for *D. obscura*, where males have comparable transmission rates to females (92% and 88% respectively) (Longdon *et al.*, 2011b). However, in both cases males infected by their fathers do not transmit the virus. Embryos infected by their fathers (i.e. when the virus is transmitted through sperm rather than eggs) have lower titres of both DAffSV and DObsSV, as has been observed for DMelSV (Brun and Plus, 1980). This likely explains why transmission through sperm is less efficient.

Viral replication following experimental injection

Sigma viruses can be transmitted between individuals in the laboratory by injecting a crude virus suspension (infected adult flies homogenized in Ringer's solution (Sullivan *et al.*, 2000) and briefly centrifuged to remove debris). Following injection, flies that are successfully infected transmit the virus as if they had been infected by their father, with only females transmitting the virus to offspring (L'Heritier, 1957, 1970).

The onset of CO_2 sensitivity following injection depends on the initial dose of the virus, and can be around 5–20 days at 20°C (Brun and Plus, 1980). Following injection with the virus the number of infectious units in a fly increases until about 10–15 days post injection where it plateaus (Brun and Plus, 1980). In females, from around day 10 onwards transmission to offspring can occur in some cases, increasing until around day 20 where it begins to decline (Brun and Plus, 1980). Classically, 'infectious units' were ascertained from taking the homogenate of an infected fly, creating a dilution series and injecting it into non-infected individuals (Plus, 1955; L'Heritier, 1957).

Cross-species infections

Although DMelSV is restricted to *D. melanogaster* by virtue of its vertical transmission, it can replicate in a range of different dipterans, but not different orders of insects (Jousset, 1969). DMelSV has also been crossed into hybrid offspring with the closely related *D. simulans* (Kalmus, 1943). DMelSV injected into *D. funebris*, *D. willistoni*, *D. prosaltans* and *D. gibberosa* causes CO_2 sensitivity. In the two species examined stabilized transmission of injected DMelSV was found to occur in the closely related *D. simulans* but not the more distantly related *D. funebris* (L'Heritier, 1957).

Despite the fact that they appear to be mostly vertically transmitted, there is significant incongruence between the host and sigma virus phylogenies, which indicates that these viruses have switched between different host species during their evolutionary history (Fig. 8.1) (Longdon *et al.*, 2011c). Although there is clear evidence of host switches, there is tentative evidence that the phylogenies of the host and virus are more similar than expected by chance (Longdon *et al.*, 2011c), although more data are required to confirm this finding. This may be due to co-speciation of the virus with the host or may be explained by host shifts more frequently occurring between closely related species. It has also been suggested that parasitic mites could play a role in vectoring the virus between host species (Longdon *et al.*, 2010).

It has recently been found that most of the variation in the ability of three sigma viruses (DMelSV, DAffSV and DObsSV) to persist and replicate in a novel host species is explained by the host phylogeny (Longdon *et al.*, 2011a). From cross- inoculating these viruses into 51 different species of Drosophilidae, it was found that species more closely related to the natural host species had greater viral titres than more distantly related species. Additionally, there is a strong effect of the host phylogeny that is independent of the effect of distance from the natural host, with groups of related species having similar viral titres. This phylogenetic effect is mostly caused by variation in susceptibility to all three viruses (there is a strong phylogenetic correlation in the titres of the three viruses). Such patterns may result from the common ancestors of different host clades having acquired or lost immune or cellular components that affect general susceptibility to all sigma viruses (Jiggins and Kim, 2005; Sackton *et al.*, 2007; Havard *et al.*, 2009; Obbard *et al.*, 2009).

CO_2 paralysis symptom

Sigma virus infection causes flies to become paralysed and die after exposure to CO_2, whilst uninfected flies recover after a short period (L'Heritier and Teissier, 1937). In DMelSV the paralysis seems to be due to damage caused to nervous tissue, in particular the thoracic ganglia, with flies being injected with the virus at a site close to the thoracic ganglia developing CO_2 sensitivity in about 2 days compared with the 5- to 10-day delay observed if injected elsewhere in the thorax or abdomen (L'Heritier, 1948; Busserea, 1970a,b). Other rhabdoviruses are known to replicate in nervous tissue, with the glycoprotein (G) of rabies virus preferentially attaching to several neural cell receptors (Lentz *et al.*, 1982; Gastka *et al.*, 1996; Thoulouze *et al.*, 1998; Tuffereau *et al.*, 1998, 2001; Lyles and Rupprecht, 2007), suggesting this may be a general rhabdovirus trait. Furthermore, other rhabdoviruses injected into *Drosophila* (and other insects) also cause CO_2 paralysis (Bussereau and Contamine, 1980; Rosen, 1980; Shroyer and Rosen, 1983) and it has been noted that rhabdovirus infected aphids have reduced longevity after CO_2 exposure (Sylvester and Richardson, 1992). Recent studies have confirmed DMelSV is found at high titres in nervous tissues using immunofluorescent microscopy (Tsai *et al.*, 2008; Ammar *et al.*, 2009).

The CO_2 paralysis in sigma virus infected *D. melanogaster* is brought on by relatively short exposures to CO_2 (< 1 min) and is specific to CO_2, with other gases and injected chemicals having no effect (L'Heritier, 1948; Brun and Plus, 1980). The paralysis is also sensitive to changes in CO_2 concentration and temperature, with lower temperatures requiring lower CO_2 concentrations to cause paralysis, perhaps due to greater gas solubility at low temperatures. At close to 0°C low concentrations of CO_2 are sufficient (< 20%), increasing to 50% at 10°C, with no CO_2 induced paralysis observed above 23°C (Brun and Plus, 1980). The standard assay for infection is exposure

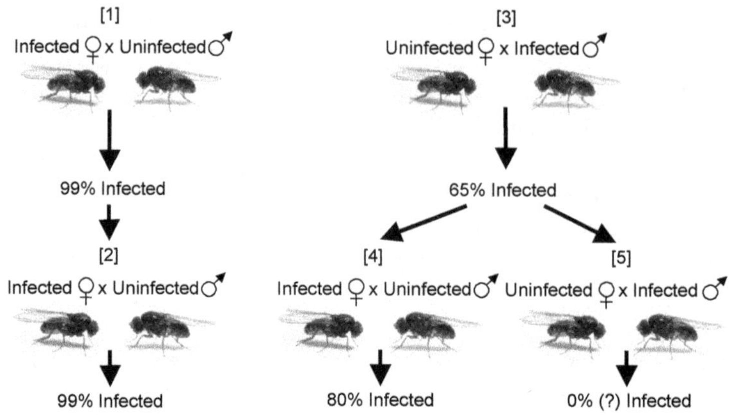

Figure 8.2 Patterns of sigma virus transmission. The rates of transmission are based on DMelSV (Fleuriet, 1988), but patterns are similar for DAffSV and DObsSV (Longdon *et al.*, 2011b) (although see main text for minor differences with DObSV). Infected females crossed to uninfected males (1) transmit the virus to close to 100% of their offspring (range of 98–100%), and are referred to as 'stabilized'. If daughters from this cross are again crossed to uninfected males (2) they too will transmit to close to 100% of their offspring. If infected males are crossed to uninfected females (3) they will transmit the virus to approximately 65% of their offspring on average (with a range of 0–100%). If infected daughters from this cross are mated to uninfected males (4) they will transmit the virus at a lower rate than daughters infected by their mothers (about 80% on average, with a range of 0–100%), and are referred to as 'unstabilized'. If infected sons from (3) are crossed to uninfected females (5) they do not transmit the virus at all, suggesting that the virus transmission through two successive male generations is rare or absent. Fly photo credit: Darren Obbard and Ben Longdon.

at 12°C for 15 minutes, with flies assessed for paralysis about 30 minutes later.

Although we know the thoracic ganglion is involved in CO_2-induced paralysis (Brun and Plus, 1980) the exact mechanism is unknown. Interestingly, CO_2 exposure will lower the pH of flies' haemolymph, and both vesicular stomatitis virus (VSV) and rabies show enhanced binding at reduced pH. Furthermore, rhabdovirus G proteins induce membrane fusion in a pH-dependent manner, suggesting a potential mechanism (Lyles and Rupprecht, 2007). However, crude attempts to manipulate fly pH by injecting acidic solutions (hydrogen cyanide, formic acid and hydrochloric acid) were found to have no effect on DMelSV infection (L'Heritier, 1948; Brun and Plus, 1980). Additionally, rhabdoviruses bind to acetylcholine receptors on nervous tissue that could also be involved in paralysis but assays of acetylcholinesterase activity, which degrades acetylcholine at synapses, found no difference between infected and uninfected flies (Brun and Plus, 1980). Furthermore, a dose of CO_2 below the threshold required to cause paralysis can protect infected *D. melanogaster* from a subsequent dose that would normally be lethal (L'Heritier, 1948). While injection of infected fly haemolymph into uninfected flies will cause CO_2 sensitivity about 10–15 days post inoculation, there is no immediate effect. This suggests that the effect is not due to mere physiological changes in the haemolymph but is induced by viral replication (L'Heritier, 1970).

It is unclear if the CO_2 sensitivity has any relevance in nature, although rotting fruit or plants, which are often *Drosophila* oviposition sites (Basden, 1954), will produce CO_2. This could have potentially sublethal effects on flies in enclosed spaces at low temperatures, under which conditions a lower concentration of CO_2 is required for paralysis (Brun and Plus, 1980). The effect of CO_2 is not limited to adults, but also causes death in larvae, although early stage embryos and pupae do not seem to be affected (L'Heritier, 1948). CO_2 paralysis has been clearly demonstrated in *D. melanogaster* and is tightly linked to sigma virus infection (Brun and Plus, 1980; Wilfert and Jiggins, 2010b), but can be less clear-cut in other species (B. Longdon personal observation), and some infected *D. obscura* do not display CO_2 paralysis (Longdon *et al.*, 2011b).

Costs of sigma virus infection to the host

Vertically transmitted pathogens rely on their hosts surviving and reproducing to be transmitted, and therefore tend to cause little harm to their hosts (i.e. they have a low virulence) (Fine, 1975). This appears to be the case with DMelSV, as infected fly stocks are generally healthy in the laboratory, but nonetheless experimentation has shown that the virus does cause some harm to the *Drosophila* host. DMelSV rapidly spreads through laboratory populations of flies kept at a low density, but if these populations were placed at high densities – so the flies have to compete for resources – the virus showed a decline in frequency (Yampolsky *et al.*, 1999). This can be explained by the virus reducing the fitness of infected flies, so they are out-competed and therefore unable to pass the infection on to future generations. Given that the virus is usually only found at low prevalence in the field (0–20%) (Fleuriet, 1988; Carpenter *et al.*, 2007; Wilfert and Jiggins, 2010a), it is likely that DMelSV is causing a fitness cost to wild flies as well. Based on the rate of expansion in lab populations, the reduction in fitness in the field has been estimated at 20–30% (Yampolsky *et al.*, 1999).

Several studies have examined the physiological basis of this fitness cost. Infected flies have reduced egg viability in the laboratory, (Seecof, 1964; Fleuriet, 1981a) and may take longer to develop from egg to adult by about 5–10 hours (Seecof, 1964). Seecof (1964) found a 20% reduction in egg viability in infected flies, but used only a single fly line. Fleuriet (1981a) used multiple fly lines and her data suggest an overall reduction in egg viability of about 10% in infected lines (averaging each line over all crosses, exact Wilcoxon rank sum test: $P = 0.049$, following re-analysis by B. Longdon). Similarly, a reduction in female fertility has been reported for another viral isolate, yet this was not due to a reduction in egg viability but because the ovarian cysts of infected individuals developed slower than uninfected ones (Jupin *et al.*, 1968). Additionally, a study under semi-natural conditions reported that virus-infected flies declined in number if over-wintered as adults (Fleuriet, 1981b). However, the patterns observed were not clear, with only five out of the six populations tested showing this trend, but even then the pattern was subjective (Fleuriet, 1981b). Studies of CO_2 sensitivity in wild populations of *D. melanogaster* show some evidence of declines in infection after winter, but the data are inadequate to be conclusive (Herforth and Westphal, 1966; Felix *et al.*, 1977). Additionally, it has been reported that DMelSV reduces flies' resistance to the fungal pathogen *Beauvaria bassiana* in a laboratory experiment (about 10% increase in mortality in DMelSV-infected flies exposed to fungi), although no suppression of the Toll pathway (which is known to be involved in anti-fungal resistance) was observed (Carpenter, 2008).

Host resistance

Given the fitness cost of DMelSV to its host, it is not surprising that *D. melanogaster* has evolved resistance mechanisms towards the virus. Most of the work on resistance has focussed on naturally occurring genetic polymorphisms that alter the susceptibility of wild flies to DMelSV infection (Gay, 1978; Bangham *et al.*, 2008). More recently, attention has also turned to the question whether there is an induced immune response to sigma virus infection.

Ref(2)P gene

The best-studied polymorphism controlling resistance to DMelSV is in the gene *ref(2)P* (Guillemain, 1953; Contamine *et al.*, 1989). Resistance is due to a mutation in the PB1 domain of the encoded protein that changes the amino acids Gln-Asn to a single Gly (Dru *et al.*, 1993; Wayne *et al.*, 1996; Bangham *et al.*, 2007). When flies are injected with certain strains of DMelSV, they are less likely to be successfully infected if they carry the resistance allele (Fig. 8.3) (Ohanessian-Guillemain, 1963; Carre-Mouka *et al.*, 2007). The polymorphism also has an effect on the rate at which flies transmit the virus to their offspring. The effect is greatest when the virus is transmitted from an infected female through eggs – only 8% of offspring homozygous for the resistance allele became infected compared to 99% of susceptible offspring (Bangham *et al.*, 2008). The resistance allele is associated with a 59% decline in male transmission in homozygous offspring (Bangham *et al.*, 2008).

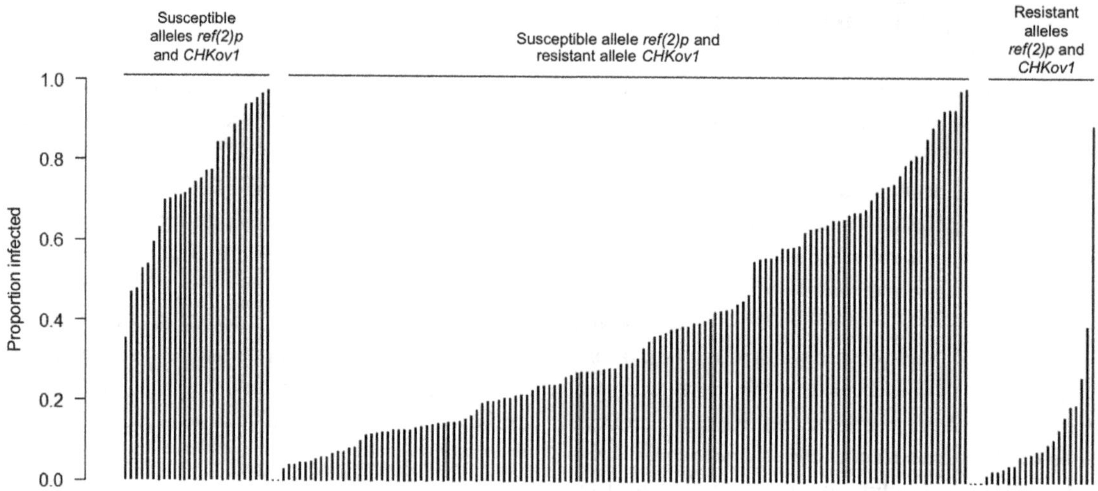

Figure 8.3 The effect of two polymorphic resistance genes on infection rates in 176 inbred lines of *Drosophila melanogaster* (Magwire *et al.*, 2011). The flies were injected with DMelSV and classified as infected or uninfected using the CO_2 sensitivity assay. Each bar represents a different inbred fly line that originated from a population in North Carolina. Details of the experiment are described in Magwire *et al.* (2011).

This polymorphism is an important factor in affecting the rate of transmission of the virus in nature. The frequency of the resistant allele can be high, ranging from 24–36% in Europe (Fleuriet and Sperlich, 1992; Wilfert and Jiggins, 2010a). Furthermore, infected female flies in the field transmit the virus to 28% fewer offspring if they carry the resistant allele (Wilfert and Jiggins, 2010a). Being able to replicate laboratory results in the field is a crucial finding if results are to be extrapolated to the real world.

The mechanism by which *ref(2)P* makes flies resistant to sigma virus is unknown. However, it has recently been discovered that this protein plays a key role in autophagy, which is an important defence mechanism against intracellular pathogens in animals (Pankiv *et al.*, 2007; Nezis *et al.*, 2008; Shelly *et al.*, 2009). Autophagy is a process by which components of the cytoplasm are encapsulated inside double membrane vesicles known as autophagosomes, which fuse with lysosomes and are then degraded (Klionsky and Ohsumi, 1999). Ref(2)P, which is called p62 in most other organisms, is an adaptor protein which selectively targets polyubiquitinated substrates for degradation by autophagy (Korolchuk *et al.*, 2009; Lamark *et al.*, 2009). In *Drosophila*, ref(2)P localizes with poly-ubiquitinated protein aggregates and accumulates in cells when autophagy is inhibited (Nezis *et al.*, 2008). The role of autophagy in innate immunity has mainly been investigated as a defence against intracellular bacteria, and here p62 interacts with ubiquitinated proteins on the bacterial surface and is essential for targeting bacteria to the autophagic pathway (Zheng *et al.*, 2009). It is now becoming clear that autophagy can also defend cells against viruses (Levine and Deretic, 2007; Shelly *et al.*, 2009). In *Drosophila*, VSV activates autophagy, protecting flies against the virus (Shelly *et al.*, 2009), while in mammalian cells other viruses are able to evade this form of immune defence by suppressing autophagy (Orvedahl *et al.*, 2007; Dreux and Chisari, 2010).

The observation that autophagy can protect flies against VSV suggests a model by which *ref(2)P* protects *Drosophila* against sigma virus infection. Because VSV and the sigma viruses are moderately close relatives (Longdon *et al.*, 2010), it is likely that autophagy also plays an antiviral role against the sigma virus. Furthermore, null mutants of *ref(2)P* are susceptible to sigma virus infection, so the resistant allele of this gene appears to have an antiviral effect rather the susceptible allele having a proviral effect (Contamine *et al.*, 1989). Co-immunoprecipitation experiments have also shown that the viral P protein interacts with ref(2)P (Wyers *et al.*, 1993). Therefore, the resistant ref(2)P isoform may protect flies against infection by targeting the sigma virus for degradation by autophagy, possibly by acting as a pattern recognition receptor that binds to viral proteins.

Ref(2)P/p62 is also involved in processes other than autophagy that could explain its role in virus resistance. p62 can target ubiquitinated cargoes for degradation by the proteasome, and this could potentially affect viral replication (Nezis *et al.*, 2008). It is also important in rel/NF-κB immune responses (Avila *et al.*, 2002), but previous work has shown that the Toll signalling pathway does not affect sigma virus titres, so it is unlikely that this is the cause of resistance (Carpenter *et al.*, 2009a).

CHKov genes

A second major-effect polymorphism (previously referred to as *ref(3)D*; Gay, 1978) that makes flies resistant to the sigma virus was recently mapped to a region containing two paralogous genes called *CHKov1* and *CHKov2* (Magwire *et al.*, 2011). It was found that three alleles of these genes occur, which confer different levels of resistance. The ancestral state is for *CHKov1* and *CHKov2* to both be intact genes, and these flies are susceptible to DMelSV. The second allele has a transposable element insertion in the protein coding sequence of *CHKov1*, and this is associated with increased resistance to infection (Fig. 8.3) (Magwire *et al.*, 2011). This insertion results in a truncated messenger RNA that encodes a far shorter protein than the susceptible allele (Aminetzach *et al.*, 2005). The third allele consists of two duplications, resulting in three copies of both the truncated allele of *CHKov1* and *CHKov2* (one of which is also truncated) (Magwire *et al.*, 2011). These flies have even greater resistance than those carrying the transposable element insertion alone. Therefore, two major mutational events that have affected the structure of these genes have led to sequential increases in resistance.

The allele carrying the transposable element (conferring a moderate level of resistance) is the commonest in North American populations, occurring at a frequency of 0.82 in North Carolina, while the most resistant allele carrying the duplications is rare and has only ever been found in a single stock collected in France (Magwire *et al.*, 2011). Analysis of the patterns of genetic variation around the *CHKov* genes has shown that natural selection has caused the resistance allele to rapidly increase in frequency (Aminetzach *et al.*, 2005), providing further evidence of the benefits it provides flies in the wild.

It is well known that pathogen resistance genes often have pleiotropic effects on other traits (Kraaijeveld and Godfray, 1997; McKean *et al.*, 2008). These pleiotropic effects tend to be harmful and it is commonly thought that resistance to pathogens is a costly trait to evolve (Kraaijeveld *et al.*, 2002). However, previous research has found that the transposable element insertion in *CHKov1* increases resistance to organophosphate insecticides (Aminetzach *et al.*, 2005). Therefore, contrary to received wisdom, this resistance allele would appear to be beneficial to flies in other ways beyond making them resistant to DMelSV. Interestingly, population genetic analyses suggest that the transposable element insertion occurred well before the use of insecticides (Aminetzach *et al.*, 2005), so it appears likely that originally this allele played a role in antiviral defence, and then it fortuitously 'preadapted' flies to the use of insecticides (Magwire *et al.*, 2011).

The reasons why these genes affect virus and pesticide resistance are unknown. Neither *CHKov1* nor *CHKov2* are upregulated in infected flies (Carpenter *et al.*, 2009a), so they do not appear to be part of an induced immune response. It has been suggested that CHKov1, which contains a choline kinase domain, might make flies resistant to organophosphates by affecting choline metabolism in general or the target of organophosphate insecticides, acetylcholine esterase (Aminetzach *et al.*, 2005). If this is the case, it is possible that it could be linked to the mechanism of virus resistance as other rhabdoviruses use acetylcholine receptors to enter cells (Lentz *et al.*, 1982; Gastka *et al.*, 1996); however, this remains highly speculative.

Other resistance genes

Although the *CHKov* genes and *ref(2)P* are the only two host resistance polymorphisms that have been identified in *D. melanogaster* at the molecular level, several more genes remain to be characterized. These include resistance genes that have been mapped that delay the onset of CO_2 sensitivity and reduce the viral titres after flies are injected with the virus (Gay, 1978; Brun and Plus, 1980). However, it seems that resistance may sometimes involve blocking the vertical transmission of the virus without impeding viral replication, as there is little genetic correlation between the rate males transmit the virus and the rate they develop CO_2 sensitivity after they have been injected (in *ref(2)P* susceptible flies) (Bangham *et al.*, 2008). Two unidentified genes, which map to the second and third chromosomes of *D. melanogaster*, have been found to reduce the rate of transmission through sperm (Gay, 1978; Bangham *et al.*, 2008).

Immune activation

Drosophila mounts potent immune responses when infected by fungi, bacteria and viruses (Lemaitre and Hoffmann, 2007; Sabin *et al.*, 2010). RNA silencing, autophagy and inflammatory signalling pathways may contribute to the antiviral immune response against *Drosophila* C virus and several entomopathogenic viruses such as Flock House virus (Sabin *et al.*, 2010). Only two studies have examined the innate immune response to DMelSV, but with conflicting results (Tsai *et al.*, 2008; Carpenter *et al.*, 2009a). Tsai *et al.* (2008) used quantitative RT-PCR to show that the sigma virus upregulated several antimicrobial peptides and two peptidoglycan recognition proteins (PGRPs). However, Carpenter *et al.* (2009a) found that the virus had very little effect on the expression of any immune system genes. The reasons why the results of the two studies differ are unclear. It is possible that only certain sigma virus genotypes activate an immune response in certain fly genotypes or conditions. For example the fly lines may have differed in their sigma virus resistance genes. It can also not entirely be ruled out that some other factor such as secondary infections with bacteria were inducing an immune response in the study of Tsai *et al.* (2008).

There is evidence for an innate immune response against VSV (which is in the sister clade to the sigma viruses; Longdon *et al.*, 2010) when injected into *D. melanogaster* (Shelly *et al.*, 2009; Mueller *et al.*, 2010). Firstly it has recently been found that VSV induces autophagy, as discussed above (Shelly *et al.*, 2009). Secondly, the RNA interference (RNAi) pathway is involved in resistance to VSV in *D. melanogaster*, with mutant flies lacking RNAi components showing increased mortality following VSV infection (Mueller *et al.*, 2010). Antiviral RNAi relies on double-stranded (ds) RNA to generate small interfering RNAs (siRNAs), but negative sense RNA viruses produce very little dsRNA compared to other RNA viruses (Obbard *et al.*, 2009). Nonetheless, while no detectable amounts of ds RNA were found in VSV-infected cells, viral siRNAs were detected (Mueller *et al.*, 2010). These siRNAs equally matched both positive and negative-sense genomic copies of the virus, suggesting that they have come from dsRNA formed between the viral genome and its replication intermediate, rather than dsRNA as a result of secondary structure in the viral genome. However, VSV is normally orally transmitted in its natural dipteran vectors (although see Tesh *et al.*, 1972) and so needs to pass through the gut wall to cause systemic infection, whereas sigma viruses are exclusively vertically transmitted in nature (Brun and Plus, 1980; Longdon *et al.*, 2011b). The immune response to horizontally and vertically transmitted viruses may differ substantially, and it remains to be seen if such induced responses occur in sigma virus-infected flies.

RNA editing

Adenosine deaminases that act on RNA (ADARs) are enzymes that introduce mutations in double stranded RNA, changing adenosine to inosine, which is converted to guanosine during reverse transcription by the virus. They are best known for 'editing' mRNAs in the central nervous system to produce different protein isoforms. However, ADARs can also cause large clusters of mutations (hypermutation) in RNA viruses in vertebrates (Carpenter *et al.*, 2009b). Recently, strains of DMelSV have been isolated with patterns of hypermutation characteristic of ADAR. Although this hypermutation could simply be an accident with little effect on viral replication, it is also possible that ADAR may be acting as an antiviral defence, by disrupting viral gene function or tagging viruses for degradation (Carpenter *et al.*, 2009b).

Coevolution and population dynamics

Drosophila melanogaster

Hosts and pathogens are engaged in a battle of adaptation and counter-adaptation, in which the spread of alleles that increase host resistance is followed by the spread of pathogen genotypes that can overcome these host defences (Dawkins and Krebs, 1979). The *ref(2)P* gene in *D. melanogaster* and the sigma virus is one of the best understood cases of this process.

The resistant allele of *ref(2)P* has arisen from an ancestral susceptible allele, and analyses of DNA sequence variation show that natural selection has caused it to recently increase in frequency in the population (Wayne *et al.*, 1996; Bangham *et al.*, 2007). Based on the degree of linkage disequilibrium with flanking markers, the resistant allele appears to have spread through the host population in the last few thousand years (Bangham *et al.*, 2007). The resistant allele is currently at frequencies of about 24% in natural populations (Wilfert and Jiggins, 2010a), so about 5% of flies will be homozygous and fully resistant (heterozygous flies only transmit the virus at rates slightly less than homozygous susceptible flies; Bangham *et al.*, 2007).

The spread of the resistant *ref(2)P* allele has led to reciprocal evolution in the viral population. In natural populations there are two viral types, one of which can infect the *ref(2)P* resistant flies ('insensitive' viruses) and one of which cannot infect resistant flies ('sensitive' viruses). These insensitive viral types have arisen relatively recently, as European isolates of DMelSV have been estimated to have shared a common ancestor about 200 years ago (Carpenter *et al.*, 2007). This suggests that the mutation that has allowed the virus to overcome host resistance is very recent, as current DMelSV isolates include both the insensitive and a small proportion of sensitive strains (about 16–27%; L. Wilfert, personal communication). The insensitive viral type then rapidly swept through natural host populations in France and Germany over a period of about ten years in the 1980s, largely replacing the original sensitive viral genotypes (Fig. 8.4) (Fleuriet and Sperlich, 1992; Fleuriet and Periquet, 1993). Therefore, this interaction appears to be a classic evolutionary arms race (Dawkins and Krebs, 1979), in which a newly evolved form of host defence rapidly failed owing to evolution in the viral population.

Fleuriet (1982) recreated this process in experimental populations of *D. melanogaster*. Initially the population contained only the susceptible *ref(2)P* allele and was infected with the

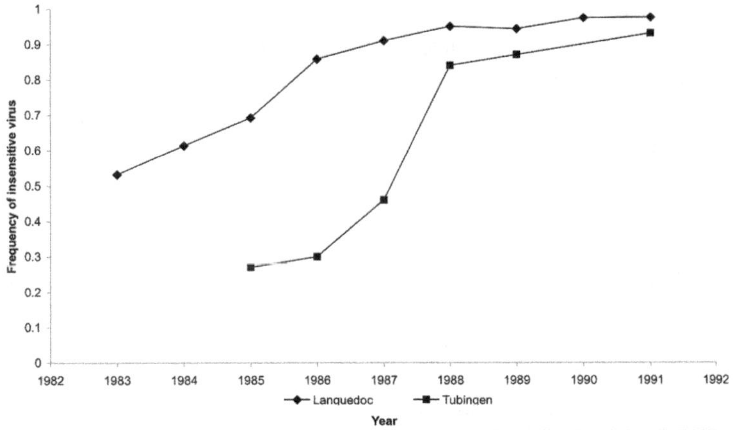

Figure 8.4 Relative frequency of the insensitive viral type (which can infect *ref(2)P* resistant flies) over time in two natural European populations of DMelSV. The figure was drawn using data from Fleuriet and Sperlich (1992) and Fleuriet and Periquet (1993).

sensitive strain of DMelSV. When resistant *ref(2)P* alleles were introduced into this population, their frequency increased to levels similar to that observed in natural populations (about 30%) and there was a decrease in the prevalence of the virus (Fleuriet, 1982). The insensitive viral strain was introduced into the population and replicating the observations in the field, replaced the original sensitive viral genotype, with the prevalence of this virus increasing (Fleuriet, 1982). This also resulted in the resistant *ref(2)P* allele decreasing in frequency, which could be due to a fitness cost of carrying this resistance allele, conveying a selective advantage to the susceptible allele in the absence of a sensitive virus. However, this result must be treated with caution. These experiments were founded with only a few flies and *ref(2)P* is in a region of the genome with low recombination, so there may have been linkage-disequilibrium between *ref(2)P* and deleterious mutations in other genes.

It has also been reported that there is a fitness cost for the virus to be infective and overcome host resistance (Fleuriet, 1999). Fleuriet (1999) attempted to examine this by measuring the frequency of competing insensitive and sensitive viral types in populations of flies lacking the *ref(2)P* resistance allele. The sensitive viral types that were affected by *ref(2)P* resistance had a slight advantage over the insensitive viral types that had evolved to evade this resistance, with about a 20% reduction in male transmission of the insensitive strain (Fleuriet, 1999). However, as this study used only one or two isolates of each of the viral types, the effect may also be due to other deleterious mutations in the viral genome, which has shown no sign of recombination (Carpenter *et al.*, 2007; L. Wilfert, unpublished data).

Drosophila obscura

In populations of *D. obscura* in the United Kingdom (UK), DObsSV shows evidence of even more rapid dynamics than have been observed for DMelSV (Longdon *et al.*, 2011b). In samples from across the UK, the DObsSV viral genomes have extremely low genetic diversity and most single nucleotide polymorphisms (SNPs) occur at a low frequency, suggesting that this virus has very recently spread though the population (Longdon *et al.*, 2011b). Using an approach based on coalescent theory to reconstruct the past population dynamics of the virus, it was estimated that in approximately the past 11 years a single viral clone has spread across the UK to infect about 40% of flies, during which time the viral population size doubled approximately every 9 months (Longdon *et al.*, 2011b). This may have resulted from the spread of the virus through a previously uninfected host population, or a selective sweep of the virus carrying an advantageous mutation through a pre-existing viral population.

Such rapid changes in prevalence are common in pathogens that spread infectiously, but are more surprising in a virus whose only mode of transmission is vertical. However, by combining a mathematical model with laboratory estimates of the vertical transmission rates of DObsSV, it was possible to show that the unusual biparental transmission of the virus through both eggs and sperm can cause such rapid increases in prevalence. Furthermore, this can occur even if the infection significantly reduces the reproductive success of infected flies (Longdon *et al.*, 2011b).

Concluding statement

To study the co-evolutionary and ecological interactions between host and parasites, studying naturally occurring host–parasite combinations is essential (Little, 2002). As they naturally infect *Drosophila,* sigma viruses have great potential as a model to study the fundamental questions relating to the genetics and evolution of viral diseases, and offer a unique opportunity to study insect–rhabdovirus interactions.

Acknowledgements

Frank Jiggins and Lena Wilfert are funded by Royal Society Research fellowships. We thank Mike Magwire for providing data and Darren Obbard for helpful comments.

References

Afzelius, B.A., Alberti, G., Dallai, R., Godula, J., and Witalinski, W. (1989). Virus-infected and rickettsia-infected sperm cells in arthropods. J. Invert. Path. 53, 365–377.

Aminetzach, Y.T., Macpherson, J.M., and Petrov, D.A. (2005). Pesticide resistance via transposition-mediated adaptive gene truncation in *Drosophila.* Science *309,* 764–767.

Ammar, E.D., Tsai, C.W., Whitfield, A.E., Redinbaugh, M.G., and Hogenhout, S.A. (2009). Cellular and molecular aspects of rhabdovirus interactions with insect and plant hosts. Ann. Rev. Entom. *54,* 447–468.

Avila, A., Silverman, N., Diaz-Meco, M.T., and Moscat, J. (2002). The *Drosophila* atypical protein kinase C-ref(2)p complex constitutes a conserved module for signaling in the toll pathway. Mol. Cell. Biol. *22,* 8787–8795.

Bangham, J., Obbard, D.J., Kim, K.W., Haddrill, P.R., and Jiggins, F.M. (2007). The age and evolution of an antiviral resistance mutation in *Drosophila melanogaster.* Proc. Biol. Sci. *274,* 2027–2034.

Bangham, J., Kim, K.W., Webster, C.L., and Jiggins, F.M. (2008). Genetic variation affecting host–parasite interactions: Different genes affect different aspects of sigma virus replication and transmission in *Drosophila melanogaster.* Genetics *178,* 2191–2199.

Basden, E.B. (1954). The distribution and biology of Drosophilidae (Diptera) in Scotland, including a new species of *Drosophila.* Trans. Royal Soc. Edinburgh *62,* 602–654.

Berkalof, A., Breglian, J., and Ohanessi, A. (1965). Demonstration of virions in *Drosophila* infected by the hereditary sigma virus. [In French.] C. R. Acad. Sci. Hebd. Seances Acad. Sci. D. *260,* 5956–5959.

Bourhy, H., Cowley, J.A., Larrous, F., Holmes, E.C., and Walker, P.J. (2005). Phylogenetic relationships among rhabdoviruses inferred using the L polymerase gene. J. Gen. Virol. *86,* 2849–2858.

Bregliano, J.C. (1970). Study of infection of germ line in female *Drosophila* infected with sigma virus. II. Evidence of a correspondance between ovarian cysts with increased virus yield and stabilized progeny. [In French.] Ann. Inst. Pasteur (Paris) *119,* 685–704.

Brun, G., and Plus, N. (1980). The viruses of *Drosophila.* In The Genetics and Biology of *Drosophila,* Ashburner, M., and Wright, T.R.F., eds (Academic Press, New York), pp. 625–702.

Buchner, P. (1965). Endosymbiosis of Animals with Plant Microorganisms (Interscience, Inc., New York).

Busserea, F. (1970a). Study of CO_2 sensitivity symptom produced by sigma virus in *Drosophila.* 2. Comparative development of efficiency of nerve centers and different organs after intraabdominal and intrathoracic inoculation. Ann. Inst. Pasteur *118,* 626–645.

Busserea, F. (1970b). Symptoms of CO_2-sensitivity induced by sigma virus in *Drosophila.*1. Effect of inoculation site on delay in symptom appearance. Ann. Inst. Pasteur *118,* 367–385.

Busserea, F., and Contamine, D. (1980). Infectivity of two rhabdoviruses in *Drosophila melanogaster*: Rabies and isfahan viruses. Ann. Virol. *131,* 3–12.

Calisher, C.H., Karabatsos, N., Zeller, H., Digoutte, J.P., Tesh, R.B., Shope, R.E., Travassos da Rosa, A.P., and St George, T.D. (1989). Antigenic relationships among rhabdoviruses from vertebrates and hematophagous arthropods. Intervirol. *30,* 241–257.

Carpenter, J.A. (2008). The evolution and genetics of *Drosophila melanogaster* and the sigma virus. PhD thesis, University of Edinburgh, Edinburgh, UK.

Carpenter, J.A., Obbard, D.J., Maside, X., and Jiggins, F.M. (2007). The recent spread of a vertically transmitted virus through populations of *Drosophila melanogaster.* Mol. Ecol. *16,* 3947–3954.

Carpenter, J.A., Hutter, S., Baines, J.F., Roller, J., Saminadin-Peter, S.S., Parsch, J., and Jiggins, F.M. (2009a). The transcriptional response of *Drosophila melanogaster* to infection with the sigma virus (Rhabdoviridae). Plos One *4,* e6838.

Carpenter, J.A., Keegan, L.P., Wilfert, L., O'Connell, M.A., and Jiggins, F.M. (2009b). Evidence for ADAR-induced hypermutation of the *Drosophila* sigma virus (Rhabdoviridae). BMC Genet. *10,* 75.

Carre-Mouka, A., Gaumer, S., Gay, P., Petitjean, A.M., Coulondre, C., Dru, P., Bras, F., Dezelee, S., and Contamine, D. (2007). Control of sigma virus multiplication by the ref(2)P gene of *Drosophila melanogaster*: An in vivo study of the PB1 domain of Ref(2)P. Genetics *176,* 409–419.

Chare, E.R., Gould, E.A., and Holmes, E.C. (2003). Phylogenetic analysis reveals a low rate of homologus recombinaion in negative-sense RNA viruses. J. Gen. Virol. *84,* 2961–2703.

Contamine, D., and Gaumer, S. (2008). Sigma Rhabdoviruses. In Encyclopedia of Virology 3rd edn, Mahy, B.W.J., and van Regenmortel, M.H.V., eds (Elsevier, Oxford, UK), 576–581.

Contamine, D., Petitjean, A.M., and Ashburner, M. (1989). Genetic resistance to viral infection: the molecular cloning of a *Drosophila* gene that restricts infection by the rhabdovirus sigma. Genetics *123,* 525–533.

Dacheux, L., Berthet, N., Dissard, G., Holmes, E.C., Delmas, O., Larrous, F., Guigon, G., Dickinson, P., Faye, O., Sall, A.A., *et al.* (2010). Application of broad-spectrum resequencing microarray for genotyping rhabdoviruses. J. Virol. *84,* 9557–9574.

Dawkins, R., and Krebs, J.R. (1979). Arms races between and within species. Proc. Biol. Sci. *205,* 489–511.

Dreux, M., and Chisari, F.V. (2010). Viruses and the autophagy machinery. Cell Cycle *9,* 1295–1307.

Dru, P., Bras, F., Dezelee, S., Gay, P., Petitjean, A.M., Pierredeneubourg, A., Teninges, D., and Contamine, D. (1993). Unusual variability of the *Drosophila melanogaster* Ref(2)P protein which controls the multiplication of sigma rhabdovirus. Genetics *133,* 943–954.

Engelstadter, J., and Hurst, G.D.D. (2009). The ecology and evolution of microbes that manipulate host reproduction. Ann. Rev. Ecol. Evol. Syst. *40,* 127–149.

Felix, R., Guzman, J., and Arellano, A. (1977). CO_2 sensitivity of Drosophilid flies from a location in the outskirts of Mexico city. Dros. Info. Service *47,* 110–112.

Fine, P.E. (1975). Vectors and vertical transmission: an epidemiologic perspective. Ann. N.Y. Acad. Sci. *266,* 173–194.

Fleuriet, A. (1981a). Comparison of various physiological traits in flies (*Drosophila melanogaster*) of wild origin, infected or uninfected by the hereditary rhabdovirus sigma. Arch. Virol. *69*, 261–272.

Fleuriet, A. (1981b). Effect of overwintering on the frequency of flies infected by the rhabdovirus sigma in experimental populations of *Drosophila melanogaster*. Arch. Virol. *69*, 253–260.

Fleuriet, A. (1982). Factors affecting the frequency of infection by the sigma virus in experimental populations of *Drosophila melanogaster*. Arch. Virol. *73*, 121–133.

Fleuriet, A. (1988). Maintenance of a hereditary virus – the sigma virus in populations of its host, *Drosophila melanogaster*. Evol. Biol. *23*, 1–30.

Fleuriet, A. (1999). Evolution of the proportions of two sigma viral types in experimental populations of *Drosophila melanogaster* in the absence of the allele that is restrictive of viral multiplication. Genetics *153*, 1799–1808.

Fleuriet, A., and Periquet, G. (1993). Evolution of the *Drosophila melanogaster* sigma virus system in natural-populations from Languedoc (Southern France). Arch. Virol. *129*, 131–143.

Fleuriet, A., and Sperlich, D. (1992). Evolution of the *Drosophila melanogaster*-sigma virus system in a natural-population from Tübingen. Theor. Appl. Genet. *85*, 186–189.

Gastka, M., Horvath, J., and Lentz, T.L. (1996). Rabies virus binding to the nicotinic acetylcholine receptor alpha subunit demonstrated by virus overlay protein binding assay. J. Gen. Virol. *77*, 2437–2440.

Gay, P. (1978). *Drosophila* genes which intervene in multiplication of sigma virus. Mol. Gen. Genet. *159*, 269–283.

Gubala, A.J., Proll, D.F., Barnard, R.T., Cowled, C.J., Crameri, S.G., Hyatt, A.D., and Boyle, D.B. (2008). Genomic characterisation of Wongabel virus reveals novel genes within the Rhabdoviridae. Virology *376*, 13–23.

Gubala, A., Davis, S., Weir, R., Melville, L., Cowled, C., Walker, P., and Boyle, D. (2010). Ngaingan virus, a macropod-associated rhabdovirus, contains a second glycoprotein gene and seven novel open reading frames. Virology *399*, 98–108.

Guillemain, A. (1953). [Discovery and localization of a gene in *Drosophila melanogaster* inhibiting the multiplication of a virus with hereditary sensitivity to carbon dioxide]. C. R. Hebd. Seances Acad. Sci. *236*, 1085–1086.

Havard, S., Eslin, P., Prevost, G., and Doury, G. (2009). Encapsulation ability: Are all *Drosophila* species equally armed? An investigation in the obscura group. Canad. J. Zoo-Rev. *87*, 635–641.

Herforth, R., and Westphal, N.J. (1966). Observations on the frequency of carbon dioxide sensitivity in a natural population of Drosophilid flies. Dros. Info. Service *41*, 79.

Jiggins, F.M., and Kim, K.W. (2005). The evolution of antifungal peptides in *Drosophila*. Genetics *171*, 1847–1859.

Jousset, F.X. (1969). Preliminary studies of *Drosophila* sigma virus proliferation in taxonomically different insects. Comptes Rendus Hebdomadaires Des Seances De L Acad. Sci. Serie D *269*, 1035–1038.

Jupin, N., Plus, N., and Fleuriet, A. (1968). Effects of a sigma virus strain on female *Drosophila* fertility. Ann. Institut Pasteur *114*, 577–595.

Kalmus, H. (1943). Transmission of susceptibility to carbon dioxide to species hybrids in *Drosophila*. Nature *152*, 692.

Katzourakis, A., and Gifford, R.J. (2010). Endogenous Viral Elements in Animal Genomes. PLos Genet. *6*, e1001191.

Klionsky, D.J., and Ohsumi, Y. (1999). Vacuolar import of proteins and organelles from the cytoplasm. Annu. Rev. Cell. Dev. Biol. *15*, 1–32.

Korolchuk, V.I., Mansilla, A., Menzies, F.M., and Rubinsztein, D.C. (2009). Autophagy inhibition compromises degradation of ubiquitin-proteasome pathway substrates. Mol. Cell *33*, 517–527.

Kraaijeveld, A.R., and Godfray, H.C. (1997). Trade-off between parasitoid resistance and larval competitive ability in *Drosophila melanogaster*. Nature *389*, 278–280.

Kraaijeveld, A.R., Ferrari, J., and Godfray, H.C. (2002). Costs of resistance in insect-parasite and insect–parasitoid interactions. Parasitol. *125 Suppl*, S71–82.

Kuwata, R., Isawa, H., Hoshino, K., Tsuda, Y., Yanase, T., Sasaki, T., Kobayashi, M., and Sawabe, K. (2011). RNA splicing in a new rhabdovirus from *Culex* mosquitoes. J. Virol. *85*, 6185–6196.

L'Heritier, P.H. (1948). Sensitivity to CO_2 in *Drosophila*. A review. Heredity *2*, 325–348.

L'Heritier, P.H. (1957). The hereditary virus of *Drosophila*. Adv. Virus Res. *5*, 195–245.

L'Heritier, P.H. (1970). *Drosophila* viruses and their role as evolutionary factors. Evol. Biol. *4*, 185–209.

L'Heritier, P.H., and Teissier, G. (1937). Une anomalie physiologique héréditaire chez la Drosophile. C. R. Acad Sci Paris *231*, 192–194.

Lamark, T., Kirkin, V., Dikic, I., and Johansen, T. (2009). NBR1 and p62 as cargo receptors for selective autophagy of ubiquitinated targets. Cell Cycle *8*, 1986–1990.

Landesdevauchelle, C., Bras, F., Dezelee, S., and Teninges, D. (1995). Gene 2 of the sigma rhabdovirus genome encodes the p-protein, and gene-3 encodes a protein related to the reverse-transcriptase of retroelements. Virology *213*, 300–312.

Lemaitre, B., and Hoffmann, J. (2007). The host defense of *Drosophila melanogaster*. Annu. Rev. Immunol. *25*, 697–743.

Lentz, T.L., Burrage, T.G., Smith, A.L., Crick, J., and Tignor, G.H. (1982). Is the acetylcholine receptor a rabies virus receptor? Science *215*, 182–184.

Levine, B., and Deretic, V. (2007). Unveiling the roles of autophagy in innate and adaptive immunity. Nat. Rev. Immunol. *7*, 767–777.

Little, T.J. (2002). The evolutionary significance of parasitism: do parasite-driven genetic dynamics occur ex silico? J. Evol. Biol. *15*, 1–9.

Longdon, B., Obbard, D.J., and Jiggins, F.M. (2010). Sigma viruses from three species of *Drosophila* form a major new clade in the rhabdovirus phylogeny. Proc. Bio. Soc. 277, 35–44.

Longdon, B., Hadfield, J.D., Webster, C.L., Obbard, D.J., and Jiggins, F.M. (2011a). Host phylogeny determines viral persistence and replication in novel hosts. PLoS Pathog. 7, e1002260.

Longdon, B., Wilfert, L., Obbard, D.J., and Jiggins, F.M. (2011b). Rhabdoviruses in two species of *Drosophila*: vertical transmission and a recent sweep. Genetics *188*, 141–150.

Longdon, B., Wilfert, L., Osei-Poku, J., Cagney, H., Obbard, D.J., and Jiggins, F.M. (2011c). Host switching by a vertically-transmitted rhabdovirus in *Drosophila*. Biol. Lett. 7, 747–750.

Lyles, D.S., and Rupprecht, C.E. (2007). Rhabdoviridae. In Fields Virology, Knipe, D.M., and Howley, P.M., eds (Lippincott Williams and Wilkins, Philadelphia, PA), pp. 1364–1394.

McKean, K.A., Yourth, C.P., Lazzaro, B.P., and Clark, A.G. (2008). The evolutionary costs of immunological maintenance and deployment. BMC Evol. Biol. *8*, 76.

Magwire, M.M., Bayer, F., Webster, C.L., Chuan, C., and Jiggins, F.M. (2011). Successive increases in the resistance of *Drosophila* to viral infection through a transposon insertion followed by a duplication. PLos Genet. 7, e1002337.

Mueller, S., Gausson, V., Vodovar, N., Deddouche, S., Troxler, L., Perot, J., Pfeffer, S., Hoffmann, J.A., Saleh, M.C., and Imler, J.L. (2010). RNAi-mediated immunity provides strong protection against the negative-strand RNA vesicular stomatitis virus in *Drosophila*. Proc. Natl. Acad. Sci. U.S.A. *107*, 19390–19395.

Nezis, I.P., Simonsen, A., Sagona, A.P., Finley, K., Gaumer, S., Contamine, D., Rusten, T.E., Stenmark, H., and Brech, A. (2008). Ref(2)P, the *Drosophila melanogaster* homologue of mammalian p62, is required for the formation of protein aggregates in adult brain. J. Cell. Biol. *180*, 1065–1071.

Obbard, D.J., Gordon, K.H., Buck, A.H., and Jiggins, F.M. (2009). The evolution of RNAi as a defence against viruses and transposable elements. Philos. Trans. R. Soc. Lond. B Biol. Sci. *364*, 99–115.

Ohanessian-Guillemain, A. (1959). Etude genetique du virus hereditaire de la drosophile (sigma) mutations et recombinaison genetique. Ann. Genet. *1*, 59–68.

Ohanessian-Guillemain, A. (1963). Etude de facteurs genetiques controlant les relations du virus sigma et de la drosophile son hote. Ann. Genet. *5*, G1–G64.

Orvedahl, A., Alexander, D., Talloczy, Z., Sun, Q., Wei, Y., Zhang, W., Burns, D., Leib, D.A., and Levine, B. (2007). HSV-1 ICP34.5 confers neurovirulence by targeting the Beclin 1 autophagy protein. Cell. Host Microbe *1*, 23–35.

Pankiv, S., Clausen, T.H., Lamark, T., Brech, A., Bruun, J.A., Outzen, H., Overvatn, A., Bjorkoy, G., and Johansen, T. (2007). p62/SQSTM1 binds directly to Atg8/LC3 to facilitate degradation of ubiquitinated protein aggregates by autophagy. J. Biol. Chem. *282*, 24131–24145.

Plus, N. (1955). Etude de la sensibilite hereditaire au CO_2 chez la drosophile.1. Multiplication du virus-Sigma et passage a la descendance apres inoculation ou transmission hereditaire. Ann. Inst. Pasteur *88*, 347–363.

Rosen, L. (1980). Carbon dioxide sensitivity in mosquitos infected with sigma, vesicular stomatitis, and other rhabdoviruses. Science *207*, 989–991.

Sabin, L.R., Hanna, S.L., and Cherry, S. (2010). Innate antiviral immunity in *Drosophila*. Curr. Opin. Immunol. *22*, 4–9.

Sackton, T.B., Lazzaro, B.P., Schlenke, T.A., Evans, J.D., Hultmark, D., and Clark, A.G. (2007). Dynamic evolution of the innate immune system in *Drosophila*. Nature Genet. *39*, 1461–1468.

Seecof, R.L. (1964). Deleterious effects on *Drosophila* development associated with sigma virus infection. Virology *22*, 142–148.

Shelly, S., Lukinova, N., Bambina, S., Berman, A., and Cherry, S. (2009). Autophagy is an essential component of *Drosophila* immunity against vesicular stomatitis virus. Immunity *30*, 588–598.

Shroyer, D.A., and Rosen, L. (1983). Extrachromosomal-inheritance of carbon-dioxide sensitivity in the mosquito *Culex quinquefasciatus*. Genetics *104*, 649–659.

Sullivan, W., Ashburner, M., and Hawley, S. (2000). *Drosophila* Protocols, 1st edn (Cold Spring Harbor Laboratory Press, New York).

Sylvester, E., and Richardson, J. (1992). Aphid-borne rhabdoviruses – relationship with their aphid vectors. In Adv. Dis. Vector Res., H. KF, ed. (Springer Verlag, New York), pp. 313–341.

Teninges, D. (1968). [Demonstration of sigma viruses in the cells of the stabilized male germinal line of *Drosophila*]. [In French.] Arch. Gesamte Virusforsch. *23*, 378–387.

Teninges, D., Bras, F., and Dezelee, S. (1993). Genome organization of the sigma rhabdovirus – 6 genes and a gene overlap. Virology *193*, 1018–1023.

Tesh, R.B., Byron, N.C., and Johnson, K.M. (1972). Vesicular stomatitis virus (Indiana serotype): Transovarial transmission by phlebotomine sandflies. Science *175*, 1477–1479.

Thoulouze, M.I., Lafage, M., Schachner, M., Hartmann, U., Cremer, H., and Lafon, M. (1998). The neural cell adhesion molecule is a receptor for rabies virus. J. Virol. 72, 7181–7190.

Tsai, C.W., McGraw, E.A., Ammar, E.D., Dietzgen, R.G., and Hogenhout, S.A. (2008). *Drosophila melanogaster* mounts a unique immune response to the rhabdovirus sigma virus. App. Env. Microbiol. *74*, 3251–3256.

Tuffereau, C., Benejean, J., Blondel, D., Kieffer, B., and Flamand, A. (1998). Low-affinity nerve-growth factor receptor (P75NTR) can serve as a receptor for rabies virus. EMBO J. *17*, 7250–7259.

Tuffereau, C., Desmezieres, E., Benejean, J., Jallet, C., Flamand, A., Tordo, N., and Perrin, P. (2001). Interaction of lyssaviruses with the low-affinity nerve-growth factor receptor p75NTR. J. Gen. Virol. *82*, 2861–2867.

Walker, P.J., Dietzgen, R.G., Joubert, D.A., and Blasdell, K.R. (2011). Rhabdovirus accessory genes. Virus Res. *162*, 110–125.

Wayne, M.L., Contamine, D., and Kreitman, M. (1996). Molecular population genetics of ref(2)P, a locus which confers viral resistance in *Drosophila*. Mol. Biol. Evol. *13*, 191–199.

Wilfert, L., and Jiggins, F.M. (2010a). Disease association mapping in *Drosophila* can be replicated in the wild. Biol. Lett. *6*, 666–668.

Wilfert, L., and Jiggins, F.M. (2010b). Host-parasite coevolution: genetic variation in a virus population and the interaction with a host gene. J. Evol. Biol. *23*, 1447–1455.

Williamson, D.L. (1959). Carbon dioxide sensitivity in *Drosophila affinis* and *Drosophila athabasca*. PhD thesis, University of Nebraska, NB.

Williamson, D.L. (1961). Carbon dioxide sensitivity in *Drosophila affinis* and *Drosophila athabasca*. Genetics *46*, 1053–1060.

Wyers, F., Dru, P., Simonet, B., and Contamine, D. (1993). Immunological cross-reactions and interactions between the *Drosophila melanogaster* ref(2)P protein and sigma rhabdovirus proteins. J. Virol. *67*, 3208–3216.

Yampolsky, L.Y., Webb, C.T., Shabalina, S.A., and Kondrashov, A.S. (1999). Rapid accumulation of a vertically transmitted parasite triggered by relaxation of natural selection among hosts. Evol. Ecol. Res. *1*, 581–589.

Zheng, Y.T., Shahnazari, S., Brech, A., Lamark, T., Johansen, T., and Brumell, J.H. (2009). The adaptor protein p62/SQSTM1 targets invading bacteria to the autophagy pathway. J. Immunol. *183*, 5909–5916.

Zhu, R.L., Lei, X.Y., Ke, F., Yuan, X.P., and Zhang, Q.Y. (2011). Genome of turbot rhabdovirus exhibits unusual non-coding regions and an additional ORF that could be expressed in fish cell. Virus Res. *155*, 495–505.

Virus–Host Protein Interactions of Plant-adapted Rhabdoviruses

9

Michael M. Goodin and Byoung-Eun Min

Abstract

Rhabdoviruses belonging to the genus *Nucleorhabdovirus* are unique in that they have not only adapted to replication in both plant and invertebrate cells but they are nucleotrophic as well. Beyond their roles as plant pathogens, these viruses also serve as probes to provide insights into plant nuclear biology. Using data gained from studies with Sonchus yellow net virus and Potato yellow dwarf virus, this review will describe protein interaction and membrane dynamics in virus-infected cells, with particular emphasis on relating these findings to the mechanism of cell-to-cell movement. To date, five host factors have been implicated in the process of moving nucleocapsids from infected cells into adjacent ones. These host factors, in concert with several virus-encoded proteins, are thought to form a cytoplasmic complex that migrates from nuclei on microtubules and membranes towards the plasmodesmata where they dock onto plasmodesmata-localized protein 1. Prior to moving cell-to-cell, these viruses induce dramatic changes in nuclear membranes, particularly the inner nuclear membrane, during establishment of a nuclear viroplasm, were nascent nucleocapsids form on ring-shaped protein complexes before accumulating in the perinuclear space.

Introduction

Plant-adapted rhabdoviruses induce dramatic changes in the localization patterns of host cell membranes and proteins during infection. Emphasis will be given here to the cytopathology of nucleorhabdoviruses, particularly Potato yellow dwarf virus (PYDV) and Sonchus yellow net virus (SYNV), as these have been the most extensively studied members of this genus. A simplified cell-to-cell transmission cycle, which will serve as the organizational basis of this review, is provided in Fig. 9.1.

In contrast to infection of animal cells by enveloped viruses, the initial entry of nucleorhabdoviruses into plant cells is a physical process mediated by the stylets of insect vectors, which are primarily aphids, leafhoppers or planthoppers. In whatever form rhabdoviruses are transmitted, it is important to remember that the minimal infectious unit of these viruses is the nucleocapsid, which is composed of the viral genomic RNA encapsidated over its entire length by the nucleocapsid protein (N) and associated with the phosphoprotein (P) and polymerase protein (L) (Dietzgen *et al.*, 2011). The process by which initially infecting nucleorhabdoviruses target the nucleus is poorly understood (Jackson *et al.*, 2005). However, it is believed that these viruses must de-envelope, if they have a membrane, on the endoplasmic reticulum (ER) or outer nuclear membrane in order to release the nucleocapsid into the cytoplasm. Given the density of nuclear localization signals present on the N protein-encapsidated nucleocapsid, it is reasonable to speculate that transport to the nucleus would occurs by an importin-α mediated process akin to the nuclear transport of complexes such as the VirD/E protein mediated transfer of DNA from agrobacterium Ti-plasmids into plant cells (Tzfira *et al.*, 2004; Lacroix *et al.*, 2006; Deng *et al.*, 2007). As is typical for monopartite negative-strand RNA viruses, biochemical evidence suggests that the

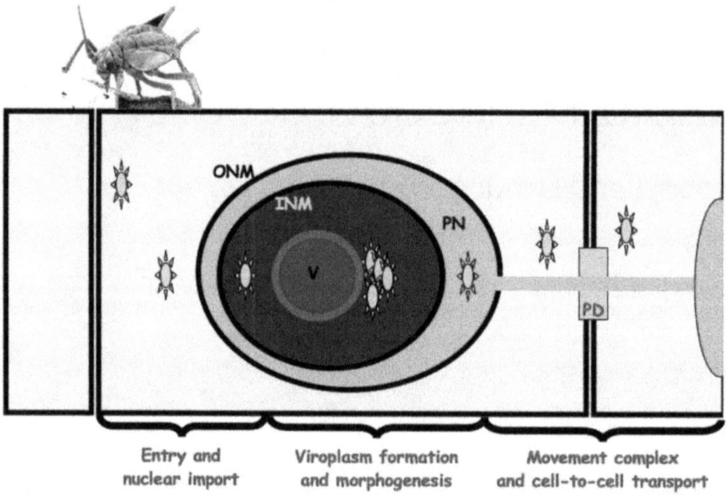

Figure 9.1 Cell-to-cell transmission of plant-adapted nucleorhabdoviruses. Initial entry into plant cells is mediated during feeding by an insect vector, most commonly a species of aphid, leafhopper or planthopper. In contrast to vertebrate-infecting rhabdoviruses, no membrane fusion events are required for viral entry, which in the case of nucleorhabdoviruses is entirely a physical process dependent upon the feeding apparatus of the insect to breach the plant cell wall. Upon entry into cells, the viruses must have their nucleocapsids imported into the nucleus, whereupon they become transcriptionally active to produce virus-specific messenger RNAs. Synthesis and nuclear import of nascent proteins results in establishment of the viroplasm where replication takes place. Mature nucleocapsids, defined as the viral genomic RNA encapsidated by the nucleocapsid (N) protein and associated with the phosphoprotein (P) and polymerase (L) protein, then accumulate in the perinuclear space. Although the precise mechanism of its formation remains equivocal, a nucleocapsid-containing cell-to-cell movement complex must be targeted towards plasmodesmata in order to establish infections in adjacent cells. ONM, outer nuclear membrane; INM, inner nuclear membrane; V, viroplasm; PN, perinucler space; PD, plasmodesmata.

nucleorhabdovirus nucleocapsids are transcriptionally competent and therefore are expected to initiate synthesis of viral mRNAs upon entry into the nucleus (Wagner and Jackson, 1997; Wang *et al.*, 2010).

Ultimately, the accumulation of viral proteins results in the formation of a nuclear matrix, called the viroplasm, which is believed to be the site of virus replication. Perhaps as a consequence of viroplasm formation, nucleorhabdoviruses induce enlargement of host cell nuclei, as well as gross modifications of the nuclear membranes during establishment of spatially separated sites of replication and morphogenesis. Nascent nucleocapsids are condensed by the matrix protein (M) and in many cases accumulate in the perinuclear space (van Beek *et al.*, 1985; Redinbaugh *et al.*, 2002). While the mechanism of budding and virion formation is poorly understood for plant-adapted rhabdoviruses, it is believed to follow the general processes utilized by other negative-strand RNA viruses (Jayakar *et al.*, 2004). The form in which nucleocapsids are transported from the nucleus into adjacent cells is equivocal. However, there is no present evidence for budding through, or breakdown of, the outer nuclear membrane as occurs for herpes viruses (Mettenleiter *et al.*, 2009). Current evidence suggests that the nucleocapsid is exported from the nucleus to form a complex that migrates on microtubule and endomembranes, ultimately docking at plasmodesmata in order to release infectious complexes into adjacent cells. Repetition of this process must result ultimately in establishing systemic infection. In the case of SYNV infections, plants experience an acute phase of infection 2–3 weeks post inoculation followed by a nearly complete 'recovery' phase, wherein virus titres and symptom severity drop dramatically. In contrast, most PYDV-infected plants will die approximately 5–6 weeks post infection, as these plants fail to recover (Ghosh *et al.*, 2008). It is remarkable that two closely related nucleorhabdoviruses can have such different effects on a common host, such as *Nicotiana benthamiana*, suggesting that the molecular details of plant–rhabdovirus interactions may differ markedly.

Nuclear import of *Nucleorhabdovirus* proteins

Nucleorhabdoviruses establish sites of replication called viroplasms in nuclei of infected cells (van Beek *et al.*, 1985; Redinbaugh *et al.*, 2002; Jackson *et al.*, 2005). As such, a critical requirement for these viruses is import of their proteins into the nucleus. Where examined, nuclear localization signals (NLSs) have been identified in nucleorhabdoviral proteins by computational analyses supported by protein localization and site-specific mutagenesis experiments (Goodin *et al.*, 2001; Tsai *et al.*, 2005; Deng *et al.*, 2007). An interesting exception to this arose during the genome sequence characterization of PYDV, for which computational analyses failed to identify NLSs in any of the seven major open reading frames (ORFs) determined for this virus (Bandyopadhyay *et al.*, 2010). However, yeast-based nuclear import assays, and live-cell imaging of GFP fusions of PYDV proteins expressed in transgenic *N. benthamiana* plants, expressing a fluorescent nuclear marker, demonstrated that all proteins predicted to be involved in replication or morphogenesis were indeed capable of localizing to nuclei. The exception was the PYDV L protein, which was too large for such analyses.

Where examined, N proteins of nucleorhabdoviruses have nuclear localization signals located in their carboxyl termini, which is required for viroplasm formation (Deng *et al.*, 2007). Nuclear import of N proteins is thought to be mediated by the nuclear import receptor importin-α (Deng *et al.*, 2007). Curiously, in all cases examined, the co-expression of N and P proteins results in their subnuclear co-localization, resulting in dramatic changes in their localization compared to when the proteins are expressed alone (Goodin *et al.*, 2001, 2008; Tsai *et al.*, 2005). Although both N and P proteins are capable of independent nuclear localization, P proteins tend to partition between the nucleus and cytoplasm, whereas the N proteins localize to the nucleus exclusively, but accumulate over the entire viroplasm. The reason for shift in localization patterns of N and P proteins when they are co-expressed is presently unknown. However, it is thought to be related to the mechanism by which the viroplasm is established (Deng *et al.*, 2007).

Modification of intranuclear membranes

Mature *Nucleorhabdovirus* particles accumulate in the perinuclear spaces of infected plant cells (van Beek *et al.*, 1985; Redinbaugh *et al.*, 2002; Ammar *et al.*, 2009). As such, these viruses induce the intranuclear accumulation of membranes, in a virus-specific manner (Fig. 9.2). The dynamics of nuclear membrane re-localization was studied live in rhabdovirus-infected cells of transgenic *N. benthamiana* plants that expressed green fluorescent protein (GFP) targeted to the ER (Goodin *et al.*, 2005). Of note is the demonstration in previous studies, that GFP targeted to the ER also accumulates in the lumen of the nuclear envelope, hereafter referred to as the perinuclear space (Ellenberg *et al.*, 1997; Ellenberg and Lippincott-Schwartz, 1999; Lippincott-Schwartz *et al.*, 1999). This has implications for interpretation of data presented below because it means that any changes in the spacing between the inner (INM) and outer nuclear membranes (ONM) will result in GFP accumulation in a larger perinuclear volume. Examination of virus-infected plant cells by confocal microscopy showed that SYNV induced the intranuclear accumulation of membranes, whereas PYDV infections resulted in the expansion of the nuclear envelope, as well as inducing intranuclear membrane accumulation. In the context of SYNV infections, virus-induced intranuclear membranes retain connectivity to the ER (Goodin *et al.*, 2007b). This conclusion was derived from experiments that demonstrated that fluorescence recovery after photobleaching (FRAP) of GFP in the nuclear envelope of uninfected nuclei was not significantly different from that in virus-induced 'spherules'. The implication of these findings is the possibility that cell-to-cell transmission of SYNV involves

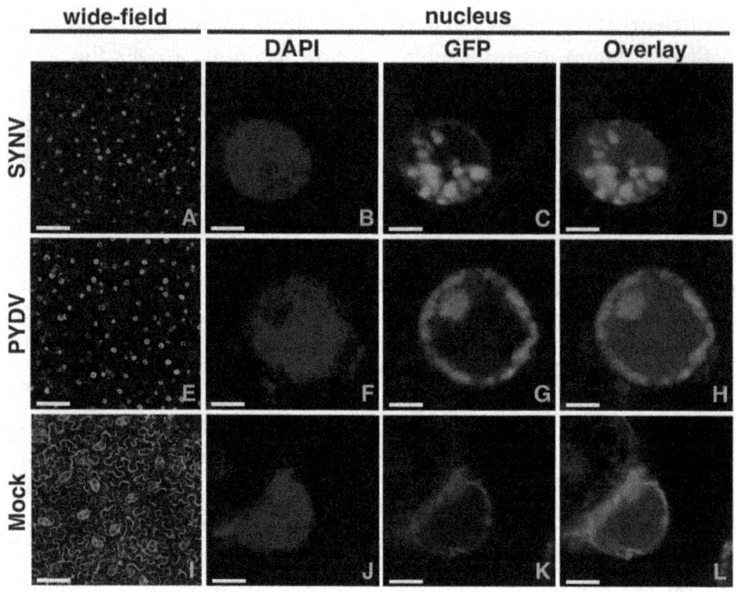

Figure 9.2 Confocal micrographs of rhabdovirus-induced nuclear inclusions in 16c (ER-targeted GFP-expressing) *N. benthamiana* plants inoculated with SYNV (A–D), PYDV (E–H) or mock-inoculated (I–L). (A,E,I) Detection of GFP fluorescence in wide-field micrographs of the abaxial surface of systemically infected leaves. Scale-bar = 200 mm. (B–D, F–H, and J–L) Detection of DAPI and GFP fluorescence, and the overlay of these images, in single nuclei of epidermal cells of SYNV-, PYDV- and mock-inoculated leaves, respectively. Scale bar = 5 mm. (A and E) Punctate loci of GFP fluorescence were observed in SYNV- and PYDV-infected leaves. (I) GFP accumulated primarily in the cytoplasm and, to a lesser degree, in nuclei of leaf epidermal cells of mock-inoculated leaves. Note that exposure times for micrographs were adjusted to provide correct acquisition of fluorescence signal from the brightest areas within cells. Therefore, the GFP fluorescence in guard cells, observed in (I) is not detected in (A) and (E). (B–D) The punctate areas of fluorescence observed in A are nuclei filled with inclusions of GFP that accumulate primarily within the area of DAPI fluorescence. (F–H) The punctate areas of fluorescence observed in E are nuclei in which inclusions of GFP accumulate primarily around the area of DAPI fluorescence and within a large internal region that excludes DAPI. (J–L) GFP fluorescence is detected primarily at the periphery and surrounding membranes of nuclei in cells of mock-inoculated leaves. Discrete inclusions of GFP were not detected within these nuclei. Reprinted from Goodin *et al.* (2005).

migration of nucleocapsids from the perinuclear space through the ER to plasmodesmata that connect adjacent cells.

In contrast to plus-strand RNA viruses that replicate in membrane-associate 'spherules' it is unlikely that rhabdoviral polymerase proteins could be responsible for the observed membrane re-localization since these proteins are typically associated only with the viral nucleocapsids and are not membrane-associated (Hammond *et al.*, 1992; Qanungo *et al.*, 2004). Instead, it appears that these membrane dynamics are mediated primarily by actions of the M proteins. PYDV-M for example, has the unique ability to re-localize the INM outside the context of virus infections (Bandyopadhyay *et al.*, 2010).

Effect of matrix proteins on membranes

From electron and light microscopic studies it appears that morphogenesis of nucleorhabdoviruses takes place on the INM, with mature nucleocapsids accumulating in the perinuclear space (Redinbaugh *et al.*, 2002; Jackson *et al.*, 2005; Goodin *et al.*, 2007b; Ammar *et al.*, 2009). However, there have been few studies that examined the mechanisms by which nucleocapsids accumulate in the perinuclear space. Recent experiments to determine the subcellular localization of PYDV proteins led to the observation that the PYDV M protein was capable of inducing the intranuclear accumulation of plant membranes outside the context of virus infections (Fig. 9.3).

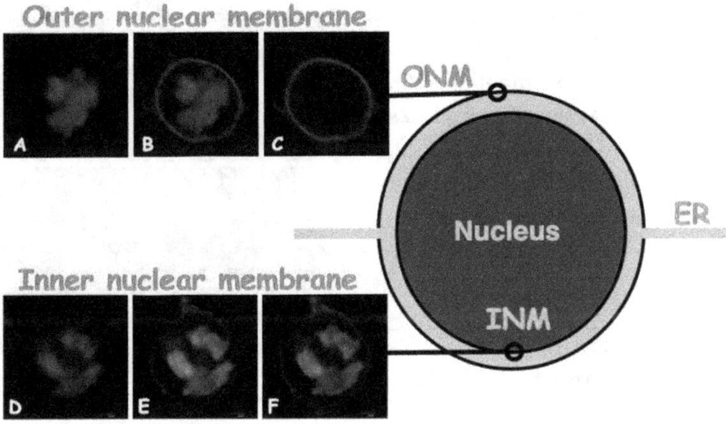

Figure 9.3 The matrix protein of PYDV induces the intranuclear accumulation of the inner nuclear membrane. (A and D) Micrographs of RFP-labelled PYDV-M in nuclei. (C) Outer nuclear membrane (ONM) marker protein fusion, GFP-WIP:1. (B) Overlay of images A and C. (F) Inner nuclear membrane (INM) marker protein fusion, GFP-LBR. E. Overlay of images A and C. All protein fusions were expressed in leaf epidermal cells of *N. benthamiana* plants following agroinfiltration. Cartoon on the right shows the relative position of the nuclear envelope (INM, ONM) to chromatin (Nucleus). Figure adapted from Goodin *et al.* (2007b).

To further define if there was selectivity for particular membranes, PYDV-M was co-expressed as a red fluorescent protein (RFP) fusion in plant cells with GFP fusions targeted to the ONM (Xu *et al.*, 2007) (Fig. 9.3A–C), or INM (Irons *et al.*, 2003; Fig. 9.3D–F). All membrane markers, including others not shown here, except WIP1-GFP (ONM marker) accumulated in the nucleus when co-expressed with RFP-M. Although the matrix proteins of both cytoplasmic and nucleotrophic rhabdoviruses have been shown to localize to nuclei (Goodin *et al.*, 2001; Glodowski *et al.*, 2002), only PYDV-M, thus far, has the ability to affect localization of nuclear membranes outside the context of virus infections. In contrast to the PYDV M protein, SYNV-M is associated with membranes only in the context of infected cells and does not alter nuclear membranes in the absence of other viral proteins (Goodin *et al.*, 2001). The association of PYDV-M with membranes may be facilitated by a YPDL late domain, located at residues 61–64 in its primary structure. This domain is also found in p9 and p6 proteins of Equine infectious anaemia virus and Human immunodeficiency virus-1, respectively. Related motifs YLDL and YMYL are found in the matrix proteins of sendai virus and niphah virus, respectively (Chen and Lamb, 2008). Importantly, these YXXL motif-containing proteins all bind the AIP1/Alix protein, which plays a role in vacuolar protein sorting (Chen and Lamb, 2008; Irie *et al.*, 2008). AIP1/Alix is a highly conserved protein with homologues in animals, fungi and plants, suggesting that a similar interaction could occur in PYDV-infected *N. benthamiana.*

FRAP analyses of RFP-M-induced intranuclear membranes in cells of transgenic plants that expressed GFP targeted to the ER showed that the GFP marker was not mobile, in contrast to the M protein. This is a particularly intriguing finding that suggests that the intranuclear membranes are highly cross-linked in a manner that traps luminal GFP, perhaps in spherules, while the cross–linking associations of PYDV-M are sufficiently transient to permit diffusion of this protein. Electron microscopy studies will be required to examine the nature of membranes cross-linked by the M protein. However, PYDV-M may be essential for formation of annulate lamellae, which could link the perinuclear space to the ER during the formation of rhabdovirus movement complexes (Steinkamp and Hoefert, 1977).

In addition to its effects on the INM, PYDV-M interacts with other PYDV proteins (Bandyopadhyay *et al.*, 2010). Of these, the observed PYDV–N/M interaction could be essential for condensation of the PYDV nucleocapsid, a process that would then be followed by budding into the perinuclear space (Martins *et al.*, 1998; Goodin *et al.*, 2007b). Interestingly, M protein also interacted with Y, the putative movement protein, suggesting that PYDV-M may form complexes similar to the ER-associated SYNV-M-containing complexes that were postulated to be rhabdoviral cell-to-cell movement complexes (Goodin *et al.*, 2007b).

Mapping protein localization in the context of virus-infected cells

Characterization of rhabdoviral proteins using *in vivo* protein expression and localization studies has been most often conducted outside the context of virus-infected cells (Goodin *et al.*, 2001; Tsai *et al.*, 2005; Bandyopadhyay *et al.*, 2010) or in virus-infected cells for a limited number of proteins (Martins *et al.*, 1998). However, it is common for viruses to induce dramatic changes in host cell membranes and to re-localize proteins to sites different to that occurring in virus-free cells. Thus, the Goodin laboratory explored the localization of autofluorescent protein fusions in SYNV-infected cells (Fig. 9.4; Goodin *et al.*, 2007b). This approach revealed novel insights into the protein and membrane dynamics in the SYNV/*N. benthamiana* pathosystem.

In order to obtain high-resolution images of membranes total internal reflectance fluorescence microscopy (TIRFM) was used instead of conventional confocal microscopy. These studies suggested that the intranuclear membranes in virus-infected cells remained connected to the endomembrane system from which they were derived. This contention was supported by FRAP analyses of GFP in the nuclear envelope and intranuclear membranes. Since there was no statistically significant difference in the FRAP kinetics of GFP in any of these loci, it was concluded that the intranuclear membranes remain contiguous with the ER and were not confined to covalently closed intranuclear membranes. This finding was significant because it allows for the possibility that the intranuclear membranes are *bona fide* sites of virion assembly and not simply alterations in host membranes that do not participate in viral biology *per se*. Contiguity of the

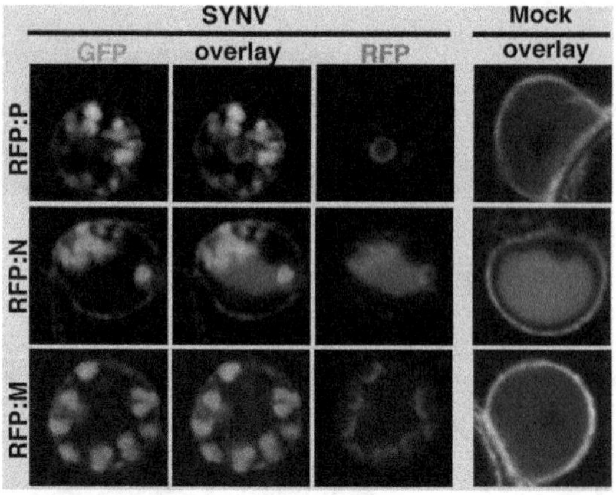

Figure 9.4 Differential localization of RFP-tagged protein fusions transiently expressed in virus-infected (SYNV) or mock-inoculated (mock) leaf epidermal cells of transgenic *N. benthamiana* expressing GFP targeted to the ER. (Top panel) RFP:SYNV-P. (Middle panel) RFP:SYNV-N. (Lower panel) RFP:SYNV-M. For virus-infected cells, panels showing localization of GFP and RFP fluorescence, and the resultant merge, are provided. Only the overlay images are shown for micrographs taken in mock-inoculated plants. Figure adapted from Goodin *et al.* (2007b).

intranuclear membranes with the ER is essential for delivery of the glycosylated SYNV-G protein to the INM from the ER and Golgi. It is noteworthy that the G protein did not induce formation of intranuclear membranes nor did co-expression of M and G, which for some negative-strand RNA viruses results in the budding of empty particles from transfected cells (Swenson *et al.*, 2004). Therefore, it was hypothesized that formation of the intranuclear membranes may require additional viral proteins and perhaps, viral RNA. Furthermore, while overexpression of viral glycoproteins commonly results in adverse cytopathic effects, such effects seen in mock-inoculated cells were absent or markedly reduced in virus-infected cells. It cannot be ruled out that following budding into the perinuclear space, contiguity with the ER might provide rhabdoviruses with a continuous conduit by which to travel cell to cell. Although the current models for cell-to-cell movement, developed primarily from studies with plus-strand RNA viruses, do not favour virus movement through ER tubules it is also true that such viruses do not accumulate in perinuclear spaces. Indeed, maize mosaic nucleorhabdovirus has been observed in ER tubules in cells of its insect vector (Herold and Munz, 1965), suggesting that this virus can migrate through the endomembrane system. An alternate means for cell-to-cell movement that also requires contiguity of the ER with virus-induced intranuclear membranes is budding of mature virions from the perinuclear space to release the condensed nucleocapsid, which could function as a movement complex. This mechanism would be akin to the bud-in bud-out, envelopment and de-envelopment of Herpes virus particles which allow this virus to move from sites of replication and assembly in the nucleus to the cytoplasm and subsequently, by additional budding events, exit from the infected cell (Mettenleiter *et al.*, 2009).

Both of the movement models above require that viral proteins be delivered to the intranuclear membranes in a manner consistent with the known function of rhabdoviral proteins in assembly (Jayakar *et al.*, 2004). To investigate this, SYNV proteins were expressed as RFP fusions in plant cells, except for the 240-kDa L protein. In contrast to expression in mock-inoculated cells, the localization patterns in virus-infected cells were consistent with models for rhabdovirus assembly and morphogenesis proposed by Green and co-workers (Green *et al.*, 2000; Green and Luo, 2006) and Jayakar *et al.* (2004). The first step in the budding process is formation of nucleocapsids by delivery of the N protein to nascent genomic-length RNAs via an N/P complex. The majority of P protein should be excluded from the RNA/N/P complex (Green and Luo, 2006) to form the mature nucleocapsid, which is in turn delivered to an M/G complex that has formed on membranes (Jackson *et al.*, 2005). Consistent with this hypothesis it was found that the majority of the P protein did not co-localize with the intranuclear membranes. In fact, this protein appears to be excluded immediately adjacent to these membranes at loci occupied by the N and M proteins. In addition to co-localizing in part with membranes, the N protein was found in a highly mobile regions in the nucleoplasm, consistent in location to the viroplasm, which is the proposed site of rhabdoviral replication (Martins *et al.*, 1998).

During the course of mapping the localization of SYNV proteins in the context of virus infections, a previously unreported complex in the cytoplasm of virus-infected cells that incorporated M protein fusions to cyan fluorescent protein (CFP) or RFP was discovered. Further analyses showed that these complexes were liberated from nuclei of virus-infected cells, which then proceeded to track on ER membranes. It has not yet been demonstrated that these complexes can be labelled with fluorescent fusions of other SYNV proteins, most notably N or P, which might indicate that nucleocapsid cores were also associated with these complexes. However, given the small amounts of P protein in virus particles relative to M and N (data not shown), it may not be possible to detect these complexes in the same manner that nucleocapsids of vesicular stomatitis virus have been labelled using GFP:P fusions (Das *et al.*, 2006). Extensive analyses failed to reveal such complexes in mock-inoculated leaves in which RFP:M was co-expressed. Further, that these complexes are ER-associated is intriguing as it suggests that G protein could also be a part of the complex. This hypothesis was born out in later studies that determined all pairwise interactions for SYNV proteins. However, validation that the observed M protein complex is the *bona fide* SYNV movement complex will require extensive characterization by electron microscopy.

Taken together, the live-cell imaging studies conducted in the context of virus-infected cells revealed the spatial relationship between viral proteins that suggests a contiguous pathway from the putative sites of viral replication to those of morphogenesis. Importantly, the ability to express autofluorescent protein fusions in the context of virus-infection represents a significant advance in our ability to study plant–rhabdovirus interactions in living cells.

Formation of SYNV nucleocapsids

Light and electron microscopy have been used to compare the structure of nuclei in rhabdovirus-infected and mock-inoculated leaf tissues. The most readily observed difference is the significant increase in the size of nuclei in virus-infected cells, with SYNV- or PYDV-infected nuclei having at least double the cross-sectional area of those in the absence of virus. Closer examination reveals that the uniform composition of the chromatin is disturbed and that electron-dense 'islands' form immediately adjacent to the nuclear envelope. Moreover, the invaginations formed by the perinuclear membrane are filled with condensed nucleocapsids. One of the most startling findings is that the SYNV-P protein accumulates in the nuclei in ring-like structures, in the context of virus infections (Fig. 9.5A). A similarly ring structure is observed by electron microscopy (EM) (Fig. 9.5B). Although immunolabelling EM has not been conducted yet to establish the composition of these nuclear rings, it is likely that the structures observed in EM and light microscopy are indeed one and the same. Examination of the rings by EM revealed that nucleocapsids condense on the perimeter of the rings (Fig. 9.5C). Presumably, these immature particles then migrate towards the perinuclear space where they accumulate. A key question arises as to where P protein rings nucleate. Unpublished protein localization experiments have ruled out that the rings nucleate on nucleoli given that P protein rings do not co-localize with nucleoli in virus-infected cells (Goodin and Anderson, unpublished). However, it is intriguing to speculate that the spatial localization of the P protein is tightly regulated and is perhaps a key element in the separation of sites of virus replication from sites of morphogenesis. Such regulation is without a doubt related not only to the formation of viral protein complexes, but also to the association of host factors as well. Thus, it is increasingly clear that detailed analyses of viral and plant protein interaction are essential for understanding the molecular basis of pathogenesis within nucleorhabdovirus pathosystems.

Protein interaction maps for plant-adapted rhabdoviruses

Comprehensive protein interaction maps, particularly when supported by protein localization data are critical for understanding molecular mechanisms that mediate viral infection and transmission processes at the systems level (Uetz and Finley, 2005; Shapira *et al.*, 2009; Bandyopadhyay *et al.*, 2010; Min *et al.*, 2010; Zilian and Maiss, 2011). They also permit identification of potential targets that can be exploited to engineer virus-resistant plants. Surprisingly, in spite of their threat to sustainable agricultural systems no high-resolution protein interaction and localization maps have been generated for any plant virus. However, an increasing amount of such data are accumulating for plant-adapted rhabdoviruses (Bandyopadhyay *et al.*, 2010; Min *et al.*, 2010). To date, protein interaction maps have been published for PYDV and SYNV, with an additional map for the *Cytorhabdovirus* lettuce necrotic yellows virus generated recently (Martin *et al.*, 2012).

Of the more interesting findings from this and related protein interaction studies is the inconsistency among interactions of nucleorhabdoviral movement proteins, which raises the possibility that each virus employs are unique strategy for cell-to-cell movement. In the simplest model, the

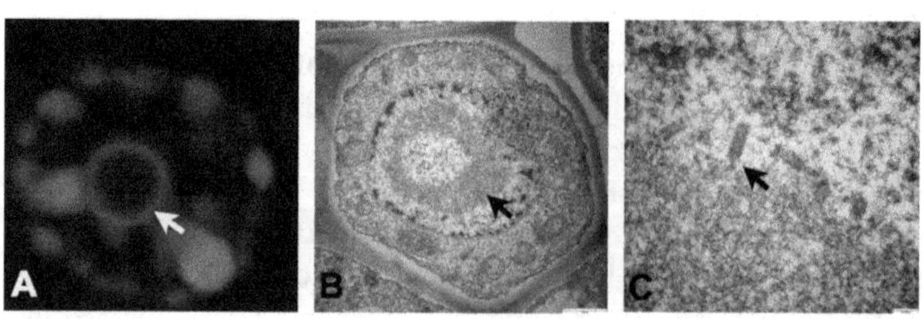

Figure 9.5 (A) Ring-shaped accumulation of SYNV-P (arrow) in nuclei of virus-infected *N. benthamiana* cells. Intranuclear membranes also accumulate in response to infections. (B) Electron micrograph of an SYNV-infected nucleus. Note formation of a ring-shaped electron dense complex (arrow). (C) Enlargement of the edge of the ring shown in B identifies the site of nucleocapsid formation (arrow). Panel A from Goodin *et al.* (2007b). Panels B and C courtesy of Tea Meulia and Lucy Stewart (unpublished).

P3 movement protein of rice yellow stunt virus (RYSV) interacts with it's cognate N protein, suggesting that the nucleocapsid is the movement complex. In contrast, the Y protein of PYDV interacts with its cognate M protein, which in turn interacts with the N protein. This suggests that a condensed nucleocapsid may function as the PYDV movement complex. An extended variation on this theme is the finding that the SYNV sc4 protein contacts its M protein via a host factor Mi7, which can be found on mobile complexes in the cytoplasm of virus-infected cells. Additionally, sc4 also interacts with the SYNV-G protein. Taken together, these interactions suggest that SYNV may move as a membrane-associated condensed nucleocapsid. Indeed, these finding are consistent with the report of M protein-containing complexes that migrate on 'rings' derived from ER membranes (Goodin *et al.*, 2007b).

A model for cell-to-cell transport for plant adapted rhabdoviruses

In order to gain an in-depth understanding of viral infection processes viral protein–protein interaction maps must be expanded to include host proteins (Uetz and Finley, 2005; Shapira *et al.*, 2009). Although it is generally established that in order to move from initially infected cells

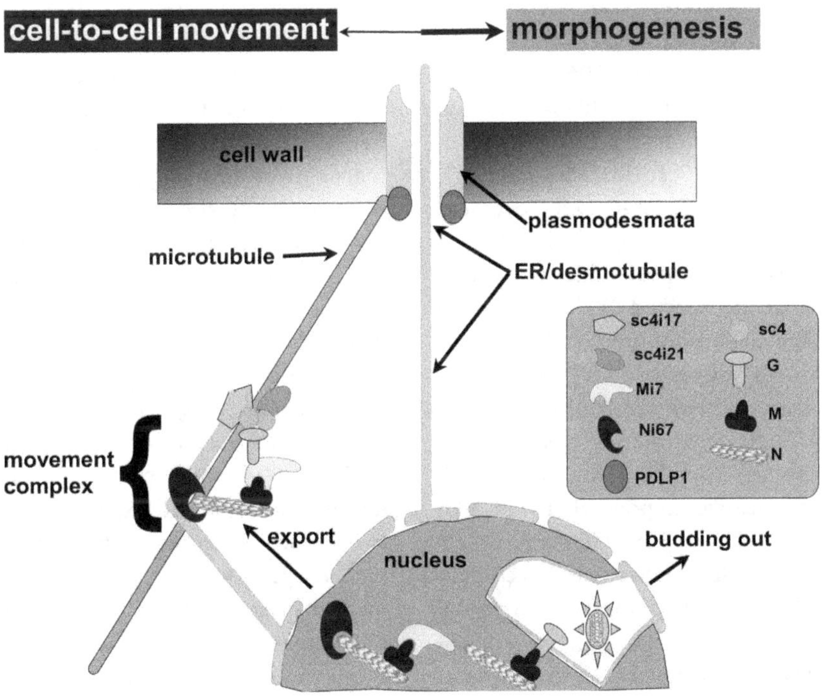

Figure 9.6 Model of the SYNV cell-to-cell movement complex. Sc4i21 and sc4i17 are a microtubule-tethered transcription activator and motor-kinesin, respectively. Ni67 is an ER-tethered transcription activator that may shuttle into the nucleus where it interacts with the SYNV-N protein and, presumably, SYNV nucleocapsids. Mi7 is also a transcription factor that interacts with the M protein, and is also probably a nuclear shuttle protein. Current data suggest that Mi7 and Ni67 could function cooperatively to export nucleocapsids from the nucleus, to establish a movement complex with sc4 and G, or its truncated derivative G*, in the cytoplasm. Nucleocapsids that are not exported bud through the inner nuclear membrane to form mature virions that accumulate in the perinuclear space. Based on EM data, it is likely that most nucleocapsids undergo morphogenesis (bold arrow, top) given that the movement complex has not been reported previously. In addition to protein interaction data shown below, evidence for an ER-associated M protein containing movement complex has been reported previously (Goodin *et al.*, 2007b). Recent live-cell imaging studies and pull-down studies have allowed the plasmodesmata-localized protein (AtPDLP1) to be added to the model, proposed by Min *et al.* (2010).

into adjacent cells plant viruses make use of the ER (Bamunusinghe *et al.*, 2011) and at least one type of protein filament of the cytoskeleton, be it actin (Prokhnevsky *et al.*, 2005) or microtubules (Epel, 2009; Sambade and Heinlein, 2009), no additional host factors have been implicated in the movement of plant-adapted rhabdoviruses.

Under the assumption that binding of host factors to the N protein is equivalent to binding to the nucleocapsid *per se*, a model for SYNV cell-to-cell movement was proposed that requires the SYNV N, M, sc4 and G proteins, or a truncated version termed G* (Min *et al.*, 2010). We believe this assumption, also made by Huang *et al.* (2005), to be valid, given that the N protein readily forms viroplasm-like complexes even in the absence of other viral proteins (Martins *et al.*, 1998; Deng *et al.*, 2007) and that N proteins can self-assemble into ribonucleoprotein complexes in the absence of viral RNA (Green and Luo, 2006). Additionally, based upon the results of high-resolution yeast two-hybrid interaction screens, at least four host factors, termed sc4i17, sc4i21, Mi7 and Ni67, have been implicated in the cell-to-cell movement of SYNV (Fig. 9.6).

A variety of web-based proteomics tools were use to characterize the selected SYNV-interacting host factors. The deduced amino acid sequence of the M protein interactor, Mi7, encodes a small 21 kDa basic protein with a pI of 9.7. Mi7 contains a DUF640 domain conserved in some transcription factors including the *Arabidopsis thaliana* Light sensitive hypocotyl 10 protein (TAIR reference AT1G78815; Genbank accession NM_106529). Mi7 was predicted to contain both a PYKKKKK nuclear localization signal (NLS) starting at amino acid 155, and a VLEFLRYLD nuclear export signal (NES) starting at amino acid residue 66. The steady-state localization pattern for Mi7 was nuclear.

The sc4 interactor sc4i17 is an 80 kDa protein with a pI of 6.0 and is 98% identical to the *N. tabacum* BY-2 kinesin-like protein 10 (Matsui *et al.*, 2001). Consistent with the predicted motor domain at amino acids 186 to 508 in the deduced protein sequence, sc4i17 localized to punctate loci on microtubules in *N. benthamiana* leaf epidermal cells.

The sc4 interactor, sc4i21, is a 53 kDa, pI 5.8, protein that contains a conserved 'no apical meristem' (NAM) domain found in some transcription factors, such as the phloem-associated Arabidopsis vascular one zinc-finger protein 1 (AtVOZ1; (Mitsuda *et al.*, 2004)), with which it shares 65% identity. Although sc4i21 contains a predicted NLS (314-KPRR-317) and NES (382-LYRLELKLVD-391) its steady-state localization was on microtubules. Interestingly, Ni67 which is an N protein interactor of 53 kDa with a pI 5.5, shares 84% sequence identity with sc4i21, with identically positioned NLS and NES sequences. However, the steady-state localization of Ni67 was in the ER.

At approximately 53 kD, it is clear that both Ni67 and sc4i21 are distinct from the 10 kDa MBF1 transcriptional coactivator that was shown to interact with p30 of *Tobacco mosaic virus* (TMV) (Matsushita *et al.*, 2002). Interestingly, both sc4i21 and Ni67 contain a NAM domain, as does the smaller 32 kDa ATAF2 that binds to the helicase domain of the TMV replicase and that is involved in suppression of systemic host defence (Wang *et al.*, 2009). Thus, to the best of our knowledge, neither the host factors reported here, nor their homologues in other plants, have been shown to participate in the biology of other plant viruses.

Min *et al.* (2010) proposed that nucleocapsids, or condensed cores, which are not targeted to the perinuclear space associate instead with Ni67 and Mi7, both of which contain functional nuclear import and predicted export signals. The use of two factors for export is attractive as it could provide a regulatory mechanism for exporting only nucleocapsids, or partially condensed cores, and not simply oligomers of the N and M proteins. It is quite interesting that VOZ1, the closest named homologue of both sc4i21 and Ni67, is expressed primarily in the phloem (Mitsuda *et al.*, 2004). It stands to reason that a nucleocapsid destined for cell-to-cell movement would associate with phloem-associated proteins, particularly if it was a non-cell autonomous protein (NCAP) with an intrinsic ability to move between cells (Oparka, 2004). Once exported, Ni67 may anchor nucleocapsids onto ER membranes bringing with it the associated M and Mi7 proteins, both of which associate with mobile ER-associated complexes

(Goodin *et al.*, 2007b and data not shown). The ER-associated complex could then be tethered to microtubules via a complex containing the motor-kinesin, sc4i17, as well as sc4i21, and a truncated, and presumably cytoplasmic, G protein. Once formed, the complete ER-associated movement complex could track on microtubules towards plasmodesmata in a manner similar to that of NCAPs. Given the punctate localization of the Ni67/sc4i21 on cell periphery, which could be plasmodesmata, we propose that upon reaching this structure, a Ni67/sc4i21 interaction could provide a mechanism to release the nucleocapsid into the adjacent cell. Some such release of the nucleocapsid from the movement complex is required in order to allow the nucleocapsid to be ferried to the nucleus, probably via an importin-α-mediated process (Deng *et al.*, 2007), to initiate the next infection cycle (Jackson *et al.*, 2005). Unpublished data from our laboratory have recently shown that the plasmodesmata localized protein 1 (PDLP1; (Amari *et al.*, 2010) interacts with sc4. This suggests that the SYNV movement complex could dock at plasmodesmata via an sc4–PDLP1 interaction (B.-E. Min and M.M. Goodin, unpublished).

The present model for SYNV cell-to-cell movement (Fig. 9.6) was established entirely on the basis of protein interaction and localization studies. Although the interactions have been validated by independent methods, an absolute requirement for these host factors in cell-to-cell movement has not been established either by VIGS or in the context of transgenic plants (Liu *et al.*, 2002; Park *et al.*, 2009) to significantly reduce the steady-state levels of host factor mRNAs, which accumulate to very low levels in both mock-inoculated and SYNV-infected plants.

The model for cell-to-cell transport of SYNV nucleocapsids is distinctly different from that proposed by Huang *et al.* (2005), who suggested that movement of RYSV required direct binding of P3 to nucleocapsids. However, these authors only investigated binding between P3 and N, whereas we have considered pairwise interactions for all SYNV proteins, except for L, which cannot be expressed or purified readily in quantities sufficient for interaction assays. Additionally, the SYNV model integrates data demonstrating the association of sc4 with G protein and its co-purification with membrane-associated nucleocapsids but not with nucleocapsids treated with organic solvents to remove membranes prior to purification (Scholthof *et al.*, 1994). It is possible that different rhabdoviruses, particularly those with dicot versus monocot hosts, could utilize different mechanisms for cell-to-cell transport.

Transcription factors that are tethered in the cytoplasm are typically membrane-associated (Kim *et al.*, 2010), and are typically members in the bZIP (basic leucine zipper) and NAC (NAM/ATAF1-2/CUC2) transcription factor families, which include AtVOZ1, Ni67 and sc4i21. At present, known cytoplasm-tethered transcription factors utilize trans-membrane anchors to mediate their retention in the ER, and are released from these membranes upon proteolytic cleavage. Ni67 is ER-associated, but lacks a predicted trans-membrane domain or signal peptide, while sc4i21 is microtubule-associated. This suggests that there may be multiple mechanisms to retain transcription factors outside of the nucleus. This contention is supported by the reports from animal and plant systems that implicate microtubules in transcriptional activation by NFkB and RNA processing (Hamada *et al.*, 2009; Jackman *et al.*, 2009). Consistent with the localization of Ni67, AtVOZ fused to tagRFP, accumulated in ER-associated complexes in *N. benthamiana* leaves (B.-E. Min M.M. and Goodin, unpublished data).

Conclusions

The plant-adapted rhabdoviruses are a particularly fascinating group of pathogens. Not only are they the agents of important diseases of agronomic importance, but their ability to replicate in both insect and plant cells is a unique property shared by a select few of all known viruses. With their nucleartrophic habit, the nucleorhabdoviruses as a group provide models for the study of plant–virus interactions, as well as nuclear functions in plant cells, which is a very understudied area.

Advances in nucleorhabdovirus research will be accelerated by resources such as transgenic marker lines (Chakrabarty *et al.*, 2007; Goodin *et al.*, 2007a), high-resolution two-hybrid libraries for isolating *N. benthamiana* proteins (Min *et al.*, 2010), as well as additional resources for this model plant that is so vital to plant virology research (Goodin *et al.*, 2008). Expanding the present protein interaction and localization maps is essential to understanding the molecular details of disease in nucleorhabdovirus pathosystems. Additionally, while light microscopy has been used to provide great advances, future studies will be increasingly dependent on electron microscopy, as much of the ultrastructure associated with rhabdovirus–plant interaction has not been defined in detail. Finally, of critical importance, given the biology of these viruses is development of research models to facilitate the parallel study of nucleorhabdoviruses in insect and plant cells.

Acknowledgements

The authors wish to acknowledge Lucy Stewart (USDA-ARS Corn and Soybean Research Unit) and Tea Meulia (Molecular and Cellular Imaging Center, The Ohio State University) for generously providing electron micrographs used in Fig. 9.5 prior to publication

References

Amari, K., Boutant, E., Hofmann, C., Schmitt-Keichinger, C., Fernandez-Calvino, L., Didier, P., Lerich, A., Mutterer, J., Thomas, C.L., Heinlein, M., *et al.* (2010). A family of plasmodesmal proteins with receptor-like properties for plant viral movement proteins. PLoS Pathog. *6*, e1001119.

Ammar, E.D., Tsai, C.W., Whitfield, A.E., Redinbaugh, M.G., and Hogenhout, S.A. (2009). Cellular and molecular aspects of rhabdovirus interactions with insect and plant hosts. Annu. Rev. Entomol. *54*, 447–468.

Bamunusinghe, D., Seo, J.K., and Rao, A.L. (2011). Subcellular localization and rearrangement of endoplasmic reticulum by brome mosaic virus capsid protein. J. Virol. *85*, 2953–2963.

Bandyopadhyay, A., Kopperud, K., Anderson, G., Martin, K., and Goodin, M. (2010). An integrated protein localization and interaction map for Potato yellow dwarf virus, type species of the genus *Nucleorhabdovirus*. Virology *402*, 61–71.

van Beek, N.A.M., Lohuis, D., Dijkstra, J., and Peters, D. (1985). Morphogenesis of Sonchus yellow net virus in cowpea protoplasts. J. Ultrastruct. Res. *90*, 294–303.

Chakrabarty, R., Banerjee, R., Chung, S.M., Farman, M., Citovsky, V., Hogenhout, S.A., Tzfira, T., and Goodin, M. (2007). pSITE vectors for stable integration or transient expression of autofluorescent protein fusions in plants: probing *Nicotiana benthamiana*–virus interactions. Mol. Plant Microbe Interact. *20*, 740–750.

Chen, B.J., and Lamb, R.A. (2008). Mechanisms for enveloped virus budding: can some viruses do without an ESCRT? Virology *372*, 221–232.

Das, S.C., Nayak, D., Zhou, Y., and Pattnaik, A.K. (2006). Visualization of intracellular transport of vesicular stomatitis virus nucleocapsids in living cells. J. Virol. *80*, 6368–6377.

Deng, M., Bragg, J.N., Ruzin, S., Schichnes, D., King, D., Goodin, M.M., and Jackson, A.O. (2007). Role of the sonchus yellow net virus N protein in formation of nuclear viroplasms. J. Virol. *81*, 5362–5374.

Dietzgen, R.G., Calisher, C.H., Kurath, G., Kuzmin, I.V., Rodriguez, L.L., Stone, D.M., Tesh, R.B., Tordo, N., Walker, P.J., Wetzel, T., *et al.* (2011). Rhabdoviridae. In Virus Taxonomy, Ninth Report of the International Committee on Taxonomy of Viruses, King, A.M.Q., Adams, M.J., Carstens, E.B., Lefkowitz, E.J., eds. (Elsevier, San Diego, CA), pp. 654–681.

Ellenberg, J., and Lippincott-Schwartz, J. (1999). Dynamics and mobility of nuclear envelope proteins in interphase and mitotic cells revealed by green fluorescent protein chimeras. Methods *19*, 362–372.

Ellenberg, J., Siggia, E.D., Moreira, J.E., Smith, C.L., Presley, J.F., Worman, H.J., and Lippincott-Schwartz, J. (1997). Nuclear membrane dynamics and reassembly in living cells: targeting of an inner nuclear membrane protein in interphase and mitosis. J. Cell. Biol. *138*, 1193–1206.

Epel, B.L. (2009). Plant viruses spread by diffusion on ER-associated movement-protein-rafts through plasmodesmata gated by viral induced host beta-1,3-glucanases. Semin. Cell Dev. Biol. *20*, 1074–1081.

Ghosh, D., Brooks, R.E., Wang, R., Lesnaw, J., and Goodin, M.M. (2008). Cloning and subcellular localization of the phosphoprotein and nucleocapsid proteins of potato yellow dwarf virus, type species of the genus *Nucleorhabdovirus*. Virus Res. *135*, 26–35.

Glodowski, D.R., Petersen, J.M., and Dahlberg, J.E. (2002). Complex nuclear localization signals in the matrix protein of vesicular stomatitis virus. J. Biol. Chem. *277*, 46864–46870.

Goodin, M., Yelton, S., Ghosh, D., Mathews, S., and Lesnaw, J. (2005). Live-cell imaging of rhabdovirus-induced morphological changes in plant nuclear membranes. Mol. Plant Microbe Interact. *18*, 703–709.

Goodin, M.M., Austin, J., Tobias, R., Fujita, M., Morales, C., and Jackson, A.O. (2001). Interactions and nuclear import of the N and P proteins of sonchus yellow net virus, a plant *Nucleorhabdovirus*. J. Virol. *75*, 9393–9406.

Goodin, M.M., Chakrabarty, R., Banerjee, R., Yelton, S., and Debolt, S. (2007a). New gateways to discovery. Plant Physiol. *145*, 1100–1109.

Goodin, M.M., Chakrabarty, R., Yelton, S., Martin, K., Clark, A., and Brooks, R. (2007b). Membrane and protein dynamics in live plant nuclei infected with Sonchus yellow net virus, a plant-adapted rhabdovirus. J. Gen. Virol. *88*, 1810–1820.

Goodin, M.M., Zaitlin, D., Naidu, R.A., and Lommel, S.A. (2008). Nicotiana benthamiana: its history and future as a model for plant–pathogen interactions. Mol. Plant Microbe Interact. *21*, 1015–1026.

Green, T.J., and Luo, M. (2006). Resolution improvement of X-ray diffraction data of crystals of a vesicular stomatitis virus nucleocapsid protein oligomer complexed with RNA. Acta Crystallogr. D Biol. Crystallogr. *62*, 498–504.

Green, T.J., Macpherson, S., Qiu, S., Lebowitz, J., Wertz, G.W., and Luo, M. (2000). Study of the assembly of vesicular stomatitis virus N protein: role of the P protein. J. Virol. *74*, 9515–9524.

Hamada, T., Itoh, T.J., Hashimoto, T., Shimmen, T., and Sonobe, S. (2009). GTP is required for the microtubule catastrophe-inducing activity of MAP200, a tobacco homolog of XMAP215. Plant Physiol. *151*, 1823–1830.

Hammond, D.C., Evans, R.K., and Lesnaw, J.A. (1992). The L protein of vesicular stomatitis virus transcription complexes is specifically photolabelled by 5-azido-uridine 5′-triphosphate, an analogue of the RNA polymerase substrate uridine 5′-triphosphate. J. Gen. Virol. *73*, 61–66.

Herold, F., and Munz, K. (1965). Electron microscopic demonstration of viruslike particles in *Peregrinus maidis* following acquisition of maize mosaic virus. Virology *25*, 412–417.

Huang, Y.W., Geng, Y.F., Ying, X.B., Chen, X.Y., and Fang, R.X. (2005). Identification of a movement protein of rice yellow stunt rhabdovirus. J. Virol. *79*, 2108–2114.

Irie, T., Nagata, N., Yoshida, T., and Sakaguchi, T. (2008). Recruitment of Alix/AIP1 to the plasma membrane by sendai virus C protein facilitates budding of virus-like particles. Virology *371*, 108–120.

Irons, S.L., Evans, D.E., and Brandizzi, F. (2003). The first 238 amino acids of the human lamin B receptor are targeted to the nuclear envelope in plants. J. Exp. Bot. *54*, 943–950.

Jackman, R.W., Rhoads, M.G., Cornwell, E., and Kandarian, S.C. (2009). Microtubule-mediated NF-kappaB activation in the TNF-alpha signaling pathway. Exp. Cell. Res. *315*, 3242–3249.

Jackson, A.O., Dietzgen, R.G., Goodin, M.M., Bragg, J.N., and Deng, M. (2005). Biology of plant rhabdoviruses. Annu. Rev. Phytopathol. *43*, 623–660.

Jayakar, H.R., Jeetendra, E., and Whitt, M.A. (2004). Rhabdovirus assembly and budding. Virus Res. *106*, 117–132.

Kim, S.G., Lee, S., Seo, P.J., Kim, S.K., Kim, J.K., and Park, C.M. (2010). Genome-scale screening and molecular characterization of membrane-bound transcription factors in Arabidopsis and rice. Genomics *95*, 56–65.

Lacroix, B., Li, J., Tzfira, T., and Citovsky, V. (2006). Will you let me use your nucleus? How Agrobacterium gets its T-DNA expressed in the host plant cell. Can. J. Physiol. Pharmacol. *84*, 333–345.

Lippincott-Schwartz, J., Presley, J.F., Zaal, K.J., Hirschberg, K., Miller, C.D., and Ellenberg, J. (1999). Monitoring the dynamics and mobility of membrane proteins tagged with green fluorescent protein. Meth. Cell. Biol. *58*, 261–281.

Liu, Y., Schiff, M., and Dinesh-Kumar, S.P. (2002). Virus-induced gene silencing in tomato. Plant J. *31*, 777–786.

Martin, K.M., Dietzgen, R.G., Wang, R.-Y., and Goodin, M.M. (2012). Lettuce necrotic yellows *Cytorhabdovirus* protein localization and interaction map and comparison with *Nucleorhabdoviruses*. J. Gen. Virol. *93*, 906–914.

Martins, C.R., Johnson, J.A., Lawrence, D.M., Choi, T.J., Pisi, A.M., Tobin, S.L., Lapidus, D., Wagner, J.D., Ruzin, S., McDonald, K., *et al.* (1998). Sonchus yellow net rhabdovirus nuclear viroplasms contain polymerase-associated proteins. J. Virol. *72*, 5669–5679.

Matsui, K., Collings, D., and Asada, T. (2001). Identification of a novel plant-specific kinesin-like protein that is highly expressed in interphase tobacco BY-2 cells. Protoplasma *215*, 105–115.

Matsushita, Y., Miyakawa, O., Deguchi, M., Nishiguchi, M., and Nyunoya, H. (2002). Cloning of a tobacco cDNA coding for a putative transcriptional coactivator MBF1 that interacts with the tomato mosaic virus movement protein. J. Exp. Bot. *53*, 1531–1532.

Mettenleiter, T.C., Klupp, B.G., and Granzow, H. (2009). Herpesvirus assembly: an update. Virus Res. *143*, 222–234.

Min, B.E., Martin, K., Wang, R., Tafelmeyer, P., Bridges, M., and Goodin, M. (2010). A host–factor interaction and localization map for a plant-adapted rhabdovirus implicates cytoplasm-tethered transcription activators in cell-to-cell movement. Mol. Plant Microbe Interact. *23*, 1420–1432.

Mitsuda, N., Hisabori, T., Takeyasu, K., and Sato, M.H. (2004). VOZ; isolation and characterization of novel vascular plant transcription factors with a one-zinc finger from Arabidopsis thaliana. Plant Cell. Physiol. *45*, 845–854.

Oparka, K.J. (2004). Getting the message across: how do plant cells exchange macromolecular complexes? Trends Plant Sci. *9*, 33–41.

Park, M.R., Park, S.H., Cho, S.Y., and Kim, K.H. (2009). *Nicotiana benthamiana* protein, NbPCIP1, interacting with Potato virus X coat protein plays a role as susceptible factor for viral infection. Virology *386*, 257–269.

Prokhnevsky, A.I., Peremyslov, V.V., and Dolja, V.V. (2005). Actin cytoskeleton is involved in targeting of a viral Hsp70 homolog to the cell periphery. J. Virol. *79*, 14421–14428.

Qanungo, K.R., Shaji, D., Mathur, M., and Banerjee, A.K. (2004). Two RNA polymerase complexes from vesicular stomatitis virus-infected cells that carry out transcription and replication of genome RNA. Proc. Natl. Acad. Sci. U.S.A. *101*, 5952–5957.

Redinbaugh, M.G., Seifers, D.L., Meulia, T., Abt, J.J., Anderson, R.J., Styer, W.E., Ackerman, J., Salomon, R., Houghton, W., Creamer, R., *et al.* (2002). Maize fine streak virus, a new leafhopper-transmitted rhabdovirus. Phytopathol. *92*, 1167–1174.

Sambade, A., and Heinlein, M. (2009). Approaching the cellular mechanism that supports the intercellular spread of Tobacco mosaic virus. Plant Signal. Behav. *4*, 35–38.

Scholthof, K.B., Hillman, B.I., Modrell, B., Heaton, L.A., and Jackson, A.O. (1994). Characterization and detection of sc4: a sixth gene encoded by sonchus yellow net virus. Virology *204*, 279–288.

Shapira, S.D., Gat-Viks, I., Shum, B.O., Dricot, A., de Grace, M.M., Wu, L., Gupta, P.B., Hao, T., Silver, S.J., Root, D.E., *et al.* (2009). A physical and regulatory map of host–influenza interactions reveals pathways in H1N1 infection. Cell *139*, 1255–1267.

Steinkamp, M.P., and Hoefert, L.L. (1977). Annulate lamellae in phloem cells of virus-infected Sonchus plants. J. Cell Biol. *74*, 111–118.

Swenson, D.L., Warfield, K.L., Kuehl, K., Larsen, T., Hevey, M.C., Schmaljohn, A., Bavari, S., and Aman, M.J. (2004). Generation of Marburg virus-like particles by co-expression of glycoprotein and matrix protein. FEMS Immunol. Med. Microbiol. *40*, 27–31.

Tsai, C.W., Redinbaugh, M.G., Willie, K.J., Reed, S., Goodin, M., and Hogenhout, S.A. (2005). Complete genome sequence and in planta subcellular localization of maize fine streak virus proteins. J. Virol. *79*, 5304–5314.

Tzfira, T., Li, J., Lacroix, B., and Citovsky, V. (2004). Agrobacterium T-DNA integration: molecules and models. Trends Genet. *20*, 375–383.

Uetz, P., and Finley, R.L., Jr. (2005). From protein networks to biological systems. FEBS Lett. *579*, 1821–1827.

Wagner, J.D., and Jackson, A.O. (1997). Characterization of the components and activity of *Sonchus* yellow net rhabdovirus polymerase. J. Virol. *71*, 2371–2382.

Wang, K., Uppalapati, S.R., Zhu, X., Dinesh-Kumar, S.P., and Mysore, K.S. (2010). SGT1 positively regulates the process of plant cell death during both compatible and incompatible plant–pathogen interactions. Mol. Plant Pathol. *11*, 597–611.

Wang, X., Goregaoker, S.P., and Culver, J.N. (2009). Interaction of the Tobacco mosaic virus replicase protein with a NAC domain transcription factor is associated with the suppression of systemic host defenses. J. Virol. *83*, 9720–9730.

Xu, X.M., Meulia, T., and Meier, I. (2007). Anchorage of plant RanGAP to the nuclear envelope involves novel nuclear-pore-associated proteins. Curr. Biol. *17*, 1157–1163.

Zilian, E., and Maiss, E. (2011). Detection of plum pox potyviral protein–protein interactions *in planta* using an optimized mRFP-based bimolecular fluorescence complementation system. J. Gen. Virol. *174*, 158–165.

Insect Vector Interaction and Transmission of Cereal-infecting Rhabdoviruses

10

Margaret G. Redinbaugh, Anna E. Whitfield and El-Desouky Ammar

Abstract

The cereal-infecting cytorhabdoviruses and nucleorhabdoviruses are transmitted to new plant hosts in a persistent, propagative manner by their leafhopper (Cicadellidae) and planthopper (Fulgoroidea) vectors. We discuss our current knowledge of virus acquisition by, replication and movement in, and transmission by their vectors. Barriers to virus acquisition and infection of specific tissues in the insect host in determining vector competence between and among vector species are beginning to be defined. Emerging information on vector genomes indicates the presence of genes with similarity to those important for rhabdovirus uptake into cells, and the innate immune system of mammals. Gaps in our knowledge of what makes a vector are identified and future directions for research are discussed.

Plant and cereal-infecting rhabdoviruses

Nucleorhabdoviruses and cytorhabdoviruses

The two genera of plant-infecting rhabdoviruses, *Cytorhabdovirus* and *Nucleorhabdovirus,* were originally classified according to the site of virion morphogenesis in host cells. In both their plant and insect hosts, cytorhabdoviruses such as barley yellow striate mosaic virus (BYSMV; see Table 10.1 for a list of virus acronyms) mature similarly to the vertebrate-infecting rhabdoviruses by budding through the endoplasmic reticulum and associated cytoplasmic cisternae where mature virus particles accumulate (reviewed in Jackson, 1987; Redinbaugh and Hogenhout, 2005). In contrast, virions of nucleorhabdoviruses, such as maize mosaic virus (MMV) and maize fine streak virus (MFSV), assemble in plant cell nuclei, and mature by budding through the inner nuclear membrane to accumulate in the perinuclear space and membrane-bound cisternae that protrude from the nucleus into the cytoplasm (Fig. 10.1a) (Ammar *et al.*, 2005). More recently, support for separation of the plant rhabdoviruses into two genera has come from phylogenetic analysis of the viral L (RNA-dependent RNA polymerase) gene sequence, with the nucleorhabdoviruses and cytorhabdoviruses forming sister clades separated from the other rhabdovirus genera (Bourhy *et al.*, 2005; Kumin *et al.*, 2009; Longdon *et al.*, 2010).

Vectors and transmission

Most plant-infecting rhabdoviruses are naturally transmitted by hemipteran insects, especially leafhoppers (Cicadellidae), planthoppers (Fulgoroidea) or aphids (Aphidoidea) (Jackson *et al.*, 2005; Redinbaugh and Hogenhout, 2005), although a few putative plant rhabdoviruses are transmitted by lacebugs (Piesmidae) or brevipalpus mites (Acarina) (Ammar *et al.*, 2009; Kitajima *et al.*, 2010). Plant rhabdoviruses are not seed transmitted, and most are not easily transmitted mechanically to their host plants (Hogenhout *et al.*, 2003; Jackson *et al.*, 2005). All characterized cereal-infecting rhabdoviruses have leafhopper or planthopper vectors (Table 10.1). However, the cereal-infecting rhabdoviruses cannot be distinguished by genus on the basis of vector taxonomy. Three members of each virus genus are transmitted by delphacid planthoppers, and two cytorhabdoviruses and four nucleorhabdoviruses are transmitted by

Table 10.1 Vector families and species for cereal-infecting rhabdoviruses.[a]

Virus genus and species[b]	Abbreviation[c]	Vectors[d]	Vector family	Crop hosts[e]	Distribution[f]	References
Cytorhabdovirus						
Barley yellow striate mosaic virus	BYSMV	*Laodelphax striatellus*	Del.	Barley, wheat, oat, rye	Mediterranean	Persley and Greber (2004a)
Maize sterile stunt virus (a strain of BYSMV)	MSSV	*Sogatella longifurcifera*	Del.	Maize	Australia	Greber (1982)
Festuca leaf streak virus	FLSV	*Javesella pellucida*	Del.	Giant fescue	Denmark Germany	Lundsgaard (1984, 1999)
Northern cereal mosaic virus[g]	NCMV	*Laodelphax striatellus, Unkanodes albifascia, U. sapporonus* and *Muellerianella fairmairei*	Del.	Oat, wheat barley, rye, annual ryegrass	Japan, China, Korea	Toriyama (2004)
Wheat Am. Striate mosaic virus[h]	WASMV	*Endria inimica* and *Elymana virescens*	Cic.	Wheat, maize	Canada USA	Seifers (2004)
Winter wheat Russian mosaic virus	WWRMV	Psammotettix alienus P. straitus	Cic.	Oat, wheat, barley, rye	Russia E. Europe	Redinbaugh (2010)
Nucleorhabdovirus						
Cereal chlorotic mottle virus	CCoMV	*Nesoclutha pallida* and *Cicadulina* spp.	Cic.	Maize, barley, oat	Australia Morocco	Persley and Greber (2004b)
Maize fine streak virus	MFSV	*Graminella nigrifrons*	Cic.	Maize	USA	Redinbaugh *et al.* (2002)
Maize mosaic virus	MMV	*Peregrinus maidis*	Del.	Maize	World-wide in tropics and sub-tropics	Falk and Tsai (1985), Ammar *et al.* (2004)
Rice yellow stunt virus	RYSV	*Nephotettix* spp.	Cic.	Rice	Asia	Hibino (1996), Signoret (2004)
Cynodon chlorotic streak virus	CCSV	*Toya propinqua* and *L. striatellus*	Del.	Bermudagrass, maize	Mediterranean	Lockhart *et al.* (1985), Lockhart (2004)
Maize Iranian mosaic virus	MIMV	*U. tanasijevici* and *L. striatellus*	Del.	Wheat, barley, maize, sorghum	Iran	Izadpanah *et al.* (1983); Izadpanah (1989)
Sorghum stunt mosaic virus	SSMV	Graminella sonora	Cic.	Sorghum, maize	USA	Mayhew and Flock (1981); Creamer *et al.* (1997)

Del., Delphacideae; Cic., Cicadellidae.
[a]Modified from Ammar and Nault (2002), Redinbaugh and Hogenhout (2005), and Ammar *et al.* (2009).
[b]Known cereal-infecting cytorhabdoviruses and nucleorhabdoviruses are listed.
[c]Known natural vectors, experimental vectors are not included.
[d]Crop hosts include: oat, *Avena sativa*; wheat, *Triticum* spp.; barley, *Hordeum vulgare*; rye, *Secale cereale*; giant

cicadellid leafhoppers. Cereal rhabdoviruses have highly specific interactions with their vectors, and most are naturally transmitted by one or a few closely related insect species (Table 10.1). Frequently, not all closely related species can act as vectors; e.g. MFSV can be transmitted by *Graminella nigrifrons* but not by the related leafhopper *Endria inimica* (Redinbaugh *et al.*, 2002), and sorghum stunt mosaic virus (SSMV) is transmitted by *G. sonora* but not by *G. nigrifrons* (Creamer *et al.*, 1997). When multiple insect vectors have been identified for a virus, some are frequently more effective vector than others. For example, *Nephotettix cincticeps* and *N. nigropictus* are efficient vectors of Rice yellow stunt virus (RYSV, synonymous with rice transitory yellows virus), but transmission efficiency by *N. virescens* is much lower (Hibino, 1996; Signoret, 2004). Similarly, *E. inimica* is an efficient vector of wheat American striate mosaic virus (WASMV), while *Elymana virescens* is less efficient (Sinha, 1970). Further, maize Iranian mosaic virus (MIMV) is transmitted efficiently by *Unkanodes tanasijevici* under laboratory conditions, but is transmitted very inefficiently by the MMV vector *Peregrinus maidis* after feeding on infected plants (Nault and Ammar, 1989). However, *P. maidis* transmits MIMV efficiently if the virus is injected into the insect's hemocel (Nault and Ammar, 1989; Ammar *et al.*, 2005).

Cereal rhabdovirus genome organization and function

All rhabdovirus genomes encode five 'core' structural proteins namely the nucleoprotein (N), phosphoprotein (P), matrix protein (M), glycoprotein (G) and large polymerase (L) protein (Dietzgen *et al.*, 2011). *Cytorhabdovirus* and *Nucleorhabdovirus* genomes encode these five core proteins plus from one to five accessory proteins, with at least one additional gene located between the viral P and M genes (Fig. 10.2) (Heaton *et al.*, 1989; Scholthof *et al.*, 1994; Wetzel *et al.*, 1994; Tanno *et al.*, 2000; Huang *et al.*, 2005; Tsai *et al.*, 2005; Walker *et al.*, 2011). At least one of the proteins encoded in this position has been proposed to be a 'movement protein' that facilitates intracellular movement between plant cells via plasmodesmata (Huang *et al.*, 2005; Jackson *et al.*, 2005; Ammar *et al.*, 2009). The cereal-infecting RYSV P3, MMV P3, MFSV P4 and Sonchus yellow net virus (SYNV) sc4 protein have secondary structures similar to members of the 30K superfamily of virus movement proteins, suggesting these proteins are important for virus cell-to-cell movement in plants (Huang *et al.*, 2005). The ability of the RYSV P3 to support spread of a movement deficient potexvirus in *Nicotiana benthamiana* further supports this notion. In addition to virus movement, at least one plant rhabdovirus protein is likely to suppress virus-induced gene silencing in the plant host. Expression of the SYNV P protein in *Nicotiana benthamiana* relieved M protein mediated suppression of green fluorescent protein expression in a manner similar to known suppressors of RNA silencing, suggesting it is also a suppressor (Jackson *et al.*, 2005). However, silencing suppressors have not been identified for cereal-infecting rhabdoviruses. Very little is known about rhabdovirus gene expression or protein function in insect hosts. However, RYSV P6 is specifically expressed in its leafhopper vector (Huang *et al.*, 2003).

Although there are significant differences in genome organization between the four sequenced cereal-infecting nucleorhabdoviruses and northern cereal mosaic virus (NCMV) (Fig. 10.2), the only completely sequenced cereal-infecting cytorhabdovirus, there is not enough genome information to distinguish between cyto- and nucleorhabdoviruses based on gene complement or function. Furthermore, genome composition and organization of two lettuce infecting cytorhabdoviruses is the same as that of two nucleorhabdoviruses, MMV and maize Iranian mosaic virus (MIMV) (Reed *et al.*, 2005;

---

fescue, *Festuca gigantea*; annual ryegrass, *Lolium multiflorum*; sorghum, *Sorghum vulgare*; maize, *Zea mays*.
[e]Geographic distribution of the virus.
[f]Synonymous with wheat rosette stunt virus.
[g]Synonymous with oat striate mosaic virus.
[h]Synonymous with rice transitory yellowing virus.

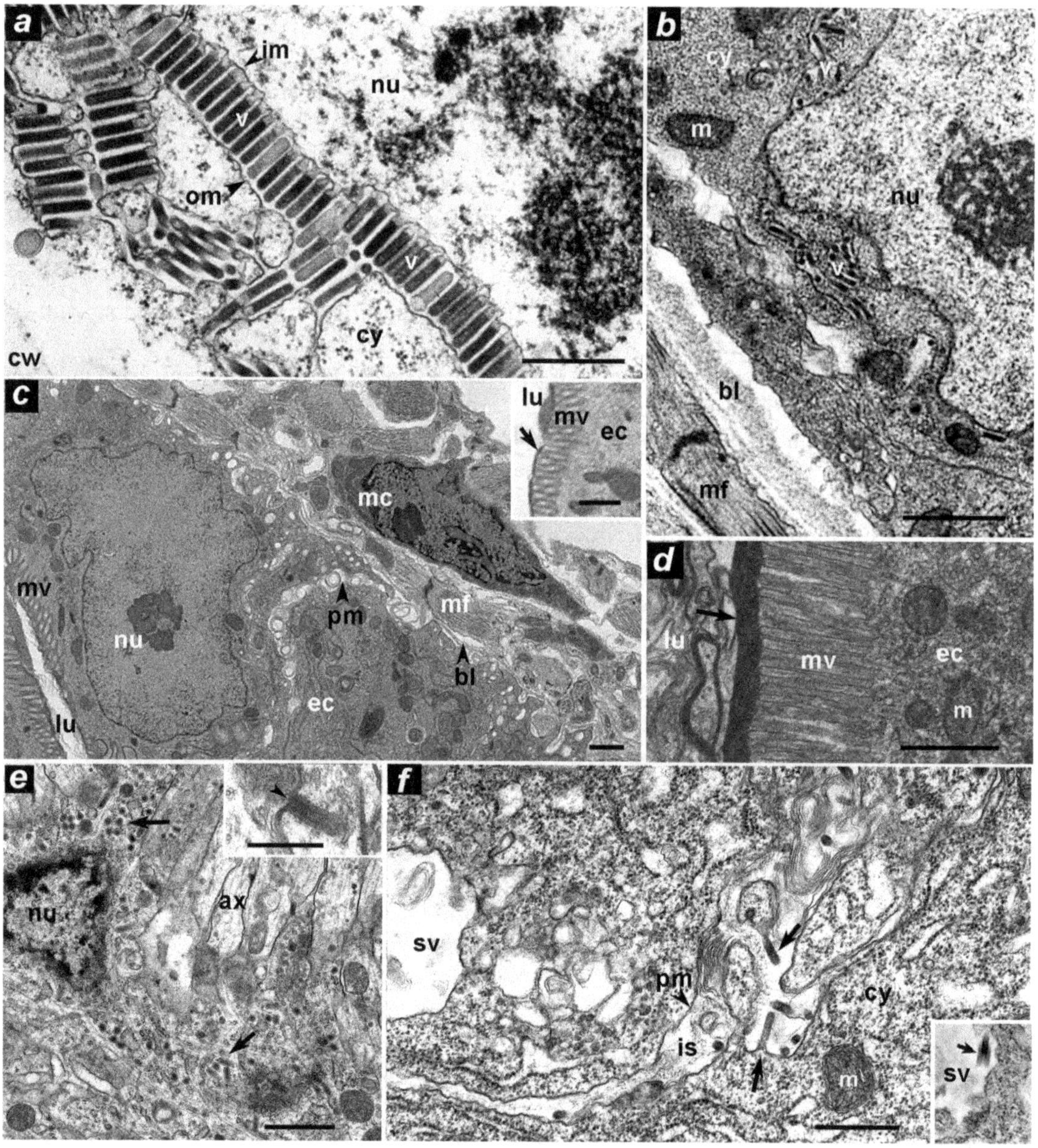

Figure 10.1 Electron micrographs of ultrathin sections in plant and insect hosts of maize mosaic virus (MMV). (a) MMV virions (v) budding through the inner nuclear membrane (im) in a maize leaf cell. (b) MMV virions (v) in cytoplasmic cisternae around the nucleus (nu) in a midgut cell of the vector *P. maidis.* (c and d) Ultrastructure of the anterior diverticulum (c) and the midgut epithelium (d) in *P. maidis,* showing the lumen (lu), microvilli (mv), peritrophic envelope (arrow in *d*), epithelial cells (ec), basal plasma membrane (pm), basal lamina (bl), and muscle cells (mc). (e) MMV virions (arrows) in the cytoplasm of nerve cells in *P. maidis*; the inset shows a virion budding through axon membranes (arrowhead). (*f*) MMV virions budding through the plasma membranes (unlabelled arrows) in secretory cells of the salivary gland in *P. maidis* and accumulating in intercellular space (is) and in secretory vesicles (sv). Other abbreviations: ax, axon; im, inner nuclear membrane; mf, muscle fibres; om, outer nuclear membrane. Scale bars: a and f, 0.5 µm; b–e, 1 µm. Inset in e: 0.25 µm. Reprinted with permission from Ammar *et al*. (2009).

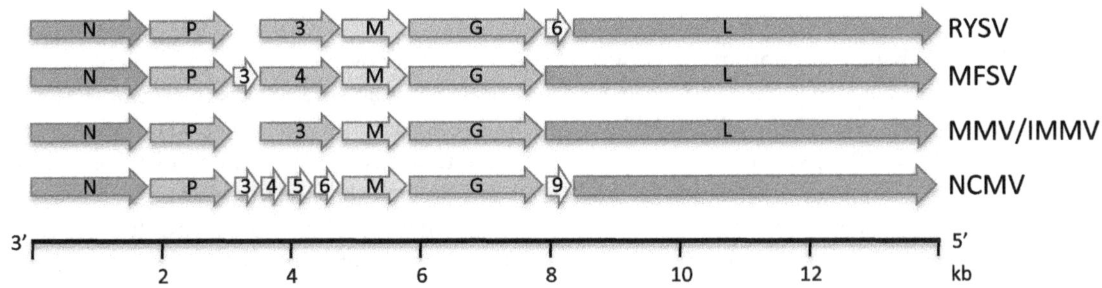

Figure 10.2 Genome organization of cereal-infecting rhabdoviruses. Genomes of the nucleorhabdoviruses rice yellow stunt virus (RYSV), maize fine streak virus (MFSV) and maize mosaic virus (MMV)/maize Iranian mosaic virus (MIMV), and the *Cytorhabdovirus* northern cereal mosaic virus (NCMV) are shown. The ca. 14 kb negative sense viral genomic RNAs encode from six to ten genes (from 3′ to 5′) including the nucleocapsid (N), phosphoprotein (P), matrix protein (M), glycoprotein (G) and large polymerase (L) structural genes. The remaining accessory genes are designated by numbers according to their position in the genome. Adapted from Redinbaugh and Hogenhout (2005).

Dietzgen *et al.*, 2006; Heim *et al.*, 2008; Massah *et al.*, 2008). Genome composition and organization also differ among nucleorhabdoviruses, with MFSV and RYSV carrying an extra gene relative to MMV and MIMV between the P and putative movement protein gene and G and L genes, respectively (Fig. 10.2).

Persistent propagative transmission

Parameters and properties

Hemipteran vectors transmit plant-infecting rhabdoviruses in a persistent, propagative manner. Thus, these viruses infect both their plant hosts and insect vectors, and after virus acquisition from infected plants, whether in the nymphal or adult stages, the vectors can transmit the virus for the remainder of their lifetime. A few plant rhabdoviruses, especially those transmitted by aphids, are transmitted vertically, i.e. transovarially, from infected female vectors to their progeny, normally at a low rate (Sylvester and Richardson, 1992). Although MMV particles were found in follicular cells of the ovaries of *P. maidis* (Ammar and Nault, 1985), there is no evidence that MMV is transovarially transmitted. Of the cereal-infecting rhabdoviruses, only WASMV has been reported as being transmitted to offspring from infected mothers (Sinha, 1970).

Several lines of evidence demonstrate that rhabdoviruses infect and replicate in the tissues of their insect vectors. Potato yellow dwarf virus (PYDV), the type member of the genus *Nucleorhabdovirus*, infected monolayers of vector *Agallia* sp. cells, but infected cells of the non-vector *Dalbulus elymanus* at a much lower rate (Hsu *et al.*, 1977; Nault, 1997). Electron and immunofluorescence microscopy of viruliferous vector insects indicated aggregated virus particles, protein products or sites of virus assembly/budding in *P. maidis, G. nigrifrons, L. striatellus* and *N. cincticeps* cells and tissues infected with MMV, MFSV, NCMV and RYSV, respectively, indicating virus multiplication in the vector (Figs. 10.1 and 10.3) (Ammar and Nault, 1985; Shikata, 1989). The increase in virus concentration in vectors following a short acquisition access period (AAP) on MMV or MFSV-infected plants, as determined by quantitative serology using ELISA, dot blot, or tissue blot assays, provides more direct evidence of rhabdovirus multiplication in insect vectors (Falk and Tsai, 1985; Todd *et al.*, 2010). Similarly, immunocytochemical analysis indicated replication of the vertebrate-infecting rhabdovirus Vesicular stomatitis virus (VSV) in its sandfly vectors (Howerth *et al.*, 2002).

Virus acquisition time thresholds by insect vectors range from less than one minute for some rhabdoviruses that infect both mesophyll and phloem tissues of their plant hosts (e.g. WASMV) to up to 15 minutes for phloem-limited viruses (e.g. RYSV) (Slykhuis, 1963; Lee, 1967; Chiu *et* al., 1968; Inoue, 1979). However, longer AAP usually lead to higher proportion of insects that acquire and/or transmit the virus, even for viruses that infect mesophyll cells (Slykhuis, 1963; Ammar et al., 1987; Creamer et al., 1997; Todd et al., 2010). Similarly, inoculation feeding time thresholds for the vectors range from a few minutes to several days, but optimal inoculation access periods (IAP) range from one to several days. As with other viruses that replicate in their vectors, rhabdoviruses undergo a latent period that ranges from several days to several weeks in the insect vector. The latent period is presumed to be the time necessary for the virus to propagate in and circulate through the vector. The process is thought to involve ingestion of virus from a virus-infected plant, infection of the alimentary canal and other tissues including the salivary glands,

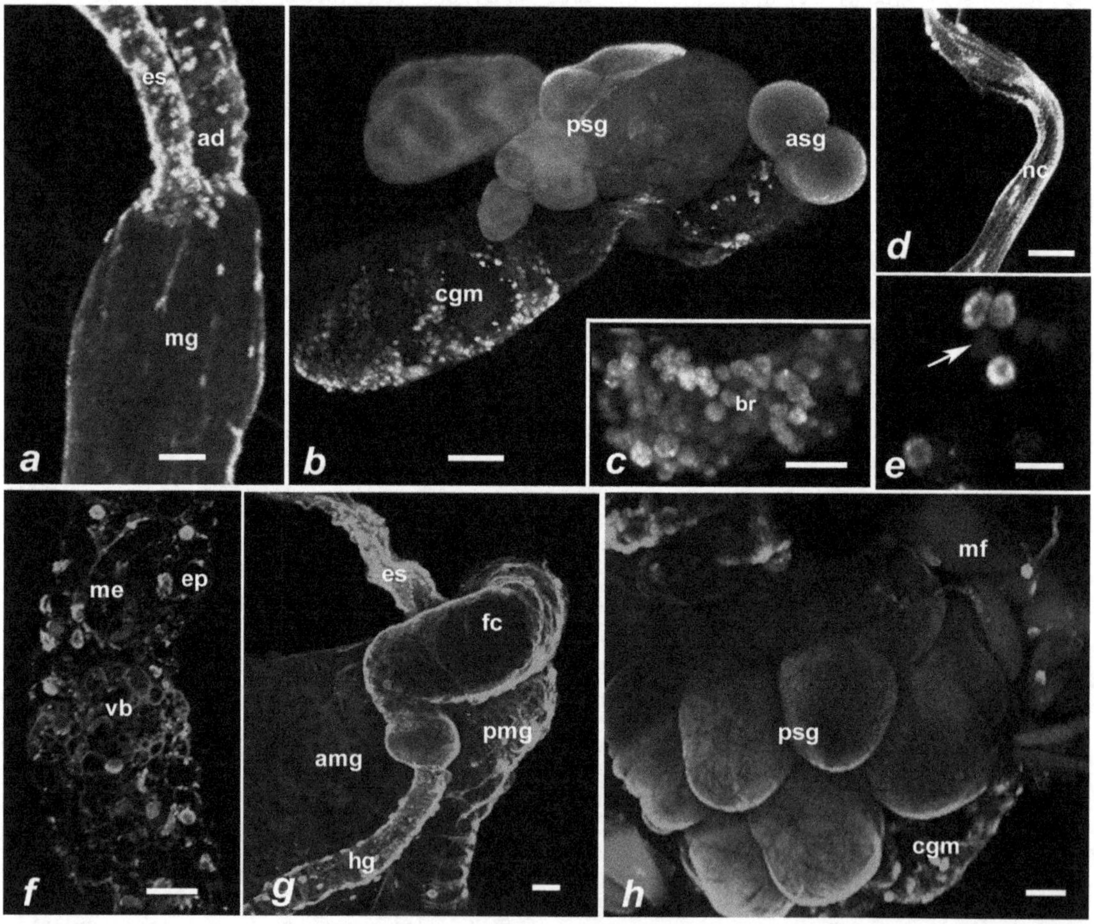

Figure 10.3 Immunofluorescence localization of maize-infecting rhabdoviruses in organs and tissues of their insect vectors. (a–e) Localization of maize mosaic virus (green fluorescence) in cells of the planthopper vector's (*Peregrinus maidis*) alimentary canal (a), salivary gland and compound ganglion (b) brain, (c) nerve cord (d) and hemocytes (e). (f–h) Localization of maize fine streak virus (green fluorescence) in cells of a maize leaf (f), and the leafhopper vector's (*Graminella nigrifrons*) alimentary canal (g), salivary gland and compound ganglion (h). Dissected organs or hemolymph smears were incubated with virus-specific antibodies, then with a secondary antibody, Alexa Fluor 488 (green), the nuclear stain propidium iodide (red), and (in a, g and h) with an actin stain, Phalloidin (blue/purple). Abbreviations: ad, anterior diverticulum; amg, anterior midgut; asg, accessory salivary gland; br, brain; cgm, compound ganglionic mass; ep, epidermis; es, oesophagus; fc, filter chamber; hg, hindgut; me, mesophyll; mf, muscle fibres; mg, midgut; nc; nerve cord; pmg, posterior midgut; psg, principal salivary gland; vb, vascular bundle. The arrow in (e) indicates the nucleus of an uninfected haemocyte. Scale bars = 40 µm, except in (c) and (e) (10 µm). Adapted from Ammar et al. (2009). A colour version of this figure can be found in the plate section at the end of the book.

and inoculation of virus into a new host plant with salivary secretions. The latent period for RYSV in *Nephotettix* spp. ranged between 3 and 66 days, and that of MFSV in *G. nigrifrons* was at least 28 days (Chiu *et al.*, 1968; Inoue, 1979; Todd *et al.*, 2010). For WASMV, the latent period was shorter in efficient vectors compared with inefficient ones (Slykhuis, 1963).

Vector competence for plant rhabdoviruses (i.e. transmission efficiency, or the percentage of a vector population that can transmit a virus to new host plants) can range from as low as 2% to as high as 100%, and may differ among vector populations, biotypes, sexes, and developmental stages (Sylvester and Richardson, 1992; Ammar, 1994; Todd *et al.*, 2010). *P. maidis* biotypes differed in their ability to transmit MMV (Ammar *et al.*, 1987), and inbred lines of *E. inimica* differed in their ability to transmit WASMV (Slykhuis, 1963). These results suggest that there are insect genetic factors that influence transmission. A higher percentage of infected individuals and a higher virus titre were found for *G. nigrifrons* when nymphs were fed on MFSV infected plants relative to those fed as adults (Todd *et al.*, 2010). This is not true for all cereal rhabdoviruses, as only small differences in acquisition of WASMV by *E. inimica* nymphs and adults were reported (Slykhuis, 1963).

Virus infection of vector cells, tissues and organs

For many years, it was presumed that after a rhabdovirus crossed the midgut cells and membranes of its insect vector, it moved into and replicated in the hemolymph prior to infecting the vector's salivary glands, where it could be excreted with the saliva and inoculated into a new plant host (Nault and Ammar, 1989; Ammar and Nault, 2002). More recently, Hogenhout *et al.* (2003) hypothesized that some rhabdoviruses might alternatively follow a neurotropic route to the salivary glands, similar to that followed by some mammalian rhabdoviruses, particularly Rabies virus (RABV). Recent research provides support for this hypothesis for the planthopper and leafhopper vectors of MMV and MFSV, respectively. Ammar and Hogenhout (2008) studied the temporal and spatial distribution of MMV in its planthopper vector by immunofluorescence labelling (Fig. 10.3 and Table 10.2). Following a 1-week AAP for *P. maidis* on MMV-infected plants, MMV spread from the anterior midgut to the oesophagus, nerve cord and nerve ganglia. This spread occurred at least 1 week before the virus was detected in hemocytes, salivary glands or other tissues. Neural tissues infected with MMV in *P. maidis* included the brain, nerve chord, compound ganglion and compound eye cells (Figs 10.1e and 10.3c,d; Table 10.2) (Ammar and Nault, 1985; Ammar and Hogenhout, 2008). Movement of MFSV through *G. nigrifrons* follows a similar pattern. Following a 1 week AAP, MFSV could not be detected in vector midgut, neural, tracheal, visceral muscle and salivary gland tissues or hemocytes (Todd *et al.*, 2010). At two weeks post access to diseased plants (PADP), MFSV was detected in the midgut, neural tissues, muscle and hemocytes of a small percentage of *G. nigrifrons* individuals (Fig. 10.3g), and this proportion increased up to 4 weeks PADP (Table 10.2). Infection in trachea was detected at 3 weeks PADP, and salivary glands at 4 weeks PADP (Fig. 10.3h). Inoculative *G. nigrifrons* were detected at the same time that MFSV was detected in the salivary glands. These results indicate that alternatives to the midgut – hemocoel – salivary gland route are possible, with MMV and MFSV infection of neural tissues occurring 1 week earlier than of hemocytes, indicating that a route from the midgut to the salivary glands via the nervous system may exist. Identification of internal barriers to rhabdovirus infection of vectors, inefficient vectors and non-vectors in different rhabdovirus – insect vector systems should provide more definitive information on the importance of various routes from the midgut to the salivary gland.

Virus replication and gene expression in insect vectors

Immunohistochemistry and transmission experiments provide strong evidence that plant-infecting rhabdoviruses replicate in their arthropod vectors. Analysis of cDNA libraries from MMV-infected *P. maidis* guts provide confirmation that virus genes are expressed in the insect vector. Of 20,771 sequenced expressed sequence tags (ESTs), 83

Table 10.2 Time course for rhabdovirus infection of vector tissues[a]

	Midgut		Neural		Muscles		Haemocytes		Tracheae		Salivary gland	
PADP[b] (weeks)	Gn	Pm	Gn	Pm	Gn	Pm	Gn	Pm	Gn	Pm	Gn	Pm
1	0	5.4	0	0	0	0	0	0	0	0	0	0
2	6.1	13.5	3.0	5.2	3.0	0	3.0	0	0	0	0	0
3	19.1	21	7.4	16.8	7.4	6.4	4.4	8.5	5.9	9.5	0	7.5
4	25.3	25.1	16.2	21.9	11.1	16	4.0	17.6	15.2	16.3	7.1	17.4
5	21.0	21.7	14.5	20.3	12.9	14.5	6.5	15.9	8.1	17.4	8.1	20.3

PADP, post access to diseased plants.
[a]Adapted from Todd *et al.* (2010) and Ammar and Hogenhout (2008).
[b]Data presented are the percentage of *Graminella nigrifrons* (Gn) or *Perigrinus maidis* (Pm) individuals positive for maize fine streak virus or maize mosaic virus as determined by immunofluorescence confocal microscopy, respectively.

were high quality MMV sequences that assembled into 7 contigs and 45 singletons (Whitfield *et al.*, 2011). Each of the six MMV genes was expressed in the gut tissue, including gene 3, the putative virus movement protein. Comprehensive characterization of the genetic diversity and abundance of MMV transcript sequences harboured by *P. maidis* will be the subject of future studies.

Virus receptors and the endocytic pathway

Based on similarities between genome composition and conservation of gene function among mammal- and plant-infecting rhabdoviruses, it is hypothesized that during arthropod vector infection plant rhabdoviruses enter cells via receptor-mediated endocytosis. Analysis of *P. maidis* gut ESTs identified putative candidate genes that may play a role in MMV binding to host membranes and entry into cells (Whitfield *et al.*, 2011). Two sequences in the *P. maidis* gut transcriptome share significant amino acid sequence similarity with the known rhabdovirus receptor molecules acetylcholine receptor (AChR) and fibronectin-like protein. The AChR protein is a transmembrane receptor-ion channel protein that has been reported to serve as a cellular receptor for RABV (Gastka *et al.*, 1996) and fibronectin-like proteins have been shown to bind to RABV glycoprotein (Tuffereau *et al.*, 1998). There is evidence that fibronectins serve as receptors for entry and replication for other rhabdoviruses such as fish-infecting novirhabdoviruses (Bearzotti *et al.*, 1999). One of these types of rhabdovirus receptors may play a role in MMV binding in *P. maidis* midgut cells. The next critical step after virion binding to a host receptor molecule is virus entry into the cell. Annotation of *P. maidis* ESTs revealed that 109 ESTs have homology to proteins associated with the endocytic pathway, indicating that pathways that play an essential role in virus entry and transport in mammals are conserved in this insect vector.

The *P. maidis* gut transcriptome had 202 *P. maidis* transcripts (about 1%) with similarity to proteins associated with insect innate immunity. Sequence comparisons between *P. maidis* gut EST collection and the currently available EST database collection for another planthopper vector (*Nilaparvata lugens*) revealed that a pathogen recognition receptor in the Imd pathway is present in both of these planthoppers (Whitfield *et al.*, 2011). This EST analysis sheds new light on immune response pathways in hemipteran guts that will be useful for further dissecting the roles of innate defence response pathways to rhabdovirus infection.

Transmission barriers and the making of vectors and non-vectors

Vectors, acquirers and non-vectors within and between species

A propagatively transmitted virus must first be acquired by the vector from an infected plant host, then move and propagate in insect vector tissues, and finally be inoculated to a new plant host. We generally define acquisition as the ability to detect the virus within the tissues of an insect after an acquisition-feeding period on diseased plants using serological (e.g. ELISA), molecular (e.g. RT-PCR) or microscopic techniques. Inoculative insects, or vectors, are defined by their ability to feed on a healthy plant and inoculate it with the virus. For many plant viruses, a much higher proportion of vector insects can acquire a virus from diseased plants than can inoculate it into new plants. With PADP greater than 3 weeks, MFSV was detected in 30% to 60% of *G. nigrifrons* individuals, but less than 10% of the insects in these experiments transmitted the virus to healthy plants (Todd *et al.*, 2010). MFSV titre was significantly higher in inoculative *G. nigrifrons* leafhoppers compared with infected but non-inoculative leafhoppers. Similarly, 28% of *P. maidis* individuals acquired MMV, but only 17% were inoculative (Ammar and Hogenhout, 2008). These results suggest there are barriers both to virus acquisition by vectors, as not all individuals acquire the virus, and barriers to virus movement within the vector to the salivary glands, as not all infected insects can transmit/inoculate the virus.

Immunofluorescence microscopy of dissected insect organs provides evidence of midgut infection, midgut escape, and salivary gland infection barriers for MMV in *P. maidis* (Fig. 10.3 and Table 10.2) (Ammar and Hogenhout, 2008). Following 1–3 weeks' AAP on diseased plants, only 28% of *P. maidis* had MMV infected midguts, and a significantly lower proportion (20%) had infected salivary glands. The proportion of MMV and MFSV infected guts were significantly higher following longer AAP for *P. maidis* and *G. nigrifrons*, suggesting that the midgut wall provides a dose-dependent bottleneck for virus acquisition (Ammar and Hogenhout, 2008; Todd *et al.*, 2010). A midgut barrier was also demonstrated by intrathoracic inoculation of MMV into *P. maidis*, which significantly increased the proportion of infected vectors (Falk and Tsai, 1985). These results, together with the relatively common increases in vector competence with longer AAP are consistent with the midgut being a major barrier to rhabdovirus acquisition by vector species.

For both *P. maidis* and *G. nigrifrons*, the proportion of vectors with virus-infected salivary glands was significantly lower than that of vectors with infected midguts, suggesting a salivary gland infection barrier (Ammar *et al.*, 2008; Todd *et al.*, 2010). However, for both MMV and MFSV, the proportions of infected salivary glands and inoculative insects are similar, suggesting that once the salivary gland becomes infected, the insect becomes inoculative, with no apparent exit barrier in these glands. However, a salivary gland exit barrier has been suggested for other propagative viruses in their vectors (Ammar, 1994). For example, more than 75% of *P. maidis* with maize stripe tenuivirus-infected salivary glands were not capable of transmitting the virus (Nault and Gordon, 1988).

The inability of a rhabdovirus to enter, replicate in or exit specific insect tissues would make it a non-vector. Inability to acquire virus was demonstrated for *P. maidis* transmission of MIMV. A very small portion of *P. maidis* transmit MIMV after feeding on infected plants, with < 2% of individuals being vectors (Ammar *et al.*, 2005). However, for planthoppers injected with an extract from MIMV-infected plants, the probability that an individual was a vector was high (62%), and similar to the rate of transmission of MMV by this vector. Preliminary studies indicate that a rhabdovirus can infect/replicate in some non-vector insects without rendering them inoculative, probably because of its failure to enter or exit the salivary glands. MFSV can be detected in the non-vector leafhoppers *Dalbulus maidis* and *Macrosteles quadrilineatus* fed on virus-infected plants, but these insects cannot transmit MFSV, and preliminary results suggest that injecting MFSV into the hemolymph did not result in transmission (Redinbaugh *et al.*, unpublished results). The factors limiting rhabdovirus replication in insect species are not understood, though replication of rhabdoviruses in a wide range of non-vectors after injection has been demonstrated. For example, the vertebrate-infecting VSV was reported to efficiently multiply in the hemolymph when injected into the planthopper *P. maidis*, vector of MMV (Lastra and Esparza, 1976). VSV appears to be a promiscuous virus with the ability to replicate in many types of invertebrate hosts including arthropods and nematodes (Wilkins *et al.*, 2005; Mueller *et al.*, 2010). Strains of the nucleorhabdovirus PYDV were able to infect cell cultures from vector and non-vector insects equally (Hsu *et al.*, 1977). However, the ability of PYDV to infect cell cultures was not unlimited, as virus strains did not infect *Dalbulus elimatus* cell cultures, or infected them at very low rates. Thus, barriers to both rhabdovirus movement and replication may play a role in determining vector competence.

Bypassing barriers

We hypothesize that vector specificity and vector competence related to transmission barriers in the insect are, at least in part, genetically controlled (Ammar, 1994; Ammar and Nault, 2002; Hogenhout *et al.*, 2003; Jackson *et al.*, 2005; Ammar *et al.*, 2009). For plant-infecting rhabdoviruses, receptors in the gut, salivary glands or other tissues may account for the high degree of host specificity that is commonly observed among invertebrates transmitting rhabdoviruses and other disease agents. Similarly, the presence and efficiency of receptors and other factors in the epithelial cells of the insect's alimentary canal, salivary glands and other tissues would control development of insect vector competence. Plant host specific manipulation of plant defence responses by factors excreted during insect feeding could also be important for controlling the efficiency of virus acquisition and/or transmission (Hogenhout and Bos, 2011). It is of great importance to reveal the true nature of these vector determinants because they offer new opportunities for disease control, including the genetic manipulation of the vector and transmission neutralization by means of vaccines or recombinant proteins expressed in transgenic plants (van den Heuvel *et al.*, 1999; Chen *et al.*, 2011; Liu *et al.*, 2011).

The nature of virus acquisition barriers

Several hypotheses have been proposed to explain the midgut infection barrier to propagative viruses in their vectors (Hardy, 1988). These include inactivation of virus by digestive enzymes in the midgut lumen, occlusion of virus by the peritrophic envelope (Fig. 10.1c), absence, reduced number or genetic variation of cellular receptor sites of virus attachment on the microvillar membrane, activation of defence responses that effectively prevent infection, and/or abortive replication of virus in midgut epithelial cells. As discussed above, infection of insect cell monolayers by both animal and plant rhabdoviruses suggests similar entry into epithelial cells by receptor-mediated endocytosis (Jackson *et al.*, 2005). While nothing is known yet about receptors for plant-infecting rhabdoviruses, VSV virions enter vertebrate and insect cell lines through coated pits and coated vesicles (Superti *et al.*, 1987), and clatherin-coated pits were recently implicated in cell entry of the *Novirhabdovirus* infectious haematopoietic necrosis virus (Liu *et al.*, 2011). Actin polymerization is involved in the transport of endocytic vesicles into the cytosol, whereas microtubules are required for the trafficking of clatherin-coated vesicles to early endosomes, late endosomes, and lysosomes. Also, inhibition of PYDV infection of vector cell monolayers by some lysosomotropic agents suggested that PYDV enters these cells by adsorptive endocytosis (Adam and Gaedigk, 1986). The rhabdoviral G protein is important both for receptor recognition in animal cells and for membrane fusion that occurs via a pH-induced conformational change (Albertini *et al.*, 2008), serving as both the viral attachment and fusion protein (Lyles and Rupprecht, 2007). Furthermore, the G protein of VSV is recognized by the *Drosophila melanogaster* innate immune system, triggering an autophagic response and resulting in decreased viral replication and virulence (Shelly *et al.*, 2009).

The nature of virus dissemination barriers

It was historically assumed that rhabdoviruses are disseminated from the midgut into other tissues of the vector through the hemolymph, based on experiments such as those described above in which MIMV and MMV are transmitted by vectors and non-vectors at higher rate following injection of virus extracts into their hemocoel. A number of humoral and cellular receptors have been identified in the insect hemolymph that recognize different microbes and other parasites (Strand, 2008). It is possible that these factors have a role in preventing dissemination of viruses from the midgut to other tissues including the salivary glands. The basal lamina between the midgut and the hemolymph (Fig. 10.1b) also seems to be a substantial barrier to the hemolymph route (Hardy, 1988). Nonetheless, recent investigations on plant and animal viruses indicate that other possible routes exist, especially through neural, tracheal, or muscle tissues, for virus movement from the vector's midgut into the salivary glands (Drolet *et al.*, 2005; Ammar and Hogenhout, 2008; Ammar *et al.*, 2009).

Salivary gland infection and escape barriers were reported for several propagative plant viruses in their vectors (Nault and Gordon, 1988; Ammar *et al.*, 2004, 2009). The mechanism(s) used by viruses for overcoming the basal laminae of the salivary gland to enter these glands is not well documented, but the neural and tracheal systems are good candidates for such mechanisms (Nagata *et al.*, 1999; Ammar and Hogenhout, 2008; Means and Passarelli, 2010). After invading and replicating in the salivary glands, the virus must escape one last physical barrier, the plasma membrane of secretory cells in the salivary glands, to enter into the saliva for introduction into plants during insect feeding. In most planthopper tissues, MMV virions bud through nuclear membranes and accumulate in the perinuclear space and in dilated cisternae in the cytoplasm (Fig. 10.1b) (Ammar and Nault, 1985) as they do in plant cells (Fig. 10.1a). However, in the secretory cells of the principal salivary gland, MMV virions bud mainly through the plasma membranes and accumulate in intercellular and extracellular spaces connected with secretory vesicles and eventually with the salivary ducts (Fig. 10.1f) (Ammar, 1985; Ammar and Nault, 1985). The ability of rhabdoviruses to bud from plasma membranes, rather than the endoplasmic reticulum or nuclear membranes of the insect salivary gland cells, is likely important for overcoming the salivary gland escape barrier. The possible role(s) of chemical and physiological conditions of the salivary secretions in enhancing or obstructing plant virus transmission by insect vectors remains largely uninvestigated. Also, the plant response to the hemipteran salivary secretions and/or to inoculated virus is just beginning to be studied (Hogenhout and Bos, 2011), and merits further investigation.

Vector responses to infection

Molecular response of vectors to rhabdoviruses

The typical insect vector sustains a persistent rhabdovirus infection, so the insect's response to rhabdovirus infection must allow the infection to persist at a non-pathogenic level. Recent analysis of *Drosophila melanogaster* response to VSV infection revealed that two branches of the insect innate immune system are activated in insect cells and flies challenged with this virus (Mueller *et al.*, 2010). Characterization of the small interfering (si)RNAs in *Drosophila* infected with VSV indicated that the rhabdovirus genome is effectively targeted by RNA inference (RNAi) and insects defective in this pathway were highly susceptible to the virus. Autophagy was also activated by exposure of *Drosophila* to VSV particles and the viral G protein was identified as the pattern-associated-molecular pattern (PAMP) recognized by the host (Shelly *et al.*, 2009). In comparison, analysis of *Drosophila* response to the vertically transmitted rhabdovirus, Sigma virus, revealed the differential regulation of a unique suite of genes: peptidoglycan recognition proteins, antimicrobial peptides, serine proteases, and ribosomal proteins (Tsai *et al.*, 2008; Carpenter *et al.*, 2009). With the development of rapid and inexpensive transcriptome sequencing technologies, the characterization of insect vector responses to virus infection is a rapidly advancing field that will enable the study of virus–vector interactions for non-model organisms, such as the vectors of plant viruses. Towards that goal, the characterization of transcriptomes for the MMV vector, *P. maidis*, has revealed 202 ESTs with putative homology to proteins associated with innate immunity at the levels of pathogen recognition, signal modulation, signal transduction, and effectors. Putative homologues associated with the antimicrobial signalling pathways Toll and Imd, and MAP kinase pathways (JNK) that lead to stress-related apoptosis and haemocyte differentiation were represented in the *P. maidis* transcriptome library. More specifically associated with antiviral defence in insects (Wang *et al.*, 2006) and plants (Lindbo *et al.*, 1993), 12 *P. maidis* gut ESTs with a putative homology to proteins involved in RNAi pathways were identified, including those involved in dsRNA recognition and degradation (Whitfield *et al.*, 2011). ESTs encoding proteins with homology to Dicer, components of the RISC holoenzyme and dsRNA binding motif-containing proteins were identified.

Behavioural responses of vectors to rhabdovirus infection

The route of MMV in *P. maidis* follows a similar path as other rhabdoviruses (Ammar and Hogenhout, 2008). For example, the initial infection of mammals by RABV occurs in muscle or epithelial cells, the virus then spreads to the central nervous system, and extensive infection of the brain is observed (Lyles and Rupprecht, 2007). RABV infection of mammals leads to the classic symptoms of hyperactivity, disorientation, and salivation. The behavioural and physiological changes in the host facilitate transmission of the virus to a new host. Similarly, VSV infection in rats results in behavioural alterations and animals that survive infection exhibit a permanent reduction in serotonin, even after the virus is cleared by the immune system (Mohammed *et al.*, 1992). VSV infection of midges (*Culicoides sonorensis*) revealed that the infection pathway in insects parallels the observed pathway in mammalian hosts (Drolet *et al.*, 2005). One study reported altered feeding behaviour of VSV-infected midges after two days post acquisition (Bennett *et al.*, 2008). It is well documented that other insect-borne viruses can also alter host behaviour and physiology (Platt *et al.*, 1997; Burand *et al.*, 2005; Vasconcelos *et al.*, 2005; Ban *et al.*, 2008; Ryabov *et al.*, 2009). Virus infection can reduce fecundity and increase mortality, alter feeding behaviours, increase locomotory activity, and increase production of winged insects. Planthoppers infected with MMV exhibit extensive infection of the brain, eyes, hemocytes, neural tissues, salivary glands, and midguts (Figs 10.1b,f and 10.3a–e; Ammar and Nault, 1985; Ammar and Hogenhout, 2008). The infection of the nervous system and brain are intriguing similarities to mammalian rhabdoviruses and research to identify behavioural and physiological changes in arthropod vectors of plant rhabdoviruses is ongoing.

Virus infection of arthropod vectors has been shown to affect their feeding behaviour (Platt *et al.*, 1997; Grimstad *et al.*, 1980). For example, tomato spotted wilt tospovirus, a plant-infecting member of the family Bunyaviridae, alters feeding behaviour of male vector insects that are infected with the virus (Stafford *et al.*, 2011). Mosquitoes infected with dengue virus (family Flaviviridae) required a longer time to feed and probing times were longer than in uninfected insects (Platt *et al.*, 1997). These mosquitoes also sustain heavy virus infection of the brain, nervous system, and salivary glands, similar to the MMV infection of *P. maidis* and MFSV infection of *G. nigrifrons*.

Diseases caused by cereal rhabdoviruses and their control

Diseases caused by cereal rhabdoviruses

Of the cereal-infecting rhabdoviruses, MMV probably causes the most significant disease worldwide. MMV is an important pathogen of maize in tropical and subtropical areas of Africa, South America, Hawaii and Australia where its vector, *P. maidis* is present (Gordon *et al.*, 1981; Greber, 1981; Thottappilly, 1992). *P. maidis* is a maize specialist, so disease is exacerbated under conditions where overlapping crops of maize are grown all year and in the presence of perennial weedy plant hosts such as foxtail (*Setaria verticilliflorum*) (Autrey, 1983; Reynaud *et al.*, 2009). RYSV is transmitted by *Nephotettix* sp. in central to southern China, Taiwan, Vietnam, Laos and Thailand (Hibino, 1996; Hiraguri *et al.*, 2010). Major outbreaks occurred in China from 1960 to 1980, but incidence has apparently decreased in recent years. Similar to MMV, the disease is found in areas where multiple rice crops are grown annually, with the disease being spread by infected vectors moving from early-planted fields to late planted fields. RYSV can occur together with rice dwarf reovirus, which is also transmitted by *N. cincticeps* and *N. nigropictus*, to cause more severe disease symptoms. Most of the other cereal-infecting rhabdoviruses cause disease regionally or locally, or within limited sets of susceptible germplasm. For example, NCMV causes disease on wheat and barley in northern Japan, Korea and temperate China (Toriyama, 2004) and MIMV causes disease on maize in the Shiraz region of Iran (Izadpanah *et al.*, 1983; Izadpanah, 1989). Similarly, cereal chlorotic mottle virus, cynodon chlorotic streak virus and sorghum stunt mosaic virus can cause disease locally in northern Europe, Morocco and California, respectively (Mayhew and Flock, 1981; Lockhart *et al.*, 1985; Lockhart, 2004; Persley and Greber, 2004b).

Virus and vector resistance in crop hosts

Genetically controlled resistance to viruses is generally identifiable in cereal crops, or their close relatives, and is considered to be an economically and environmentally sustainable approach for disease control (Kang *et al.*, 2005). Strong resistance to MMV, MFSV and maize sterile stunt virus (MSSV) has been reported (Greber, 1982; Ming *et al.*, 1997; Redinbaugh *et al.*, 2002). Resistance to MSSV appears to be conferred by a frequent dominant allele(s), as few maize lines are susceptible, and F_1 plants from crosses of susceptible and resistant lines are resistant. MMV resistance is conferred by a major QTL, *Mv1*, that maps close to the chromosome 3 centromere in the inbred line Hi31 (Ming *et al.*, 1997). Lines resistant to MFSV were identified (Redinbaugh *et al.*, 2002), and preliminary evidence indicates a resistance locus on chromosome 2 in the inbred line Oh1VI (Redinbaugh *et al.*, unpublished results). Much less is known about rhabdovirus resistance in small grains, likely owing to the fact that these viruses do not have geographically widespread economic impact. Nonetheless, resistant and partially resistant cultivars are used to control the cytorhabdoviruses winter wheat Russian mosaic virus, NCMV and WASMV (Toriyama, 2004; Brockus *et al.*, 2008). At least one report of pathogen-derived resistance for cereal rhabdoviruses has been described, with rice plants expressing RYSV RNAs being resistant to virus infection (Fang *et al.*, 1996).

In addition to deploying virus-resistant crops, removing virus reservoirs, including weeds and vectors can be effective for controlling virus diseases (Hibino, 1996; Sinha, 1970). Because rhabdoviruses are persistently transmitted, plant resistance to vector feeding can be effective for disease control. Maize inbred lines with resistance to *P. maidis* developed less disease when insect pressure was low (Dintinger *et al.*, 2005). In rice, cultivars with resistance to leafhoppers, including the RTYV vector *N. virescens*, have been developed and deployed (Hibino, 1996). The mechanisms associated with plant host resistance to leafhopper and planthopper vectors are not defined, and may provide insights for control of rhabdovirus diseases.

Future directions

From the discussion above, it is clear that the basic parameters for transmission of rhabdoviruses to cereals and other plants, as well as serological, molecular and microscopic evidence indicate that leafhopper and planthopper vectors of rhabdoviruses are also hosts of these viruses. This mode of transmission provides a basis for considering control of the vector and/or transmission process as means for disease control, in contrast to non-persistently transmitted viruses. We are beginning to characterize barriers within insect hosts to vector competence, and to develop the molecular information on vectors that will be required to define molecular factors required for an insect to be a host and/or a vector of rhabdoviruses. However, transcriptome information is currently available for only one rhabdovirus vector. Complementary information is required for other vector and non-vector insects. Reverse genetics systems for both the vectors and viruses would allow us to identify insect and pathogen genes required for virus acquisition, replication and transmission by vectors. Together with currently available microscopic and analytical systems, functional genomics will allow us to identify specific molecular factors required for virus entry into gut cells, movement through the vector and exit from the salivary glands, and provide potential targets at each site for control of diseases caused by rhabdoviruses and other viruses that are propagatively transmitted by vectors.

References

Adam, G., and Gaedigk, K. (1986). Inhibition of potato yellow dwarf virus infection in vector cell monolayers by lysosomotropic agents. J. Gen. Virol. 67, 2775–2780.

Albertini, A., Roche, S., Lepault, J., Bressanelli, S., and Gaudin, Y. (2008). Glycoprotein structures of rhabdovirus. Virologie *12*, 407–418.

Ammar, E. (1985). Internal morphology and ultrastructure of leafhoppers and planthoppers. In Leafhoppers and Planthoppers, Nault, L.R., and Rodriguez, J.G., eds (Wiley, New York), pp. 121–162.

Ammar, E., Meulia, T., Özbek, E., and Hogenhout, S.A. (2004). Assembly and accumulation sites of maize mosaic virus (Rhabdoviridae) in plant host and insect vector using transmission electron and confocal laser scanning microscopy. In Current Issues on Multidisciplinary Microscopy Research and Education, Méndez-Vilas A., and Labajos-Broncano, L., eds (Formatex Publishing, Badajoz, Spain), pp. 56–64.

Ammar, E.D. (1994). Propagative transmission of plant and animal viruses by insects: factors affecting vector specificity and competence, Vol. 10 (Springer-Verlag, New York:).

Ammar, E.D., and Nault, L.R. (1985). Assembly and accumulation sites of maize mosaic virus in its planthopper vector. Intervirol. *24*, 33–41.

Ammar, E.D., and Nault, L.R. (2002). Virus transmission by leafhoppers, planthoppers and treehoppers (Auchenorrhyncha, Homoptera). Adv. Bot. Res. *36*, 141–167.

Ammar, E.D., and Hogenhout, S.A. (2008). A neurotropic route for maize mosaic virus (Rhabdoviridae) in its planthopper vector *Peregrinus maidis*. Virus Res. *131*, 77–85.

Ammar, E.D., Gingery, R.E., and Nault, L.R. (1985). 2 types of inclusions in maize infected with maize stripe virus. Phytopathol. *75*, 84–89.

Ammar, E.D., Gingery, R.E., and Nault, L.R. (1987). Interactions between maize mosaic and maize stripe viruses in their insect vector, *Peregrinus maidis*, and in maize. Phytopathol. *77*, 1051–1056.

Ammar, E.D., Gomez-Luengo, R.G., Gordon, D.T., and Hogenhout, S.A. (2005). Characterization of maize Iranian mosaic virus and comparison with Hawaiian and other isolates of maize mosaic virus (Rhabdoviridae). J. Phytopathol. *153*, 129–136.

Ammar, E.D., Tsai, C.W., Whitfield, A.E., Redinbaugh, M.G., and Hogenhout, S.A. (2009). Cellular and molecular aspects of rhabdovirus interactions with insect and plant hosts. Annu. Rev. Entomol. *54*, 447–468.

Autrey, L.J.C. (1983). Maize mosaic and other maize virus diseases in the islands of the western Indian Ocean. International Maize Virus Disease Colloquium and Workshop, Gordon D.T., Knoke J.K., Nault, L.R., and Ritter, R.M., eds. OARDC: Wooster, OH, pp. 167–181.

Ban, L.P., Ahmed, E., Ninkovic, V., Delp, G., and Glinwood, R. (2008). Infection with an insect virus affects olfactory behaviour and interactions with host plant and natural enemies in an aphid. Entomol. Experiment. Applic. *127*, 108–117.

Bearzotti, M., Delmas, B., Lamoureux, A., Loustau, A.M., Chilmonczyk, S., and Bremont, M. (1999). Fish rhabdovirus cell entry is mediated by fibronectin. J. Virol. *73*, 7703–7709.

Bennett, K.E., Hopper, J.E., Stuart, M.A., West, M., and Drolet, B.S. (2008). Blood-feeding behavior of vesicular stomatitis virus infected *Culicoides sonorensis* (Diptera: Ceratopogonidae). J. Med. Entomol. *45*, 921–926.

Bourhy, H., Cowley, J.A., Larrous, F., Holmes, E.C., and Walker, P.J. (2005). Phylogenetic relationships among rhabdoviruses inferred using the L polymerase gene. J. Gen. Virol. *86*, 2849–2858.

Brockus, W.W., Bowden, R.L., Hunger, R.M., Morrill, W.L., Murray, T.D., and Smiley, R.W. (2008). Compendium of Wheat Diseases (APS Press, St. Paul, MN).

Burand, J.P., Tan, W.J., Kim, W.J., Nojima, S., and Roelofs, W. (2005). Infection with the insect virus Hz-2v alters mating behavior and pheromone production in female *Helicoverpa zea* moths. J. Insect Sci. *5*, 6.

Carpenter, J., Hutter, S., Baines, J.F., Roller, J., Saminadin-Peter, S.S., Parsch, J., and Jiggins, F.M. (2009). The transcriptional response of *Drosophila melanogaster* to infection with the sigma virus (Rhabdoviridae). PLoS One 4, e6838.

Chen, A.Y.S., Walker, G.P., Carter, D., and Ng, J.C.K. (2011). A virus capsid component mediates virion retention and transmission by its insect vector. Proc. Natl. Acad. Sci. U.S.A. *108*, 16777–16782.

Chiu, R.J., Jean, J.H., Chen, M.H., and Lo, T.C. (1968). Transmission of transitory yellowing virus of rice by 2 leafhoppers. Phytopathol. *58*, 740–745.

Creamer, R., He, X., and Styer, W.E. (1997). Transmission of sorghum stunt mosaic rhabdovirus by the leafhopper vector, *Graminella sonora* (Homoptera: Cicadellidae). Plant Dis. *81*, 63–65.

Dietzgen, R.G., Callaghan, B., Wetzel, T., and Dale, J.L. (2006). Completion of the genome sequence of lettuce necrotic yellows virus, type species of the genus *Cytorhabdovirus*. Virus Res. *118*, 16–22.

Dietzgen, R.G., Calisher, C.H., Kurath, G., Kuzmin, I.V., Rodriguez, L.L., Stone, D.M., Tesh, R.B., Tordo, N., Walker, P.J., Wetzel, T., *et al.* (2011). Family Rhabdoviridae. In Virus Taxonomy. Ninth Report of the International Committee on Taxonomy of Viruses, King, A.M.Q., Adams, M.J., Carstens, E.B., and Lefkowitz, E.J., eds (Elsevier Academic Press, Oxford, UK), pp. 686–714.

Dintinger, J., Boissot, N., Chiroleu, F., Hamon, P., and Reynaud, B. (2005). Evaluation of maize inbreds for maize stripe virus and maize mosaic virus resistance: Disease progress in relation to time and the cumulative number of planthoppers. Phytopathol. *95*, 600–607.

Drolet, B.S., Campbell, C.L., Stuart, M.A., and Wilson, W.C. (2005). Vector competence of *Culicoides sonorensis* (Diptera: Ceratopogonidae) for vesicular stomatitis virus. J. Med. Entomol. *42*, 409–418.

Falk, B.W., and Tsai, J.H. (1985). Serological detection and evidence for multiplication of maize mosaic virus in the planthopper, *Peregrinus maidis*. Phytopathol. *75*, 852–855.

Fang, R.-X., Zhu, H.-T., Wang, Q., Mang, K.-Q., Gao, D.-M., Qin, W.-S., Zhang, L., Cao, S.-Y., Tian, W.-Z., and Li, L.-C. (1996). Construction of transgenic rice plant resistant to rice yellow stunt virus, a plant rhabdovirus. Rice Genetics III Proceedings of the Third International Rice Genetics Symposium (IRRI), Manila, the Phillipines. pp. 201–205.

Gastka, M., Horvath, J., and Lentz, T.L. (1996). Rabies virus binding to the nicotinic acetylcholine receptor alpha subunit demonstrated by virus overlay protein binding assay. J. Gen. Virol. *77*, 2437–2440.

Gordon, D.T., Bradfute, O.E., Gingery, R.E., Knoke, J.K., Nault, L.R., and Scott, G.E. (1981). Introduction: history, geographical distribution, pathogen characteristics and economic importance. In Virus and Virus-like Diseases of Maize in the United States, Knoke, J.K., Gordon, D.T., and Scott, G.E., eds (OARDC, Wooster, OH), pp. 1–18.

Greber, R.S. (1981). Maize stripe disease in Australia. Aust. J. Agric. Res. *32*, 27–36.

Greber, R.S. (1982). Maize sterile stunt – a delphacid transmitted rhabdovirus disease affecting some maize genotypes in Australia. Aust. J. Agric. Res. *33*, 13–23.

Grimstad, P.R., Ross, Q.E., and Craig, G.B. (1980). *Aedes triseriatus* (Diptera, Culicidae) and La Crosse virus 2. Modification of mosquito feeding behavior by virus infection. J. Med. Entomol. *17*, 1–7.

Hardy, J.L. (1988). Susceptibility and resistance of vector mosquitoes. In The Arboviruses: Epidemiology and Ecology, Monath, T.P., ed (CRC Press, Boca Raton, FL), pp. 87–126.

Heaton, L.A., Hillman, B.I., Hunter, B.G., Zuidema, D., and Jackson, A.O. (1989). Physical map of the genome of sonchus yellow net virus, a plant rhabdovirus with 6 genes and conserved gene junction sequences. Proc. Natl. Acad. Sci. U.S.A. *86*, 8665–8668.

Heim, F., Lot, H., Delecolle, B., Bassler, A., Krczal, G., and Wetzel, T. (2008). Complete nucleotide sequence of a putative new *Cytorhabdovirus* infecting lettuce. Arch. Virol. *153*, 81–92.

van den Heuvel, J., Hogenhout, S.A., and van der Wilk, F. (1999). Recognition and receptors in virus transmission by arthropods. Trends Microbiol. *7*, 71–76.

Hibino, H. (1996). Biology and epidemiology of rice viruses. Annu. Rev. Phytopathol. *34*, 249–274.

Hiraguri, A., Hibino, H., Hayashi, T., Shimizu, T., Uehara-Ichiki, T., Omura, T., and Sasaya, T. (2010). Complete sequence analysis of rice transitory yellowing virus and its comparison to rice yellow stunt virus. Arch. Virol. *155*, 243–245.

Hogenhout, S.A., and Bos, J.I.B. (2011). Effector proteins that modulate plant–insect interactions. Curr. Opin. Plant Biol. *14*, 422–428.

Hogenhout, S.A., Redinbaugh, M.G., and Ammar, E.D. (2003). Plant and animal rhabdovirus host range: a bug's view. Trends Microbiol. *11*, 264–271.

Howerth, E.W., Mead, D.G., and Stallknecht, D.E. (2002). Immunolocalization of vesicular stomatitis virus in black flies (*Simulium vittatum*). In Domestic Animal/Wildlife Interface: Issue for Disease Control, Conservation, Sustainable Food Production, and Emerging Diseases (Annals of the New York Academy of Sciences, New York), pp. 340–345.

Hsu, H.T., McBeath, J.H., and Black, L.M. (1977). Comparative susceptibilities of cultured vector and non-vector leafhopper cells to 3 plant viruses. Virology *81*, 257–262.

Huang, Y.W., Geng, Y.F., Ying, X.B., Chen, X.Y., and Fang, R.X. (2005). Identification of a movement protein of rice yellow stunt rhabdovirus. J. Virol. *79*, 2108–2114.

Inoue, H. (1979). Transmission efficiency of rice transitory yellowing virus by the green rice leafhoppers, *Nephotettix* spp. (Hemiptera, Cicadellidae). Appl. Entomol. Zool. *14*, 123–126.

Izadpanah, K. (1989). Purification and serology of the Iranian maize mosaic rhabdovirus. J. Phytopathol. *126*, 43–50.

Izadpanah, K., Ahmadi, A.A., Parvin, S., and Jafari, S.A. (1983). Transmission, particle-size and additional hosts of the rhabdovirus causing maize mosaic in Shiraz, Iran. J. Phytopathol. *107*, 283–288.

Jackson, A.O. (1987). Biology, structure, and replication of plant rhabdoviruses. In The Rhabdoviruses, Wagner, R.R., ed. (Plenum Press, New York), pp. 427–508.

Jackson, A.O., Dietzgen, R.G., Goodin, M.M., Bragg, J.N., and Deng, M. (2005). Biology of plant rhabdoviruses. Annu. Rev. Phytopathol. *43*, 623–660.

Kang, B.C., Yeam, I., and Jahn, M.M. (2005). Genetics of plant virus resistance. Annu. Rev. Phytopathol. *43*, 581–621.

Kitajima, E.W., Rodrigues, J.C.V., and Freitas-Astua, J. (2010). An annotated list of ornamentals naturally found infected by Brevipalpus mite-transmitted viruses. Scientia Agricola *67*, 348–371.

Kumin, I.V., Novella, I.S., Dietzgen, R.G., Padhi, A., and Rupprecht, C.E. (2009). The rhabdoviruses: Biodiversity, phylogenetics, and evolution. Infect. Genet. Evol. *9*, 541–553.

Lastra, J.R., and Esparza, J. (1976). Multiplication of vesicular stomatitis virus in leafhopper *Perigrinus maidis* (Ashm), a vector of a plant rhabdovirus. J. Gen. Virol. *32*, 139–192.

Lee, P.E. (1967). Morphology of wheat striate mosaic virus and its location in infected cells. Virology *33*, 84–94.

Lindbo, J.A., Silva-Rosales, L., Proebsting, W.M., and Dougherty, W.G. (1993). Induction of a highly specific antiviral state in transgenic plants: implications for regulation of gene expression and virus resistance. Plant Cell *5*, 1749–1759.

Liu, H.B., Liu, Y., Liu, S.L., Pang, D.W., and Xiao, G.F. (2011). Clathrin-mediated endocytosis in living host cells visualized through quantum dot labeling of infectious hematopoietic necrosis virus. J. Virol. *85*, 6252–6262.

Lockhart, B. (2004). Cynodon chlorotic streak virus. In Virus and Virus Diseases of Poaceae (Gramineae), Lapierre, H., and Signoret, P., eds (INRA, Paris), pp. 624–625.

Lockhart, B.E.L., Khaless, N., Elmaataoui, M., and Lastra, R. (1985). Cynodon chlorotic streak virus, a previously undescribed plant rhabdovirus infecting bermuda grass and maize in the Mediterranean area. Phytopathol. *75*, 1094–1098.

Longdon, B., Obbard, D.J., and Jiggins, F.M. (2010). Sigma viruses from three species of *Drosophila* form a major new clade in the rhabdovirus phylogeny. Proc. R. Soc. B Biol. Sci. *277*, 35–44.

Lundsgaard, T. (1984). Comparison of festuca leaf streak virus antigens with those of 3 other rhabdoviruses infecting the *Gramineae*. Intervirol. *22*, 50–55.

Lundsgaard, T. (1999). *Javesella pellucida* (F.) is a vector of festuca leaf streak virus (FLSV, Genus *Cytorhabdovirus*). J. Plant Dis. Protect. *106*, 545–549.

Lyles, D.S., and Rupprecht, C.E. (2007). Rhabdoviridae. In Fields Virology, Knipe, D.M., and Howley, P.M., eds (Lippincott Williams and Wilkins, Philadelphia, PA), pp. 1363–1408.

Massah, A., Izadpanah, K., Afsharifar, A.R., and Winter, S. (2008). Analysis of nucleotide sequence of Iranian maize mosaic virus confirms its identity as a distinct nucleorhabdovirus. Arch. Virol. *153*, 1041–1047.

Mayhew, D.E., and Flock, R.A. (1981). Sorghum stunt mosaic. Plant Dis. *65*, 84–86.

Means, J.C., and Passarelli, A.L. (2010). Viral fibroblast growth factor, matrix metalloproteases, and caspases are associated with enhancing systemic infection by baculoviruses. Proc. Natl. Acad. Sci. U.S.A. *107*, 9825–9830.

Ming, R., Brewbaker, J.L., Pratt, R.C., Musket, T.A., and McMullen, M.D. (1997). Molecular mapping of a major gene conferring resistance to maize mosaic virus. Theor. Appl. Genet. *95*, 271–275.

Mohammed, A.K.H., Maehlen, J., Magnusson, O., Fonnum, F., and Kristensson, K. (1992). Persistent changes in behavior and brain-serotonin during aging in rats subjected to infant nasal virus-infection. Neurobiol. Aging *13*, 83–87.

Mueller, S., Gausson, V., Vodovar, N., Deddouche, S., Troxler, L., Perot, J., Pfeffer, S., Hoffmann, J.A., Saleh, M.C., and Imler, J.L. (2010). RNAi-mediated immunity provides strong protection against the negative-strand RNA Vesicular stomatitis virus in *Drosophila*. Proc. Natl. Acad. Sci. U.S.A. *107*, 19390–19395.

Nagata, T., Inoue-Nagata, A.K., Smid, H.M., Goldbach, R., and Peters, D. (1999). Tissue tropism related to vector competence of *Frankliniella occidentalis* for tomato spotted wilt tospovirus. J. Gen. Virol. *80*, 507–515.

Nault, L.R. (1997). Arthropod transmission of plant viruses: a new synthesis. Ann. Entomol. Soc. Amer. *90*, 521–541.

Nault, L.R., and Ammar, E. (1989). Leafhopper and planthopper transmission of plant-viruses. Annu. Rev. Entomol. *34*, 503–529.

Nault, L.R., and Gordon, D.T. (1988). Multiplication of maize stripe virus in *Peregrinus maidis*. Phytopathol. *78*, 991–995.

Persley, D.M., and Greber, R.S. (2004a). Maize sterile stunt. In Viruses and Virus Diseases of the Poaceae (Gramineae), Lapierre, H., ed. (INRA, Paris), pp. 670–671.

Persley, D.M., and Greber, R.S. (2004b). Cereal chlorotic mottle. In Viruses and Virus Diseases of the Poaceae (Gramineae), Lapierre, H., ed. (INRA, Paris), pp. 620–621.

Platt, K.B., Linthicum, K.J., Myint, K.S.A., Innis, B.L., Lerdthusnee, K., and Vaughn, D.W. (1997). Impact of dengue virus infection on feeding behavior of *Aedes aegypti*. Amer. J. Trop. Med. Hyg. *57*, 119–125.

Redinbaugh, M. (2010). Winter wheat Russian mosaic virus. In Compendium of Wheat Diseases and Pests, 3rd edn, Brockus, W., ed. (APS Press, St. Paul, MN), p. 118.

Redinbaugh, M.G., and Hogenhout, S.A. (2005). Plant rhabdoviruses. In The World of Rhabdoviruses, Fu, Z.F., ed. (Springer, New York), pp. 143–163.

Redinbaugh, M.G., Seifers, D.L., Meulia, T., Abt, J.J., Anderson, R.J., Styer, W.E., Ackerman, J., Salomon, R., Houghton, W., Creamer, R., *et al.* (2002). Maize fine streak virus, a new leafhopper-transmitted rhabdovirus. Phytopathol. *92*, 1167–1174.

Reed, S.E., Tsai, C.W., Willie, K.J., Redinbaugh, M.G., and Hogenhout, S.A. (2005). Shotgun sequencing of the negative-sense, RNA genome of the rhabdovirus maize mosaic virus. J. Virol. Meth. *129*, 91–96.

Reynaud, B., Delatte, H., Peterschmitt, M., and Fargette, D. (2009). Effects of temperature increase on the epidemiology of three major vector-borne viruses. Eur. J. Plant Pathol. *123*, 269–280.

Ryabov, E.V., Keane, G., Naish, N., Evered, C., and Winstanley, D. (2009). Densovirus induces winged morphs in asexual clones of the rosy apple aphid, *Dysaphis plantaginea*. Proc. Natl. Acad. Sci. U.S.A. *106*, 8465–8470.

Scholthof, K.B.G., Hillman, B.I., Modrell, B., Heaton, L.A., and Jackson, A.O. (1994). Characterization and detection of sc4 – a 6th gene encoded by sonchus yellow net virus. Virology *204*, 279–288.

Seifers, D.L. (2004). Wheat American striate mosaic virus. In Viruses and Virus Diseases of Poaceae (Gramineae), Lapierre, H., ed. (INRA, Paris, France), pp. 589–590.

Shelly, S., Lukinova, N., Bambina, S., Berman, A., and Cherry, S. (2009). Autophagy is an essential component of *Drosophila* immunity against vesicular stomatitis virus. Immunity *30*, 588–598.

Shikata, E. (1989). Cytopathological changes in leafhopper vectors of plant viruses. In Leafhopper Vectors and Plant Disease Agents, Maramorosch, K., and Harris, K.F., eds (Academic Press, New York), pp. 309–325.

Signoret, P. (2004). Rice transitory yellowing/rice yellow stunt. In Viruses and Virus Diseases of Poaceae (Gramineae), Lapierre, H., ed. (INRA, Paris, France), pp. 524–527.

Sinha, R.C. (1970). *Elymana virescens*, a newly described vector of wheat striate mosaic virus. Can. Plant Dis. Surv. *50*, 118–120.

Slykhuis, J.T. (1963). Vector and host relations of North American wheat striate mosaic virus. Can. J. Bot. *41*, 1171–1185.

Stafford, C.A., Walker, G.P., and Ullman, D.E. (2011). Infection with a plant virus modifies vector feeding behavior. Proc. Natl. Acad. Sci. U.S.A. *108*, 9350–9355.

Strand, M.R. (2008). The insect cellular immune response. Insect Sci. *15*, 1–14.

Superti, F., Seganti, L., Ruggeri, F.M., Tinari, A., Donelli, G., and Orsi, N. (1987). Entry pathway of vesicular stomatitis virus into different host cells. J. Gen. Virol. *68*, 387–399.

Sylvester, E.S., and Richardson, J. (1992). Aphid-borne rhabdoviruses-relationship with their aphid vectors. In Advances in Disease Vector Research, Harris, K.F., ed. (Springer-Verlag, New York), pp. 313–341.

Tanno, F., Nakatsu, A., Toriyama, S., and Kojima, M. (2000). Complete nucleotide sequence of northern cereal mosaic virus and its genome organization. Arch. Virol. *145*, 1373–1384.

Thottappilly, G. (1992). Plant virus diseases of importance of African Agriculture. J. Phytopathol. *134*, 265–288.

Todd, J.C., Ammar, E.D., Redinbaugh, M.G., Hoy, C., and Hogenhout, S.A. (2010). Plant host range and leafhopper transmission of maize fine streak virus. Phytopathol. *100*, 1138–1145.

Toriyama, S. (2004). Northern cereal mosaic. In Viruses and Virus Diseases of Poaceae (Gramineae), Lapierre, H., ed. (INRA, Paris, France), pp. 578–580.

Tsai, C.W., Redinbaugh, M.G., Willie, K.J., Reed, S., Goodin, M., and Hogenhout, S.A. (2005). Complete genome sequence and in planta subcellular localization of maize fine streak virus proteins. J. Virol. *79*, 5304–5314.

Tsai, C.W., McGraw, E.A., Ammar, E.D., Dietzgen, R.G., and Hogenhout, S.A. (2008). *Drosophila melanogaster* mounts a unique immune response to the rhabdovirus sigma virus. Appl. Envir. Microbiol. *74*, 3251–3256.

Tuffereau, C., Benejean, J., Alfonso, A.M.R., Flamand, A., and Fishman, M.C. (1998). Neuronal cell surface molecules mediate specific binding to rabies virus glycoprotein expressed by a recombinant baculovirus on the surfaces of lepidopteran cells. J. Virol. 72, 1085–1091.

Vasconcelos, S.D., Hails, R.S., Speight, M.R., and Cory, J.S. (2005). Differential crop damage by healthy and nucleopolyhedrovirus-infected *Mamestra brassica* L. (Lepidoptera: Noctuidae) larvae: a field examination. J. Invert. Pathol. *88*, 177–179.

Walker, P.J., Dietzgen, R.G., Joubert, A.A., and Blasdell, K.R. (2011). Rhabdovirus accessory genes. Virus Res. *162*, 110–125.

Wang, X.-H., Aliyari, R., Li, W.-X., Li, H.-W., Kim, K., Carthew, R., Atkinson, P., and Ding, S.W. (2006). RNA interference directs innate immunity against viruses in adult *Drosophila*. Science *312*, 452–454.

Wetzel, T., Dietzgen, R.G., and Dale, J.L. (1994). Genomic organization of lettuce necrotic yellows rhabdovirus. Virology *200*, 401–412.

Whitfield, A.E., Rotenberg, D., Aritua, V., and Hogenhout, S.A. (2011). Analysis of expressed sequence tags from maize mosaic rhabdovirus-infected gut tissues of *Peregrinus maidis* reveals the presence of key components of insect innate immunity. Insect Mol. Biol. *20*, 225–242.

Wilkins, C., Dishongh, R., Moore, S.C., Whitt, M.A., Chow, M., and Machaca, K. (2005). RNA interference is an antiviral defence mechanism in *Caenorhabditis elegans*. Nature *436*, 1044–1047.

Recent Advances in the Characterization of Animal Rhabdoviruses

11

Aneta Gubala

Abstract

Recent molecular characterization studies of taxonomically unassigned animal rhabdoviruses have uncovered a large and genetically diverse group within the family *Rhabdoviridae*, informally referred to as the 'dimarhabdovirus supergroup'. These viruses contain intricate genome structures, novel open reading frames (ORFs) in addition to the typical five (N–P–M–G–L) rhabdovirus genes and other interesting features previously not described in other representatives of the family. Consequently, the family awaits the classification of two new genera, Tibrovirus and Sigmavirus. The Hart Park, Tupaia–Oak Vale and Sinistar groups of viruses, and several independent representative viruses, await additional characterization to establish their positions within the family. Whilst an array of new and valuable information has been generated, more than 70 animal rhabdoviruses remain fundamentally uncharacterized. These viruses unquestionably conceal a great abundance of new information that will further sculpt our understanding of this important family.

Introduction

Viruses of the family *Rhabdoviridae* infect a variety of hosts such as mammals (including marsupials), fish, birds, reptiles, insects, crustaceans and plants. Rhabdoviruses are responsible for significant diseases of public health concern and cause substantial impact on livestock production, agriculture and wildlife, worldwide (Wagner, 1987; Lyles and Rupprecht, 2007). Different rhabdoviruses have evolved to exploit various modes of transmission, including transmission by arthropods, through direct contact, through gametes and through water. This remarkable diversity and ecological adaptability is complemented by recent discoveries of increasingly complex genomes within the *Rhabdoviridae*.

Until the sequencing of the first ephemerovirus genomes in the early 1990s there were no rhabdoviruses of terrestrial vertebrates known to contain more than the typical five N–P–M–G–L genes. Prior to this, only rhabdoviruses of fish and plants were believed to contain additional genes, suggested to be relics from ancestral viruses. The discovery of large and complex genomes in the ephemeroviruses initially made them outliers among the animal rhabdoviruses and raised important questions regarding rhabdovirus evolution. Facilitated by advancements in next-generation technologies for *de novo* virus sequencing, recent characterization studies have revealed that genome complexity is a widespread phenomenon of animal rhabdoviruses. This chapter outlines recent discoveries that arose from the molecular characterization studies of diverse viruses of terrestrial animals. The important evolutionary questions that are raised as a consequence and the impacts to the taxonomy of the family are discussed. Viruses that require further study are also considered.

Historical perspective

The impacts of rhabdoviruses on humans and animals have been evident since antiquity. Rabies features in ancient Greek, Roman and Egyptian mythology (Steele and Fernandez, 1991; Lyles and Rupprecht, 2007). The earliest record of rabies-like illness in canines was made by Democritus in around 900 BC. In 1885 Louis Pasteur developed the first post-exposure vaccine using a rabies virus (RABV) preparation derived from the dried spinal cords of infected rabbits. In the early 1930s the bacteriologist Joseph Lennox Pawan from Trinidad was the first to propose that rabies could be spread from bats to humans and animals, which was initially regarded as implausible by his peers. Molecular characterization studies culminated in the completion of the RABV genome sequence (Tordo *et al.*, 1986a,b, 1988). Isolation of rabies-related viruses in Africa, Europe, Australia and Asia shaped the current delineation of the genus *Lyssavirus*.

Vesicular stomatitis viruses (VSV) are found predominantly in the Americas. The first description of VSV infection was done in 1916 in the USA following an epidemic in cattle and horses; however, reports of an illness with similar symptoms in horses date back to the 1860s during the American Civil War (Whelan, 2008). The significance of the disease in domestic animals (primarily cattle, horses, pigs) pressed research forward at a rapid pace. VSV was the first rhabdovirus to be observed by electron microscopy (Chow *et al.*, 1954), and in the 1960s a flurry of immunological, biochemical, ultrastructural and epidemiological studies followed. VSV was subsequently at the centre of the molecular biology boom in the 1970s, with numerous genetic studies exploiting modern developments in nucleotide sequencing. By the early 1980s, understanding of VSV genetics surpassed that of any other rhabdovirus. Today many different facets of VSV infection are well characterized, and currently VSIV (vesicular stomatitis Indiana virus) serves as the prototype virus for the mononegavirales.

The arthropod-transmitted bovine ephemeral fever virus (BEFV), the causative agent of a disabling and sometimes lethal disease in cattle and water buffalo, is thought to be an ancient virus endemic to Africa and southern Asia. The disease was first reported from east Africa in the late nineteenth century and the disease was first identified in Australia in 1936 (Seddon, 1938; St George, 1988; Walker, 2005). As cattle and buffalo are not native in Australia, BEFV is believed to have entered from the Indonesian Archipelago and manifested following the commencement of cattle breeding programmes in northern Australia. The Americas and Pacific Island countries have been free of BEFV to date. Although the disease was reproduced experimentally on several occasions by intravenous injection of blood collected from cattle during the febrile stage of the disease, six decades passed before the virus was finally isolated from bovines in Africa (van der Westhuizen, 1967). In 1974 the first isolation of BEFV from an insect (*Culicoides* spp.) was reported in Kenya (Davies and Walker, 1974) and numerous isolations from bovine bloods and different species of midge and mosquito have since been made in Africa, Australia, Japan, China, Taiwan and Israel. Despite this, definitive identification of the principal arthropod vector(s) has remained elusive. Owing to the impact of BEFV on beef and dairy industries, the virus and its close relatives became subjects of detailed molecular studies in the late 1980s and early 1990s (Walker *et al.*, 1991, 1992; Wang and Walker, 1993; Wang *et al.*, 1994, 1995; McWilliam *et al.*, 1997). Within the ephemerovirus genomes a complex region between the G and L genes was identified, consisting of several novel ORFs of unknown function, including an intriguing second non-structural glycoprotein gene. The genus *Ephemerovirus* was formally established in 1994 (Wunner *et al.*, 1995).

The far-reaching effects of disease caused by fish rhabdoviruses on farmed and wild, freshwater and marine fish species have been well documented since the 1950s (Tordo *et al.*, 2005; Kurath and Winton, 2008). Following the development of fish cell lines in the 1960s, understanding of fish rhabdovirus infections progressed significantly and two distinct groups of fish rhabdoviruses were identified: the genus *Novirhabdovirus*, representatives of which are highly diverged from all other rhabdoviruses, and the VSV-like fish viruses (Kurath and Winton, 2008). Recently, Spring viraemia of carp virus was the first fish virus species to be assigned to the genus *Vesiculovirus* (Carstens, 2010). A new distinct clade (informally named the Sinistar group) contains a growing list of viruses of marine and fresh water fish and also appears to have a close relationship to, but is distinct from, the genus *Vesiculovirus* (Dacheux *et al.*, 2010; Talbi *et al.*, 2011).

Plant rhabdoviruses have been widely studied since the 1960s. The genera *Cytorhabdovirus* and *Nucleorhabdovirus* were established in 1994 to accommodate two distinct and genetically diverse groups of plant rhabdoviruses based on the site of virus maturation: in the cytoplasm or nucleus of the infected cells, respectively (Wunner *et al.*, 1995). More than 90 putative plant rhabdoviruses have been described to date; however, molecular characterization of the majority is incomplete or lacking (Jackson *et al.*, 2008).

Many of the recently characterized rhabdoviruses were isolated by arbovirus monitoring programmes. These programmes were established in the 1950s in previously unexplored tropical regions of Africa, the Americas, Australia and South East Asia. Their aim was to provide a means for monitoring the distribution and ecology of endemic and exotic arboviruses of medical and veterinary importance, and their arthropod vectors. Viruses were isolated from terrestrial vertebrate and hematophagous arthropod trapping programmes, which were often established in association with sentinel animals. Concurrent advancements in laboratory and field techniques for virus isolation and serological identification played a vital role in the identification of many new viruses. Rhabdoviruses were isolated primarily from mosquitoes, midges and sandflies, but also from other animals including cattle, birds, bats, reptiles, rodents, amphibians, ticks and humans. Presently there are over 90 taxonomically unassigned animal rhabdoviruses.

Characterization methodologies

Traditional techniques for virus isolation and identification

Prior to the 1970s, virus isolation routinely involved intracerebral inoculation of suckling mice using clarified supernatant obtained after the homogenization of pools of arthropods, or from clinical material collected from ill or deceased animals. Mice, which showed signs of paralysis, were euthanized and their brains were used to inoculate additional mice in order to amplify the virus.

This approach has been almost completely superseded by *in vitro* cell culture methods. Commonly used cell lines for isolation and routine propagation of terrestrial vertebrate rhabdoviruses include baby hamster kidney (BHK-21), African green monkey kidney epithelial (Vero) or mosquito (*Aedes albopictus* C6/36) (St George, 1985; Calisher *et al.*, 1988). Following publication of the first viral electron micrographs in the mid 1950s, electron microscopy became the gold standard for the classification of viruses into families. Other important tests used for primarily classification included acridine orange staining for differentiation of viruses with single and double stranded genomes, and sodium deoxycholate sensitivity for differentiation of enveloped and non-enveloped viruses. Electron microscopy is still useful for the primary classification of new viruses that resist identification by molecular methods such as PCR or serology. The distinctive bullet or cone-shaped virions feature an envelope and are typically 100–430 nm in length and 45–100 nm in diameter. Rhabdoviruses of plants are bacilliform in appearance and tend to be longer than those of animals. It was suggested that virion size can be influenced by the size of the viral genome (Lazzarini *et al.*, 1981; Schnell *et al.*, 1996). This characteristic may provide rhabdoviruses with the flexibility to accommodate additional genes.

Until recently, serological tests including complement fixation, fluorescent antibody assay, virus neutralization test and haemagglutination inhibition were the primary tools for assessing relationships between rhabdoviruses. Pioneering work in the 1970s and 1980s produced a framework for the taxonomic assignment of new rhabdoviruses and demonstrated that many of the isolated rhabdoviruses could not be assigned to existing genera. (Tesh *et al.*, 1983; Brown, 1987; Calisher *et al.*, 1989). Serological surveys mapped the geographical distributions and identified potential hosts for some rhabdoviruses, although definitive causal links with disease could not be made for the majority of these viruses. Serological approaches remain vital tools for primary identification and comparison of relationships, but also have their limitations. For example, cross-reactivity between viruses of different genera, possibly due to shared epitopes within the highly conserved N protein (Wang and Walker, 1993; Wang *et al.*, 1995), has in the past led to the incorrect tentative grouping of some viruses. As such, recognition of the importance of genetic characterization has grown progressively in the last two decades.

Sanger sequencing

Until the 1970s, rhabdovirus characterization was primarily based on structural, morphogenic, physico-chemical and antigenic characteristics. Following the intensive molecular characterization studies of VSV and RABV in the 1970s and 1980s, the 1990s was a time of a rapid advancement in the genetic characterization of other rhabdoviruses. In the 1990s complete genome sequences were reported for viruses representing the genera *Cytorhabdovirus, Nucleorhabdovirus, Novirhabdovirus* and *Ephemerovirus,* for the first time enabling comprehensive genetic comparisons of the different members of the family.

The inherent difficulty of sequencing viruses lies in the nature of the association of viruses with their host cells. Until recently, approaches have required purification of the virus from host cell genetic material prior to performing conventional sequencing (Sanger *et al.*, 1977). These types of approach can be cumbersome, time consuming and unpredictable, particularly for novel viruses.

In recent years, PCR-based methods such as degenerate PCR, sequence-independent single primer amplification (SISPA), degenerate oligonucleotide primed PCR (DOP-PCR), random PCR and rolling circle amplification have been useful for the generation of sequence information for novel viruses (Bexfield and Kellam, 2011). One of the most successful recent approaches is subtractive hybridization, which has been used to obtain full genome sequences for at least eight novel negative sense ssRNA viruses, including four rhabdoviruses (Bowden *et al.*, 2001; Miller *et al.*, 2003; Jack *et al.*, 2005; Gubala *et al.*, 2008, 2010, 2011; Lambeth *et al.*, 2009). Despite these achievements, the major limitation of the published approaches is the slow pace of conventional sequencing. These methods are becoming rapidly superseded by 'next-generation' sequencing technologies.

Next-generation high-throughput sequencing

The commercial availability of next-generation high-throughput sequencing since 2005 has had a major impact on virus discovery and molecular characterization. A good demonstration of the power of high throughout sequencing for pathogen discovery occurred in 2008 following an outbreak of haemorrhagic fever in southern Africa (Briese *et al.*, 2009). Within 72 hours of receipt of clinical samples (serum and liver tissue), over 50% of the genome of a novel arenavirus (subsequently named Lujo virus) had been obtained with pyrosequencing using the 454 method. Combining traditional approaches with pyrosequencing has also proved extremely effective. Following subtractive hybridization, a pooled sample of three novel rhabdoviruses was subjected to 454 sequencing, simultaneously yielding complete genome sequences for two of the viruses (Ngaingan and Tibrogargan viruses) and half of the genome of the third virus (Oak Vale virus) (Gubala, 2009; Gubala *et al.*, 2010, 2011). Other examples include Durham rhabdovirus, two strains of Moussa virus and two strains of Oak Vale virus, which were sequenced by random amplification or DOP-PCR and 454 pyrosequencing (Quan *et al.*, 2010, 2011; Allison *et al.*, 2011). These approaches take a fraction of the time of conventional approaches to obtain full genome sequences. However, it is usually still necessary to confirm the sequences of genome termini by other methods, for example, RACE (rapid amplification of cDNA ends).

Ongoing developments and improvements are expediting faster, more accurate and more cost effective sequence generation. Younger platforms such as the SOLiD system (Applied Biosystems) and the Illumina sequencers offer increased scaleability and higher accuracy when sequencing through homopolymers and repetitive sequences. The gigabases of sequence data that these technologies are capable of generating generally surpass the level that is required for the sequencing of small RNA virus genomes. Therefore, efforts are under way to maximize output by pooling and multiplexing. The power of next-generation sequencing and improved bioinformatic tools can compensate to a large extent for the presence of contaminating host genetic material in the sample, effectively enabling the analysis of less pure virus preparations. Genomes of Ord River and Harrison Dam rhabdoviruses (described later in this chapter) and the ephemeroviruses Kotonkan and Obodhiang were sequenced in a multiplex format of pooled viruses from several different families using the Illumina genetic analyser (Blasdell *et al.*, 2012; Gubala, unpublished). These samples were prepared using a fast, crude method without the need for any enrichment or amplification of viral genetic material.

De novo virus genome sequencing has been greatly facilitated by the latest advancements in sequencing technologies. For example, speed and throughput of sequencing are no longer the bottleneck in the discovery process. Instead, new challenges lie in the analysis and biological interpretation of gigabases of genetic sequence data. Establishing Koch's postulates to provide proof of disease causation is rapidly becoming a greater challenge with the increasing ease of sequencing virus genes directly from clinical material (Rivers, 1937; Bexfield and Kellam, 2011).

Rhabdovirus genome organization and common features

The negative-sense single stranded RNA genomes of rhabdoviruses have several similarities with those of viruses from the families *Paramyxoviridae, Filoviridae* and *Bornaviridae* within the order *Mononegavirales* (Tordo *et al.*, 1992; Dietzgen *et al.*, 2011). These viruses typically contain helical nucleocapsids that consist of (–) ssRNA genome templates that are intimately associated with N protein molecules, and RNA-dependant RNA-polymerases (L) that are responsible for primary transcription, 5′-capping and 3′-polyadenylation of RNA. These viruses have a similar gene order, and genes typically contain a single short 3′ promoter region, short terminal untranslated regions (UTR) and short intergenic regions (IR).

Rhabdovirus genomes are generally between 11–15 kb in length and contain at least five open reading frames (ORFs) coding for structural proteins in the order N–P–M–G–L. The N protein of rhabdoviruses is tightly associated with genomic RNA to form a ribonucleocapsid that acts as a template for transcription and replication by the L and P proteins. The multifunctional M protein plays a diverse range of roles including assembly of the virion by interacting with the ribonucleocapsid, lipid membranes and the cytoplasmic tails of the G glycoproteins, which form trimers that project from the surface of the virion. The terminal regions of the genome consist of a 3′ leader and 5′ trailer sequence. These sequences play vital roles in gene transcription, genome replication, encapsidation of RNA by the N protein, binding of the L polymerase and assembly of infectious particles (Pattnaik *et al.*, 1995; Whelan *et al.*, 2004; Lyles and Rupprecht, 2007). Each ORF is typically flanked by transcription start and stop signals that are generally highly conserved within each virus. The transcription stop signals typically contain a conserved tetra-nucleotide sequence followed by a run of U_7, which the polymerase reiteratively transcribes to create the mRNA polyA tail. Each ORF is usually transcribed as an independent mRNA species; however an increasing number of rhabdoviruses appear to employ polycistronic transcription for some ORFs, possibly as a means for controlling transcript abundance or as a result of adaptation to cell culture. Levels of gene transcription decrease progressively by approximately 25–30% as the polymerase moves downstream of each gene (Iverson and Rose, 1981).

Genome features of viruses from the six rhabdovirus genera

Although the prototype rhabdovirus contains only five genes, in the order N–P–M–G–L, many rhabdoviruses contain additional ORFs or other unique features, as described below (refer to Fig. 11.1).

Vesiculoviruses

Viruses of the genus *Vesiculovirus* contain the smallest genomes of all rhabdoviruses (~ 11.2 kb, VSV) and have the prototype genome structure, coding for the five structural proteins N, P, M, G and L only. However, an additional overlapping ORF within the P gene, termed C, is found in all vesiculoviruses, similar to viruses in the

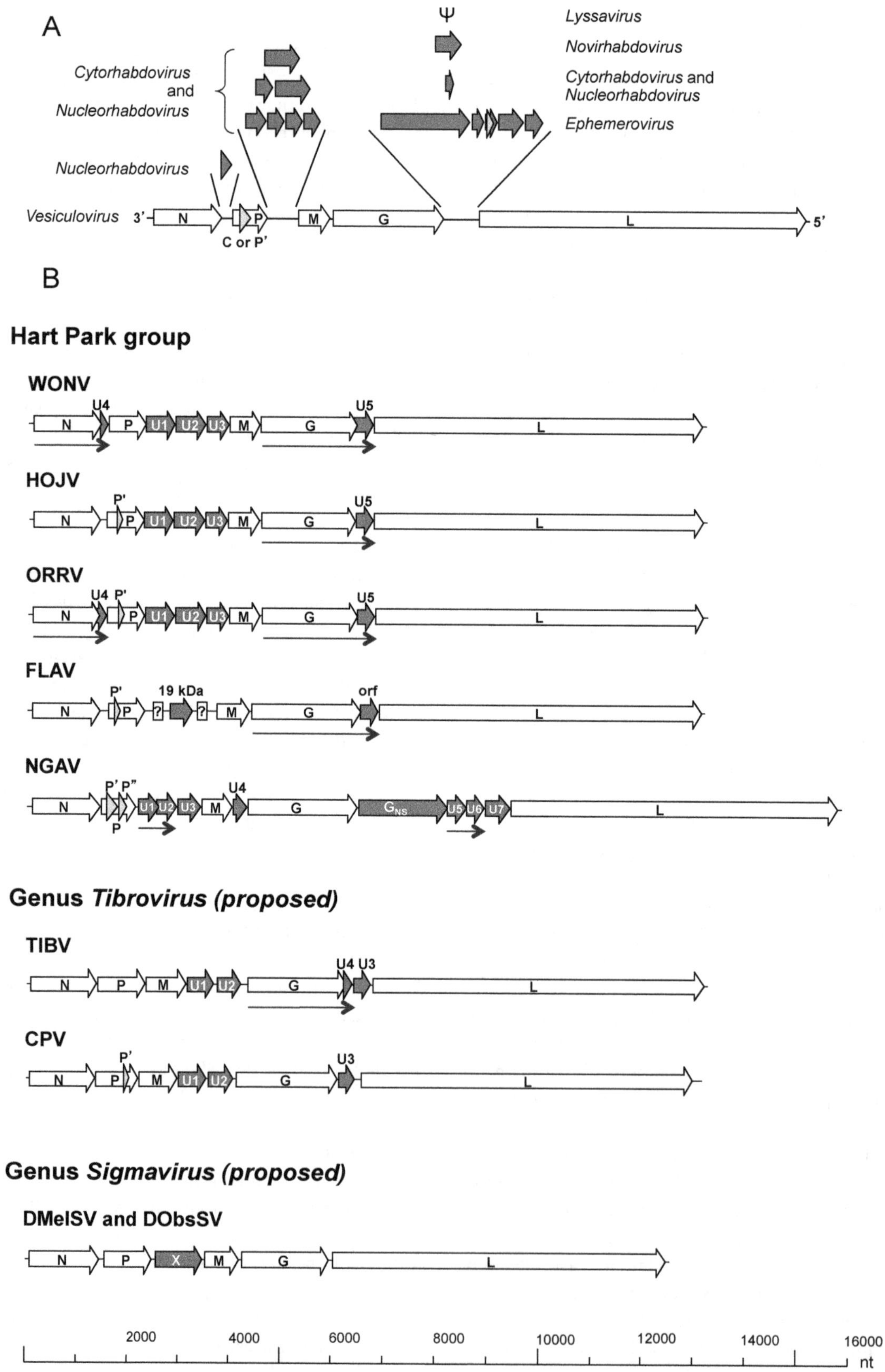
A
Ψ
Lyssavirus
Novirhabdovirus
Cytorhabdovirus and Nucleorhabdovirus
Cytorhabdovirus and Nucleorhabdovirus
Ephemerovirus
Nucleorhabdovirus
Vesiculovirus
3'
N
P
M
G
L
5'
C or P'
B
Hart Park group
WONV
U4
U5
U1 U2 U3
HOJV
P'
ORRV
FLAV
19 kDa
orf
?
NGAV
P' P"
GNS
U5 U6 U7
Genus Tibrovirus (proposed)
TIBV
U4 U3
CPV
Genus Sigmavirus (proposed)
DMelSV and DObsSV
X
2000
4000
6000
8000
10000
12000
14000
16000
nt

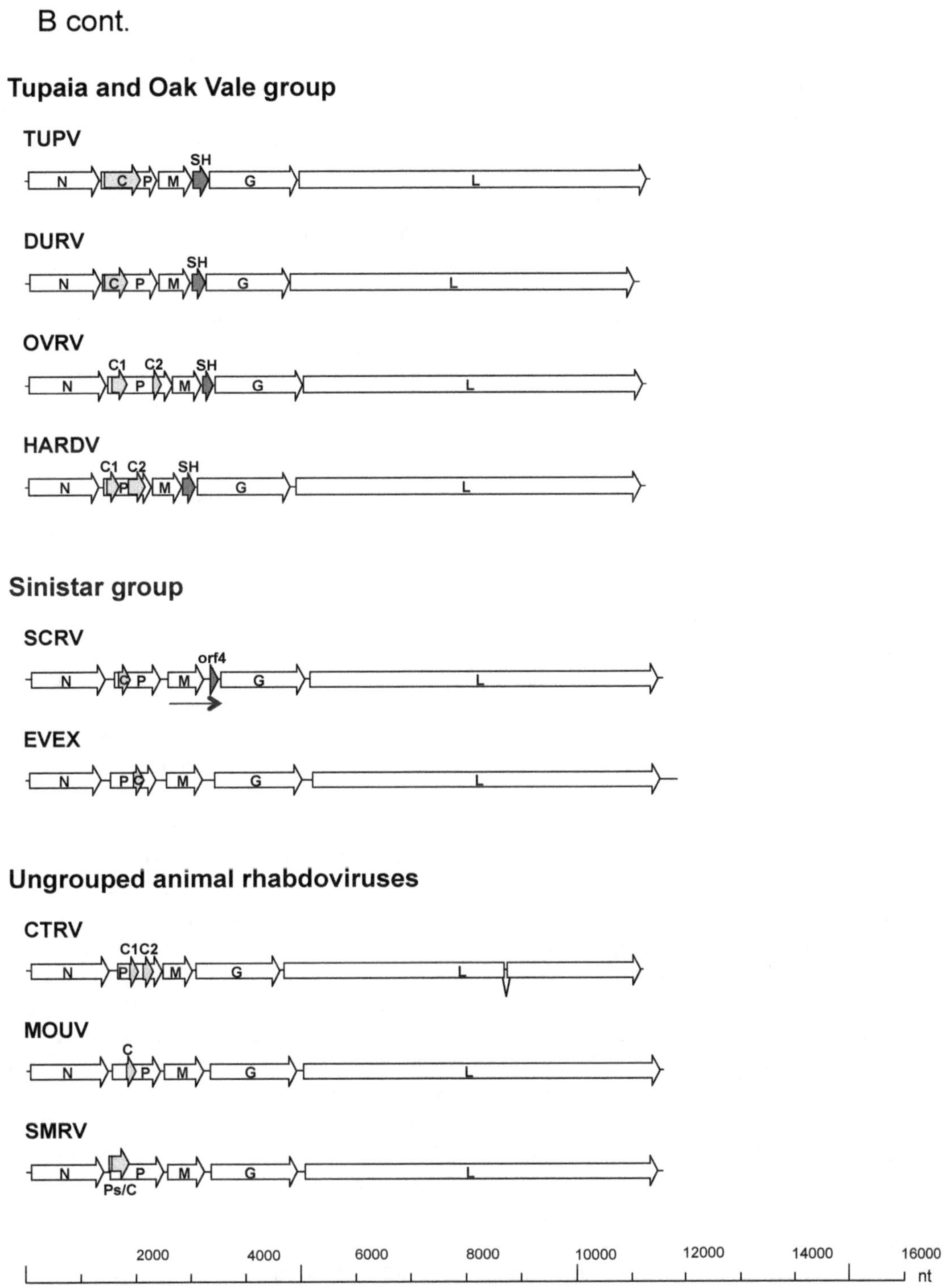

Figure 11.1 Genome organization of rhabdoviruses. (A) A representation of additional ORFs (shaded) present in the genomes of viruses of the six genera of the *Rhabdoviridae*. Viruses of the genus *Vesiculovirus* contain the prototype genome with five genes coding for structural proteins and an additional ORF overlapping the P gene. (B) Genome organization of unassigned animal rhabdoviruses of the dimarhabdovirus supergroup, including the sigma viruses. Additional genes are shaded. An intron in the L gene of CTRV is indicated. Thin arrows below ORFs indicate tentative bi-cistronic transcription owing to the absence of apparent transcription stop and start sequences.

family *Paramyxoviridae*. Many rhabdoviruses are now known to contain such ORFs (also variably referred to as P′ in some viruses). Studies of the C protein and the shorter isoform C′ of VSIV and VSNJV (vesicular stomatitis New Jersey virus) have demonstrated that they increase viral transcription by acting on the viral polymerase but that their absence does not adversely affect viral replication *in vitro* (Spiropoulou and Nichol, 1993; Kretzschmar *et al.*, 1996; Peluso *et al.*, 1996). These proteins have been proposed to contribute to transmission by insect vectors. The analogous paramyxovirus C proteins have a variety of functions in viral pathogenesis: inhibition of viral RNA synthesis, counteracting the innate immune response of the host cell, inhibition of virus-induced apoptosis, facilitating virus budding and they are essential for maintenance of negative-sense RNA genomes within virus particles (Irie *et al.*, 2008). Furthermore, the VSV M gene encodes two additional smaller in-frame translation products, referred to as M2 and M3 (Jayakar and Whitt, 2002). Both of these have been found to have a significant contribution to the cytopathology of the virus.

Lyssaviruses

Similar to vesiculoviruses, lyssaviruses code for only five structural proteins, but in addition lyssaviruses contain a long non-coding region between the G and L genes (within the long 3′ UTR of the G mRNA). The RABV G-L intergenic region (historically referred to as Ψ, or Psi) of several vaccine strains contains sequences resembling transcription start and stop signals, but no intact ORF, and has been thus proposed to be a remnant of an ancestral gene (Tordo *et al.*, 1986b). In other lyssaviruses (including wild-type RABV isolates) this region does not contain sequences resembling transcription initiation and termination signals (Kuzmin *et al.*, 2008). Although this region does not appear to be protein coding and its removal is not deleterious in cell culture or for pathogenesis in mice, its absence can have a negative effect on neuro-invasiveness of some strains (Schnell *et al.*, 1996; Ceccaldi *et al.*, 1998; Faber *et al.*, 2004). From another suggestion, this region could act as a structural spacer that regulates transcription (Ceccaldi *et al.*, 1998). Uniquely, WCBV contains a 180-nt ORF within the 3′UTR of the G gene, potentially encoding a small 60-aa protein, but this region does not contain any independent transcription initiation and termination signals and it is still unknown if this protein is translated (Kuzmin *et al.*, 2008).

Novirhabdoviruses

The fish-infecting novirhabdoviruses encode an additional non-virion (NV), non-structural protein between the G and L genes, which was documented in the infected cells (Kurath *et al.*, 1985; Kurath and Leong, 1985). It has been demonstrated that the NV protein is not essential for replication in cell culture, but its absence can impair virus growth (Biacchesi *et al.*, 2000). NV of infectious haematopoietic necrosis virus (IHNV) and viral haemorrhagic septicaemia virus (VHSV) seems to have a crucial role in infection of trout and perch, respectively (Thoulouze *et al.*, 2004; Ammayappan *et al.*, 2010). In contrast, the NV protein of snakehead rhabdovirus (SHRV) is not obligatory for pathogenesis in zebra fish and its absence does not appear to adversely affect viral replication in cultured fish cells (Johnson *et al.*, 2000; Alonso *et al.*, 2004). This apparent difference is suggested a result of the adaptation of SHRV to warm water (Biacchesi, 2011). The hypothesis that this region also has a spacing function that regulates genome transcription remains to be tested (Johnson *et al.*, 2000; Alonso *et al.*, 2004). The successful interchangeability of NV between IHNV and VHSV, despite only a 31% amino acid homology between them, suggests similar roles for both proteins (Thoulouze *et al.*, 2004). Recent results suggest that NV may have an anti-apoptotic function at the early stages of infection (Ammayappan and Vakharia, 2011). ORFs that overlap the P gene are also found in viruses of this genus.

Cytorhabdoviruses and nucleorhabdoviruses

The plant-infecting rhabdoviruses from the genera *Cytorhabdovirus* and *Nucleorhabdovirus* typically contain at least one additional gene between P and M, believed to code for a protein that facilitates movement of virus complexes between cells. The most convincing evidence of this function has been demonstrated for the rice yellow stunt virus (RYSV) P3 protein. In several viruses these proteins are known to be membrane or cell wall associated, and many share considerable amino acid sequence similarity with movement proteins of viruses from other families (Scholthof *et al.*, 1994; Goodin *et al.*, 2002; Huang *et al.*, 2005; Revill *et al.*, 2005; Dietzgen *et al.*, 2006; Goodin *et al.*, 2007; Heim *et al.*, 2008; Bandyopadhyay *et al.*, 2010). Genome of the maize fine streak virus (MFSV) codes for two proteins between P and M, which are similar in size to other putative movement proteins; however, their different subcellular localization could be indicative of different roles (Tsai *et al.*, 2005). Genome of northern cereal mosaic virus (NCMV) contains four ORFs between P and M which are transcribed in infected planthoppers, but further studies are required to determine their expression, subcellular localization and putative roles (Tanno *et al.*, 2000).

Genomes of some plant rhabdoviruses also contain an additional ORF between G and L. The RYSV P6 protein is associated with virions in infected leafhoppers, but has not been detected in leaf tissues (Huang *et al.*, 2003). A similarity between P6 and the amino terminal region of the L polymerase could suggest an involvement of P6 in virus replication (Huang *et al.*, 2003). The additional ORF between the G and L ORFs in the genome of strawberry crinkle virus (SCV) and NCMV is of unknown function (Tanno *et al.*, 2000; Dietzgen *et al.*, 2006). Potato yellow dwarf virus (PYDV) is the only rhabdovirus known to contain a gene (X) between the N and P genes. This gene codes for a small acidic protein, which is localized to the nuclei in yeast cells (Bandyopadhyay *et al.*, 2010). Studies of the plant rhabdovirus accessory proteins have recently been comprehensively reviewed (Jackson *et al.*, 2005; Kormelink *et al.*, 2011; Walker *et al.*, 2011).

Orchid fleck virus (OFV) and citrus leprosis virus (CiLV) were originally classified as rhabdoviruses based on their rhabdovirus-like morphology. Their genomic characterization, however, exposed bi-partite genomes. This is a considerable deviation from the rhabdovirus genome (and general concept of the order *Mononegavirales*), even if OFV displays similarity to the nucleorhabdoviruses at the gene sequence level and has a similar subcellular distribution (Kondo *et al.*, 2006). CiLV is phylogenetically distinct from viruses of other families (Locali-Fabris *et al.*, 2006). Currently there is no taxonomic provision for either virus. Considering the large number of tentative plant rhabdoviruses described in the literature (90+), their further characterization may expose that a significant proportion does not conform to the family *Rhabdoviridae*.

Ephemeroviruses

The most extensively studied ephemeroviruses are BEFV and Adelaide river virus (ARV). Ephemeroviruses have large genomes (~15 kb) that contain a second, non-structural glycoprotein gene (G_{NS}) located immediately downstream of the structural glycoprotein (G), and several additional small accessory ORFs of undefined function (α1, α2, α3, β and γ in BEFV; α1, α2, and β in ARV) (Walker *et al.*, 1991, 1992; Wang and Walker, 1993; Wang *et al.*, 1994; McWilliam *et al.*, 1997). Transcription of several of these ORFs appears to be on polycistronic mRNAs, which is postulated to be a means for modulation of the relative product abundance, or a consequence of virus adaptation to cell culture (Wang and Walker, 1993; Wang *et al.*, 1994; McWilliam *et al.*, 1997). Until the recent sequencing of Ngaingan virus (NGAV) genome (see below), the G_{NS} protein was a defining characteristic found only in the ephemeroviruses.

The ephemerovirus accessory proteins are subjects of ongoing study and have recently been comprehensively reviewed by Walker *et al.* (2011). The BEFV G_{NS} protein can be detected in infected mammalian and insect cell culture, but does not appear to be incorporated into virions, at least in mammalian cells (Walker *et al.*, 1991; Hertig *et al.*, 1996). The antigenicities of the BEFV G_{NS} and G proteins are significantly different (Hertig *et al.*, 1996; Johal *et al.*, 2008). The small novel proteins have a suggested involvement in pathogenesis, modulation of the host response to infection, or in the regulation of transcription and replication (Dhillon *et al.*, 2000; Walker *et al.*, 2011). The small hydrophobic α1 proteins of BEFV and ARV have characteristics suggestive of viroporins (McWilliam *et al.*, 1997; Gonzalez and Carrasco, 2003). However, these proteins have so far not been detected in infected cell culture, suggesting that they may be present in small amounts or not constitutively expressed *in vitro*. The β and γ proteins of ephemeroviruses appear to be variably expressed and detected in infected cells and virions (Walker *et al.*, 2011). Attenuation of expression by point mutations or corruption of transcription control signals (resulting in polycistronic mRNA transcription) of these proteins in cell culture suggests they may have greater roles in animal infection (Walker *et al.*, 2011). The G-L region of the recently sequenced genome of Obodhiang virus (OBOV) appears to reflect the genome structure of ARV, and the genomes of Kimberley (KIMV) and Berrimah viruses (BRMV) appear to resemble the genome structure of BEFV (Walker *et al.*, 2011; Blasdell *et al.*, 2012). The 15,870-nt genome of kotonkan virus (KOTV) contains an additional accessory gene, δ, that is absent from the other ephemeroviruses. The cognate protein has been detected in infected cells and preliminary data suggests it may play a role in inhibiting the innate anti-viral immune response (Walker *et al.*, 2011; Blasdell *et al.*, 2012).

Characterization of unassigned animal rhabdoviruses

Phylogenetic studies of partial N and L genes

New information generated through phylogenetic studies of partial N and L protein sequences has enabled the provisional grouping of numerous unclassified animal rhabdoviruses (Bourhy *et al.*, 2005; Kuzmin *et al.*, 2006; Dacheux *et al.*, 2010). The study of Bourhy *et al.* (2005), the largest phylogenetic study of unclassified rhabdoviruses at the time, analysed short fragments (~474 nt; ~158-aa) of a highly conserved region within block III of the L polymerase that were amplified using degenerate primers. The 20 unclassified or tentatively assigned viruses analysed in this study originate from Australia, Europe, Africa, Japan and the Americas. A succeeding investigation generated analogous sequence data for several additional viruses (Dacheux *et al.*, 2010). The study of Kuzmin and colleagues analysed a highly conserved 915-nt region and the deduced 305-aa region of the N protein in seven African and American rhabdoviruses that are antigenically cross-reactive with lyssaviruses and vesiculoviruses (Kuzmin *et al.*, 2006). The majority of the viruses in these studies were found to have distinct lineages in the vicinity of, but independent from, the existing genera *Ephemerovirus* and *Vesiculovirus*. The informal name 'dimarhabdovirus supergroup' (dipteran-mammal associated rhabdovirus) has been proposed to encompass this diverse group of animal viruses (Bourhy *et al.*, 2005).

Although useful new information was generated from these studies, it is apparent that the classification of these viruses cannot be resolved using only partial gene sequences. Presently, parallel comparisons between many of the viruses cannot be made owing to lack of analagous data. The comprehensive characterization of complete genomes enables more accurate analyses and greater scrutiny of phylogenetic relationships. Table 11.1 lists the unassigned animal rhabdoviruses that have been recent subjects of characterization studies and Fig. 11.1 depicts their genome structures.

Hart Park group

The Hart Park group was established on the basis of two serologically related viruses, Hart Park virus (HPV) and Flanders virus (FLAV) (Whitney, 1964; Johnson, 1965; Tesh *et al.*, 1983). Both viruses have been isolated numerous times from mosquitoes and birds throughout the USA, they are capable of causing fatal encephalitis in suckling mice, and FLAV causes haemorrhage in chick embryos (Jenson *et al.*, 1967; Karabatsos, 1985). A near-complete genome for FLAV was posted on GenBank in 2002; however, HPV remains uncharacterized at the genetic level.

Wongabel virus (WONV) and NGAV are tentatively assigned to the Hart Park group based on their phylogenetic relationship to FLAV (Fig. 11.2) (Bourhy *et al.*, 2005; Gubala *et al.*, 2008, 2010). WONV has a similar genome organization to FLAV (Fig. 11.1), containing three additional putative genes between P and M, an ORF encoding a small hydrophobic protein within the 3′ non-coding region of the G gene, and a small ORF capable of coding a putative 49-aa protein within

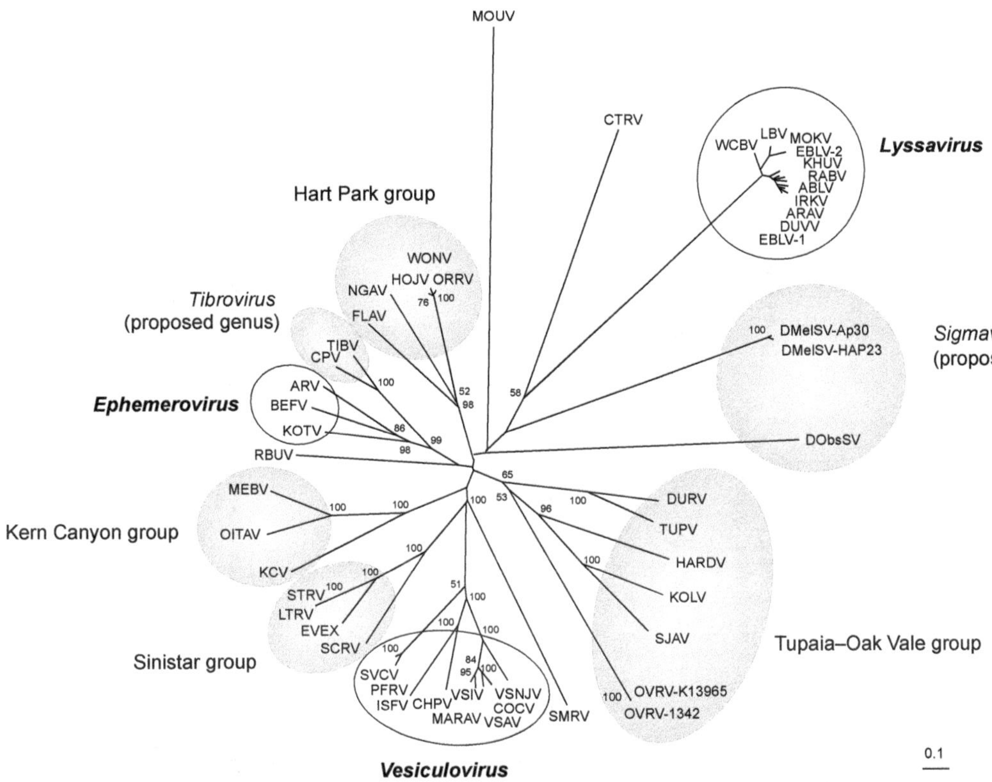

Figure 11.2 A neighbour-joining tree of partial N protein sequences (~300 aa) of viruses from, and in proximity to, the dimarhabdovirus supergroup. Groups of unassigned viruses are highlighted by the shading. Bootstrap values above 50 from 100 resampled data sets are overlaid on the tree (values within the genus *Lyssavirus* are not depicted for clarity).

the N gene 3′ UTR (which is absent from FLAV) (Gubala *et al.*, 2008). WONV was isolated from biting midges (*Culicoides austropalpalis*) collected in the Wongabel State Forest on the Atherton Tablelands of northern Queensland. Midges of this species have been observed to have a feeding preference for birds, and neutralizing antibodies to this virus have been detected in sea birds collected off the Great Barrier Reef, suggesting that WONV may have an avian reservoir host (Muller *et al.*, 1981; Humphery-Smith *et al.*, 1991). No neutralizing antibodies have been detected in human sera from the island residents. A broad survey of over 1,600 animal sera (including cattle, buffalo, pigs, wild and domestic birds, wallabies, kangaroos, wallaroos, horses, rats and pythons) collected between 1997 and 2007 in the Northern Territory, Australia, did not conclusively identify any significant seroprevalence to WONV (Gubala, 2009). Low level neutralization was observed in cattle and wallabies, possibly indicative of infection with another serologically related virus.

NGAV was isolated from biting midges (*Culicoides brevitarsis* and *Culicoides actoni*) that were collected at Mitchell River Aboriginal Community (currently known as Kowanyama) in northern Queensland, Australia (Doherty *et al.*, 1973; Kay *et al.*, 1978). The NGAV genome is one of the largest (15.7 kb) genomes within the *Rhabdoviridae* and arguably the most complex (Fig. 11.1). Despite the inference that NGAV is most closely related to WONV and FLAV at the sequence level of the structural proteins, the genome organization of NGAV is significantly different. The NGAV genome contains 15 ORFs (Gubala *et al.*, 2010). Additional genes are located in three different intergenic regions. Three novel ORFs are located in the P-M region like in WONV,

Table 11.1 Unassigned animal rhabdoviruses ('?' represents unknown)

Virus	Acronym	Prototype strain	Isolated from	Antibody detected in	Origin	Year of first isolation	Sequencing completed
Tentative groups with completed genomes							
Hart Park group							
Flanders	FLAV	61–7484	*Culiseta melanura*, *Culex pipans*, *Cx. restuans*, *Cx. salinarius*, *Cx. territans*, *Cx. tarsalis*, birds: Starling liver (bird), *Seirus aurocapillus*	Wild birds	New York, USA	1961	Genome
Wongabel	WONV	CS264	*Culicoides austropalpalis*	Sea birds, low level neutralization in cattle and wallaby	QLD, Australia	1979	Genome
Holmes Jungle[a]	HOJV	DPP1163	*Cx. annulirostris*	Cattle, buffalo, human	NT, Australia	1987	Genome
Ord River[a]	ORRV	OR1023	*Cx. annulirostris*	–	Western Australia	1976	Genome
Ngaingan	NGAV	MRM14556	*C. brevitarsis*/*C. actoni*	Wallabies, wallaroos, kangaroos, cattle	QLD, Australia	1970	Genome
Hart Park	HPV	AR 70	*Cx. tarsalis*, *Cx. restuans*, *Cx. nigripalpus*, *Cx. pipiens*, *Cs. melanura* and the following birds: *Xanthocephalus xantho*, *Passer domesticus*, *Agelaius tricolor*, *Carpodacus mexicanus*, *Seiurus aurocapillus*	Birds	California, USA	1955	none
Possible members							
Parry Creek	PCRV	OR189	*Cx. annulirostris*	–	Western Australia	1973	Part L
Kamese	KAMV	MP 6186	*Cx. annulioris, Cx. pruina, Cx. trigripes*	Humans	Uganda	1967	Part L
Mossuril	MOSV	SAAr 1995	*Cx. sitiens*, *Cx. neavei*, *Aedes abnormalis* and other *Culex* spp., birds: *Adropadus virens*, *Coliuspasser macrourus*	Humans, baboon	Mozambique	1959	Part L
Bangoran	BGNV	DakArB 2053	*Cx. perfuscus*, *Turdus libonyanus* (bird: brain)	–	C Afr Rep	1969	Part L
Porton	PORV	S 1643	*Mansonia uniformis*	–	Malaysia (Sarawak)	?	Part L
Mosqueiro	MQOV	BeAr185559	*Cx. portesi*, *Wyemyia* spp., *Mansonia* spp.	–	Brazil	1970	none

Genus Tibrovirus (proposed)							
Tibrogargan	TIBV	CS132	*C. brevitarsis*, asymptomatic bovine	Cattle, buffalo	QLD, Australia	1976	Genome
Coastal Plains	CPV	DPP53	Unsymptomatic bovine	Cattle, buffalo	NT, Australia	1981	Genome
Bivens Arm	BAV	UF 10	*C. insignis*	Cattle, buffalo	Florida, USA	1982	Part L
Genus Sigmavirus (proposed)							
Drosophila melanogaster sigma virus	SIGMAV and DMelSV-HAP23	HAP-23	*D. melanogaster*	–	unknown	ca. 1937	Genome
Drosophila melanogaster sigma virus	DMelSV-Ap30	Ap-30	*D. melanogaster*	–	USA	2005	Genome
Drosophila obscura sigma virus[a]	DObsSV	n/a	*D. obscura*	–	UK	2007	Genome
Drosophila affinis sigma virus[a]	DAffSV	n/a	*D. affinis*	–	USA	2007	L
Drosophila ananassae sigma virus[a]	DAnaSV	n/a	*D. ananassae*	–	Kenya	2010	Part L
Drosophila immigrans sigma virus[a]	DImmSV	n/a	*D. immigrans*	–	Germany	2010	Part L
Drosophila tristis sigma virus[a]	DTriSV	n/a	*D. tristis*	–	UK	2010	Part L
Mucina stabulans sigma virus[a]	MStaSV	n/a	*Mucina stabulans*	–	UK	2010	Part L
Tupaia and Oak Vale group							
Tupaia	TUPV	TRV1591	Tupaia belangeri	–	Germany (ex. Thailand)	?	Genome
Oak Vale	OVRV-CS1342	CS1342	*Cx. edwardsi*, *Ae. vigilax*	Feral pigs, low level neutralization in cattle	Australia	1981	Genome
Oak Vale	OVRV-K13965	K13965	*Anopheles annulipes* s.l.	–	Australia	1993	Genome
Durham	DURV	n/a	*Fulica americana* (bird)	–	North Carolina, USA	2005	Genome
Harrison Dam[a]	HARDV	CS75	*Cx. annulirostris*	–	NT, Australia	1975	Genome

Table 11.1 Continued

Virus	Acronym	Prototype strain	Isolated from	Antibody detected in	Origin	Year of first isolation	Sequencing completed
Sinistar group							
Siniperca chuatsi rhabdovirus	SCRV	?	*Siniperca chuatsi* (mandarin fish)	–	China	1999	Genome
Eel virus European X	EVEX	?	*Anguilla anguilla* (European eel)	–	Europe/Japan	1977	Genome
Sea trout rhabdovirus	STRV	28/97	*Salmo trutta trutta* (sea trout)	–	Sweden	1996	Part genome
Lake trout rhabdovirus[a]	LTRV	903/87	*Salmo trutta lacustris* (brown trout)	–	Finland	1987	Part genome
Starry flounder rhabdovirus[a]	SFRV	?	*Platichthys stellatus* (starry flounder)	–	USA	2000	Part L
Ungrouped							
Culex tritaeniorhynchus rhabdovirus	CTRV	TY	*Cx. tritaeniorhynchus*	–	Japan	2010?	Genome
Moussa virus	MOUV	D24, C23	*Cx. Decens, Culex* spp.	–	Côte d'Ivoire	2004	Genome
Scophthalmus maximus rhabdovirus	SMRV	?	*Scophthalmus maximus* (cultured turbot fish)	–	China	2006	Genome
Tentative groups without completed genomes							
Kern Canyon and Le Dantec groups							
Kern Canyon	KCV	M 206	*Myotis yumanensis* (bat)	–	California, USA	1956	Part N and L
Mount Elgon bat	MEBV	BP 846	*Rhinolophus hildebrandtii eloquens* (bat)	–	Kenya	1964	Part N
Oita	OITAV	296/1972	*Rhinolophus cornutus* (bat)	–	Japan	1972	N
Fukuoka	FUKAV	FUK-11	*C. punctalis*, *Cx. tritaeniorhynchus*, febrile calves	Cattle	Japan	1982	Part L
Nkolbisson	NKOV	YM 31–65	Mosquitoes: *Eretmapodites leucopus*, *Er. chrysogaster*, *Eretmapodites* spp., *Cx. telesilla*, *Ae. mutilus*, *Ae. argenteopunctatus*, *Ae. cumminsi*	–	Cameroon, Ivory Coast	1965	Part L

Barur	BARV	6235	*Rattus rattus wroughtoni* (rat), *Ma. uniformis*, ticks: *Haemaphysalis intermedia* (from goats), *Rhipicephalus pulchellus* (from camels)	–	India, Kenya, Somalia	1962	none
Le Dantec	LDV	DakHD 763	Human (hospitalized)	Ill human travelling from Nigeria	Senegal	1965	Part L
Keuraliba	KEUV	DakAnD 5314	*Tatera, Mastomys* and *Taterillus* (rodents)	–	Senegal	1968	Part L
Kolongo and Sandjimba groups							
Kolongo	KOLV	DakAnB 1094d	*Euplectes afra* and *Plesiositagra cucullata* (birds)	–	C Afr Rep	1970	Part N and L
Sandjimba	SJAV	DakAnB 373d	*Acrocephalus schoenobaenus* (bird)	–	C Afr Rep	1970	Part N and L
Possible members				–			
Nasoule	NASV	AnB 4289	*Andropadus virens* (bird: blood)	–	C Afr Rep	1973	Part L
Garba	GARV	DakAnB 439a	*Corythornis cristata* (bird), *Nectarinia pulchella* (bird: brain)	–	C Afr Rep	1970	Part L
Bimbo	BBOV	DakAnB 1054d	*Euplectes afra* (bird)	–	C Afr Rep	1970	Part L
Ouango	OUAV	DakAnB 1582a	*Sitagra melanocephala* (bird)	–	C Afr Rep	1970	Part L
Boteke	BTKV	DakArB 1077	*Ma. maculipennis*	–	C Afr Rep	1968	Part L
Bahia Grande group							
Bahia Grande	BGV	TB4–1054	*Ae. sollicitans, Anopheles* spp., *Psorophora* spp., *Culex* spp.	Humans, cattle, sheep, reptiles and wild mammals	Texas, USA	1974	Part L
Muir Springs	MSV	76V-23524	*Aedes* spp.	–	Colorado, USA	1976	Part L
Reed Ranch	RRV	TB4–222	*Cx. salinarius*	–	Texas, USA	1974	none
Almpiwar group							
Almpiwar	ALMV	MRM4059	*Ablepharus boutonii virgatus* (skink)	Same species of skink, cattle, horse, sheep, kangaroo, bandicoot, various birds, human	QLD, Australia	1966	Part L
Charleville	CHVV	Ch9824, Ch9847	*Phlebotomus* spp. (sandfly), *Gehyra australis* (lizard), *Lasiohelea* spp. (biting midge)	Human	QLD, Australia	1969	Part L

Table 11.1 Continued

Virus	Acronym	Prototype strain	Isolated from	Antibody detected in	Origin	Year of first isolation	Sequencing completed
Humpty Doo	HDOOV	CSIRO79, CSIRO80	*C. marksi* and *Lasiohelea* spp.	–	NT, Australia	1974	Part L
Bracorhabdoviruses							
Itacaiunas	ITAV	BE AR 427036	*Culicoides* spp.	–	Amazon, Brazil	1984	Part N
Curionopolis	CURV	BE AR 440009	*Culicoides* spp.	*Nasua nasua* (aardvark), *Cebus apella* (capuchin monkey)	Amazon, Brazil	1985	Part N
Sawgrass group							
Sawgrass	SAWV	64A1247	*Dermacentor variabilis* (tick from raccoon), *Haemaphysalis leporis-palusis* (tick)	–	Florida, USA	1964	none
Connecticut	CNTV	Ar-1152–78	*Ixodes (Ixodes) dentatus* (tick) from *Sylvilagus floridanus* (Eastern Cottontail rabbit)	*Sylvilagus floridanus*	Connecticut, USA	1978	none
New Minto	NMV	0579	*Haemaphysalis leporispalustris* (tick) from *Lepus americanus* (asymptomatic snowshoe hare)	–	Alaska, USA	1971	none
Timbo group							
Timbo	TIMV	BeAn 41787	*Ameiva ameiva ameiva* (lizard)	Same species of lizard and other reptiles	Brazil	1962	none
Chaco	CHOV	BeAn 42217	*Ameiva ameiva ameiva* and *Kentropyx calcaratus* (lizards)	–	Brazil	1962	none
Sena Madureira	SMV	BeAn 303197	*Ameiva ameiva ameiva* (lizard)	–	Brazil	1976	none
Marco (serologically unrelated)	MCOV	BeAn 40290	*Ameiva ameiva ameiva* (lizard)	–	Brazil	1962	none
Mataryia group							
Matariya[a]	none	An 147761	*Sylvia curraca* and *Sylvia borin* (birds)	–	Egypt	1961	none
Burg el Arab[a]	none	An 378262	*Sylvia curraca* and *Sylvia borin* (birds)	–	Egypt	1962	none
Ungrouped							
Rochambeau	RBUV	CaAr 16102	*Coquillettidia albicosta* (mosquitoes)	–	French Guiana	1973	Part N

Tentative animal rhabdoviruses							
Aruac	ARUV	TRVL 9223	Mosquitoes: *Trichoprosopon theobaldi* (mosquitoes), *Wyeomyia* spp., *Psorophora ferox*, *Phoniomyia* spp., *Culex* spp., *Sabethes chloropterus*	–	Trinidad	1955	none
Blue crab	BCV	?	?	–	?	?	none
Cuiaba virus	CUIV	BeAn 227841	*Bufo marinus* (toad)	–	Brazil	1972	none
DakArK 7292	DAKV-7292	DakArK 7292	*Amblyomma variegatum* (tick)	–	Guinea	?	none
Entamoeba	ENTV	?	*Entamoeba histolytica* and *Entamoeba invadens* (protazoa collected from human and snake)	–	UK	1974	none
Farmington	FARV	CT-114	Unknown wild bird	–	Connecticut, USA	1969	none
Gossas	GOSV	DakAnD 401	*Tadarida* spp. (bat)	–	Senegal	1964	none
Harlingen	HARV	?	*Culex* spp.	–	?	?	none
Iriri	IRIV	?	*Lutzomyia* spp. (sandfly)	–	?	?	none
Inhangapi virus	none	BeAr 177325	*Lutzomyia flaviscutellata*	*Proechimys guyannensis* (spiny rat), *Proechumys* spp., *Oryzomys capito goeldii* (semi-aquatic rodent), *Coendu* spp. (porcupine)	Brazil	1969	none
Joinjakaka	JOIV	MK7837	*Culicines* spp.	Cattle	Papua New Guinea		none
Kannamangalam	KANV	VRC 66415	*Corvus splendens* (house crow)	–	India	1963	none
Koolpinyah	KOOLV	DPP883	Cattle	Cattle	NT, Australia	1985	none
Landjia	LJAV	DakAnB 769d	*Riparia paludicola* (bird)	–	C Afr Rep	1970	none
Little Lilly Creek[a]	LLCV	OR559	*Cx. annulirostris*	–	Western Australia	1974	none
Manitoba	MNTBV	Mn 936–77	*Cx. tarsalis*	–	Manitoba, Canada	1977	none
Morreton	MRTV	?	*Lutzomyia* spp.	–	?	?	none
Navarro	NAVV	Cali 874	*Cathartes aura* (vulture: spleen)	–	Colombia	1956	none

Table 11.1 Continued

Virus	Acronym	Prototype strain	Isolated from	Antibody detected in	Origin	Year of first isolation	Sequencing completed
Poona virus[a]	none	733646	*Phlebotomus* spp.	–	India	?	none
Rio Grande cichlid	RGRCV	?	?	–	?	?	none
Rhode Island virus[a]	none	RI-166	*Columbia livia* (pigeon: brain)	–	Rhode Island, USA	2000	none
Sripur	SRIV	733646	*Sergentomyia* spp. (sandfly)	–	India	1973	none
Sweetwater Branch	SWBV	UF-11	*C. insignis*; (may be closely related to BAV)	–	Florida, USA	1982	none
Walkabout Creek[a]	WACV	CS1056	*C. austropalpalis*	–	QLD, Australia	1981	none
Xiburema	XIBV	BeAr 362159	*Sabethes (Sbn) intermedius* (mosquito)	–	Brazil	1977	none
Yata	YATAV	DakArB 2181	*M. uniformis*	–	C Afr Rep	1969	none

C Afr Rep, Central African Republic; NT, Northern Territory; QLD, Queensland. *Ae.*, *Aedes*; *C.*, *Culicoides*; *Cs.*, *Culiseta*; *Cx.*, *Culex*; *Ma.*, *Mansonia*.
[a]Not listed by the ICTV.

and one novel gene is located in the M-G region. Between the G and L genes NGAV has a second non-structural glycoprotein (G_{NS}) gene followed by three additional novel ORFs. The latter region is reminiscent of the ephemerovirus genomes. Two additional ORFs overlap the P gene. The G_{NS} protein is analogous in sequence and secondary structure to the G_{NS} proteins of ephemeroviruses and appears to have the same origin, but it is unclear whether this indicates a common ancestor or if this gene is a recent acquisition (Gubala *et al.*, 2010). NGAV is the only rhabdovirus outside of the genus *Ephemerovirus* known to contain a G_{NS} gene, which creates a taxonomic challenge. Two independent serological surveys of Australian livestock and wildlife identified macropods (wallabies, kangaroos, and wallaroos) and possibly cattle as primary hosts for the virus (Doherty *et al.*, 1973; Gubala *et al.*, 2010). At the antigenic level NGAV shares no obvious relationship to the ephemeroviruses or any of the tentative Hart Park viruses (Calisher *et al.*, 1989; Gibbs *et al.*, 1989). At present NGAV is unique within the family. Currently, FLAV, WONV and NGAV are recognized as three independent viral species within *Rhabdoviridae*, without inclusion in any established genera (Dietzgen *et al.*, 2012).

The recent genomic sequencing of Holmes Jungle virus (HOJV) and Ord River virus (ORRV) suggests that they have a very close relationship to WONV (Gubala *et al.*, unpublished). These three viruses are similar phylogenetically (Fig. 11.2) and contain almost identical genome structures (Fig. 11.1). The main differences are that the HOJV genome does not contain a small ORF overlapping the 3′ region of the N gene, and that the genomes of both HOJV and ORRV contain a putative small C ORF that overlaps the P gene, which is absent from WONV. The available short fragment of the L protein of Parry Creek virus (PCRV) similarly suggests a close relationship to WONV (Bourhy *et al.*, 2005), but this needs confirmation by additional sequencing. PCRV was isolated from the same species of mosquito as HOJV and ORRV (*Culex annulirostris*) (Liehne *et al.*, 1976, 1981). ORRV and PCRV have unknown reservoir hosts but HOJV, or a closely related virus, may be present in cattle (Weir, 2002). Although historical records suggest that all of these viruses are distinct from each other, detailed serological comparisons and ecological studies would be valuable for accuracy in taxonomic demarcation.

Kamese virus (KAMV), Mossuril virus (MOSV) and Mosqueiro virus (MQOV) are proposed members of the Hart Park group based on their antigenic similarity to HPV and FLAV (Karabatsos *et al.*, 1973; Calisher *et al.*, 1989). Short L protein sequence data recently generated for KAMV and MOSV support their placement in the group; the two viruses form a clade with FLAV (Dacheux *et al.*, 2010). KAMV was isolated from *Culex annulioris* in Uganda and MOSV was isolated from *Culex sitiens* collected in Mozambique (Kokernot *et al.*, 1962; Anon., 1970). Numerous isolations of MOSV have also been made in central Africa from various species of mosquito (Karabatsos, 1985). Evidence suggests that both viruses are capable of infecting humans, and MOSV antibodies have also been found in a baboon. MQOV was isolated in Brazil from *Culex portesi* and from mosquitoes of other species but no genetic information exists for this virus (Karabatsos, 1985; Calisher *et al.*, 1989). Comparisons of short L protein sequences suggest that Bangoran virus (BGNV), isolated from *Culex perfuscus* and from the brain of a bird in the Central African Republic, and Porton virus (PORV), isolated from *Mansonia uniformis* in Malaysia, may also belong to the Hart Park group (Dacheux *et al.*, 2010), but there is no serological or other support for this suggestion. On antigenic grounds the Hart Park group has in the past been proposed to constitute a new genus (Calisher *et al.*, 1989). The genetic data obtained to date support that FLAV, WONV, HOJV and ORRV may represent a new genus, but the placement of NGAV into this genus is questionable.

Proposed genus Tibrovirus

The proposed genus Tibrovirus consists of two proposed species, Tibrogargan virus and Coastal Plains virus. Complete genomes have been sequenced from both viruses (Gubala *et al.*, 2011). Phylogenetic analyses demonstrate that Tibrogargan virus (TIBV) and Coastal Plains virus (CPV) form an independent monophyletic group with limited relationship to viruses of the genus *Ephemerovirus* (Fig. 11.2). TIBV was isolated from a pool of *Culicoides brevitarsis* midges near Peachester in South Eastern Queensland, Australia (Cybinski *et al.*, 1980), and later from the blood of a healthy bovine (T. D. St George, unpublished). CPV was isolated from the blood of an asymptomatic steer at the Coastal Plains Research Station (currently known as Beatrice Hill), Northern Territory, Australia (Cybinski and Gard, 1986). A third candidate, Bivens Arm virus (BAV), is serologically and genetically (based on a 408-nt L gene fragment) closely related to TIBV, sharing 94% sequence identity (Gibbs *et al.*, 1989; Allison *et al.*, 2011). Further sequencing of BAV will be necessary to support its phylogenetic relationship to TIBV and CPV. BAV was isolated from *Culicoides insignis* in Florida, USA (Gibbs *et al.*, 1989; Tuekam *et al.*, 1991).

The genomic sequencing of TIBV and CPV has revealed a unique genome structure within the *Rhabdoviridae* (Gubala *et al.*, 2011). Each virus contains two additional putative genes between the M and G genes, and one gene between the G and L genes (Fig. 11.1). Like many other rhabdoviruses, CPV also contains an overlapping ORF with the P gene, which is absent from TIBV. The TIBV genome contains an additional ORF, coding for a small putative 55-aa hydrophobic protein, which overlaps the G ORF and the 3′ G gene UTR similarly to the way that the WONV U5 protein is coded. The three corresponding novel proteins of TIBV and CPV are highly diverged from each other and from other known proteins at the sequence level, but significant conservation of predicted secondary structure elements suggests conserved biological functions (Gubala *et al.*, 2011).

At the antigenic level these three viruses form a distinct group within the *Rhabdoviridae* (Tesh *et al.*, 1983; Cybinski and Gard, 1986; Calisher *et al.*, 1989; Gibbs *et al.*, 1989). Although no clinical disease has been observed in association with either virus, antibodies to TIBV and CPV are found in a high proportion of cattle and buffalo across northern Australia, and antibodies to BAV are widely spread in cattle and water buffalo throughout Florida and the Caribbean (Cybinski *et al.*, 1980; Gibbs *et al.*, 1989; Tuekam *et al.*, 1991; Gubala *et al.*, 2011). The discovery of BAV in the USA initially fuelled speculations that it had entered the Americas through infected cattle imported from Australia via Trinidad, but subsequent analyses showed that its presence in cattle predated the arrival of the imported buffalo (Gibbs *et al.*, 1989; Tuekam *et al.*, 1991). Antibody titres of cattle in Florida at times were higher to TIBV than BAV, leading to the suggestion that several serologically related viruses may be co-circulating in cattle. Although there is no conclusive evidence whether BAV was a recent introduction or not, it is still believed to have origins in Asia where water buffalo are native (Gibbs *et al.*, 1989). If this were the case, it is interesting that ephemeroviruses, which are likewise present throughout Asia, have not similarly spread to the Americas. Perhaps this is indicative of differences in vector compatibility or in the mode of transmission of these two distinct groups of viruses.

Proposed genus Sigmavirus

In 1937 *Drosophila* spp. (fruit flies) were found with paralysing and irreversible CO_2 sensitivity that was hereditary and could be experimentally transmitted to resistant *Drosophila* flies using extracts from sensitive flies (Contamine and Gaumer, 2008). In 1965, the causative agent, sigma virus, was described. The sigmaviruses are unusual in that they are transmitted vertically through the eggs and sperm of their hosts. Originally found only in *Drosophila* spp., they have since been shown to infect other representatives of *Diptera*, and are the only known group within the *Rhabdoviridae* that have solely Dipteran hosts (Longdon *et al.*, 2010, 2011c). These unique characteristics are reflected in their high sequence divergence from other rhabdoviruses. Until recently, the majority of studies focused on sigma virus of *D. melanogaster* (DMelSV) for which the genomes of two strains have been sequenced (Teninges *et al.*, 1993; Longdon *et al.*, 2010). The genome of the related sigma virus of *D. obscura* (DObsSV) has also been sequenced, and a complete L gene of sigma virus of *D. affinis* (DAffSV), is also available. Partial gene sequences have also been generated for other sigma viruses, including the first sigmavirus with a host outside the genus *Drosophila* (*Muscina stabulans*) (Longdon *et al.*, 2010, 2011c). Past phylogenetic studies have produced variable branching of these viruses, but the most recent phylogenetic analyses of L genes suggested that the sigma viruses form a 'sister clade' to the dimarhabdoviruses (Longdon *et al.*, 2010). Owing to the high level of divergence of the sigmaviruses, L genes appear to be more appropriate than N genes for their evolutionary comparisons (Longdon *et al.*, 2010).

Between the P and M genes the sigmaviruses contain an additional gene (X) coding for a putative PP3 protein (DMelSV, 298 aa; DObsSV, 321 aa). The roles of the PP3 proteins remain a mystery – but they contain regions similar to those found in RNA polymerases, topoisomerases, signal proteins and toxin molecules. At best, the presence of signal peptides and *N*-glycosylation sites could suggest that they are secreted glycoproteins (Landes-Devauchelle *et al.*, 1995; Walker *et al.*, 2011).

Tupaia and Oak Vale group

The genomes of Tupaia virus (TUPV), Oak Vale virus (OVRV), Durham virus (DURV) and Harrison Dam virus (HARDV) suggest that these viruses are highly divergent at the sequence level but have a clear monophyletic origin (Fig. 11.2). A distinguishing feature found in only these four rhabdoviruses is the presence of a gene between the M and G genes, which putatively codes for a small hydrophobic protein (SH) (Fig. 11.1) (Springfeld *et al.*, 2005; Gubala, 2009; Allison *et al.*, 2011; Quan *et al.*, 2011; Gubala *et al.*, unpublished). All four viruses also contain one or two C ORFs that overlap the P proteins. Despite the high divergence of these four viruses at the sequence level, bioinformatic comparisons of the OVRV and TUPV G proteins further support a common ancestor in that the G proteins contain a unique pattern of cysteine residue conservation (Gubala, 2009) that is different from the previously described cysteine bonding model for animal rhabdoviruses (Walker and Kongsuwan, 1999). In addition, the OVRV and TUPV G proteins contain short cytoplasmic tails compared to other animal rhabdoviruses and both viruses contain atypically large P proteins. These characteristics suggest a different lineage from the ephemeroviruses, tibroviruses and Hart Park viruses.

TUPV, the first rhabdovirus found to contain an additional gene between the M and G genes, was isolated from the hepatocellular carcinoma cells of a moribund tree shrew (*Tupaia belangeri*) which was imported from Thailand and resided in Germany for six years prior to the development of illness (Kurz *et al.*, 1986; Springfeld *et al.*, 2005). This virus is currently recognized as a species within the *Rhabdoviridae* without assignment into any established genus (1). In 2005 a new rhabdovirus, DURV, was isolated from the brain of an ataxic, native bird (American coot) in Durham County, North Carolina, USA (Allison *et al.*, 2011). The N proteins of these two viruses share 57% sequence identity.

OVRV (prototype strain CS1342) was isolated in December 1981 (T. D. St George, unpublished) and subsequently nine times between January and February 1984 from *Culex edwardsi* mosquitoes collected near Peachester in south-east Queensland (Cybinski and Muller, 1990). Although it was isolated in association with an outbreak of BEFV in cattle, the serological and insect data collectively suggest that OVRV is not a cattle virus (*Cx. edwardsi* rarely feeds on cattle). However, the ability of OVRV to propagate well through mammalian cell culture, insect cell culture and live mosquitoes suggests that likely it is an arbovirus. OVRV was speculated to circulate between mosquitoes and an avian host, but attempts to demonstrate compatibility with birds by performing replication studies in cattle egrets suggest this is unlikely the case (Cybinski and Muller, 1990). Concurrently, OVRV was isolated from a pool of *Aedes (Ochlerotatus) vigilax* mosquitoes (which feed on a broad range of animals including humans and birds) collected at Darwin, Northern Territory, in January 1984 (Calisher *et al.*, 1989; Weir *et al.*, 1996). Nine years later in May 1993, an isolation of OVRV (strain K13965) was made from *Anopheles annulipes* mosquitoes collected in the Kununurra region in the north of Western Australia (500 km west of Darwin) during a routine annual university survey of mosquitoes and arboviruses (Quan *et al.*, 2011). The genomes of both strains are sequenced and have an overall genome identity of 93%, with N demonstrating the highest (100%), and P (92%) and SH (93%) demonstrating the lowest nucleotide identities. Serological surveys suggest that OVRV may have a considerable presence in the feral pig population of northern Australia and, to a lesser extent, in domestic pigs (Gubala, 2009). Low virus neutralization titres are also observed in cattle. As such, OVRV could have widespread implications for livestock industries in Australia. HARDV was isolated from *Culex annulirostris* at Beatrice Hill, near Darwin, and its ecology remains unresolved (Standfast *et al.*, 1984).

Klamath virus (KLAV), isolated from a meadow vole (*Microtus montanus*) in Klamath County, Oregon, USA (Murphy *et al.*, 1972), has a close relationship to TUPV and DURV based on comparisons of limited L protein sequences and serological analyses (Calisher *et al.*, 1989; Allison *et al.*, 2011). Partial L and N gene phylogenetic analyses suggest that Kolongo and Sandjimba viruses (described below) may also be related to this group, but weak branching support demonstrates the need for further sequencing (Fig. 11.2) (Kuzmin *et al.*, 2006; Dacheux *et al.*, 2010; Quan *et al.*, 2011). There is anecdotal evidence that the Almpiwar group (described below) may also join this clade, albeit distantly (Bourhy *et al.*, 2005). A lack of supporting sequence data for all these viruses summons caution in drawing conclusions about their relationships at the present time.

Sinistar group

The tentatively named Sinistar group (name derived from Siniperca and starry flounder) consists of a growing number of fish rhabdoviruses that are related to viruses from the genus *Vesiculovirus*. These fish viruses form a distinct cluster that may be grounds for the consideration of a new genus (Kurath and Winton, 2008; Dacheux *et al.*, 2010; Talbi *et al.*, 2011). This group is distinct from the fish-infecting spring viraemia of carp virus (SVCV, a recently assigned member to the genus *Vesiculovirus)* and the tentative vesiculovirus pike fry rhabdovirus (PFRV). The provisional members of the Sinistar group are: *Siniperca chuatsi* rhabdovirus (SCRV), starry flounder rhabdovirus (SFRV), eel virus X (EVEX, currently tentatively assigned to the genus *Vesiculovirus*), sea trout rhabdovirus (STRV), lake trout rhabdovirus 903/87 (LTRV), and a number of perch viruses isolated in Europe with partially sequenced N and G genes (Johansson *et al.*, 2001, 2002; Betts *et al.*, 2003; Mork *et al.*, 2004; Tao *et al.*, 2008; Talbi *et al.*, 2011). The *Scophthalmus maximus* rhabdovirus (SMRV), isolated from deceased turbot, also has a clear phylogenetic relationship to viruses from the genus *Vesiculovirus* but it does not seem to belong to the Sinistar group (Zhang *et al.*, 2007; Zhu *et al.*, 2011).

All these viruses have genome structures similar to the vesiculoviruses, with the exception of SCRV, which uniquely contains an ORF putatively coding for a small 45-aa hydrophobic protein within the 3′ non-coding region of the M gene. This protein resists detection in cell culture (Tao *et al.*, 2008). None of the Sinistar group viruses contain an NV protein between the G and L genes like their fish-infecting cousins of the genus *Novirhabdovirus*, but like many other rhabdoviruses, SCRV, EVEX and SMRV contain C ORFs that overlap the P gene (Tao *et al.*, 2008; Zhu *et al.*, 2011).

Moussa virus

Moussa virus (MOUV) was isolated several times from mosquitoes (*Culex decens* and other *Culex* spp.) collected in the vicinity of coffee plantations near the Taï National Park, Ivory Coast (Quan *et al.*, 2010). The genomes of isolates C23 (originally known as Taï Forest rhabdovirus) and D24 demonstrate an overall 96% nt identity. Both isolates demonstrate the prototype rhabdovirus genome organization containing the genes N–P–M–G–L, and contain an additional small C ORF overlapping the P gene putatively coding for a 56-aa C protein. Despite initial impressions that MOUV shares a close relationship with TIBV, which was inferred using a short TIBV L protein fragment available at that time (Bourhy *et al.*, 2005; Quan *et al.*, 2010), N protein phylogeny suggests that MOUV is distanced from all other dimarhabdoviruses (Fig. 11.2). Serological surveys of humans and animals to determine possible exposure and ecology of the virus remain to be studied. Currently MOUV is recognized as a species within the *Rhabdoviridae*, without assignment into any established genus (Dietzgen *et al.*, 2011).

Culex tritaeniorhynchus rhabdovirus

Culex tritaeniorhynchus rhabdovirus (CTRV) was isolated during the course of routine surveillance for Japanese encephalitis virus in Japan (Kuwata *et al.*, 2011). L gene of CTRV atypically contains a 76-nt intron, which suggests that this virus may require RNA splicing for mRNA maturation. This intron contains the expected elements suggestive of eukaryotic RNA splicing, and mature mRNAs without the intron sequence can be detected in cells. Replication of the virus within nuclei and very elongated virion morphology are characteristics that are more reminiscent of the plant-adapted nucleorhabdoviruses rather than the dimarhabdoviruses. RNA splicing is a highly unusual observation within the order *Mononegavirales*; indeed only viruses of the family *Bornaviridae* are known to undergo this process. Because RNA splicing is unprecedented in the *Rhabdoviridae*, the origin and significance of this phenomenon is not clear, but it is speculated to influence the level of expression of the L protein (Kuwata *et al.*, 2011). Owing to its high divergence, the phylogeny of CTRV is difficult to decipher, but it may share an ancestry with the sigmaviruses (Kuwata *et al.*, 2011). Like the sigmaviruses, CTRV may be a Dipteran virus.

Tentative groups of viruses with partially sequenced genes

Kern Canyon and Le Dantec groups

Viruses within the Kern Canyon and Le Dantec groups are less well characterized at the genetic level. Antigenic information and limited genetic evidence suggest that these two groups may be closely related but more sequence information is required to confirm this supposition (Bourhy *et al.*, 2005; Kuzmin *et al.*, 2006; Dacheux *et al.*, 2010; Allison *et al.*, 2011).

The Kern Canyon group, based on N gene phylogenetics (~ 1000 nt), consists of Kern Canyon virus (KCV), Mount Elgon bat virus (MEBV) and Oita virus (OITAV) (Kuzmin *et al.*, 2006). KCV was isolated from insectivorous bats (*Myotis yumanensis*) in California (Murphy and Fields, 1967) and MEBV was isolated from an insectivorous bat (*Rhinolophus hildebrandtii eloquens*) from Kenya (Metselaar *et al.*, 1969). MEBV can multiply in experimentally infected mosquitoes, and can be transmitted to suckling mice by mosquitoes. Serological tests comparing MEBV with known at that time animal rhabdoviruses failed to show significant cross-reactivity, although this virus was originally tentatively placed into the genus *Vesiculovirus* (Murphy *et al.*, 1970). OITAV was isolated from another insectivorous bat (*Rhinolophus cornutus*) in Japan during a survey for natural hosts of Japanese encephalitis virus (Iwasaki *et al.*, 2004). Although all three viruses were isolated from insectivorous bats, they originate from distant regions of the world (North America, Africa and Japan). It has thus been suggested that they could be specifically associated with bats, and transmitted by broad-ranging blood-feeding insects (such as mosquitoes), bat ectoparasites, or possibly through directly between bats (Kuzmin *et al.*, 2006).

Through serology it was inferred that Barur virus (BARV), Fukuoka virus (FUKAV), and Nkolbisson virus (NKOV) are antigenically related to KCV (Calisher *et al.*, 1989). BARV was originally isolated in India from rats (*Rattus rattus wroughtoni*) during studies into encephalomyocarditis disease (Ghosh and Rajagopalan, 1973). The virus was subsequently isolated from ticks in India, and from mosquitoes and ticks in Africa (Johnson *et al.*, 1977; Butenko *et al.*, 1981). Pathogenicity studies in green monkeys revealed that BARV causes death within 5–6 days after intracerebral inoculation, with lesions typically observed in the brain (Abramova *et al.*, 1987). No genetic sequence data exist for this virus. NKOV was isolated from forest mosquitoes (*Eretmapodites leucopus*), *Aedes* spp. and *Culex* spp. in Cameroon, and has subsequently been isolated from *Culicidea* spp. in the Ivory Coast and from humans in the Central African Republic (Salaun *et al.*, 1969; Bourhy *et al.*, 2008). FUKAV was first isolated from *Culicoides punctalis* biting midges, *Culex tritaeniorhynchus* mosquitoes, and subsequently from calves with fever and leucopoenia in Japan (Noda *et al.*, 1992). Partial L protein sequence analysis suggests a close relationship between FUKAV and NKOV (Dacheux *et al.*, 2010), but this requires further corroboration based on additional genetic information.

The Le Dantec group, according to serology and partial L protein analysis, consists of Le Dantec virus (LDV) and Keuraliba virus

(KEUV) (Tesh *et al.*, 1983; Cropp *et al.*, 1985; Dacheux *et al.*, 2010). LDV was originally isolated from a human with febrile illness from Senegal in 1965 (Cropp *et al.*, 1985). It is believed that a second person was exposed to this virus in 1969 through the bite of an insect that had arrived in London on a boat from Nigeria. This patient was hospitalized for 6 weeks with delirium (Woodruff *et al.*, 1977). Seven years after exposure, antibodies to LDV were detected in this person. It has been suggested that the patient's Parkinson's disease-like symptoms, which developed after the insect bite, could be attributed to this virus, although this remains unconfirmed. Complement fixation tests revealed that LDV is most closely related to KEUV, which has been isolated from different rodents (genera *Tatera, Mastomys,* and *Taterillus*) from Senegal (Tesh *et al.*, 1983; Cropp *et al.*, 1985).

Kolongo, Sandjimba, and related viruses

Kolongo virus (KOLV) and Sandjimba virus (SJAV) are representatives of a distinct monophyletic group within the dimarhabdovirus supergroup (Bourhy *et al.*, 2005; Kuzmin *et al.*, 2006; Dacheux *et al.*, 2010). KOLV and SJAV were isolated from birds in the Central African Republic (Karabatsos, 1985). Only minor one-way serological reactivity is present between these two viruses, however, KOLV can be neutralized by antibody generated to the lyssavirus MOKV, which initially placed KOLV and SJAV in close relationship to the lyssaviruses (Calisher *et al.*, 1989). Genetic analyses suggest otherwise, indicating that they may share a monophyletic origin with the Tupaia-Oak Vale group (Fig. 11.2) (Bourhy *et al.*, 2005; Kuzmin *et al.*, 2006; Dacheux *et al.*, 2010). Nasoule virus (NASV), Garba virus (GARV), Bimbo virus (BBOV), Ouango virus (OUAV) and Boteke virus (BTKV) may also belong to this clade according to comparisons of limited L protein sequences (Dacheux *et al.*, 2010). Of these, only NASV demonstrates a serological relationship to SJAV (Calisher *et al.*, 1989). All of these viruses have been isolated from birds with the exception of BTKV, which was isolated from mosquitoes (Karabatsos, 1985).

Bahia Grande group

The tentative Bahia Grande group consists of Bahia Grande virus (BGV), Muir Springs virus (MSV) and Reed Ranch virus (RRV). At the antigenic level the three viruses are related to each other and distinct from other rhabdoviruses (Kerschner *et al.*, 1986). Only short fragments of the L genes have been sequenced for BGV and MSV, and their comparison suggested that these viruses may constitute another independent clade (Allison *et al.*, 2011). BGV and RRV were isolated from salt-marsh mosquitoes (*Aedes sollicitans* and *Culex salinarius*, respectively) in Texas, and MSV was isolated from *Aedes* spp. in Colorado (Kerschner *et al.*, 1986). A serological survey showed that humans, cattle, sheep, reptiles and wild mammals from southern Texas had neutralizing antibodies to BGV (Kerschner *et al.*, 1986).

Almpiwar group

According to partial L protein phylogenetics this tentative group presently consists of the Australian viruses Almpiwar virus (ALMV), Charleville virus (CHVV), and Humpty Doo virus (HDOOV) (Bourhy *et al.*, 2005).

ALMV was isolated from the skink *Ablepharus boutonii virgatus* at Mitchell River Aboriginal Community (currently known as Kowanyama) in northern Queensland. Antibodies to ALMV were detected in other skinks of the same species and the virus was subsequently isolated on two further occasions from these lizards (Karabatsos, 1985). In cell culture the virus optimally replicates at 30°C, which further supports the suggestion that it is a reptilian virus (Monath *et al.*, 1979). Although ALMV has never been isolated from arthropods, it is an assumed arbovirus on its ability to multiply in experimentally infected mosquitoes (Carley *et al.*, 1973). Limited evidence suggests the presence of neutralizing antibodies to ALMV in sera from several different vertebrates including humans (Doherty *et al.*, 1970).

CHVV was isolated numerous times from sandflies (*Phlebotomus* spp.) collected near Charleville in southern Queensland (Karabatsos, 1985). Isolations of CHVV were also made from the heart, liver and lung of a lizard (*Gehyra australis*) and from a pool of biting midges (*Lasiohelea* spp.) collected at Mitchell River (Doherty *et al.*,

1973). Similarly to ALMV, CHVV is also an assumed arbovirus based on its ability to multiply in experimentally infected mosquitoes (Carley *et al.*, 1973). Weak serum neutralization of CHVV by one of 30 tested human sera has been reported (Doherty *et al.*, 1973). Although both viruses were isolated from reptiles (of the order *Squamata*) they do not share a serological relationship with each other. Complement fixation analyses of ALMV and CHVV with known at the time Australian arboviruses found no relationships.

HDOOV was isolated from biting midges (*Culicoides marksi* and *Lasiohelea* spp.) that were collected at Beatrice Hill near Darwin (Standfast *et al.*, 1984). Serologically this virus has no close relatives, and no potential hosts have been identified.

Bracorhabdoviruses

Itacaiunas virus (ITAV) and Curionopolis virus (CURV) were isolated from *Culicoides* spp. in Brazil (Diniz *et al.*, 2006). Analysis of a very short N protein fragment of the carboxy-terminal region suggests that ITAV and CURV are related to each other, albeit distantly, and highly diverged from all other rhabdoviruses, sharing closest similarity with the vesiculovirses (25–45% aa identity with different vesiculoviruses). Based on this analysis, and on serological data demonstrating lack of cross-reactivity with any other rhabdovirus, it has been tentatively suggested that the two viruses may constitute a new genus '*Bracorhabdovirus*' (Brazilian Amazonian *Culicoides* rhabdoviruses). However, a severe deficiency in supporting data means that the phylogenetic and moreover taxonomic placement of these viruses remains uncertain until sufficient information is collected.

Rochambeau virus

Rochambeau virus (RBUV), isolated from mosquitoes (*Coquillettidia albicosta*) in the French Guiana, was initially suggested to be a lyssavirus by serology (Calisher *et al.*, 1989). However, the available partial N gene sequence of RBUV is more similar to that of ephemeroviruses and Hart Park viruses than of lyssaviruses (Kuzmin *et al.*, 2006). In general, although RBUV is clearly a dimarhabdovirus, it is not closely related to any known virus based on the available data, and further studies are needed to resolve its phylogeny, ecology and pathobiology.

Serogroups of viruses for which there is no genetic information

Sawgrass group

The Sawgrass serogroup consists of Sawgrass virus (SAWV), Connecticut virus (CNTV) and New Minto virus (NMV). All three viruses were isolated from ticks and are related to each other at the antigenic level (Main and Carey, 1980; Calisher *et al.*, 1989). The first isolation of SAWV was made from *Dermacentor variabilis* ticks collected from a raccoon in Florida, USA, and it was later also isolated from *Haemaphysalis leporispalusis* ticks (Sather *et al.*, 1970; Wellings *et al.*, 1972). SAWV was, similarly to BARV (see Kern Canyon group), tested for pathogenicity in green monkeys; however, it did not appear to have pathogenicity for these animals (Abramova *et al.*, 1987). NMV was isolated from *Haemaphysalis leporispalusis* ticks that were collected from snowshoe hares in central Alaska (Ritter *et al.*, 1978). CNTV was isolated from *Ixodes dendatus* ticks removed from eastern cottontail rabbits in Connecticut, USA. Neutralizing antibodies to this virus have been detected in cottontail rabbits from the region, suggesting that ticks may play a role in its transmission (Main and Carey, 1980).

Timbo group

The Timbo serogroup consists of Timbo virus (TIMV), Chaco virus (CHOV) and Sena Madureira virus (SMV). TIMV and CHOV were isolated from *Ameiva ameiva ameiva* lizards during a study of reptilian viruses in Brazil, and CHOV was also isolated from *Kentropyx calcaratus* lizards during the same study (Causey *et al.*, 1966). These two viruses are antigenically closely related but distinguishable in serological tests (Causey *et al.*, 1966; Monath *et al.*, 1979; Calisher *et al.*, 1989). SMV was likewise isolated from *Ameiva ameiva ameiva* lizards in Brazil during a different study, and is antigenically related to TIMV and CHOV (Calisher *et al.*, 1989). These viruses do not share antigenic relationships with the Almpiwar group viruses ALMV and CHVV, which were also isolated from reptiles.

A serologically distinct virus, Marco virus (MCOV) was also isolated from *Ameiva ameiva ameiva* lizards during the same study as TIMV and CHOV. This virus has a possible distant relationship to the unassigned Navarro virus (NAVV) by serology. CHOV, TIMV and MCOV have been shown to multiply in experimentally infected mosquitoes, suggesting that they are arthropod-transmitted (Causey *et al.*, 1966).

Matariya group

This serogroup consists of Matariya virus and Burg el Arab virus, both isolated from *Sylvia curraca* birds in Egypt and neither is yet confirmed to be a rhabdovirus. GARV, isolated from *Corythornis cristata* birds in the Central African Republic (putatively related to the Kolongo and Sandjimba viruses), is serologically related to Mataryia and Burg el Arab viruses (Karabatsos, 1985; Calisher *et al.*, 1989; Dacheux *et al.*, 2010).

Serologically ungrouped viruses

There are many poorly characterized, tentative animal rhabdoviruses based on the two most comprehensive serological comparisons performed to date (Tesh *et al.*, 1983; Calisher *et al.*, 1989) (Table 11.1). No sequence data have yet been generated for these viruses, and very little is known about their lineages within the family or about their ecology. In several cases, the history of isolation was not well documented. Owing to their novel nature, further characterization can be anticipated to reveal valuable new information.

Evolution of complex rhabdovirus genomes

How do novel genes arise?

The mechanisms by which additional genes arise in rhabdoviruses are not clear. Until the discovery of novel genes in the ephemeroviruses, only rhabdoviruses of plants and fish were known to contain more than five genes. According to the molecular evolution theory, viruses evolve towards simpler, more efficient genomes. The lyssavirus long G–L intergenic region was considered an example of gene attrition with a bias towards a simpler, easier to maintain, genome (Tordo *et al.*, 1986b). The recent discovery of numerous additional genes in an increasing list of terrestrial vertebrate rhabdoviruses, however, suggests that this is not the whole story. The question seems to be whether these genes are remnants from ancestral viruses that somehow confer an evolutionary advantage, or whether rhabdoviruses are capable of readily gaining or losing genes under selective pressure to facilitate adaptation to new environments.

One of the proposed modes by which new genes arise is via duplication of upstream genes, through a mechanism similar to the way defective interfering particles are created (Lazzarini *et al.*, 1981; Perrault, 1981; Walker *et al.*, 1992). Generally however, with the exception of the G_{NS} proteins, the additional genes of rhabdoviruses and their protein products show no significant sequence homology to any other known genes or proteins, which would suggest that if they did arise by gene duplication they subsequently undergo considerable changes. The G_{NS} gene (in ephemeroviruses and NGAV), which contains significant sequence similarity to the structural G genes and deduced proteins, could have arisen through a gene duplication process. Similarly, the β and γ genes of BEFV and the U1/U2/U3 genes of WONV and FLAV may have arisen through the same process (McWilliam *et al.*, 1997; Walker *et al.*, 2011). Some similarity in support of duplication has also been observed between the RYSV (plant rhabdovirus) P6 protein, which is encoded between the G and L genes, and the amino terminal region of the L polymerase (Huang *et al.*, 2003). Along the same lines, the 112-aa NGAV U6 protein demonstrates 39% identity with the amino-terminal region of the upstream G_{NS} protein (Gubala *et al.*, 2010).

The possibility that new genes arise through recombination is traditionally considered to be unlikely in negative-sense ssRNA viruses (Domingo and Holland, 1997), but its occurrence could be largely underestimated. Recombination, or gene transfer, could offer an immediate advantage to the virus, unlike duplication, which would need to stand the test of time before a gene evolves a new function (Walker *et al.*, 2011). Evidence does support the occurrence of lateral gene transfer amongst viruses from other families, but this has not yet been reported for any viruses of the order *Mononegavirales* (Walker *et al.*, 2011). The similarity in the crystal structure of the VSV G protein and the herpesvirus gB protein could suggest that mononegavirales are capable of acquiring genes from their environment, for example through copying exogenous mRNA during genome synthesis (Roche *et al.*, 2006). Homologous recombination could be a mode by which rhabdoviruses exchange genes during mixed infection (Walker *et al.*, 1992).

The presence of the G_{NS} gene in NGAV is particularly intriguing. Previously known to only occur in the ephemeroviruses, the implication of the presence of G_{NS} in a virus that is phylogenetically more closely related to the Hart Park group, is unknown (Fig. 11.2). Like the other Hart Park viruses, NGAV also contains additional genes between the P and M genes, but none of the known Hart Park viruses contain G_{NS}. NGAV also contains an additional gene between the M and G genes, like the Tupaia–Oak Vale group and the tibroviruses (Fig. 11.1). The high homology between the G_{NS} genes of NGAV and the ephemeroviruses could suggest a common ancestral virus that conceivably contained many additional genes. Alternatively, NGAV may have acquired additional genes from the environment, which hypothetically could occur during virus replication in a mixed infection. Co-infection with NGAV and ephemeroviruses could occur in nature; these viruses circulate in the same ecosystem, they have been isolated from the same species of arthropod (*Culicoides brevitarsis*) and they probably share a greater vertebrate host range than is currently known. Rhabdoviruses may have an under-appreciated capacity to evolve from dynamic populations by gaining and losing genes in response to selective pressures and environmental adaptation (Walker *et al.*, 1992, 2011; Wang and Walker, 1993; Wang *et al.*, 1994; McWilliam *et al.*, 1997; Gubala *et al.*, 2010).

Patterns in genome organization

Rhabdoviruses exploit overlapping, or overprinting, as a means for encoding additional proteins within existing genes. This may be advantageous in that it does not compromise genome size and does not adversely alter the abundance of downstream transcripts. These proteins can only be translated by internal initiation on bicistronic mRNAs, through a still poorly understood mechanism. The most widely overprinted gene in the *Rhabdoviridae* is P which, of all the structural proteins, is the most amenable to change and is the least conserved in sequence. The ORFs that overlap P, commonly referred to as C in vesiculoviruses and RABV, or P′ in ephemeroviruses, vary in size and characteristics but are most often between 30 to 80 aa in length and are usually basic proteins (Spiropoulou and Nichol, 1993; Kretzschmar *et al.*, 1996; Peluso *et al.*, 1996; Springfeld *et al.*, 2005).

Many of the recently characterized rhabdoviruses contain ORFs that overlap the P gene but in most cases their expression has not yet been studied, they have different characteristics and their significance is unknown. SMRV, however, seems to contain two isoforms of the C protein (118 and 110 aa) (Zhu *et al.*, 2011). The protein can be detected in infected fish cell culture, but attempts to distinguish the two isoforms have not been successful. EVEX and SCRV from the Sinistar group contain putative C ORFs coding for a 65-aa and 72-aa protein, respectively, but expression studies have not yet been reported (Tao *et al.*, 2008). The Hart Park group viruses contain ORFs that overlap P, coding for proteins ranging in size from 30 to 68 aa, with the exception of WONV. NGAV contains two separate ORFs overlapping with P (68 and 58 aa). CPV, but not TIBV (proposed genus *Tibrovirus*), and the distantly related MOUV, each contain one ORF encoding a putative 33 aa and 56 aa protein, respectively. Viruses of the Tupaia – Oak Vale group likewise code for putative C proteins, but the 221-aa TUPV and 136-aa DURV C proteins are considerably larger than those generally found in rhabdoviruses and are more similar in size to the large paramyxovirus C proteins (typically 150–220 aa in length) (Springfeld *et al.*, 2005; Allison *et al.*, 2011). OVRV and HARDV P genes on the other hand contain two smaller ORFs overlapping P (OVRV, 91 and 48 aa; HARDV, 74 and 102 aa). The two shorter ORFs of OVRV and HARDV could have evolved through truncation of a larger progenitor ORF such as the one found in the related TUPV.

Unlike most C proteins, which are basic, the two OVRV C proteins are acidic.

WONV, HOJV, ORRV (Hart Park group) and TIBV contain ORFs that overlap the 3′ non-coding region of the G gene, coding for putative small hydrophobic proteins that could be viroporins (Gubala *et al.*, 2008, 2011, unpublished). The 3′ non-coding regions of the N genes of WONV and ORRV uniquely contain an ORF that putatively codes for a small hydrophobic and highly acidic protein (49 aa and 75 aa, respectively). The 3′ non-coding region of the M gene of the fish-infecting SCRV contains an ORF capable of coding a putative 45-aa membrane anchored protein that could not been detected by western blotting (Tao *et al.*, 2008).

Increasing evidence suggests that different rhabdoviruses encode additional proteins on independent genes that are interspersed between the typical five genes. Additional genes are most commonly interposed in two locations: between the P and M, and between G and L genes (Tordo *et al.*, 1992). In plant rhabdoviruses the P-M region commonly codes for a protein that facilitates cell-to-cell movement. Some plant viruses contain more than one gene in this location, such as NCMV, which contains four, and MFSV, which contains two. In the sigmaviruses the PP3 glycoprotein is encoded in this region. The Hart Park group viruses putatively encode three novel proteins in this location. The G–L intergenic region is the most commonly utilized location for additional genes (Walker *et al.*, 2011). This region is used by ephemeroviruses, novirhabdoviruses, some plant rhabdoviruses, and many of the recently characterized rhabdoviruses. The non-structural glycoprotein (G_{NS}) gene of the ephemeroviruses and NGAV are found here, as are numerous other small, uncharacterized ORFs. ORFs coding for small, viroporin-like proteins are often found in this location.

The occurrence of additional genes between the M and G genes is uncommon; however, several rhabdoviruses have recently been reported to contain additional genes in this location. Viruses of the Tupaia and Oak Vale group (TUPV, OVRV, DURV and HARDV) contain a gene coding for a small hydrophobic protein (SH) in this location. The distantly related NGAV contains one gene (U4), and TIBV and CPV (proposed genus *Tibrovirus*) contain two genes (U1 and U2) in this region. As this region does not seem to be widely utilized by many rhabdoviruses, it may be that additional genes in this location have an adverse effect on genome stability, replication or on the ratio of the envelope-associated M and G proteins.

Roles and characteristics of the accessory rhabdovirus proteins

The characteristics and putative functions of accessory rhabdovirus proteins have recently been comprehensively reviewed (Walker *et al.*, 2011). Whilst some of the accessory proteins have been widely studied, such as the C proteins of vesiculoviruses, the NV proteins of novirhabdoviruses and the putative movement proteins of plant rhabdoviruses, the majority of recently identified novel proteins have not yet been detected in cell culture. Several of the ephemerovirus accessory proteins can be detected in cell culture and in virions, but their detection has been challenging (Walker *et al.*, 1991, 2011; Wang and Walker, 1993; Wang *et al.*, 1994; McWilliam *et al.*, 1997). There is some evidence suggesting the expression of novel proteins in FLAV-infected cells, and possibly also in WONV-infected cells, but the small novel proteins of NGAV have resisted detection in mammalian cell culture although G_{NS} appeared to be present (Boyd and Whitaker-Dowling, 1988; Gubala *et al.*, 2008, 2010). In TUPV-infected cells the cognate mRNA species for the SH protein can be detected, but attempts to detect the protein remain to be performed (Springfeld *et al.*, 2005). The challenges in detecting the accessory proteins in cell culture suggest that they are probably of greater relevance in the animal host. Lack of similarity to known proteins, small size, charge, low abundance and absence of suitable immunological reagents have hindered the characterization of the putative, accessory proteins.

These proteins are usually small (~100 to 200 aa) and a few are extremely small (~30 aa). At the sequence level they are generally unrelated to known proteins, but some contain regions of high homology to other proteins that are probably of functional significance. Several studies have inferred that their roles may be accessory to infection in that they are not essential but may enhance replication and transcription, alter host innate immune responses, modulate apoptotic pathways and other cellular functions, and facilitate cell-to-cell movement (Walker *et al.*, 2011). Preliminary data suggest that the WONV U3 protein may be involved in the modulation of apoptosis (Walker *et al.*, 2011). Some of the additional proteins have characteristics resembling viroporins, or carrier proteins, which are small, dispensable proteins that may have roles in the formation of pores in membranes to facilitate passage of ions and small molecules during infection. Viroporins are also known to interact with the cell vesicle system and may be involved in glycoprotein trafficking (Gonzalez and Carrasco, 2003). The small hydrophobic α1 proteins of BEFV and ARV have characteristics suggestive of viroporins (McWilliam *et al.*, 1997), as do the U5 proteins of WONV, HOJV, ORRV, the U6 protein of NGAV and the U3 proteins in TIBV and CPV.

The SH proteins found in viruses of the Tupaia–Oak Vale group, are named by analogy to the SH proteins of paramyxoviruses but their functions remain untested. In paramyxoviruses the SH proteins are non-essential but are involved in blocking apoptosis and in decreasing the cytopathic effect (He *et al.*, 1998, 2001; Lin *et al.*, 2003). Varying predicted characteristics and high sequence divergence of the TUPV, OVRV, DURV and HARDV SH proteins make it difficult to decipher whether they have analogous roles. The OVRV and TUPV SH proteins are predicted to contain transmembrane regions and signal peptides, suggesting transfer to membranes but in contrast no signal peptides are predicted in the DURV or HARDV SH proteins (Springfeld *et al.*, 2005; Gubala, 2009, unpublished; Allison *et al.*, 2011). OVRV SH is predicted to be a type II transmembrane protein, whereas TUPV SH is predicted as type I (Springfeld *et al.*, 2005; Gubala, 2009). The SH proteins of TUPV and HARDV share a highly conserved centrally located region that does support a common function, even though the two viruses are highly diverged. NGAV also contains a gene (U4) between M and G coding for a putative 81-aa protein, which like the SH proteins of DURV and HARDV contains a putative transmembrane region but no signal peptide (Gubala *et al.*, 2010). The two putative proteins of TIBV and CPV (U1 and U2), that are likewise located between M and G, do not contain transmembrane regions and are hence likely to be functionally different (Gubala *et al.*, 2011).

It is likely that the accessory rhabdovirus genes have roles in the invertebrate host. Viruses that do not rely exclusively on arthropod transmission tend to have fewer genes (for example lyssaviruses, novirhabdoviruses, vesiculoviruses). Arthropods and plants are known to employ RNA silencing (or RNA interference; RNAi) as a vital means for protecting themselves against viral infection (Li and Ding, 2006; Ding and Voinnet, 2007; Kormelink *et al.*, 2011). In response, viruses produce viral suppressors of RNA silencing (VSRs), which are remarkably diverse, non-structural proteins that are often encoded by novel ORFs that overlap other genes. Although identified in a variety of plant, insect and fungus viruses, studies of VSRs remain to be reported in the *Rhabdoviridae*. Recent studies of the effects of mutations on sigma virus genes revealed that the X gene, which codes for the accessory PP3 protein, is the most highly tolerant to ADAR-induced hypermutations of all the genes (Carpenter *et al.*, 2009). ADAR molecules likewise play a role in the immune response by altering dsRNA and may share a complex interconnection with RNAi in the antiviral immune response (Carpenter *et al.*, 2009). Some viruses are known to exploit the ADAR mechanism to augment viral protein functions from, for example, virus replication to virion assembly (Sato *et al.*, 2001). A high tolerance to host-induced mutations is one possible explanation for the high divergence observed between the accessory proteins of related viruses. There appear to be far fewer constraints on sequence conservation, which may also suggest that their functions are more reliant on secondary structure than sequence conservation. Low sequence conservation and a flexibility to undergo frequent changes may allow these proteins to adapt and respond to antiviral immune responses of different kinds of hosts.

Mechanisms for controlling gene transcription and protein expression

The utilization of different intergenic regions as sites for additional genes invariably affects the abundance of downstream transcripts and has important implications for genome stability. The insertion of additional genes between N and P disrupts a critical N:P protein ratio. Doing this in VSV was shown to impair viral replication by over an order in magnitude (Iverson and Rose, 1981; Wertz *et al.*, 2002). After just two passages, VSV is able to overcome the highly unfavourable experimental insertion of an additional gene into the N–P junction by introducing a mutation that abolishes transcription termination, resulting in a bicistronic mRNA (Wertz *et al.*, 2002). Conversely, the insertion of genes in the remaining three intergenic regions has no adverse effect on replication. This explains the relatively rare occurrence of genes between N and P; indeed the only known rhabdovirus to contain a gene in this location is the potato-infecting PYDV (Bandyopadhyay *et al.*, 2010). Of all the possible genome locations, the favourable usage of the G–L intergenic region for locating novel genes has the lowest impact on downstream transcription, as the only affected downstream gene is L, which is required at comparatively low amounts (Tordo *et al.*, 1992). This suggests that novel genes in this location are also transcribed at comparatively low levels. It can thus be presumed that the utilization of the P–M intergenic region produces the highest possible abundance of transcripts without an exorbitant impact on virus replication that would otherwise occur if a gene was inserted between N and P.

In addition to using the most appropriate location in the genome to control relative transcript abundance, viruses may introduce subtle changes to their transcription control sequences, resulting in polymerase read-through and the production of bi- or polycistronic mRNAs. This is prominently observed in the ephemeroviruses (Wang and Walker, 1993; Wang *et al.*, 1994; McWilliam *et al.*, 1997). Several of the recently characterized dimarhabdoviruses also appear to exhibit variable control sequences. Polycistronic transcription could be a means by which a virus with additional genes can balance transcription efficiency of a large and complex genome, but it is presently unknown if this phenomenon occurs in nature or to what extent it is driven by adaptation to cell culture. The efficiency of protein translation from polycistronic mRNAs also remains to be explored. Recent studies of RABV transcription have demonstrated that the nature of the polymerase has a greater bearing on polycistronic transcription than do the transcription control sequences (Morimoto *et al.*, 2011). Study of a highly passaged strain found that, even although the transcription control sequences were conserved, a high proportion of transcripts were polycistronic and that the ratios of transcripts and the estimated protein abundance varied greater than expected based on the stop-start monocistronic model. It is believed that adaptation of virus to its particular host has a marked impact on the activity of the polymerase and plays an important role in transcription. Transcription likely varies considerably between different viruses and strains and may be much more complicated than is currently appreciated.

Point mutations reported in some of the novel ORFs of ephemeroviruses and CPV (tibrovirus), producing prematurely abridged proteins, is a phenomenon that is likely to occur during adaptation to cell culture (Gubala *et al.*, 2011; Walker *et al.*, 2011). Studies of the arbovirus Rift Valley fever (family *Bunyaviridae*) demonstrated that large deletions in the NSs gene (non-essential virulence factor) were introduced after only ten passages in the same cell line (mammalian and mosquito) (Moutailler *et al.*, 2011). After 30 passages in the same cell line the virus lost virulence for mice, but alternation of the two cell lines maintained an intact NSs gene and virulence. In view of these results, host alternation may have an important role in the stability of large and complex genomes. Regular host alternation (or a range of different hosts) may be a major contributing factor to the existence of additional genes.

Transmission of rhabdoviruses

The ways in which different rhabdoviruses are transmitted and the vectors involved is a subject that requires further study. VSV is regarded an arthropod-borne virus; however, the means by which it is transmitted remains incompletely understood. The classical criterion of arthropod-transmitted viruses is transmission by insects through feeding on infected blood, but VSV has not been isolated from blood and viraemia in mammalian hosts has not been detected. The virus can be transmitted to mice by experimentally infected haematophagous insects but it can also be mechanically transmitted between flies while co-feeding on non-viraemic animals and by direct contact between mammals (Tesh *et al.*, 1987; Mead *et al.*, 2000). A study demonstrated that migratory grasshoppers are capable of amplifying and mechanically transmitting VSV between animals by eating plants that are contaminated with material shed from VSV-infected livestock (Drolet *et al.*, 2009). The virus is transovarially transmitted to the progeny of infected sandflies (as are the Drosophila-infecting sigma viruses) (Tesh *et al.*, 1972).

The transmission of BEFV is considerably different from VSV. Attempts to transmit BEFV between cattle through direct contact of nasal and ocular discharges or saliva failed to produce infection, however, experimental injection of infected blood into healthy cattle can produce disease (Mackerras *et al.*, 1940; St George, 1990). The major difference between BEFV and the vesiculoviruses and lyssaviruses is that the virus causes viraemia and can be isolated from blood. It is believed that transmission occurs through infected blood only and not through direct or mechanical contact. BEFV has been isolated from a number of different species of mosquito and midge (Walker, 2005). It is uncertain if one or both of these insect groups are responsible for transmitting BEFV, but it has been proposed that mosquitoes are the more likely candidates (St George, 2008). Mosquitoes have a capacity to directly insert their mouthparts into the venules of their prey, which enables the virus to bypass the lymphatic system. The lymphatic system does not appear to be infiltrated by the virus, which intuitively precludes transmission by midges which would require virus to pass through the lymphatic system before entering the blood stream (St George, 2008). This is furthermore supported by the observation that BEFV has been found in the saliva of experimentally infected mosquitoes but not in midges (St George, 2008). It is believed that the mosquito *Culex annulirostris* plays an important role in the transmission of BEFV in Australia, however epidemiological studies suggest that *Culicoides* spp. may also play a role in what might be a complex transmission cycle (Muller and Standfast, 1986; Walker, 2008). Detailed transmission studies are required to answer these important questions.

Plant rhabdoviruses from the genera *Cytorhabdovirus* and *Nucleorhabdovirus* are transmitted by phytophagous (plant-feeding) arthropods such as aphids, leafhoppers and planthoppers (Jackson *et al.*, 2008). Although some plant rhabdoviruses can also be transmitted mechanically, this is not believed to be a significant means of transmission. The phylogeny of plant-adapted rhabdoviruses is partly congruent on the phylogeny of their arthropod hosts, suggesting that the arthropod may have an important driving influence on rhabdovirus evolution (Bandyopadhyay *et al.*, 2010).

The fish-infecting novirhabdoviruses are transmitted horizontally between fish through water and vertically from adult to progeny through egg-associated transmission (Kurath and Winton, 2008). The transmission of novirhabdoviruses by fish ectoparasites in nature remains elusive, but the fish-infecting vesiculovirus SVCV can be experimentally vectored by carp lice even though it is primarily waterborne (Pfeil-Putzien, 1978; Ahne *et al.*, 2002; Kurath and Winton, 2008). In further support of this, rhabdovirus-like sequences have been identified in fish lice (Longdon *et al.*, 2010).

Sigmaviruses are transmitted vertically through sperm and eggs of their dipteran hosts. A 'sweep' of DObsSV infection in *D. obscura* in the UK, with an estimated 39% natural infection rate, has recently been documented (Longdon *et al.*, 2011b). Vertical transmission appears to facilitate very rapid changes in virus prevalence in insects. Although the primary mode of transmission of sigmaviruses is vertical, rare horizontal transfer events (host switching) between distantly related host species have been suggested to play important roles in long-term persistence of the virus in an evolutionary timescale (Longdon *et al.*, 2011a, 2011c). There are no other groups of rhabdoviruses identified with solely insect hosts, but it is envisaged that rhabdoviruses probably infect insects more frequently than is currently known (Longdon *et al.*, 2010).

Lyssaviruses are known to be transmitted only through direct contact (most prominently through a bite of an infected mammal). These viruses are not known to be transmitted by insects, however, MOKV has been shown to replicate in insect cell culture, although it has never been isolated from insects (Buckley, 1975). Lyssaviruses are all known to infect bats. The only other rhabdoviruses isolated from bats are from the Kern Canyon group (MEBV, KCV and OITAV).

Several unclassified rhabdoviruses have been isolated from ticks: BARV (with a possible antigenic relationship to viruses of the Kern Canyon group) and the three Sawgrass group viruses (CNTV, SAWV and NMV). The detection of neutralizing antibodies to CNTV in cottontail rabbits, the same species from which ticks containing CNTV were collected, is possible evidence suggesting a tick-mammalian maintenance cycle (Main and Carey, 1980). No other rhabdoviruses are known to originate from ticks, and as such the role of ticks as hosts or vectors of rhabdoviruses has not yet been corroborated. There is no genetic sequence data for any of the tick rhabdoviruses.

Many animal rhabdoviruses have been isolated from haematophagous insects and a number of them have been demonstrated to multiply effectively in insect cells and experimentally infected mosquitoes (Carley *et al.*, 1973). In conjunction with serological data from terrestrial vertebrates, these viruses are assumed to be essentially arboviruses. But the isolation of a virus from a haematophagous arthropod does not automatically imply that the arthropod can effectively amplify and transmit the virus. It is feasible that different kinds of vectors, for example both mosquitoes and midges, may be involved in the transmission of a virus. This is currently poorly understood but conceivably could be facilitated by the novel genes observed in so many of the newly characterized rhabdoviruses. Controlled transmission studies to associate vectors with disease are complex and logistically challenging; however, the identification of vectors and virus circulation patterns are the essential tools that feed the development of early warning systems and the implementation of appropriate control strategies.

Impacts on taxonomy

The currently recognized six rhabdovirus genera are distinguished using a range of characteristics (Dietzgen *et al.*, 2011). Some viruses from different genera are known to be serologically cross reactive, therefore other factors like genome organization, phylogenetic analyses of the structural N, G and L proteins, the replication site within cells (cytoplasmic or nuclear), host range and other biological data contribute to the classification of rhabdoviruses into genera and species.

The recently characterized novel rhabdoviruses demonstrate that complexity within the basic genomic structure of the animal rhabdovirus is more common than previously anticipated. Currently there is no provision for these new viruses within the taxonomical structure of the *Rhabdoviridae*. New genera have been recently proposed to encompass the sigmaviruses (genus Sigmavirus) and the Tibrogargan group (genus Tibrovirus) but at the present moment the taxonomy of many other viruses remains unresolved owing to either their divergence and novel features, or insufficient data. The impending characterization of new viruses will feed the development of new criteria for additional genera and may facilitate the demarcation of existing genera, but at the same time it may create indistinct interfaces between genera.

Phylogenetic studies using partial L and N genes and proteins (Bourhy *et al.*, 2005; Kuzmin *et al.*, 2006; Dacheux *et al.*, 2010; Quan *et al.*, 2010, 2011; Allison *et al.*, 2011) have been valuable for initial categorization of viruses, but uncertainty of branching and the commonly observed inconsistent placement of viruses highlight the benefits of interpreting virus relationships in the context of the entire genome. Analysis of genomes exposes an array of information that is hidden in the shadows of single-gene phylogenetics. A prominent example of this is NGAV, the relationship of which to the Hart Park group may seem clear based on phylogenetic studies of the structural proteins, but at the genomic structure level NGAV is strikingly different. The presence of additional genes may need to be considered when evolutionary relationships between different viruses are analysed; however, their significance in taxonomy cannot be established without the resolution of their functions, and importantly without unravelling if their existence is owed to an ancestral virus or if they are a more recent acquisition.

The ancestry of the *Rhabdoviridae* is perplexing. Since a large proportion of rhabdoviruses are arthropod-transmitted, one school of thought suggests that the ancestral rhabdovirus was an arthropod virus (insect or crustacean), which henceforth evolved to infect a range of different host species (Hogenhout *et al.*, 2003; Bourhy *et al.*, 2008). The fish-infecting genus *Novirhabdovirus*, however, also has an ancient lineage within the family and therefore greater focus on rhabdoviruses of aquatic arthropods, such as ectoparasites and crustaceans, may help shape our understanding of the progenitor rhabdovirus (Longdon *et al.*, 2010).

Future perspectives

The recently characterized rhabdoviruses have provided a valuable insight into the genetic complexity and flexibility of the *Rhabdoviridae*. These studies have revealed large diversity at the sequence and genome structure levels, which is reflected in the wide geographical distribution and remarkable ecological adaptation of this family. Without a doubt, advancements in nucleic acid sequencing will enable even faster data acquisition, and in the near future genomes of many unassigned rhabdoviruses will be sequenced. It can be anticipated that their characterization will reveal a cornucopia of unexpected features. Technical advancements will likely involve the development of improved approaches to *de novo* virus sequencing directly from clinical material and from their arthropod hosts. This will importantly minimize adaptation of virus to cell culture and subsequently produce data that are reflective of the virus in its most natural state.

There is a void in knowledge of the biology of most of the newly characterized animal rhabdoviruses, and of those awaiting characterization. Importantly, future studies will need to be directed towards identifying their hosts, vectors, and demonstrating their role in clinical disease. At the molecular level, functional studies of novel proteins will have an important bearing on the understanding of the virus and host life cycle, and on the understanding of virus evolution. The interchangeability of analogous accessory proteins between different viruses using reverse genetic approaches could provide valuable insights into their functions. It is possible that other unexplored ORFs within highly conserved structural genes could have biological significance. Similarly to VSV, other rhabdoviruses could play beneficial roles in medicine, for example in the treatment of cancer. In addition they could be developed as live-attenuated vaccine vectors for human disease, or the novel proteins could be adapted as trafficking proteins for localized drug delivery.

Although the diversity seen to date within the *Rhabdoviridae* seems large at the present moment, it is without doubt a view distorted by the intense study of a minority of strains. Most rhabdoviruses that have been studied to date have surfaced owing to economic and public health drivers. In order to advance knowledge the next step is the exploration of previously unstudied hosts and new ecological spaces. In view of the rapid advancements in *de novo* virus sequencing technologies and ongoing studies of different rhabdoviruses, it can be expected that our knowledge and appreciation of this diverse and complex family will continue to grow.

References

Abramova, L.N., Terskikh, I.I., Gromashevskii, V.L., and L'Vov D.K. (1987). Comparative characteristics of the pathogenicity of 2 viruses of the family *Rhabdoviridae* for monkeys. Vopr. Virusol. *32*, 357–359.

Ahne, W., Bjorklund, H.V., Essbauer, S., Fijan, N., Kurath, G., and Winton, J.R. (2002). Spring viremia of carp (SVC). Dis. Aquat. Organ. *52*, 261–272.

Allison, A.B., Palacios, G., Travassos da Rosa, A., Popov, V.L., Lu, L., Xiao, S.Y., DeToy, K., Briese, T., Lipkin, W.I., Keel, M.K., *et al.* (2011). Characterization of Durham virus, a novel rhabdovirus that encodes both a C and SH protein. Virus Res. *155*, 112–122.

Alonso, M., Kim, C.H., Johnson, M.C., Pressley, M., and Leong, J.A. (2004). The NV gene of snakehead rhabdovirus (SHRV) is not required for pathogenesis, and a heterologous glycoprotein can be incorporated into the SHRV envelope. J. Virol. *78*, 5875–5882.

Ammayappan, A., and Vakharia, V.N. (2011). Nonvirion protein of novirhabdovirus suppresses apoptosis at the early stage of virus infection. J. Virol. *85*, 8393–8402.

Ammayappan, A., Kurath, G., Thompson, T.M., and Vakharia, V.N. (2010). A reverse genetics system for the Great Lakes strain of viral hemorrhagic septicemia virus: the NV gene is required for pathogenicity. Mar. Biotechnol. (NY) *13*, 672–683.

Anon. (1970). Kamese (KAM) strain. MP 6186. Am. J. Trop. Med. Hyg. *19* (Suppl.), 1145–1146.

Bandyopadhyay, A., Kopperud, K., Anderson, G., Martin, K., and Goodin, M. (2010). An integrated protein localization and interaction map for potato yellow dwarf virus, type species of the genus *Nucleorhabdovirus*. Virology *402*, 61–71.

Betts, A.M., Stone, D.M., Way, K., Torhy, C., Chilmonczyk, S., Benmansour, A., and de Kinkelin, P. (2003).

Emerging vesiculo-type virus infections of freshwater fishes in Europe. Dis. Aquat. Organ. *57*, 201–212.

Bexfield, N., and Kellam, P. (2011). Metagenomics and the molecular identification of novel viruses. Vet. J. *190*, 191–198.

Biacchesi, S. (2011). The reverse genetics applied to fish RNA viruses. Vet. Res. *42*, 12.

Biacchesi, S., Thoulouze, M.I., Bearzotti, M., Yu, Y.X., and Bremont, M. (2000). Recovery of NV knockout infectious hematopoietic necrosis virus expressing foreign genes. J. Virol. *74*, 11247–11253.

Blasdell, K.R., Voysey, R., Bulach, D., Joubert, D.A., Tesh, R.B., Boyle, D.B., and Walker, P.J. (2012). Kotonkan and Obodhiang viruses: African ephemeroviruses with large and complex genomes. Virology *425*, 143–153.

Bourhy, H., Cowley, J.A., Larrous, F., Holmes, E.C., and Walker, P.J. (2005). Phylogenetic relationships among rhabdoviruses inferred using the L polymerase gene. J. Gen. Virol. *86*, 2849–2858.

Bourhy, H., Gubala, A., Weir, R., and Boyle, D. (2008). Animal rhabdoviruses. In Encyclopedia of Virology, Mahy, B.W.J., and van Regenmortel, M.H.V., eds (Elsevier Academic Press, Oxford), pp. 111–121.

Bowden, T.R., Westenberg, M., Wang, L.F., Eaton, B.T., and Boyle, D.B. (2001). Molecular characterization of Menangle virus, a novel paramyxovirus which infects pigs, fruit bats, and humans. Virology *283*, 358–373.

Boyd, K.R., and Whitaker-Dowling, P. (1988). Flanders virus replication and protein synthesis. Virology *163*, 349–358.

Briese, T., Paweska, J.T., McMullan, L.K., Hutchison, S.K., Street, C., Palacios, G., Khristova, M.L., Weyer, J., Swanepoel, R., Egholm, M., *et al.* (2009). Genetic detection and characterization of Lujo virus, a new hemorrhagic fever-associated arenavirus from southern Africa. PLoS Pathog. *5*, e1000455.

Brown, F. (1987). The family *Rhabdoviridae*: general description and taxonomy. In The Rhabdoviruses, Wagner, R.R., ed. (Plenum Press, New York), pp. 1–8.

Buckley, S.M. (1975). Arbovirus infection of vertebrate and insect cell cultures, with special emphasis on Mokola, Obodhiang, and kotonkan viruses of the rabies serogroup. Ann. N.Y. Acad. Sci. *266*, 241–250.

Butenko, A.M., Gromashevsky, V.L., L'Vov D.K., and Popov, V.F. (1981). First isolations of Barur virus (*Rhabdoviridae*) from ticks (Acari: Ixodidae) in Africa. J. Med. Entomol. *18*, 232–234.

Calisher, C.H., Shope, R.E., and Walton, T.E. (1988). Cell cultures for diagnosis of arbovirus infections in livestock and wildlife. Meth. Cell Sci. *11*, 157–163.

Calisher, C.H., Karabatsos, N., Zeller, H., Digoutte, J.P., Tesh, R.B., Shope, R.E., Travassos da Rosa, A.P., and St George, T.D. (1989). Antigenic relationships among rhabdoviruses from vertebrates and hematophagous arthropods. Intervirology *30*, 241–257.

Carley, J.G., Standfast, H.A., and Kay, B.H. (1973). Multiplication of viruses isolated from arthopods and vertebrates in Australia in experimentally infected mosquitoes. J. Med. Entomol. *10*, 244–249.

Carpenter, J.A., Keegan, L.P., Wilfert, L., O'Connell, M.A., and Jiggins, F.M. (2009). Evidence for ADAR-induced hypermutation of the *Drosophila* sigma virus (*Rhabdoviridae*). BMC Genet. *10*, 75.

Carstens, E.B. (2010). Ratification vote on taxonomic proposals to the International Committee on Taxonomy of Viruses (2009). Arch. Virol. *155*, 133–146.

Causey, O.R., Shope, R.E., and Bensabath, G. (1966). Marco, Timbo, and Chaco, newly recognized arboviruses from lizards of Brazil. Am. J. Trop. Med. Hyg. *15*, 239–243.

Ceccaldi, P.E., Fayet, J., Conzelmann, K.K., and Tsiang, H. (1998). Infection characteristics of rabies virus variants with deletion or insertion in the pseudogene sequence. J. Neurovirol. *4*, 115–119.

Chow, T.L., Chow, F.H., and Hanson, R.P. (1954). Morphology of vesicular stomatitis virus. J. Bacteriol. *68*, 724–726.

Contamine, D., and Gaumer, S. (2008). Sigma Rhabdoviruses. In Encyclopedia of Virology, Mahy, B.W.J., and van Regenmortel, M.H.V., eds (Elsevier Academic Press, Oxford), pp. 576–581.

Cropp, C.B., Prange, W.C., and Monath, T.P. (1985). LeDantec virus: identification as a rhabdovirus associated with human infection and formation of a new serogroup. J. Gen. Virol. *66*, 2749–2754.

Cybinski, D.H., and Gard, G.P. (1986). Isolation of a new rhabdovirus in Australia related to Tibrogargan virus. Aust. J. Biol. Sci. *39*, 225–232.

Cybinski, D.H., and Muller, M.J. (1990). Isolation of arboviruses from cattle and insects at two sentinel sites in Queensland, Australia, 1979–85. Aust. J. Zool. *38*, 25–32.

Cybinski, D.H., St George, T.D., Standfast, H.A., and McGregor, A. (1980). Isolation of Tibrogargan virus, a new Australian rhabdovirus, from *Culicoides brevitarsis*. Vet. Microbiol. *5*, 301–308.

Dacheux, L., Berthet, N., Dissard, G., Holmes, E.C., Delmas, O., Larrous, F., Guigon, G., Dickinson, P., Faye, O., Sall, A.A., *et al.* (2010). Application of broad-spectrum resequencing microarray for genotyping rhabdoviruses. J. Virol. *84*, 9557–9574.

Davies, F.G., and Walker, A.R. (1974). The isolation of ephemeral fever virus from cattle and *Culicoides* midges in Kenya. Vet. Rec. *95*, 63–64.

Dhillon, J., Cowley, J.A., Wang, Y., and Walker, P.J. (2000). RNA polymerase (L) gene and genome terminal sequences of ephemeroviruses bovine ephemeral fever virus and Adelaide River virus indicate a close relationship to vesiculoviruses. Virus Res. *70*, 87–95.

Dietzgen, R.G., Callaghan, B., Wetzel, T., and Dale, J.L. (2006). Completion of the genome sequence of lettuce necrotic yellows virus, type species of the genus *Cytorhabdovirus*. Virus Res. *118*, 16–22.

Dietzgen, R.G., Calisher, C.H., Kurath, G., Kuzmin, I.V., Rodriguez, L.L., Stone, D.M., Tesh, R.B., Tordo, N., Walker, P.J., Wetzel, T., *et al.* (2011). Rhabdoviridae. In Virus Taxonomy: Ninth Report of the International Committee on the Taxonomy of Viruses, King, A.M.Q., Adams, M.J., Carstens, E.B., and Lefkowitz, E.J., eds (Elsevier, Oxford, UK), pp. 654–681.

Ding, S.W., and Voinnet, O. (2007). Antiviral immunity directed by small RNAs. Cell *130*, 413–426.

Diniz, J.A., Nunes, M.R., Travassos da Rosa, A.P., Cruz, A.C., de Souza, W., Medeiros, D.B., Chiang, J.O., and Vasconcelos, P.F. (2006). Characterization of two new rhabdoviruses isolated from midges (*Culicoides* spp.) in the Brazilian Amazon: proposed members of a new genus, Bracorhabdovirus. Arch. Virol. *151*, 2519–2527.

Doherty, R.L., Whitehead, R.H., Wetters, E.J., Gorman, B.M., and Carley, J.G. (1970). A survey of antibody to 10 arboviruses (Koongol group, Mapputta group and ungrouped) isolated in Queensland. Trans. R. Soc. Trop. Med. Hyg. *64*, 748–753.

Doherty, R.L., Carley, J.G., Standfast, H.A., Dyce, A.L., Kay, B.H., and Snowdon, W.A. (1973). Isolation of arboviruses from mosquitoes, biting midges, sandflies and vertebrates collected in Queensland, 1969 and 1970. Trans. R. Soc. Trop. Med. Hyg. *67*, 536–543.

Domingo, E., and Holland, J.J. (1997). RNA virus mutations and fitness for survival. Annu. Rev. Microbiol. *51*, 151–178.

Drolet, B.S., Stuart, M.A., and Derner, J.D. (2009). Infection of *Melanoplus sanguinipes* grasshoppers following ingestion of rangeland plant species harboring vesicular stomatitis virus. Appl. Environ. Microbiol. 75, 3029–3033.

Faber, M., Pulmanausahakul, R., Nagao, K., Prosniak, M., Rice, A.B., Koprowski, H., Schnell, M.J., and Dietzschold, B. (2004). Identification of viral genomic elements responsible for rabies virus neuroinvasiveness. Proc. Natl. Acad. Sci. U.S.A. *101*, 16328–16332.

Ghosh, S.N., and Rajagopalan, P.K. (1973). Encephalomyocarditis virus activity in *Mus booduga* (Gray) in Barur Village (1961–1962), Sagar KFD area, Mysore State, India. Indian J. Med. Res. *61*, 989–991.

Gibbs, E.P., Calisher, C.H., Tesh, R.B., Lazuick, J.S., Bowen, R., and Greiner, E.C. (1989). Bivens arm virus: a new rhabdovirus isolated from *Culicoides insignis* in Florida and related to Tibrogargan virus of Australia. Vet. Microbiol. *19*, 141–150.

Gonzalez, M.E., and Carrasco, L. (2003). Viroporins. FEBS Lett. *552*, 28–34.

Goodin, M.M., Dietzgen, R.G., Schichnes, D., Ruzin, S., and Jackson, A.O. (2002). pGD vectors: versatile tools for the expression of green and red fluorescent protein fusions in agroinfiltrated plant leaves. Plant J. *31*, 375–383.

Goodin, M.M., Chakrabarty, R., Yelton, S., Martin, K., Clark, A., and Brooks, R. (2007). Membrane and protein dynamics in live plant nuclei infected with sonchus yellow net virus, a plant-adapted rhabdovirus. J. Gen. Virol. *88*, 1810–1820.

Gubala, A. (2009). Characterisation of novel viruses isolated from vertebrates and insects in Australia. PhD thesis. School of Chemistry and Molecular Biosciences, University of Queensland, Brisbane, Australia.

Gubala, A., Davis, S., Weir, R., Melville, L., Cowled, C., Walker, P., and Boyle, D. (2010). Ngaingan virus, a macropod-associated rhabdovirus, contains a second glycoprotein gene and seven novel open reading frames. Virology *399*, 98–108.

Gubala, A., Davis, S., Weir, R., Melville, L., Cowled, C., and Boyle, D. (2011). Tibrogargan and Coastal Plains rhabdoviruses: genomic characterization, evolution of novel genes and seroprevalence in Australian livestock. J. Gen. Virol. *92*, 2160–2170.

Gubala, A.J., Proll, D.F., Barnard, R.T., Cowled, C.J., Crameri, S.G., Hyatt, A.D., and Boyle, D.B. (2008). Genomic characterisation of Wongabel virus reveals novel genes within the *Rhabdoviridae*. Virology *376*, 13–23.

He, B., Leser, G.P., Paterson, R.G., and Lamb, R.A. (1998). The paramyxovirus SV5 small hydrophobic (SH) protein is not essential for virus growth in tissue culture cells. Virology *250*, 30–40.

He, B., Lin, G.Y., Durbin, J.E., Durbin, R.K., and Lamb, R.A. (2001). The SH integral membrane protein of the paramyxovirus simian virus 5 is required to block apoptosis in MDBK cells. J. Virol. *75*, 4068–4079.

Heim, F., Lot, H., Delecolle, B., Bassler, A., Krczal, G., and Wetzel, T. (2008). Complete nucleotide sequence of a putative new cytorhabdovirus infecting lettuce. Arch. Virol. *153*, 81–92.

Hertig, C., Pye, A.D., Hyatt, A.D., Davis, S.S., McWilliam, S.M., Heine, H.G., Walker, P.J., and Boyle, D.B. (1996). Vaccinia virus-expressed bovine ephemeral fever virus G but not G_{NS} glycoprotein induces neutralizing antibodies and protects against experimental infection. J. Gen. Virol. *77*, 631–640.

Hogenhout, S.A., Redinbaugh, M.G., and Ammar el, D. (2003). Plant and animal rhabdovirus host range: a bug's view. Trends Microbiol. *11*, 264–271.

Huang, Y., Zhao, H., Luo, Z., Chen, X., and Fang, R.X. (2003). Novel structure of the genome of *Rice yellow stunt virus*: identification of the gene 6-encoded virion protein. J. Gen. Virol. *84*, 2259–2264.

Huang, Y.W., Geng, Y.F., Ying, X.B., Chen, X.Y., and Fang, R.X. (2005). Identification of a movement protein of rice yellow stunt rhabdovirus. J. Virol. *79*, 2108–2114.

Humphery-Smith, I., Cybinski, D.H., Byrnes, K.A., and St George, T.D. (1991). Seroepidemiology of arboviruses among seabirds and island residents of the Great Barrier Reef and Coral Sea. Epidemiol. Infect. *107*, 435–440.

Irie, T., Nagata, N., Yoshida, T., and Sakaguchi, T. (2008). Paramyxovirus sendai virus C proteins are essential for maintenance of negative-sense RNA genome in virus particles. Virology *374*, 495–505.

Iverson, L.E., and Rose, J.K. (1981). Localized attenuation and discontinuous synthesis during vesicular stomatitis virus transcription. Cell 23, 477–484.

Iwasaki, T., Inoue, S., Tanaka, K., Sato, Y., Morikawa, S., Hayasaka, D., Moriyama, M., Ono, T., Kanai, S., Yamada, A., *et al.* (2004). Characterization of Oita virus 296/1972 of *Rhabdoviridae* isolated from a horseshoe bat bearing characteristics of both lyssavirus and vesiculovirus. Arch. Virol. *149*, 1139–1154.

Jack, P.J., Boyle, D.B., Eaton, B.T., and Wang, L.F. (2005). The complete genome sequence of J virus reveals a unique genome structure in the family *Paramyxoviridae*. J. Virol. *79*, 10690–10700.

Jackson, A.O., Dietzgen, R.G., Goodin, M.M., Bragg, J.N., and Deng, M. (2005). Biology of plant rhabdoviruses. Annu. Rev. Phytopathol. *43*, 623–660.

Jackson, A.O., Dietzgen, R.G., Fang, R.X., Goodin, M.M., Hogenhout, S.A., Deng, M., and Bragg, J.N. (2008). Plant rhabdoviruses. In Encyclopedia of Virology, Mahy, B.W.J., and van Regenmortel, M.H.V., eds (Elsevier Academic Press, Oxford), pp. 187–196.

Jayakar, H.R., and Whitt, M.A. (2002). Identification of two additional translation products from the matrix (M) gene that contribute to vesicular stomatitis virus cytopathology. J. Virol. *76*, 8011–8018.

Jenson, A.B., Rabin, E.R., Wende, R.D., and Melnick, J.L. (1967). A comparative light and electron microscopic study of rabies and Hart Park virus encephalitis. Exp. Mol. Pathol. *7*, 1–10.

Johal, J., Gresty, K., Kongsuwan, K., and Walker, P.J. (2008). Antigenic characterization of bovine ephemeral fever rhabdovirus G and G_{NS} glycoproteins expressed from recombinant baculoviruses. Arch. Virol. *153*, 1657–1665.

Johansson, T., Nylund, S., Olesen, N.J., and Bjorklund, H. (2001). Molecular characterisation of the nucleocapsid protein gene, glycoprotein gene and gene junctions of rhabdovirus 903/87, a novel fish pathogenic rhabdovirus. Virus Res. *80*, 11–22.

Johansson, T., Ostman-Myllyoja, L., Hellstrom, A., Martelius, S., Olesen, N.J., and Bjorklund, H. (2002). A novel fish rhabdovirus from sweden is closely related to the Finnish rhabdovirus 903/87. Virus Genes *25*, 127–138.

Johnson, B.K., Shockley, P., Chanas, A.C., Squires, E.J., Gardner, P., Wallace, C., Simpson, D.I., Bowen, E.T., Platt, G.S., Way, H., *et al.* (1977). Arbovirus isolations from mosquitoes: Kano Plain, Kenya. Trans. R. Soc. Trop. Med. Hyg. *71*, 518–521.

Johnson, H.N. (1965). Diseases derived from wildife. Calif. Health *23*, 35–39.

Johnson, M.C., Simon, B.E., Kim, C.H., and Leong, J.A. (2000). Production of recombinant snakehead rhabdovirus: the NV protein is not required for viral replication. J. Virol. *74*, 2343–2350.

Karabatsos, N., ed. (1985). The American Committee on Arthropod-borne Viruses: International Catalog of Arboviruses Including Certain Other Viruses of Vertebrates, 3rd edn (American Society for Tropical Medicine and Hygiene, San Antonio, TX).

Karabatsos, N., Lipman, M.B., Garrison, M.S., and Mongillo, C.A. (1973). The morphology, morphogenesis, and serological characterization of the rhabdoviruses Navarro, Kwatta, and Mossuril. J. Gen. Virol. *21*, 429–433.

Kay, B.H., Boreham, P.F.L., Dyce, A.L., and Standfast, H.A. (1978). Blood feeding of biting midges (Diptera: Ceratopogonidae) at Kowanyama, Cape York Peninsula, North Queensland. J. Aust. Ent. Soc. *17*, 145–149.

Kerschner, J.H., Calisher, C.H., Vorndam, A.V., and Francy, D.B. (1986). Identification and characterization of Bahia Grande, Reed Ranch and Muir Springs viruses, related members of the family *Rhabdoviridae* with widespread distribution in the United States. J. Gen. Virol. *67*, 1081–1089.

Kokernot, R.H., McIntosh, B.M., Worth, C.B., and De Sousa, J. (1962). Isolation of viruses from mosquitoes collected at Lumbo, Mozambique. II. Mossuril virus, a new virus isolated from the *Culex* (*culex*) *sitiens* Wiedemann group. Am. J. Trop. Med. Hyg. *11*, 683–684.

Kondo, H., Maeda, T., Shirako, Y., and Tamada, T. (2006). Orchid fleck virus is a rhabdovirus with an unusual bipartite genome. J. Gen. Virol. *87*, 2413–2421.

Kormelink, R., Garcia, M.L., Goodin, M., Sasaya, T., and Haenni, A.L. (2011). Negative-strand RNA viruses: The plant-infecting counterparts. Virus Res. *162*, 184–202.

Kretzschmar, E., Peluso, R., Schnell, M.J., Whitt, M.A., and Rose, J.K. (1996). Normal replication of vesicular stomatitis virus without C proteins. Virology *216*, 309–316.

Kurath, G., and Leong, J.C. (1985). Characterization of infectious hematopoietic necrosis virus mRNA species reveals a nonvirion rhabdovirus protein. J. Virol. *53*, 462–468.

Kurath, G., and Winton, J. (2008). Fish Rhabdoviruses. In Encyclopedia of Virology, Mahy, B.W.J., and van Regenmortel, M.H.V., eds (Elsevier Academic Press, Oxford), pp. 221–227.

Kurath, G., Ahern, K.G., Pearson, G.D., and Leong, J.C. (1985). Molecular cloning of the six mRNA species of infectious hematopoietic necrosis virus, a fish rhabdovirus, and gene order determination by R-loop mapping. J. Virol. *53*, 469–476.

Kurz, W., Gelderblom, H., Flugel, R.M., and Darai, G. (1986). Isolation and characterization of a tupaia rhabdovirus. Intervirology *25*, 88–96.

Kuwata, R., Isawa, H., Hoshino, K., Tsuda, Y., Yanase, T., Sasaki, T., Kobayashi, M., and Sawabe, K. (2011). RNA splicing in a new rhabdovirus from *Culex* mosquitoes. J. Virol. *85*, 6185–6196.

Kuzmin, I.V., Hughes, G.J., and Rupprecht, C.E. (2006). Phylogenetic relationships of seven previously unclassified viruses within the family *Rhabdoviridae* using partial nucleoprotein gene sequences. J. Gen. Virol. *87*, 2323–2331.

Kuzmin, I.V., Wu, X., Tordo, N., and Rupprecht, C.E. (2008). Complete genomes of Aravan, Khujand, Irkut and West Caucasian bat viruses, with special attention to the polymerase gene and non-coding regions. Virus Res. *136*, 81–90.

Lambeth, L.S., Yu, M., Anderson, D.E., Crameri, G., Eaton, B.T., and Wang, L.F. (2009). Complete genome sequence of Nariva virus, a rodent paramyxovirus. Arch. Virol. *154*, 199–207.

Landes-Devauchelle, C., Bras, F., Dezelee, S., and Teninges, D. (1995). Gene 2 of the sigma rhabdovirus genome encodes the P protein, and gene 3 encodes a protein related to the reverse transcriptase of retroelements. Virology *213*, 300–312.

Lazzarini, R.A., Keene, J.D., and Schubert, M. (1981). The origins of defective interfering particles of the negative-strand RNA viruses. Cell *26*, 145–154.

Li, F., and Ding, S.W. (2006). Virus counterdefense: diverse strategies for evading the RNA-silencing immunity. Annu. Rev. Microbiol. *60*, 503–531.

Liehne, P.F., Stanley, N.F., Alpers, M.P., and Liehne, C.G. (1976). Ord River arboviruses – the study site and mosquitoes. Aust. J. Exp. Biol. Med. Sci. *54*, 487–497.

Liehne, P.F., Anderson, S., Stanley, N.F., Liehne, C.G., Wright, A.E., Chan, K.H., Leivers, S., Britten, D.K., and Hamilton, N.P. (1981). Isolation of Murray Valley encephalitis virus and other arboviruses in the Ord River Valley 1972–1976. Aust. J. Exp. Biol. Med. Sci. *59*, 347–356.

Lin, Y., Bright, A.C., Rothermel, T.A., and He, B. (2003). Induction of apoptosis by paramyxovirus simian virus 5 lacking a small hydrophobic gene. J. Virol. *77*, 3371–3383.

Locali-Fabris, E.C., Freitas-Astua, J., Souza, A.A., Takita, M.A., Astua-Monge, G., Antonioli-Luizon, R., Rodrigues, V., Targon, M.L., and Machado, M.A. (2006). Complete nucleotide sequence, genomic organization and phylogenetic analysis of Citrus leprosis virus cytoplasmic type. J. Gen. Virol. *87*, 2721–2729.

Longdon, B., Obbard, D.J., and Jiggins, F.M. (2010). Sigma viruses from three species of *Drosophila* form a major new clade in the rhabdovirus phylogeny. Proc. Biol. Sci. *277*, 35–44.

Longdon, B., Hadfield, J.D., Webster, C.L., Obbard, D.J., and Jiggins, F.M. (2011a). Host phylogeny determines viral persistence and replication in novel hosts. PLoS Pathog. *7*, e1002260.

Longdon, B., Wilfert, L., Obbard, D.J., and Jiggins, F.M. (2011b). Rhabdoviruses in two species of *Drosophila*: vertical transmission and a recent sweep. Genetics *188*, 141–150.

Longdon, B., Wilfert, L., Osei-Poku, J., Cagney, H., Obbard, D.J., and Jiggins, F.M. (2011c). Host-switching by a vertically transmitted rhabdovirus in *Drosophila*. Biol. Lett. *7*, 747–750.

Lyles, D.S., and Rupprecht, C.E. (2007). Rhabdoviridae. In Fields Virology, Knipe, D.M., and Howley, P.M., eds (Lippincott Williams and Wilkins, Philadelphia, PA), pp. 1364–1394.

Mackerras, I.M., Mackerras, M.J., and Burnet, F.M. (1940). Experimental studies of ephemeral fever in Australian cattle. Aust. Commonw. Sci. Ind. Res. Bull. *136*, 1–116.

McWilliam, S.M., Kongsuwan, K., Cowley, J.A., Byrne, K.A., and Walker, P.J. (1997). Genome organization and transcription strategy in the complex G_{NS}-L intergenic region of bovine ephemeral fever rhabdovirus. J. Gen. Virol. *78*, 1309–1317.

Main, A.J., and Carey, A.B. (1980). Connecticut virus: a new Sawgrass group virus from *Ixodes dentatus* (Acari: Ixodidae). J. Med. Entomol. *17*, 473–476.

Mead, D.G., Ramberg, F.B., Besselsen, D.G., and Mare, C.J. (2000). Transmission of vesicular stomatitis virus from infected to noninfected black flies co-feeding on nonviremic deer mice. Science *287*, 485–487.

Metselaar, D., Williams, M.C., Simpson, D.I., West, R., and Mutere, F.A. (1969). Mount Elgon bat virus: a hitherto undescribed virus from *Rhinolophus hildebrandtii eloquens* K. Anderson. Arch. Virol. *26*, 183–193.

Miller, P.J., Boyle, D.B., Eaton, B.T., and Wang, L.F. (2003). Full-length genome sequence of Mossman virus, a novel paramyxovirus isolated from rodents in Australia. Virology *317*, 330–344.

Monath, T.P., Cropp, C.B., Frazier, C.L., Murphy, F.A., and Whitfield, S.G. (1979). Viruses isolated from reptiles: identification of three new members of the family *Rhabdoviridae*. Arch. Virol. *60*, 1–12.

Morimoto, K., Kawai, A., Sato, Y., and Ohkubo, A. (2011). A unique transcription mode of rabies virus high egg passage-Flury strain detected in infected baby hamster kidney-21 cells. Microbiol. Immunol. *55*, 558–564.

Mork, C., Hershberger, P., Kocan, R., Batts, W., and Winton, J. (2004). Isolation and characterization of a rhabdovirus from starry flounder (*Platichthys stellatus*) collected from the northern portion of Puget Sound, Washington, USA. J. Gen. Virol. *85*, 495–505.

Moutailler, S., Roche, B., Thiberge, J.M., Caro, V., Rougeon, F., and Failloux, A.B. (2011). Host alternation is necessary to maintain the genome stability of Rift Valley fever virus. PLoS Negl. Trop. Dis. *5*, e1156.

Muller, M.J., and Standfast, H.A. (1986). Vectors of ephemeral fever group viruses. In Arbovirus Research in Australia Proceedings 4th Symposium 6–9 May, St George, T.D., Kay, B.H., and Blok, J., eds (CSIRO/QIMR, Brisbane, Australia), pp. 295–298.

Muller, M.J., Murray, M.D., and Edwards, J.A. (1981). Blood-sucking midges and mosquitoes feeding on mammals at Beatrice Hill, N.T. Aust. J. Zool. *29*, 573–588.

Murphy, F.A., and Fields, B.N. (1967). Kern Canyon virus: electron microscopic and immunological studies. Virology *33*, 625–637.

Murphy, F.A., Shope, R.E., Metselaar, D., and Simpson, D.I. (1970). Characterization of Mount Elgon bat virus, a new member of the rhabdovirus group. Virology *40*, 288–297.

Murphy, F.A., Johnson, H.N., Hareison, A.K., and Shope, R.E. (1972). Ultrastructural characterization of a new member of the Rhabdovirus group-Klamath virus. Arch. virol. *37*, 323–331.

Noda, M., Inaba, Y., Banjo, M., and Kubo, M. (1992). Isolation of Fukuoka virus, a member of the Kern Canyon serogroup viruses of the family *Rhabdoviridae*, from cattle. Vet. Microbiol. *32*, 267–271.

Pattnaik, A.K., Ball, L.A., LeGrone, A., and Wertz, G.W. (1995). The termini of VSV DI particle RNAs are sufficient to signal RNA encapsidation, replication, and budding to generate infectious particles. Virology *206*, 760–764.

Peluso, R.W., Richardson, J.C., Talon, J., and Lock, M. (1996). Identification of a set of proteins (C′ and C) encoded by the bicistronic P gene of the Indiana serotype of vesicular stomatitis virus and analysis of their effect on transcription by the viral RNA polymerase. Virology *218*, 335–342.

Perrault, J. (1981). Origin and replication of defective interfering particles. Curr. Top. Microbiol. Immunol. *93*, 151–207.

Pfeil-Putzien, C. (1978). [Experimental transmission of spring viraemia of carp through carp lice (*Angulus foliaceus*)]. Zentralbl. Veterinarmed. B *25*, 319–323.

Quan, P.L., Junglen, S., Tashmukhamedova, A., Conlan, S., Hutchison, S.K., Kurth, A., Ellerbrok, H., Egholm, M., Briese, T., Leendertz, F.H., *et al.* (2010). Moussa virus: a new member of the *Rhabdoviridae* family isolated from *Culex decens* mosquitoes in Cote d'Ivoire. Virus Res. *147*, 17–24.

Quan, P.L., Williams, D.T., Johansen, C.A., Jain, K., Petrosov, A., Diviney, S.M., Tashmukhamedova, A., Hutchison, S.K., Tesh, R.B., Mackenzie, J.S., *et al.* (2011). Genetic characterization of K13965, a strain of Oak Vale virus from Western Australia. Virus Res. *160*, 206–213.

Revill, P., Trinh, X., Dale, J., and Harding, R. (2005). Taro vein chlorosis virus: characterization and variability of a new nucleorhabdovirus. J. Gen. Virol. *86*, 491–499.

Ritter, D.G., Calisher, C.H., Muth, D.J., Shope, R.E., Murphy, F.A., and Whitfield, S.G. (1978). New Minto virus: a new rhabdovirus from ticks in Alaska. Can. J. Microbiol. *24*, 422–426.

Rivers, T.M. (1937). Viruses and Koch's Postulates. J. Bacteriol. *33*, 1–12.

Roche, S., Bressanelli, S., Rey, F.A., and Gaudin, Y. (2006). Crystal structure of the low-pH form of the vesicular stomatitis virus glycoprotein G. Science *313*, 187–191.

Salaun, J.J., Rickenbach, A., Bres, P., Brottes, H., Germain, M., Eouzan, J.P., and Ferrara, L. (1969). The Nkolbisson virus (YM 31-65), a new prototype of arbovirus isolated in Cameroun. Ann. Inst. Pasteur. Paris *116*, 254–260.

Sanger, F., Nicklen, S., and Coulson, A.R. (1977). DNA sequencing with chain-terminating inhibitors. Proc. Natl. Acad. Sci. U.S.A. *74*, 5463–5467.

Sather, G.E., Lewis, A.L., Jennings, W., Bond, J.O., and Hammon, W.M. (1970). Sawgrass virus: a newly described arbovirus in Florida. Am. J. Trop. Med. Hyg. *19*, 319–326.

Sato, S., Wong, S.K., and Lazinski, D.W. (2001). Hepatitis delta virus minimal substrates competent for editing by ADAR1 and ADAR2. J. Virol. *75*, 8547–8555.

Schnell, M.J., Buonocore, L., Kretzschmar, E., Johnson, E., and Rose, J.K. (1996). Foreign glycoproteins expressed from recombinant vesicular stomatitis viruses are incorporated efficiently into virus particles. Proc. Natl. Acad. Sci. U.S.A. *93*, 11359–11365.

Scholthof, K.B., Hillman, B.I., Modrell, B., Heaton, L.A., and Jackson, A.O. (1994). Characterization and detection of sc4: a sixth gene encoded by sonchus yellow net virus. Virology *204*, 279–288.

Seddon, H.R. (1938). The spread of ephemeral fever (three-day sickness) in Australia in 1936–37. Aust. Vet. J. *14*, 90–101.

Spiropoulou, C.F., and Nichol, S.T. (1993). A small highly basic protein is encoded in overlapping frame within the P gene of vesicular stomatitis virus. J. Virol. *67*, 3103–3110.

Springfeld, C., Darai, G., and Cattaneo, R. (2005). Characterization of the Tupaia rhabdovirus genome reveals a long open reading frame overlapping with P and a novel gene encoding a small hydrophobic protein. J. Virol. *79*, 6781–6790.

St George, T.D. (1985). Studies on the pathogenesis of bovine ephemeral fever in sentinel cattle. I. Virology and serology. Vet. Microbiol. *10*, 493–504.

St George, T.D. (1988). Bovine ephemeral fever. In The Arboviruses: Epidemiology and Ecology, Monath, T.P., ed. (CRC Press, Boca Raton, FL), pp. 71–86.

St George, T.D. (1990). Bovine ephemeral fever virus. In Virus infections of vertebrates, Dinter, Z., and Morein, B., eds (Elsevier, London, UK), pp. 405–415.

St George, T.D. (2008). Evidence that mosquitoes are the vectors of bovine ephemeral fever virus. In Arbovirus Research in Australia Proceedings 10th Symposium 15–19 September, Ryan, P., Aaskov, J., and Russell, R., eds (QIMR, Brisbane, Australia), pp. 161–164.

Standfast, H.A., Dyce, A.L., St George, T.D., Muller, M.J., Doherty, R.L., Carley, J.G., and Filippich, C. (1984). Isolation of arboviruses from insects collected at Beatrice Hill, Northern Territory of Australia, 1974–1976. Aust. J. Biol. Sci. *37*, 351–366.

Steele, J.H., and Fernandez, P.J. (1991). History of rabies and global aspects. In The Natural History of Rabies, Baer, G.M., ed. (CRC Press, Boca Raton, FL), pp. 1–28.

Talbi, C., Cabon, J., Baud, M., Bourjaily, M., de Boisseson, C., Castric, J., and Bigarre, L. (2011). Genetic diversity of perch rhabdoviruses isolates based on the nucleoprotein and glycoprotein genes. Arch. Virol. *156*, 2133–2144.

Tanno, F., Nakatsu, A., Toriyama, S., and Kojima, M. (2000). Complete nucleotide sequence of northern cereal mosaic virus and its genome organization. Arch. Virol. *145*, 1373–1384.

Tao, J.J., Zhou, G.Z., Gui, J.F., and Zhang, Q.Y. (2008). Genomic sequence of mandarin fish rhabdovirus with an unusual small non-transcriptional ORF. Virus Res. *132*, 86–96.

Teninges, D., Bras, F., and Dezelee, S. (1993). Genome organization of the sigma rhabdovirus: six genes and a gene overlap. Virology *193*, 1018–1023.

Tesh, R.B., Chaniotis, B.N., and Johnson, K.M. (1972). Vesicular stomatitis virus (Indiana serotype): transovarial transmission by phlebotomine sandlies. Science *175*, 1477–1479.

Tesh, R.B., Travassos Da Rosa, A.P., and Travassos Da Rosa, J.S. (1983). Antigenic relationship among rhabdoviruses infecting terrestrial vertebrates. J. Gen. Virol. *64*, 169–176.

Tesh, R.B., Boshell, J., Modi, G.B., Morales, A., Young, D.G., Corredor, A., Ferro de Carrasquilla, C., de Rodriguez, C., Walters, L.L., and Gaitan, M.O. (1987). Natural infection of humans, animals, and phlebotomine sand flies with the Alagoas serotype of vesicular stomatitis virus in Colombia. Am. J. Trop. Med. Hyg. *36*, 653–661.

Thoulouze, M.I., Bouguyon, E., Carpentier, C., and Bremont, M. (2004). Essential role of the NV protein of *Novirhabdovirus* for pathogenicity in rainbow trout. J. Virol. *78*, 4098–4107.

Tordo, N., Poch, O., Ermine, A., and Keith, G. (1986a). Primary structure of leader RNA and nucleoprotein genes of the rabies genome: segmented homology with VSV. Nucleic Acids Res. *14*, 2671–2683.

Tordo, N., Poch, O., Ermine, A., Keith, G., and Rougeon, F. (1986b). Walking along the rabies genome: is the large G-L intergenic region a remnant gene? Proc. Natl. Acad. Sci. U.S.A. *83*, 3914–3918.

Tordo, N., Poch, O., Ermine, A., Keith, G., and Rougeon, F. (1988). Completion of the rabies virus genome sequence determination: highly conserved domains among the L (polymerase) proteins of unsegmented negative-strand RNA viruses. Virology *165*, 565–576.

Tordo, N., de Haan, P., Goldbach, R., and Poch, O. (1992). Evolution of negative-stranded RNA genomes. In Seminars in Virology, Koonin, E., ed. (Saunders Scientific Publications, London, UK), pp. 341–357.

Tordo, N., Benmansour, A., Calisher, C., Dietzgen, R.G., Fang, R.-X., Jackson, A.O., Kurath, G., Nadin-Davis, S., Tesh, R.B., and Walker, P.J. (2005). Rhabdoviridae. In Virus Taxonomy: Eighth Report of the International Committee on the Taxonomy of Viruses, Fauquet, C.M., Mayo, M.A., Maniloff, J., Desselberger, U., and Ball, L.A., eds (Elsevier Academic Press, San Diego, TX), pp. 623–644.

Tsai, C.W., Redinbaugh, M.G., Willie, K.J., Reed, S., Goodin, M., and Hogenhout, S.A. (2005). Complete genome sequence and in planta subcellular localization of maize fine streak virus proteins. J. Virol. *79*, 5304–5314.

Tuekam, T., Greiner, E.C., and Gibbs, E.P. (1991). Seroepidemiology of Bivens Arm virus infections of cattle in Florida, St. Croix and Puerto Rico. Vet. Microbiol. *28*, 121–127.

Wagner, R.R. (1987). Rhabdovirus biology and infection: an overview. In The Rhabdoviruses, Wagner, R.R., ed. (Plenum Press, New York), pp. 9–74.

Walker, P.J. (2005). Bovine ephemeral fever in Australia and the world. Curr. Top. Microbiol. Immunol. *292*, 57–80.

Walker, P.J. (2008). Bovine ephemeral fever virus. In Encyclopedia of Virology, Mahy, B.W.J., and Van Regenmortel, M.H.V., eds (Elsevier Academic Press, Oxford), pp. 354–362.

Walker, P.J., and Kongsuwan, K. (1999). Deduced structural model for animal rhabdovirus glycoproteins. J. Gen. Virol. *80*, 1211–1220.

Walker, P.J., Byrne, K.A., Cybinski, D.H., Doolan, D.L., and Wang, Y.H. (1991). Proteins of bovine ephemeral fever virus. J. Gen. Virol. *72*, 67–74.

Walker, P.J., Byrne, K.A., Riding, G.A., Cowley, J.A., Wang, Y., and McWilliam, S. (1992). The genome of bovine ephemeral fever rhabdovirus contains two related glycoprotein genes. Virology *191*, 49–61.

Walker, P.J., Dietzgen, R.G., Joubert, D.A., and Blasdell, K.R. (2011). Rhabdovirus accessory genes. Virus Res. *162*, 110–125.

Wang, Y., and Walker, P.J. (1993). Adelaide river rhabdovirus expresses consecutive glycoprotein genes as polycistronic mRNAs: new evidence of gene duplication as an evolutionary process. Virology *195*, 719–731.

Wang, Y., McWilliam, S.M., Cowley, J.A., and Walker, P.J. (1994). Complex genome organization in the G_{NS}-L intergenic region of Adelaide River rhabdovirus. Virology *203*, 63–72.

Wang, Y., Cowley, J.A., and Walker, P.J. (1995). Adelaide River virus nucleoprotein gene: analysis of phylogenetic relationships of ephemeroviruses and other rhabdoviruses. J. Gen. Virol. *76*, 995–999.

Weir, R. (2002). Classification and identification of viruses isolated from mosquitoes in the Northern Territory, 1982–1992, using a range of techniques. PhD thesis. Faculty of Medicine, University of Sydney, Sydney, Australia.

Weir, R.P., Hyatt, A.D., Calisher, C.H., and Whelan, P.I. (1996). New records of arboviruses isolated from mosquitoes in the Northern Territory 1982–1992. In Arbovirus Research in Australia Proceedings 7th Symposium 25–29 November, Kay, B.H., Brown, M.D., and Aaskov, J.G., eds (CSIRO/QIMR, Brisbane, Australia), pp. 311–321.

Wellings, F.M., Lewis, A.L., and Pierce, L.V. (1972). Agents encountered during arboviral ecological studies: Tampa Bay area, Florida, 1963 to 1970. Am. J. Trop. Med. Hyg. *21*, 201–213.

Wertz, G.W., Moudy, R., and Ball, L.A. (2002). Adding genes to the RNA genome of vesicular stomatitis virus: positional effects on stability of expression. J. Virol. *76*, 7642–7650.

van der Westhuizen, B. (1967). Studies on bovine ephemeral fever. I. Isolation and preliminary characterization of a virus from natural and experimentally produced cases of bovine ephemeral fever. Onderstepoort J. Vet. Res. *34*, 29–40.

Whelan, S.P. (2008). Vesicular stomatitis virus. In Encyclopedia of Virology, Mahy, B.W.J., and van Regenmortel, M.H.V., eds (Elsevier Academic Press, Oxord), pp. 291–299.

Whelan, S.P., Barr, J.N., and Wertz, G.W. (2004). Transcription and replication of nonsegmented negative-strand RNA viruses. Curr. Top. Microbiol. Immunol. *283*, 61–119.

Whitney, E. (1964). Flanders strain, an arbovirus newly isolated from mosquitoes and birds of New York State. Am. J. Trop. Med. Hyg. *13*, 123–131.

Woodruff, A.W., Ansdell, V.E., and Bowen, E.T. (1977). Le Dantec virus infection in a patient who had not been to West Africa. Br. Med. J. *2*, 1632–1633.

Wunner, W.H., Calisher, C.H., Dietzgen, R.G., Jackson, A.O., Kitajima, E.W., Lafon, M., Leong, J.C., Nichol, S., Peters, D., Smith, J.S., and Walker, P.J. (1995). *Rhabdoviridae*. In Virus Taxonomy: Sixth Report of the International Committee on the Taxonomy of Viruses, F.A. Murphy, C.M. Fauquet, D.H.L. Bishop, S.A. Ghabrial, A.W. Jarvis, G.P. Martelli, M.A. Mayo, and M.D. Summers, eds. (Wien, Austria, Springer-Verlag), pp. 275–288.

Zhang, Q.Y., Tao, J.J., Gui, L., Zhou, G.Z., Ruan, H.M., Li, Z.Q., and Gui, J.F. (2007). Isolation and characterization of *Scophthalmus maximus* rhabdovirus. Dis. Aquat. Organ. *74*, 95–105.

Zhu, R.L., Lei, X.Y., Ke, F., Yuan, X.P., and Zhang, Q.Y. (2011). Genome of turbot rhabdovirus exhibits unusual non-coding regions and an additional ORF that could be expressed in fish cell. Virus Res. *155*, 495–505.

Experimental Evolution of Rhabdoviruses

12

Isabel S. Novella and John B. Presloid

Abstract

Rhabdoviruses have been preferred models for studies on the evolution of RNA viruses. Experiments in the laboratory can seldom, if ever, reproduce the complexity of natural environments, but because we can control and manipulate individual components, we can assess how each of these components affects (or not) viral evolution, including population size, environmental changes, co-infection, host responses, antiviral molecules etc. Experimental evolution is becoming increasingly sophisticated as technological advances provide the tools that we need to answer some novel and some old-standing questions, and to test extensively theoretical models.

Introduction

Rhabdoviruses have been preferred models for studies on the evolution of RNA viruses for decades. Experiments in the laboratory can seldom, if ever, reproduce the complexity of natural environments, but because we can control and manipulate individual components, we can assess how each of these components affects (or not) viral evolution, including population size, environmental changes, co-infection, host responses, antiviral molecules etc. Experimental evolution is becoming increasingly sophisticated as technological advances provide the tools that we need to answer some novel and some old-standing questions, and to test extensively theoretical models. Much of what we know about the biology and replication of Mononegavirales was first learned with rabies and vesicular stomatitis virus (VSV), and further extended or extrapolated to the viruses of other families. VSV has been used extensively to test evolutionary theory and to learn the rules governing the evolution or mutation-prone RNA viruses (Domingo *et al.*, 1997).

Models of experimental evolution of rhabdoviruses

Most of the work published to date explores the evolutionary behaviour of VSV in cell culture or infectious haematopoietic necrosis virus (IHNV) *in vivo*. Both models have some features in common but there are some substantial differences.

The VSV model was developed by John Holland and colleagues (Holland *et al.*, 1991). Fitness changes are determined by competition between two populations that differ in their sensitivity to a monoclonal antibody I1. The original wild type (WT) was plated in the presence of I1 to produce monoclonal antibody-resistant mutants (MARMs). There are several mutations that mediate resistance to I1 but are otherwise neutral, and these can be used as genetic markers in competition experiments to measure fitness. Briefly, WT and MARM are mixed and the mixture used for infections of a cell monolayer. Ratios of incoming and progeny virus are measured by plaque assay in the presence and absence of monoclonal antibody, and changes in the ratios can be used to calculate the relative contribution to the next generation. More fit viruses will produce more progeny.

The IHNV model was developed by Gael Kurath and collaborators (Troyer *et al.*, 2008). In this case the two competitors are strains from different genotypes (genogroups or subgenogroups) and quantitation is done in a two-step process. The first step consists of the generation of cDNA from both using non-specific primers, and in the second step each genotype is amplified independently using specific primers. Each genotype produces a cDNA band of a different size after restriction analysis, and these can be separated in an agarose gel and quantified by densitometry. Competitions can be done in cell culture or in live animals, usually rainbow trout or sockeye salmon. It is particularly important to note that transmission is one of the fitness components that affects the final result.

Even though both systems are set up to promote competition for resources, this competition does not seem to take place during IHNV infection of fish (Wargo *et al.*, 2010). Thus, the growth curves that each competitor produce are identical, independent of the presence or absence of a competitor. In other words, competitions are not required to assess relative fitness. In contrast, work with VSV shows that results from independent growth curves, even when extremely detailed, cannot always serve as predictors of relative fitness (Novella *et al.*, 2004). A corollary is that virus titre may also fail to correlate with fitness, so higher titres will not always correspond to the strain with the highest relative fitness, and similar titres may be found in viruses that differ in fitness.

Survival in the wild

Fitness assays help to understand the dynamics of survival in natural populations of IHNV. This virus infects salmonids and is endemic in North America, between California and Alaska (Fig. 12.1). Phylogenetic analysis of natural isolates led to the recognition of three genogroups, U, M and L (Kurath *et al.*, 2003). The M genogroup includes several subgroups referred to as MA-MF (Troyer and Kurath, 2003) and has been involved in frequent epidemics. In the past decades, the viruses from the M genogroup spread to Europe, where they cause periodic outbreaks of IHNV infection (Enzmann *et al.*, 2005).

The overarching hypothesis under consideration is that *in vivo* fitness is the major factor determining whether genogroups (or subgenogroups) co-circulate or not. Thus, the co-circulating subgenogroups may have a similar fitness. Competition between the members of MB and MC subgenogroups, and competitions between MB and MD subgenogroups, showed no variations in fitness beyond the error of the method (Troyer *et al.*, 2008). There was a substantial degree of variation among individual fish, consistent with sizeable bottleneck during transmission, perhaps influenced by the outbred nature of the rainbow trout used as the host, but overall results indicated that each progenitor contributed equally to the next generation.

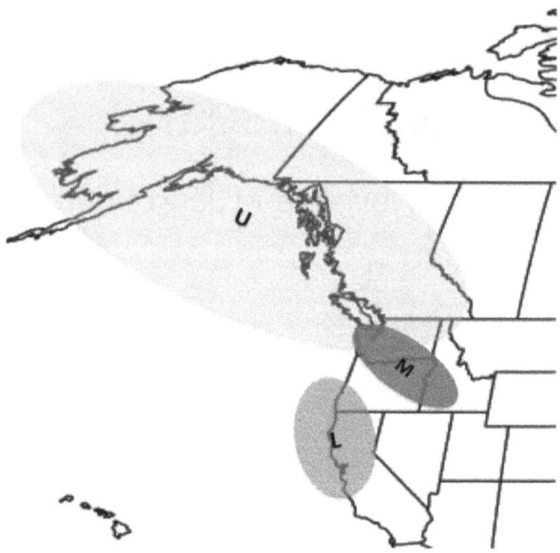

Figure 12.1 Geographical distribution of IHNV in North America. Light grey indicates the distribution of U genogroup, intermediate grey corresponds to genogroup L, and dark grey represents genogroup M.

Not all M genogroup isolates showed the same fitness, and when differences were observed, they correlated well with virulence in rainbow trout and sockeye salmon (Kurath *et al.*, 2006; Wargo *et al.*, 2010; Penaranda *et al.*, 2011). Detailed analysis of replication dynamics indicated that multiple life-history traits differed between the high virulence/high fitness isolate and the low virulence/low fitness isolate, including entry, replication and shedding (Wargo and Kurath, 2011). These studies illustrate how we can combine traditional analysis and fitness measurements in the laboratory to understand virus survival in the wild. While one has to be aware of the limitations intrinsic to any laboratory analysis, this type of work can help us to go beyond sequence analyses, which was the only method available until recently.

Fitness, adaptability and robustness

Several factors determine long-term survival or extinction, including availability of genetic variation, random events and selection. Fitness is important for short-term survival, but selection can operate at other levels. Once populations reach a fitness peak, robustness becomes critical. Robustness is defined as the ability of a genome to incorporate mutations without changing its fitness. More robust genomes have a relatively high neutral mutation rate and low deleterious or beneficial mutation rates. A population of relatively low robustness may lose a competition against a less fit, but more robust virus because of the progeny, which is known as 'survival of the flattest' (Fig. 12.2). This is a prediction of quasispecies theory (Eigen, 1992, 1996; Wilke, 2005) that has been supported experimentally, for rhabdoviruses specifically (de la Torre and Holland, 1990), but also for other RNA viruses (Burch and Chao, 2000) and digital organisms (Wilke *et al.*, 2001). A high fitness genome with low robustness will produce progeny that will suffer a substantial fitness loss, because most mutations are deleterious (Duarte *et al.*, 1994). In contrast, a robust genome will produce progeny most of which will have the same fitness as its progenitor. Thus, after sufficient number of replication cycles, the relative fitness of both progenies will change, and the genome occupying the flat peak will survive.

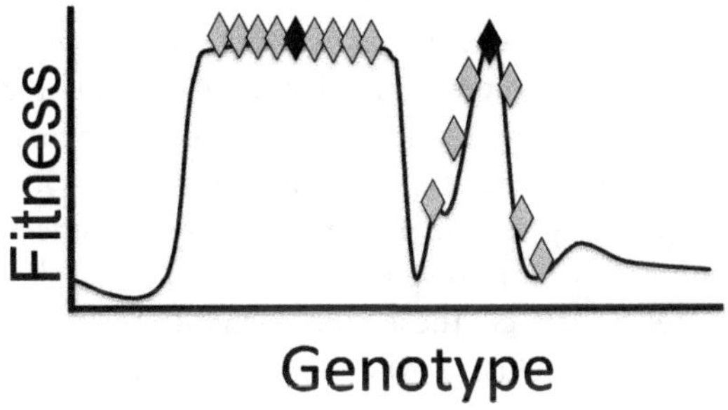

Figure 12.2 Two-dimensional fitness landscape. Each peak has a progenitor (black diamond) that can produce progeny (grey diamonds) with multiple mutations. The flatness of the peak on the left (robustness) allows this progeny to move along the x-axis without fitness loss. In contrast, the sharpness of the peak on the right results in substantial fitness loss as additional mutations move the genotype away from the progenitor.

The last factor determining long-term survival is adaptability, or the ability of a population to increase in fitness. Adaptability and robustness are inversely correlated when the changes in the latter are the result of changes in the beneficial mutation rate, but this may or may not be the case. In some literature authors use the term *evolvability* to refer to adaptability, but because evolution may take place without fitness increase (or fitness change of any type) we propose to use *adaptability* to refer to evolution with fitness increase.

Because under quasispecies theory selection favours more robust genomes, one would predict that robustness will be lost to certain degree if selection is relaxed. Two environmental factors result in relaxed selection: high multiplicity of infection (MOI) and bottlenecks. During high MOI multiple virions in the inoculum may co-infect the same cell, so resources can be shared and there will be complementation. Because genomes and their products cannot recognize whether they are matched or not (e.g. whether a specific genomic sequence coded for a specific protein), there will be all possible interactions, and because nature only targets functions, some high-fitness genomes paired with low-fitness products will be eliminated, while low-fitness genomes paired with high-fitness products will survive. Unfortunately, survival of low robustness genomes during high MOI infection is difficult to study because of the generation of defective interfering particles (Holland, 1991), but experimental evolution of other viral systems has shown that high MOI indeed relaxes selection and allows the survival of genomes with low robustness (Turner *et al.*, 2005).

The second mechanism that minimizes selection is repeated genetic bottlenecking, which promotes accumulation of mutations due to random sampling. Because a small population size diminishes or eliminates competition, viruses that are not particularly fit can survive. Most of the mutations will be in fact deleterious and result in fitness losses, a process known as Muller's ratchet (Muller, 1964). Experiments with VSV provided the first demonstration of Muller's ratchet in eukaryotic viruses (Duarte *et al.*, 1992). Regardless of their effect on fitness, mutations accumulated during bottleneck change the genetic background on which new mutations arise and, not surprisingly, have an effect in the robustness and adaptability of viral populations (Novella, 2004; Novella *et al.*, 2010). It is surprising that little changes (only two or three nucleotides in a 11 Kb genome) are sufficient to cause a difference in robustness (Novella and Ebendick-Corpus, 2004).

The robustness of VSV has been measured in two ways. First, isolating random clones, measuring their fitness, and determining distributions compared to that of the population (Duarte *et al.*, 1994; Novella *et al.*, 2010). Changes in the neutral mutation rate can affect the deleterious mutation rate, the beneficial mutation rate, or both. Loss of robustness due to an increased deleterious mutation rate will appear as a shift in the clonal distribution towards lower fitness. If the loss results in increased deleterious and beneficial mutation rates, it appears as a widening of the distribution regardless of average fitness (Fig. 12.3). While both outcomes are possible, VSV has only produced shifts in the distribution. The second approach to identify changes in robustness included mutation-accumulation experiments, during which several replicas of repeated bottlenecks were carried out, and fitness changes were measured at the end (Novella *et al.*, 2010). A third potential method to assess robustness in VSV is measure of the plaque size (Sanjuan *et al.*, 2007), although individual values cannot be normalized by the population average.

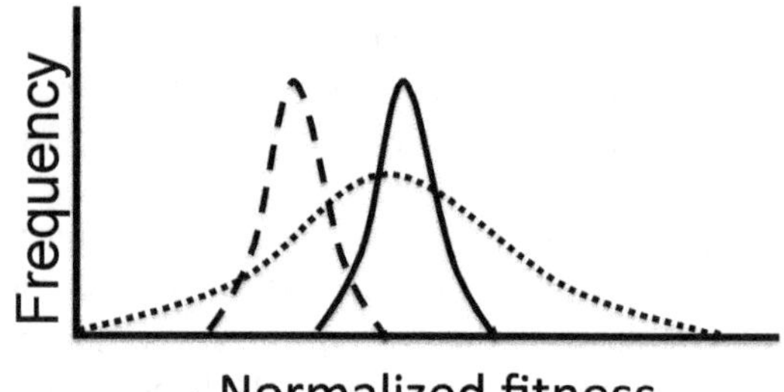

Figure 12.3 Changes in robustness result in differences in fitness distributions. The solid line represents a fitness distribution of a given population. If selection is relaxed, the expected loss of robustness can produce two types of changes. First, it can result in increased deleterious and beneficial mutation rates at the cost of the neutral mutation rate; in other words, the distribution is expanded (dotted line). Second, it can result in increased deleterious mutation rate at the cost of the neutral and beneficial mutations rates; in other words, the distribution shifts to the left (dashed line).

Bottleneck effects on VSV robustness can be observed at different times in history. Sometimes loss of robustness (and adaptability) can be measure in the progeny populations. In such cases low robustness translates into fitness distributions that show a decreased fitness of clones compared to the population average (Novella *et al.*, 2010), and low adaptability is observed either as inability to increase fitness at all, or as a slow speed of adaptation (Novella, 2004; Novella *et al.*, 2010). Even in the instances where bottlenecked viruses can replicate as well as wild type viruses, the long-lasting effects of bottlenecks can be observed as decreased robustness of the high-fitness viral populations. Thus, in long-term experiments, where beneficial variation is exhausted (e.g. the beneficial mutation rate is zero), the viruses that have gone through repeated bottlenecks will be outcompeted by non-bottlenecked counterparts, because the latter have a lower deleterious mutation rate and a higher neutral mutation rate (Novella *et al.*, 2010).

Gene sequencing of evolved VSV populations that differed in adaptability facilitated distinguishing between the two molecular mechanisms that explain this defect: either low adaptability is the result of fewer beneficial mutations, or the result of mutations with weaker beneficial effect. These two mechanisms are not mutually exclusive. The results of determining the number of accumulated mutations showed no evidence that VSV strains with adaptability defects accumulated fewer mutations than controls, and, if anything, it was the opposite. Thus, adaptability is compromised because the appearing beneficial mutations are not as beneficial as those in control populations (Novella *et al.*, 2010).

Evolution during host alternation

Arboviruses are viruses that have mammalian hosts and arthropod vectors, and this group includes many important human pathogens such as Dengue virus, yellow fever virus, West Nile virus, and others. Arbovirus evolution has received extensive attention in the past decades, and VSV has served as an ideal surrogate model. In addition, arboviruses in general and VSV in particular have been useful tools for testing ecological theory related to selection of specialists and generalists.

In natural populations, arbovirus evolution (as measured by genomic changes per year) tends to be slow when compared to the speed of evolution in other RNA viruses, such as influenza or HIV (Webster *et al.*, 1992). Scott and co-workers (1994) hypothesized that this is the result of trade-offs incurred during replication in such disparate hosts as an arthropod and a vertebrate, so viruses had to compromise to be able to replicate in both environments (Weaver *et al.*, 1993). This idea was consistent with the observation that repeated passages of arboviruses in insect cells often results in poor replication and/or fitness loss in mammalian cells (Peleg, 1971; Singh *et al.*, 1971; Taylor and Marshall, 1975; Novella *et al.*, 1995a). In the particular case of VSV, these trade-offs are fairly dramatic during long-term persistent infection of sand fly cells (Novella *et al.*, 1995a; Zarate and Novella, 2004), but there is no trade-off when the virus is recovered at maximum titre and transferred to a new monolayer (Novella *et al.*, 1999b). Furthermore, under these conditions the virus that adapts to the vector cells concomitantly increases its fitness in the bypassed mammalian environment, and, not surprisingly, during alternation the virus increases fitness in both vector and host cells (Novella *et al.*, 1999a). Work with other arboviruses confirmed that alternation usually results in adaptation to both environments and does not explain evolutionary stability (reviewed in Novella *et al.*, 2011). VSV provided the only exception to this rule: when the alternation included a persistent regime for sand fly cell infection, the virus behaved as if it were replicating persistently in sand fly cells only (Zarate and Novella, 2004). This result is consistent with the view of Rodriguez and co-workers, who proposed that VSV is essentially an insect virus (Letchworth *et al.*, 1999). The mammalian hosts are required for long-term survival of the virus because of inefficient vertical transmission (Tesh *et al.*, 1972), but they provide little or nothing beyond a physical support where one sand fly can deposit the virus and another can pick it up (Mead *et al.*, 2000).

Many mammalian species and cell types are susceptible to VSV infection and this allowed a detailed analysis of the consequences of host switching. Comparison of evolution in homogeneous environment and heterogeneous environments is important to understand speciation. As indicated earlier, adaptation to a single environment may result in compromised survival in other environments. This is the cost of specialization. In contrast, adaptation to multiple environments may be the result of a compromise that makes the population capable of survive in all these environments, but unable to compete against a specialist in any of them. That is the cost of generalization (Kassen, 2002).

Costs can come in two different ways. First, as a trade-off, which occurs when adaptation to one environment results on fitness loss in alternative environments. Second, the cost can be associated with the shape of landscapes, so there is no fitness loss, and may be even certain fitness increases, but not as large as the increase achieved by a specialist. Analysis of evolution in different mammalian and insect cell types indicates that VSV is usually a generalist regardless of regime, and that the cost of generalization is mild, and mostly dominated by differences in fitness landscapes, not trade-offs (Smith-Tsurkan *et al.*, 2010). Interestingly, these results disproved some of the assumptions of ecological theory, namely that heterogeneous environments select for generalists while homogeneous environment select specialists. VSV typically increases in fitness during replication in any susceptible cell type (Llewellyn *et al.*, 2002; Novella *et al.*, 1995a,b; Turner *et al.*, 2010). How these strains behave in the original host (if there was a host change) or in bypassed hosts is unpredictable, but overall there is a good chance of concurrent adaptation (Novella *et al.*, 2011).

Trade-offs during VSV replication in mammalian acute infection or sand fly persistent infection were asymmetrical and only observed in persistent populations (Novella *et al.*, 1995a; Zarate and Novella, 2004). The molecular basis of the cost can be studied using gene sequence analysis. Trade-offs can be the result of two processes: antagonistic pleiotropy and mutation accumulation. Antagonistic pleiotropy is the result of mutations that are selected in one environment (because they are beneficial in that environment), but they are deleterious in other environment(s). During mutation accumulation, genomic changes are acquired owing to drift or hitch hiking, and while these changes have no effect in the selective environment, they are deleterious in other environments. Results of such experiments suggested that mutation accumulation has a significant contribution to fitness costs in VSV (Turner and Elena, 2000; Remold *et al.*, 2008), but one must be careful interpreting the genetic changes identified in the different populations, because the phenotypes of the strains and their characterization as specialists or generalists may be not well established (Smith-Tsurkan *et al.*, 2010).

There is a good evidence that antagonistic pleiotropy mediates trade-offs (Presloid *et al.*, 2008). Several mutations affecting different protein functions have been identified as likely contributors to antagonistic pleiotropy under different regimes, including matrix and glycoprotein mutations (Presloid *et al.*, 2008; Remold *et al.*, 2008). However, the most interesting ones are those that occur in the regulatory sequences, and in particular promoters (Presloid *et al.*, 2008). Analysis of the mutations accumulated during adaptation, and after fitness plateaus in sand fly persistence, resulted in the identification of genetic changes associated with replication, including the polymerase gene(s) and promoters. Promoter mutations are particularly remarkable because they consist of large insertions (several dozen nucleotides) of duplicated sequences (Novella *et al.*, 2007; Presloid *et al.*, 2008), which are relative rare among the negative-strand RNA viruses.

Ecological theory also predicts the effect of environmental heterogeneity on selection of other genetic parameters such as population variance. Because of the failure to use heterogeneity to select for specialists or generalists, we do not expect that viral population variance will correlate with the type of regime as expected. Indeed, we have evidence that there is no correlation between variance and (1) selective regime (homogeneous vs. heterogeneous) or (2) population behaviour (specialist vs. generalist) (Smith-Tsurkan *et al.*, 2010; Smith-Tsurkan *et al.*, unpublished results). Furthermore, and in contrast to expectations, VSV selected in a heterogeneous environment demonstrated lower adaptability than VSV selected in a homogeneous environment (Alto and Turner, 2010).

Experimental testing of theoretical models

Viruses provide excellent systems to test theoretical models of evolution. It is important, however, to take into consideration the specific biological features of the model before expanding the applicability of results: some of the rules we learn will be applicable across the biosphere, other will have more restrictive applications (microbes, asexual organisms, viruses, negative strand RNA viruses, rhabdoviruses, VSV). Muller's ratchet is one example of a theory demonstrated first with phage (Chao, 1990), confirmed with VSV (Duarte *et al.*, 1992) and with extensive application. The evolutionary behaviour of rhabdoviruses has led to the development of theoretical models by several groups (Tsimring *et al.*, 1996; Rouzine *et al.*, 2003). In turn, these models have inspired new experiments that help refine the models (Dutta *et al.*, 2008; Novella *et al.*, 2008).

Solitary wave

Rouzin and colleagues developed solitary wave theory to describe the evolution of asexual viruses at high mutation rates (Rouzine *et al.*, 2003). This model is based on Tsimring's work (Tsimring *et al.*, 1996) which assumes two levels of adaptation: the first consists of selection within the population to produce the fixation of beneficial alleles, while the second level consists of the continuous repetition of this process to result in gradual accumulation of mutations. One important feature of solitary wave theory is that it assumes that beneficial variation is not limiting, which makes this model different from others, such as clonal interference (Gerrish and Lenski, 1998). The initial model assumed the need for a lag period during which beneficial variation is generated (Rouzine *et al.*, 2003). Experimental test of this model produced results that fit remarkably well with the computer simulations (Novella *et al.*, 1995a,c; Dutta *et al.*, 2008). This model is interesting for several reasons, and specifically because it opens that possibility of estimating important biological parameters that are technically difficult or impossible to measure directly in experiments, such as the average selective value of mutations or the number of sites under selection. The main change in the modified model was the elimination of the lag period at the beginning of adaptation. The results and simulations clearly indicated that as soon as the virus replicates enough to make a plaque (clone), beneficial variation arises, and that beneficial variation drives the initial phase of adaptation (Dutta *et al.*, 2008). Indeed, simulations under mutation rates of zero still produced fitness increases during the first 10–15 passages, which represent the first level of adaptation in Tsimring and colleagues' model (Tsimring *et al.*, 1996).

Further experimental evidence that preexisting variation is crucial during early adaptation comes from the behaviour of VSV that had been adapted to persistence, with or without alternating replication in the mammalian host. VSV clones that are debilitated after Muller's ratchet recover fitness usually by compensation and occasionally by back mutation. In contrast, when persistent VSV is allowed to recover in mammalian cells, all or most mutations accumulated during adaptation to persistence are lost, and a new set of mutations seems to be acquired. These new mutants can be found at low frequency in the alternating or persistent populations and are presumably maintained either by the step of mammalian replication during alternation or by complementation during strict persistence (Novella *et al.*, 2007).

Complementation

Complementation takes place when a virus interacts with a product that is not encoded in its genome, and this used to be a good tool in the early times of virology research to understand and map mutations. Complementation often occurs during natural infections, usually during co-infection of individual cells, and can have significant effects in the outcome of infection and in the evolution of viruses. Complementation is responsible for the accumulation of defective interfering particles during passages at high MOI (Holland, 1991), and it may be the mechanism that led to the generation of segmented and multipartite viruses (Domingo *et al.*, 1997). In general, complementation will promote diversity by hiding deleterious genomes from elimination by natural selection, but the outcome will depend on the type of interaction between a mutated genome and other viral and cellular products. Minimizing natural selection will result in extended survival of deleterious genomes. Mutations in any function that can be trans-complemented results in the extended survival of the mutant, but the dynamics of the competition depend on the function of the mutated nucleotide (Fig. 12.4). We will discuss some of the potential models.

Model 1 corresponds to a mutation in a non-complementable trait, such as a cis-acting element. By definition, complementation cannot operate independently on the level of co-infection or the relative frequency of each competitor, and there will be no density- or frequency- dependent selection. In a competition, the deleterious mutant will decrease in frequency at a constant rate, which will be determined by the selective disadvantage conferred by the mutation.

Model 2 corresponds to a mutation that affects entry. Initially the inferior mutant cannot take advantage of more fit competitor product because the virions cannot physically exchange proteins. Thus, in the first round of infection complementation cannot operate and density-dependent selection will not be observed, and, unless the receptor is limiting, there will be no frequency-dependent selection either. However, in high MOI competitions, once the two competitors replicate inside the same cell during the first passage, there will be phenotypic mixing-and-hiding, and virions of the deleterious mutant will have in their surface superior protein that will help them improve their competitive ability. Thus, if the competition is continued at high MOI, after the first passage there will be a deceleration in the slope of mutant frequency

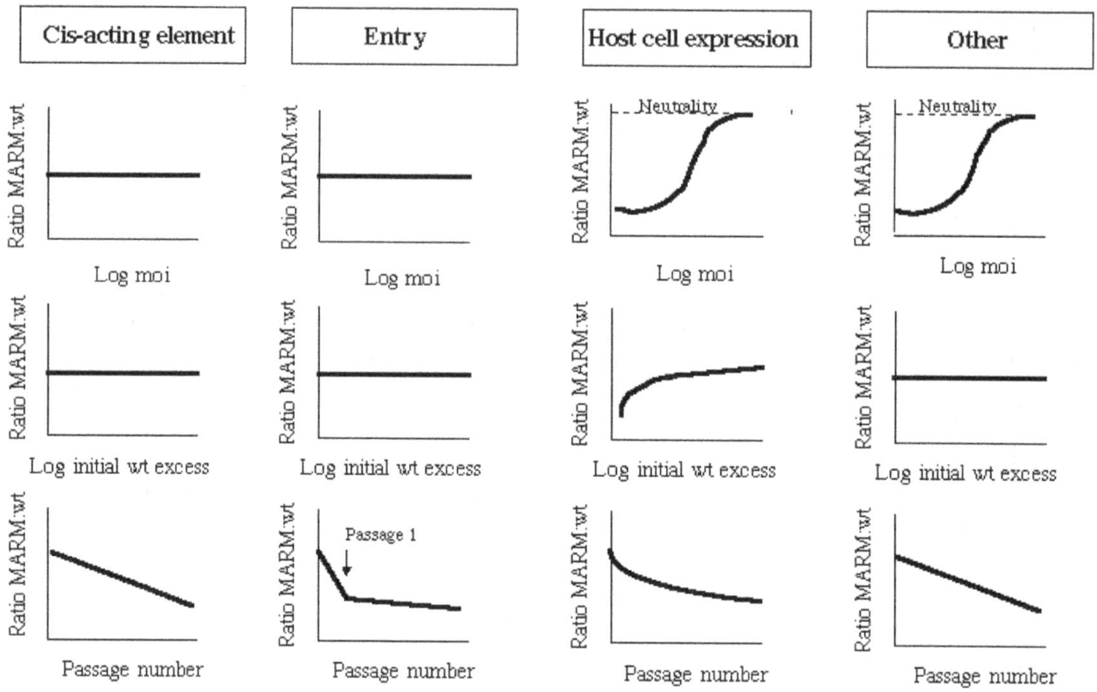

Figure 12.4 Theoretical models can be used to determine the effect of a deleterious mutation on the viral replication cycle. Differences in density-dependent selection (top), frequency-dependent selection at high MOI (middle), or mutant frequency decay during standard competitions (bottom) are expected depending on the function of the mutation.

decrease (Wilke and Novella, 2003). Previously published VSV studies support this model. The frequency of antibody-resistant mutants in viral populations can be greatly underestimated if viral preparations are neutralized before infection. Mutants arise by mutations of WT genomes, but as they leave the cell they will be surrounded by WT protein, which is sensitive to antibodies. However, if plaque assays are done by adding the antibody to the agar overlay, all mutants will be able to make plaques, while WT virus will be unable to form a plaque owing to the presence of the antibody (Holland *et al.*, 1991).

Model 3 reflects the behaviour of a mutant that has an effect on the host cell environment, such as interferon induction/repression. Perhaps these mutations can be complemented in the first round of infection, and this complementation will operate more efficiently the higher the MOI is. The shape of the curve presented at the top of Fig. 12.4 is explained as follows: At low and very low MOI the level of co-infection is negligible, so complementation never operates. At one point the level of co-infection starts being significant, and complementation can operate. Once the MOI is very high, all the cells will be co-infected so additional virus input will not have any effect. Frequency-dependent selection is expected because an excess of the WT will fully trigger (or inhibit) cellular functions, bringing the fitness of the mutant to that of its competitor. For this reason, regular competitions will show a curving in the MARM: WT ratio decay.

Many mutants will be best described by Model 4, based on the averaging of fitness in co-infected cells (rather than bringing up the fitness of the deleterious mutant to that of WT, as in Model 3). In this case we expect density-dependent selection during the first round of infection (as in Model 3), but not frequency-dependent selection. In addition, the loss of the deleterious mutant will take place at a constant rate. MARM N, a mutant with a defect in the L component of the RNA-dependent-RNA polymerase, belongs to this group.

We have done similar work with six

recombinant VSV strains with rearranged gene orders. These rearrangements resulted in changes in mRNA concentration for each viral protein, and all had deleterious effects. However, even the most deleterious of the rearranged strains behave as neutral when the competitions were done at high MOI, and the relationship between MOI and fitness is qualitatively similar to that described for MARM N (Novella *et al.*, 2004). The fact that all the deleterious mutants we tested achieved neutrality at high MOI indicates that viral proteins are unrestrictedly shared during VSV replication.

These models may need fine-tuning, particularly if extended to other viruses. For instance, there may be some restriction to the sharing of viral products or mosaic particles may need to be considered.

References

Alto, B.W., and Turner, P.E. (2010). Consequences of host adaptation for performance of vesicular stomatitis virus in novel thermal environments. Evol. Ecol. *24*, 299–315.

Burch, C.L., and Chao, L. (2000). Evolvability of an RNA virus is determined by its mutational neighbourhood. Nature *406*, 625–628.

Chao, L. (1990). Fitness of RNA virus decreased by Muller's ratchet. Nature *348*, 454–455.

Domingo, E., Menendez-Arias, L., and Holland, J.J. (1997). RNA virus fitness. Rev. Med. Virol. *7*, 87–96.

Duarte, E.A., Clarke, D.K., Moya, A., Domingo, E., and Holland, J.J. (1992). Rapid fitness losses in mammalian RNA virus clones due to Muller's ratchet. Proc. Natl. Acad. Sci. U.S.A. *89*, 6015–6019.

Duarte, E.A., Novella, I.S., Ledesma, S., Clarke, D.K., Moya, A., Elena, S.F., Domingo, E., and Holland, J.J. (1994). Subclonal components of consensus fitness in an RNA virus clone. J. Virol. *68*, 4295–4301.

Dutta, R.N., Rouzine, I.M., Smith, S.D., Wilke, C.O., and Novella, I.S. (2008). Rapid adaptive amplification of preexisting variation in an RNA virus. J. Virol. *82*, 4354–4362.

Eigen, M. (1992). Steps Towards Life (Oxford University Press, New York).

Eigen, M. (1996). On the nature of virus quasispecies. Trends Microbiology *4*, 216–218.

Enzmann, P.J., Kurath, G., Fichtner, D., and Bergmann, S.M. (2005). Infectious hematopoietic necrosis virus: monophyletic origin of European isolates from North American Genogroup M. Dis. Aquat. Organ. *66*, 187–195.

Gerrish, P.J., and Lenski, R.E. (1998). The fate of competing beneficial mutations in an asexual population. Genetica *103*, 127–144.

Holland, J.J. (1991). Defective viral genomes. In Fundamental Virology, Knipe, D.M., and Howley, P.M., eds (Raven Press, New York), pp. 151–165.

Holland, J.J., delaTorre, J.C., Clarke, D.C., and Duarte, E. (1991). Quantitation of relative fitness and great adaptibility of clonal populations of RNA viruses. J. Virol. *65*, 2960–2967.

Kassen, R. (2002). The experimental evolution of specialists, generalists, and the maintenance of diversity. J. Evol. Biol. *15*, 173–190.

Kurath, G., Garver, K.A., Troyer, R.M., Emmenegger, E.J., Einer-Jensen, K., and Anderson, E.D. (2003). Phylogeography of infectious haematopoietic necrosis virus in North America. J. Gen. Virol. *84*, 803–814.

Kurath, G., Garver, K.A., and Batts, W.N. (2006). Virulence comparisons of infectious hematopoietic necrosis virus U and M genogroups in sockeye salmon and rainbow trout. J. Aquat. Anim. Health *18*, 232–243.

Letchworth, G.J., Rodriguez, L.L., and Barrera, J.D.C. (1999). Vesicular stomatitis. Vet. J. *157*, 239–260.

Llewellyn, Z.N., Salman, M.D., Pauszek, S., and Rodriguez, L.L. (2002). Growth and molecular evolution of vesicular stomatitis serotype New Jersey in cells derived from its natural insect-host: evidence for natural adaptation. Virus Res. *89*, 65–73.

Mead, D.G., Ramberg, F.B., Besselsen, D.G., and Mare, C.J. (2000). Transmission of vesicular stomatitis virus from infected to noninfected black flies co-feeding on nonviremic deer mice. Science *287*, 485–487.

Muller, H.J. (1964). The Relation of Recombination to Mutational Advance. Mutat. Res. *106*, 2–9.

Novella, I.S. (2004). Negative effect of genetic bottlenecks on the adaptability of vesicular stomatitis virus. J. Mol. Biol. *336*, 61–67.

Novella, I.S., and Ebendick-Corpus, B.E. (2004). Molecular basis of fitness loss and fitness recovery in vesicular stomatitis virus. J. Mol. Biol. *342*, 1423–1430.

Novella, I.S., Clarke, D.K., Quer, J., Duarte, E.A., Lee, C.H., Weaver, S.C., Elena, S.F., Moya, A., Domingo, E., and Holland, J.J. (1995a). Extreme fitness differences in mammalian and insect hosts after continuous replication of vesicular stomatitis virus in sandfly cells. J. Virol. *69*, 6805–6809.

Novella, I.S., Duarte, E.A., Elena, S.F., Moya, A., Domingo, E., and Holland, J.J. (1995b). Exponential increases of RNA virus fitness during large population transmissions. Proc. Natl. Acad. Sci. U.S.A. *92*, 5841–5844.

Novella, I.S., Elena, S.F., Moya, A., Domingo, E., and Holland, J.J. (1995c). Size of genetic bottlenecks leading to virus fitness loss is determined by mean initial population fitness. J. Virol. *69*, 2869–2872.

Novella, I.S., Hershey, C.L., Escarmis, C., Domingo, E., and Holland, J.J. (1999a). Lack of evolutionary stasis during alternating replication of an arbovirus in insect and mammalian cells. J. Mol. Biol. *287*, 459–465.

Novella, I.S., Quer, J., Domingo, E., and Holland, J.J. (1999b). Exponential fitness gains of RNA virus populations are limited by bottleneck effects. J. Virol. *73*, 1668–1671.

Novella, I.S., Ball, L.A., and Wertz, G.W. (2004). Fitness analyses of vesicular stomatitis strains with rearranged genomes reveal replicative disadvantages. J. Virol. *78*, 9837–9841.

Novella, I.S., Ebendick-Corpus, B.E., Zarate, S., and Miller, E.L. (2007). Emergence of mammalian cell-adapted vesicular stomatitis virus from persistent infections of insect vector cells. J. Virol. *81*, 6664–6668.

Novella, I.S., Dutta, R.N., and Wilke, C.O. (2008). A linear relationship between fitness and the logarithm of the critical bottleneck size in vesicular stomatitis virus populations. J. Virol. *82*, 12589–12590.

Novella, I.S., Presloid, J.B., Zhou, T., Smith-Tsurkan, S.D., Ebendick-Corpus, B.E., Dutta, R.N., Lust, K.L., and Wilke, C.O. (2010). Genomic evolution of vesicular stomatitis virus strains with differences in adaptability. J. Virol. *84*, 4960–4968.

Novella, I.S., Presloid, J.B., Smith, S.D., and Wilke, C.O. (2011). Specific and non-specific host adaptation during arboviral experimental evolution. J. Mol. Microbiol. Biotechnol. *21*, 59–69.

Peleg, J. (1971). Growth of viruses in arthropod cell cultures: applications. I. Attenuation of Semliki Forest (SF) virus in continuously cultured Aedes aegypti mosquito cells (Peleg) as a step in production of vaccines. Curr. Top. Microbiol. Immunol. *55*, 155–161.

Penaranda, M.M.D., Wargo, A.R., and Kurath, G. (2011). In vivo fitness correlates with host-specific virulence of Infectious hematopoietic necrosis virus (IHNV) in sockeye salmon and rainbow trout. Virology *417*, 312–319.

Presloid, J.B., Ebendick-Corpus, B.E., Zarate, S., and Novella, I.S. (2008). Antagonistic pleiotropy involving promoter sequences in a virus. J. Mol. Biol. *382*, 342–352.

Remold, S.K., Rambaut, A., and Turner, P.E. (2008). Evolutionary genomics of host adaptation in vesicular stomatitis virus. Mol. Biol. Evol. *25*, 1138–1147.

Rouzine, I.M., Wakeley, J., and Coffin, J.M. (2003). The solitary wave of asexual evolution. Proc. Natl. Acad. Sci. U.S.A. *100*, 587–592.

Sanjuan, R., Cuevas, J.M., Furio, V., Holmes, E.C., and Moya, A. (2007). Selection for robustness in mutagenized RNA viruses. PLOS Genetics *3*, e93.

Scott, T.W., Weaver, S.C., and Mallampalli, V.L. (1994). Evolution of mosquito-borne viruses. In Evolutionary Biology of Viruses, Morse, S.S., ed. (Raven Press, New York), pp. 293–324.

Singh, K.R., Bhat, U.K., and Paul, S.D. (1971). Adaptation of Singh's *Aedes albopictus* (ATC-15) cell line to media with calf and goat sera and their susceptibility to infection with some arboviruses. Indian J. Med. Res. *59*, 31–34.

Smith-Tsurkan, S.D., Novella, I.S., and Wilke, C.O. (2010). Incongruent fitness landscapes, not tradeoffs, dominate the adaptation of vesicular stomatitis virus to novel host types. J. Gen. Virol. *91*, 1484–1493.

Taylor, W.P., and Marshall, I.D. (1975). Adaptation studies with Ross river virus: laboratory mice and cell cultures. J. Gen. Virol. *28*, 59–72.

Tesh, R.B., Chaniotis, B.N., and Johnson, K.M. (1972). Vesicular stomatitis virus (Indiana serotype): transovarial transmission by phlebotomine sandflies. Science *175*, 1477–1479.

de la Torre, J.C., and Holland, J.J. (1990). RNA virus quasispecies populations can suppress vastly superior mutant progeny. J. Virol. *64*, 6278–6281.

Troyer, R.M., and Kurath, G. (2003). Molecular epidemiology of infectious hematopoietic necrosis virus reveals complex virus traffic and evolution within southern Idaho aquaculture. Dis. Aquat. Organ. *55*, 175–185.

Troyer, R.M., Garver, K.A., Ranson, J.C., Wargo, A.R., and Kurath, G. (2008). *In vivo* virus growth competition assays demonstrate equal fitness of fish rhabdovirus strains that co-circulate in aquaculture. Virus Res. *137*, 179–188.

Tsimring, L.S., Levine, H., and Kessler, D.A. (1996). RNA virus evolution via a fitness-space model. Phys. Rev. Lett. *76*, 4440–4443.

Turner, P.E., and Elena, S.F. (2000). Cost of host radiation in an RNA virus. Genetics *156*, 1465–1470.

Turner, P.E., Montville, R., Froissart, R., Remold, S.K., and Tenaillon, O. (2005). Evolution of mutational robustness in an RNA virus. PLoS Biol. *3*, 1939–1945.

Turner, P.E., Morales, N.M., Alto, B.W., and Remold, S.K. (2010). Role of evolved host breadth in the initial emergence of an RNA virus. Evolution *64*, 3273–3286.

Wargo, A.R., and Kurath, G. (2011). *In Vivo* fitness associated with high virulence in a vertebrate virus is a complex trait regulated by host entry, replication, and shedding. J. Virol. *85*, 3959–3967.

Wargo, A.R., Garver, K.A., and Kurath, G. (2010). Virulence correlates with fitness in vivo for two M group genotypes of infectious hematopoietic necrosis virus (IHNV). Virology *404*, 51–58.

Weaver, S.C., Hagenbaugh, A., Bellew, L.A., Netesov, S.V., Volchkov, V.E., Chang, G.J., Clarke, D.K., Gousset, L., Scott, T.W., Trent, D.W., and *et al.* (1993). A comparison of the nucleotide sequences of eastern and western equine encephalomyelitis viruses with those of other alphaviruses and related RNA viruses. Virology *197*, 375–390.

Webster, R.G., Bean, W.J., Gorman, O.T., Chambers, T.M., and Kawaoka, Y. (1992). Evolution and ecology of influenza A viruses. Microbiol. Rev. *56*, 152–179.

Wilke, C.O. (2005). Quasispecies theory in the context of population genetics. BMC Evol. Biol. *5*, 44.

Wilke, C.O., and Novella, I.S. (2003). Phenotypic mixing and hiding may contribute to memory in viral quasispecies. BMC Microbiol. *3*, 11.

Wilke, C.O., Wang, J.L., Ofria, C., Lenski, R.E., and Adami, C. (2001). Evolution of digital organisms at high mutation rates leads to survival of the flattest. Nature *412*, 331–333.

Zarate, S., and Novella, I.S. (2004). Vesicular stomatitis virus evolution during alternation between persistent infection in insect cells and acute infection in mammalian cells is dominated by the persistence phase. J. Virol. *78*, 12236–12242.

Rabies Prevention and Control: Advances and Challenges

13

Charles E. Rupprecht and Dennis Slate

Abstract

Rabies prevention and control techniques improved greatly over the twentieth century. Gradual understanding of the aetiology, transmission details, immunological facets, and inter-specific dynamics among mammalian reservoirs revolutionized current intervention theory and practice. Technical progress appears even more notable given the importance of the global distribution of lyssaviruses, the discovery of more than a dozen causative agents of rabies, a broad mammalian host range, and a case fatality greater than any conventional infectious disease. Implementation of pure, potent, safe and efficacious vaccines produced major impacts on the interruption of viral transmission to humans and perpetuation among animals. Use of safe and effective vaccine-laden baits targeting wild carnivores led to substantial case reductions in Europe and North America. Continued vigilance, enhanced de-centralized surveillance, rapid diagnosis, viral characterization, and implementation of contingency actions, minimize the chance for disease re-introduction, which can thwart programme success. New biologics, baits, and efficient distribution techniques are adaptable for vaccination programmes targeting free-ranging carnivores over diverse landscapes. Although not a candidate for eradication at this time, human rabies is readily preventable, canine rabies can be eliminated, and wildlife rabies can be controlled. Dynamic multidisciplinary partnerships and international advocacy have created new hope in the face of neglect for many affected human and animal populations in developing regions. Optimization of prevention and control methods, for application along the human, domestic animal, and wildlife interface, continue to make rabies a model for comparison among other emerging zoonoses, despite the considerable burden and economic challenges faced.

Introduction

The *Rhabdoviridae* consists of more than 200 viruses associated with plants, invertebrates, and vertebrates, many of which have not been well characterized, either at the individual molecular level or epidemiologically within reservoir populations, making discussions on broad prevention and control of agents in this Family extremely difficult (Walker, 2005; Drolet *et al.*, 2009; Enzmann *et al.*, 2010; Kim *et al.*, 2011). Given the great diversity of described agents and hosts, any thorough consideration of a more holistic treatment of the *Rhabdoviridae* regarding relevant interventions against the diseases caused by these viruses as a whole would be beyond the scope of a simple review. Rather, this chapter will focus upon prevention and control of viruses in the Genus *Lyssavirus,* arguably the most important rhabdovirus taxonomic group from the standpoint of global human and veterinary medicine. More than a dozen lyssaviruses have been described since the 1950s (Kuzmin *et al.*, 2010; Marston *et al.*, 2012). All are highly neurotropic agents that cause the acute, progressive zoonotic encephalomyelitis known as rabies. Simply considered, after *Lyssavirus* replication in the central nervous system (CNS), virions are excreted intermittently for days to weeks in the saliva of clinically ill subjects, and transmitted routinely via a bite (Jackson and Wunner, 2007). The incubation period, from exposure to disease onset, is extremely variable, predicated in part by viral dose and site of inoculation, and usually lasting several weeks to months, but may range from days to years. In contrast to the prolonged incubation, the morbidity period is much shorter, usually on the order of days, before the subject dies. The fearsome clinical syndrome, often characterized as hydrophobia in humans and maniacal behaviour in animals, is a prominent historical hallmark of this disease over several millennia. Regardless of which *Lyssavirus* is involved, basic biomedical and epizootiological features of the disease appear similar. This chapter will largely concentrate on the type species of the genus, Rabies virus, given its broad taxonomic and geographical distribution and practical applied public health, veterinary, and wildlife significance (Rupprecht *et al.*, 2006).

With the exception of Antarctica, rabies is found on all inhabited continents (Fig. 13.1). This global pattern is reflected in the perpetuation of several different lyssaviruses in widely distributed mammalian hosts. Although not all extant species have been diagnosed rabid, all mammals are believed to be susceptible to infection. Within the Class *Mammalia*, the primary reservoirs responsible for disease perpetuation reside in the orders Carnivora and the Chiroptera. The present distribution of rabies is associated with the natural zoogeographical occurrence of major taxa over geological time, superimposed with the more recent historical anthropogenic translocation of domestic species and wildlife during the last several centuries (Talbi *et al.*, 2010; Vos *et al.*, 2011). Because disease re-introduction may occur rapidly from infected to non-affected areas, the current distribution of rabies may be reported and maintained electronically at several sites, such as those maintained by the World Health Organization (WHO), the Pan American Health Organization (PAHO), and information related to travellers' health by the Centers for Disease Control and Prevention (CDC), among others (http://wwwnc.cdc.gov/travel).

Unlike other viral diseases, such as polio, rinderpest and smallpox, rabies is not a candidate for true global eradication at this time (Rupprecht *et al.*, 2008). However, human rabies is preventable, canine rabies is a major problem that can be eliminated, and wildlife rabies can be controlled regionally. Also, although the terms are often used interchangeably, and there are areas of overlap, prevention activities tend to be proactive before

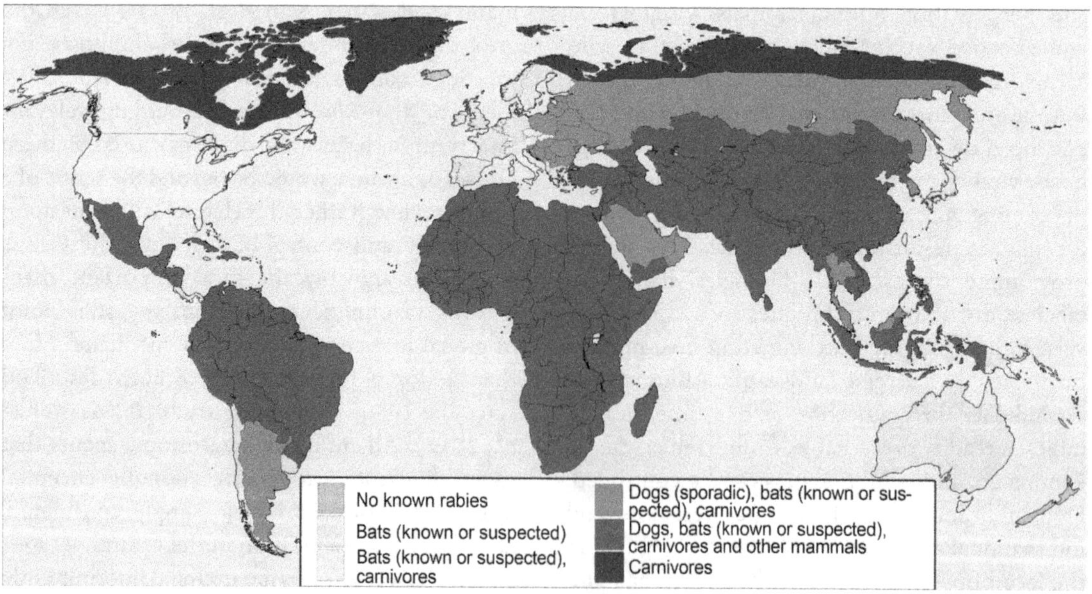

Figure 13.1 Global rabies distribution by reservoir. (Map redesigned courtesy of I.V. Kuzmin). A colour version of this figure can be found in the plate section at the end of the book.

an infectious event occurs, rather than necessitating a control response (Table 13.1). To maintain such a global vision on actual elimination, prevention, and control requires active collaborations. Multiple groups provide health communications resources on multiple infectious disease topics such as rabies, and include the CDC (http://www.cdc.gov/), FAO, Food and Agriculture Organization of the United Nations (http://www.fao.org/), the OIE, World Organization for Animal Health (http://www.oie.int/), and the WHO (http://www.who.int/en/).

The objective of this chapter is to review topics related to the prevention and control of rabies in humans, domestic animals, and wildlife, especially focusing upon published recommendations, reports, and research from the past five years, and to provide a perspective on available opportunities and remaining challenges (Table 13.2).

Human rabies prevention

As a zoonosis, public health programmes targeting rabies may be conceptualized via a barrier-systems approach to managing the interactions among different mammalian species (Fig. 13.2). Hence, rabies in humans is preventable by promotion of behaviours to avoid exposures to suspect rabid animals, and administration of emergency actions if such exposure occurs. Timely and appropriate prophylaxis prevents the occurrence of rabies in

Table 13.1 Comparison of a 'One Health' promotion consideration versus programmatic reactions to rabies

Prevention plans	Control actions
Primary public education	Public service/media announcements
Professional biomedical training	Emergency operations
International entry requirements	Local movement restrictions/prohibitions
Home maintenance	Remedial animal exclusion
Personal protection equipment	Wound care
Occupational pre-exposure vaccination	Human post-exposure prophylaxis
Responsible pet ownership	Stray animal capture
Leash laws	Mandatory confinement
Animal reproductive planning	Shelter management
Veterinary vaccinations	Treatment/euthanasia practices

Table 13.2 Selected examples of potential solutions to perceived problems in rabies prevention and control

Challenge	Opportunity
Increased human exposures	Integrated primary education on animal bite avoidance
High human vaccine costs	Optimized intradermal vaccine delivery
Complicated vaccine schedules	Harmonized, simplified, rational evidence-based boosters
Low global supply of RIG	Production of anti-glycoprotein monoclonal antibodies
Underserved Amazon populations	Pre-exposure vaccination in paediatric schedules
Canine 'overpopulation'	Population management, including immuno-contraception
Lack of political will for control	Renewed advocacy and local champions in One Health
Diverse phylogroup II viruses	Pan-lyssavirus vaccines to broaden vaccine cross-reactivity
Free-ranging wild carnivores	Strategic use of modern oral vaccine-laden bait programmes
Bats as primary reservoirs	Novel design of ecologically sound interventions
Virus–host shifts	Predict, detect, and respond to disease emergence
Extreme rabies case fatality	Focus on pathogenesis, anti-viral drugs, and animal models

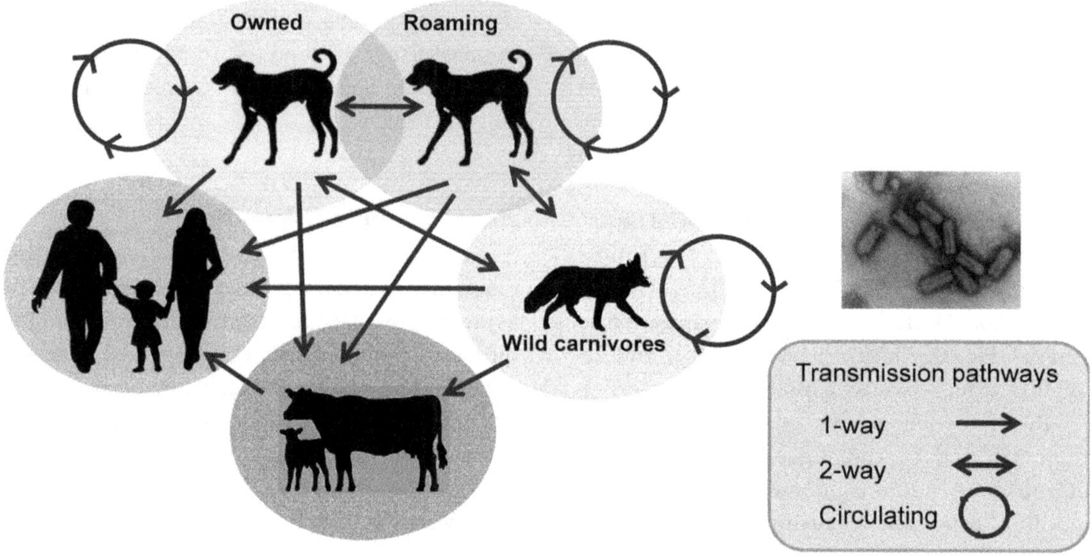

Figure 13.2 Schematic rabies perpetuation and transmission.

humans, but is unnecessary if other basic public health and veterinary programmes are operative for functional application to other species. For example, continued attention on practical and accurate health education about rabies, domestic animal vaccination, responsible pet care, modern and humane animal population management, prompt diagnostic follow-up, and environmental remediation can minimize the advent of animal exposures, alleviate risks after animal exposure, and obviate the need for any rabies prophylaxis in humans (NASPHV, 2011). Often, such basic outreach is lacking or minimal at best, among both professional and lay groups, so it is understandable how exposures occur and mismanagement compounds errors downstream of an event, if systems are lacking or dysfunctional (Salahuddin *et al.*, 2011). Similarly, the actual burden of rabies, either measured in the tens of millions of dead animals (i.e. even one densely populated cave of free-tailed bats may produce tens of thousands of rabid conspecifics over a single season) or exposed humans (i.e. some estimates range from 50,000–70,000 fatal cases annually) is largely unknown, with resulting broad confidence intervals as to the realized numbers (Knobel *et al.*, 2005). As with many neglected conditions, rabies suffers from a cycle of indifference, with laboratory-based surveillance frequently non-existent for a basic 'disease of nature', and a lack of cooperation among often competing health sectors, relying upon the other for essential responsibility (Fig. 13.3). Put simply, the greatest costs of rabies lay in the public health area associated with *Homo sapiens* and rabies prophylaxis, but as the dog is not an economic commodity (unlike livestock), the agricultural sector is often less than engaged with pet vaccination, which essentially protects people. Rabies is the only routine diagnostic test applied to animals which directly determines medical intervention in an exposed person. Clearly, improving basic laboratory-based surveillance is one of the most critical facets that require improvement in developing countries (Dürr *et al.*, 2008).

Rabies biologics for human prophylaxis

Rabies biologics for humans include two basic types of products – vaccines and rabies immune globulins (RIG). The former are used for both pre-exposure (immunization prior to viral exposure) and post-exposure prophylaxis (PEP), whereas the latter is only used once during PEP (Briggs, 2011; Both *et al.*, 2012). From the time of Pasteur until about the mid-twentieth century, all human rabies vaccines were based upon virus

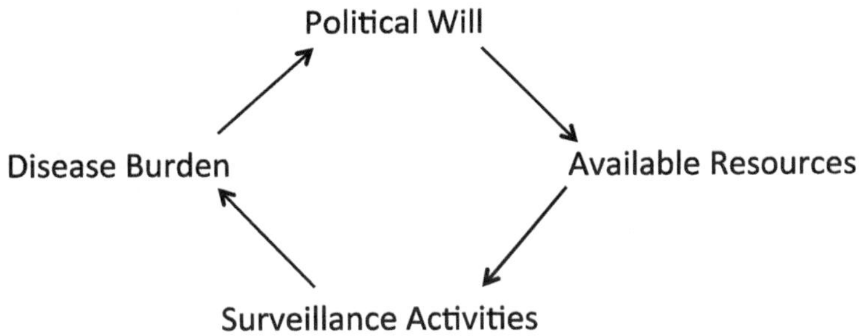

Figure 13.3 The surveillance cycle of indifference.

grown in the brains of inoculated animals (Wu *et al.*, 2011). Over the past 30 years, several much safer and potent cell culture vaccines (CCV) have been licensed for use in humans, including: human diploid cell vaccine (HDCV), purified chick embryo cell vaccine (PCECV), purified Vero cell vaccine (PVRV), and purified duck embryo vaccine (PDEV). Such modern CCV induce an active immune response that includes the rapid production of rabies virus neutralizing antibodies (VNA). This active antibody response requires approximately 7–10 days to develop, and detectable rabies VNA generally persist for several years (Fayaz *et al.*, 2011). Typically, a vaccination series is initiated and completed with a single CCV product, based upon published data and label claims. However, there are no clinical trials that document a change in efficacy or the frequency of adverse reactions when a vaccine series is initiated with one product (e.g. HDCV), but completed with another (e.g. PCECV). In contrast to CCV, the administration of RIG is intended to provide an immediate supply of rabies VNA until the production of active immunity in response to vaccine administration. Use of RIG provides a rapid immunity that persists for only a short time (half-life of several days to weeks), with a basic strategy to intervene prudently, before viral access to the nervous system (Fig. 13.4). Unfortunately, modern human rabies biologics are not always widely available or affordable for use as originally marketed. Hence, reduced schedules (i.e. lessening the absolute number of prime-boost vaccine doses applied over time) and alternative sites of application (i.e. intradermal) are two strategies employed to make modern CCV more accessible to people at highest risk in dog rabies enzootic areas (Wilde and Lumlertdacha, 2011). Monoclonal antibodies, produced in humans or in other species, offer several theoretical advantages over RIG, and are under current clinical development as another technological application expected over the next five years (Smith *et al.*, 2011). Modern rabies CCV protect against all known variants of rabies virus, but not against all lyssaviruses. New lyssaviruses will continue to emerge, as shown by the recent descriptions in Europe and Africa (Kuzmin *et al.*, 2010). Conventional rabies PEP may offer little or no protection against these new viruses. Predictably, vaccine protection may be inversely related to the genetic distance of any new isolates. Traditional fixed rabies virus preparations could be broadened for cross reactivity, to lessen the potential impact of these new lyssaviruses on human and domestic animal health (Smith *et al.*, 2011; Wu *et al.*, 2011).

Rationale for prophylaxis

Both the Advisory Committee on Immunization Practices (ACIP) and WHO recommend that human PEP should include timely and thorough wound cleansing to remove debris and chemically inactivate any deposited virus, followed by local infiltration with RIG, and prompt vaccination with a CCV (WHO, 2005; CDC, 2008). Human rabies biologics are critical life-saving resources that are periodically in limited supply, so a careful assessment of the proven benefit, and potential harm, associated with administering PEP should be conducted in each situation involving a likely exposure (Briggs, 2011). The relative medical benefits (or harm from adverse effects) of these

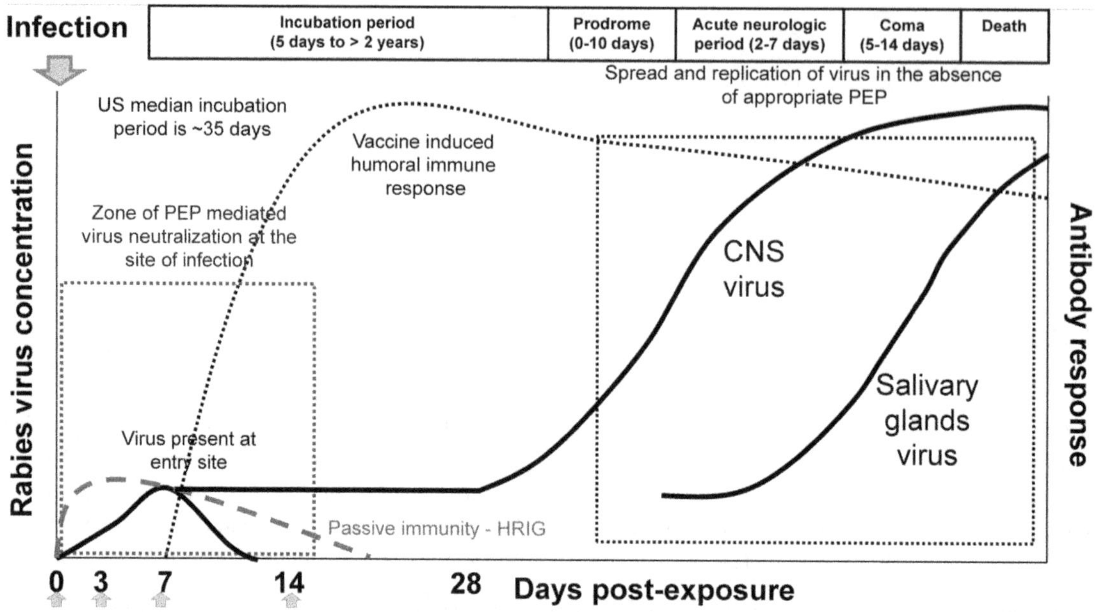

Figure 13.4 The imposition of prophylaxis upon rabies virus pathogenesis.

biologics differ based on the relative risk of infection. Recommendations regarding PEP are dependent in part upon: the type of viral exposure; epizootiology of animal rabies where the event occurred; species of animal; and circumstances of the incident. The decision to initiate PEP also depends on the availability of the suspect animal for observation or laboratory diagnosis. Circumstances surrounding each encounter with a suspect animal vary widely, from severe dog bites to the head of a child to an elderly person that may awake to a dead bat in bed. Clinicians should seek assistance from experienced public health officials for evaluating human exposures or determining the need for management in situations that are not routine.

Rabies exposures

Lyssaviruses perpetuate in animal populations by infected bites and non-bite exposures. Humans become infected most often through interactions directly with rabid animals (Brito *et al.*, 2011; Hossain *et al.*, 2011). The most dangerous route of virus exposure is through the bite of a rabid animal, particularly on the head and face. Exposures by rabid wolves are especially dangerous, given the ferocity of such attacks. An exposure might also occur if the virus, from saliva or other potentially infectious material, such as neural tissue, is introduced into fresh, open cuts in skin or onto mucous membranes (i.e. non-bite exposure). Often, farmers and veterinarians may be exposed in this way when examining livestock for evidence of a foreign body in the mouth or throat of a choking animal. Indirect contact, such as petting or handling an animal, contact with blood, urine or faeces, and contact of saliva with intact skin, do not constitute exposures, and do not require PEP. When true exposure occurs, the likelihood of infection varies with the nature and extent of that exposure. In contrast to the more obvious bites by carnivores, exposures to bats deserve special assessment (Messenger *et al.*, 2003). Bats may pose a greater risk of infecting humans under certain circumstances that might be considered inconsequential from a human perspective (i.e. a minor bite or lesion). Human-to-human transmission occurs almost exclusively owing to organ or tissue transplantation (Maier *et al.*, 2010).

Bites

Lyssaviruses are highly neurotropic and routinely animal bites deposit virus-laden saliva at the level of the neuromuscular junction for reception and peripheral viral entry (Dietzschold *et al.*, 2008). Any penetration of the skin by the teeth of a rabid animal constitutes a bite exposure. All bites, regardless of anatomical site, or perceived degree of trauma, represent a risk (CDC, 2012). The risk of transmission varies in part with the species of biting animal, the anatomic site of the bite, and the severity of the wound. Bites by bats may inflict rather minor injury compared to larger bodied carnivores, and such lesions may be more difficult to detect (CDC, 2011).

Non-bite exposures

Nearly all cases are due to an animal bite, and non-bite exposures rarely cause rabies (Dimaano *et al.*, 2011). Occasional reports of non-bite viral transmission suggest that such exposures require assessment to determine if there is reason to consider human prophylaxis. Non-bite exposures of highest risk include recipients of corneas, solid organs, and vascular tissue transplanted from patients who died of rabies, and exposure to aerosolized virus (Jackson and Wunner, 2007). At least two cases of rabies were attributed to aerosol exposures in laboratories, and two cases have been attributed to possible airborne exposures in caves containing millions of bats, although other explanations cannot be excluded (Gibbons, 2002). The contamination of open wounds or abrasions (including scratches) or mucous membranes with saliva or other potentially infectious material (such as neural tissue) from a rabid animal also constitutes a non-bite exposure. Lyssaviruses are inactivated readily by desiccation, ultraviolet radiation, and other physical and chemical factors, and do not persist in the environment. Non-bite exposures, other than organ or tissue transplants, have almost never been proven to cause rabies. In theory, it may be possible for someone to become infected with rabies by drinking raw milk or milk products from an infected animal. However, no human or animal has been reported to develop rabies in this manner and no studies have been published that demonstrated the presence of virus in bovine milk (CDC, 1999). While not recommended, healthy people who consume raw milk from a rabid animal appear to be at minimal risk of developing rabies. Milk that has been pasteurized presents no risk for virus transmission.

Rabies epidemiology, exposure circumstances, and species characteristics

The relative risk of infection varies broadly throughout the world, and not all interactions with animals warrant prophylaxis (Table 13.3). The risk of viral transmission in different scenarios has been estimated recently to range from > 1 in 20 to < 1 in a million (Vaidya *et al.*, 2010). For example, an unprovoked domestic animal exposure may be more likely than a provoked attack to indicate that an animal is rabid, although one may not always easily differentiate such behaviours. Generally, bites inflicted after attempts to remove food from an apparently healthy animal or disturb its young may be regarded as provoked actions. Other factors to consider when evaluating potential rabies exposures include the overall epidemiology of rabies in the area (e.g. presumed 'free', ongoing outbreak, low level enzootic), the biting animal's history (e.g. feral, owned and well maintained) and current health status (e.g. abnormal behaviour, signs of encephalitis), and the potential for the animal to be exposed to rabies virus (e.g. presence of an unexplained wound or history of exposure to a rabid animal). A dog, cat, or ferret with a history of current vaccination (i.e. no significant gaps in vaccination coverage) is unlikely to become infected with rabies, although no vaccine provides complete efficacy (Thiptara *et al.*, 2011). Even after an initial rabies vaccination, young or naïve animals remain at risk for the development of rabies owing to potential exposures prior to vaccination, before adequate induction of immunity during the month after vaccination. Other factors will determine the ultimate outcome after immunization, and a history of vaccination is not identical to a protected animal (David *et al.*, 2010).

Knowledge of the exposure date and domestic pet status is also useful in human PEP considerations. Rabies virus tends to be excreted intermittently in rabid dog, cat, and ferret saliva concomitant with, or a few days prior to illness (Niezgoda *et al.*, 1998; Zhang *et al.*, 2008). Any healthy domestic dog, cat, or ferret in a rabies enzootic region that exposes a person should be confined and observed for 10 days. Animals that remain healthy 10 days after

Table 13.3 Rabies post-exposure prophylaxis (PEP) guide by animal source (adapted from current ACIP/WHO recommendations)

Animal type	Evaluation and disposition of animal	PEP recommendations
Dogs, cats and ferrets	Healthy and available for ~10 days' observation	Persons should not begin prophylaxis unless animal develops clinical signs of rabies[a]
	Rabid or suspected rabid	Immediately begin PEP
	Unknown (e.g. escaped)	Consult public health officials
Reservoir carnivores (foxes, raccoons, skunks, mongoose, etc.); bats[d]	Regarded as rabid unless animal proven negative by laboratory tests[b]	Consider immediate PEP[c]
Livestock, small rodents (rabbits and hares), large rodents (woodchucks and beavers), and other mammals	Consider individually	Consult public health officials. Bites of squirrels, hamsters, guinea pigs, gerbils, chipmunks, rats, mice, other small rodents, rabbits, and hares almost never require PEP

[a]During the 10-day observation period, begin PEP at the first sign of rabies in a dog, cat, or ferret that has bitten someone. If the animal exhibits clinical signs of rabies, it should be euthanized immediately and tested.
[b]The animal should be euthanized and tested as soon as possible. Holding for observation is not recommended.
[c]Modern PEP should be initiated as soon as possible following exposure to such wildlife unless the animal is available for diagnosis and public health authorities are facilitating expeditious laboratory testing or it is already known that brain material from the animal has tested negative. In addition to the availability of the animal for testing, the logistics of specimen preparation and transfer to a laboratory, and arrangement of prompt diagnostic testing, other factors that might influence the urgency of decision-making regarding initiation of PEP prior to diagnostic results include the species of the animal, the general appearance and behaviour of the animal, whether the encounter was provoked by the presence of a human, and the severity and location of bites. Discontinue vaccine if an appropriate laboratory diagnostic test (i.e. the direct fluorescent antibody test) is negative.
[d]Any bite by a bat is considered a risk for rabies acquisition.

a bite are not excreting virus in saliva, and are not considered infectious at the time of the bite. Ideally, all domestic dogs, cats and ferrets kept as pets should be currently vaccinated against rabies, but should still be observed after contact, because vaccine failures have occurred (Murray *et al.*, 2009). Any illness during confinement should be evaluated by a qualified veterinarian and reported immediately to the local public health unit. If signs of rabies develop, PEP of the bite victim should begin. The animal should be euthanized, its head removed and shipped under refrigeration, for examination by a qualified diagnostic laboratory. If the biting animal is a stray or unwanted, the animal should either be confined and observed for 10 days or euthanized immediately and submitted for rabies virus diagnosis. Besides dogs, cats, and ferrets, little information is available about the excretion of virus in other domestic species. Management of domestic mammals such as horses, swine and ruminants that expose a person depends on the species of animal involved, the circumstances of the biting incident, the epizootiology of rabies in the area, the current health status of the animal responsible for the bite, and the potential that the animal was exposed to rabies. The length of time between rabies virus appearing in the saliva and onset of clinical signs is unknown for wildlife. In particular, wild carnivores and bats that bite humans should be euthanized and tested as soon as possible. Holding for observation is not acceptable. After exposure to wildlife (including hybrids) in which rabies is suspected, immediate human PEP is warranted in most circumstances.

Wildlife

Foxes, jackals, mongoose, raccoons, and skunks are wild carnivores most often found infected with rabies (Jackson and Wunner, 2007). Rabid bats have been documented throughout the Americas, Africa, and Eurasia (Banyard *et al.*, 2011; Rupprecht *et al.*, 2011). Recent data suggest that transmission of lyssaviruses can occur from minor, seemingly underappreciated, or unrecognized bites from bats, and laboratory data support a hypothesis that bat rabies virus variants associated with certain species have biological characteristics that might allow a higher likelihood of infection after superficial inoculation, such as into cells of epidermal origin (Dietzschold *et al.*, 2000; van Thiel *et al.*, 2009).

Human and domestic animal contact with wildlife should be minimized. Wildlife should never be handled by untrained and unvaccinated persons. All bites by such wildlife should be considered exposures to virus. Prophylaxis should be initiated as soon as possible following exposure, unless (1) the animal is available for immediate diagnosis and public health authorities are facilitating expeditious laboratory testing or (2) it is already known that brain tissue from the animal has tested negative. In addition to the availability of the animal for diagnostic testing, the logistics of specimen preparation and transfer to a laboratory, and arrangement of prompt diagnostic testing, other factors that might influence the urgency of decision-making regarding the initiation of prophylaxis prior to diagnostic results include the species of the animal, the general appearance and behaviour of the animal, whether the encounter was provoked by the presence of a human, and the severity and location of bites. Suggestive clinical signs of rabies among wildlife cannot be interpreted reliably. Any such animal that exposes a person should be euthanized as soon as possible (without unnecessary damage to the head, such as from a gunshot wound) and the brain should be submitted for rabies diagnosis. If the results of testing are negative, prophylaxis is not necessary. Unlike carnivores and bats, small mammals, such as rodents, are not reservoirs of rabies. Most rodents (e.g. squirrels, chipmunks, rats, mice, hamsters, guinea pigs, and gerbils) and lagomorphs (e.g. rabbits and hares) are almost never found rabid and have not been known to transmit rabies to humans (Childs *et al.*, 1997). In cases involving rodents, the public health department should be consulted before a decision is made to initiate prophylaxis (Eidson *et al.*, 2005). Wildlife should not be kept as pets. Humans who work with wildlife in licensed research facilities or accredited zoological parks should be educated on preventing bites and should receive pre-exposure vaccination. Rabies exposures to these animal handlers may require booster vaccinations in lieu of euthanasia and testing of the animal depending on employment requirements, conservation management guidelines, and public health recommendations.

Bat exposures

Despite a low rate of infection among bats that are tested for lyssaviruses during routine public health surveillance, potential rabies exposure from contact with a bat requires a thorough evaluation and community education to prevent exposures (Carroll *et al.*, 2011; Kikuti *et al.*, 2011; Robertson *et al.*, 2011). The risk of rabies resulting from an encounter with a bat may be difficult to determine for several reasons: a more limited injury can be inflicted by a bat bite (in contrast to more obvious wounds caused by the bite of carnivores); an often inaccurate recall of a bat encounter that might have occurred several weeks or months earlier; and evidence that some bat-related viruses may be more likely to result in infection after inoculation into superficial epidermal layers. For these reasons, direct contact with a bat should be evaluated for an exposure, unless the person can be reasonably certain a bite, scratch, or mucous membrane exposure did not occur. Rabies PEP should be administered when exposure is documented, unless the bat is available and is proven negative by diagnostic testing. In addition, finding a bat in the same room as a person who might be unaware that a bite or direct contact had occurred (e.g. a deeply sleeping person awakens to find a bat in the room, or an adult witnesses a bat in the room with a previously unattended child, mentally disabled person, or intoxicated person) should be considered an exposure, unless rabies is ruled out by diagnostic testing of the bat, or circumstances suggest it is unlikely that an exposure took place. Other household members may be assessed for potential exposure to the bat. However, unless these persons meet the above-described criteria, they should not be considered exposed. General circumstances that would make it unlikely that an undetected exposure occurred include: the simple observation of bats roosting or flying at the top of a large room or in a room open to the outdoors; the observation of bats outdoors or in a setting where bats may normally be present; or, in situations in which the use of protective covers such as mosquito netting would reasonably be expected to preclude unnoticed contact. In all instances of potential human exposures involving bats, the bat in question should be safely collected, if possible, and submitted for rabies diagnosis. Based upon surveillance data, most submitted bats will not be rabid and such timely diagnostic assessments rules out the need for PEP (Blanton *et al.*, 2011). Clustering of human cases associated with bat exposures has not been reported, with the exception of outbreaks associated with predation by vampire bats (Carvalho-Costa *et al.*, 2012).

Basic rabies post-exposure prophylaxis practices

Administration of rabies post-exposure prophylaxis (PEP) is a medical urgency, and medical decisions must not be delayed (Briggs, 2011). Incubation periods of greater than one year have been reported, so when a documented exposure has occurred, prophylaxis is indicated regardless of the length of the delay, provided that compatible clinical signs of rabies are not already present in the patient (CDC, 2008). Rabies PEP should always include thorough wound care, and administration of both RIG and vaccine in the naïve patient (Table 13.4). Optimal treatment of bite wounds includes the recognition of immediate serious injury, management of bacterial infection, and approaches that yield the best possible cosmetic results. For many wounds, immediate gentle irrigation with water or a dilute water povidone-iodine solution will markedly decrease the risk of infection. Wound cleansing is especially important in rabies PEP, and in animal studies thorough wound cleansing alone without other prophylaxis reduced the likelihood of infection. As with all contaminated wounds, consideration should also be given to tetanus prevention. Decisions regarding the use of specific antibiotics and wound closure should be individualized based on the animal responsible for the bite, size and location of the wounds, and time interval since occurrence. Suturing should be avoided to limit the introduction of virus into deeper tissues, if possible, and only considered after wound cleansing and infiltration of RIG. The combination

Table 13.4 Conventional rabies intramuscular PEP schedule (adapted from current ACIP/WHO recommendations)

Vaccination status	Treatment	Regimen[a]
Not previously vaccinated	Wound cleansing	All PEP should begin with immediate thorough cleansing of all wounds with soap and water
	RIG	Administer 20 IU/kg body weight. If anatomically feasible, the full dose should be infiltrated into and around the wound(s) and any remaining volume should be administered i.m. at an anatomical site distant from vaccine administration. Also, RIG should not be administered in the same syringe as vaccine. Because RIG might partially suppress active production of antibody, no more than the recommended dose should be given
	Vaccine	CCV 1.0 ml, i.m. (deltoid area[b]), one each on days 0[c], 3, 7, 14
Previously vaccinated[d]	Wound cleansing	All PEP should begin with immediate thorough cleansing of all wounds with soap and water
	RIG	RIG should not be administered
	Vaccine	CCV 1.0 ml, i.m. (deltoid area[b]), one each on days 0[c] and 3

CCV, modern cell culture vaccine from the WHO pre-approved list; RIG, rabies immune globulin; i.m., intramuscular.
[a]These regimens are applicable for all age groups, including children.
[b]The deltoid area is an acceptable site of vaccination for adults and older children. For younger children, the outer aspect of the thigh may be used. Vaccine should not be administered in the gluteal area.
[c]Day 0 is the day the first dose of vaccine is administered.
[d]Any person with a history of pre-exposure vaccination with CCV; prior PEP with CCV; or previous vaccination with any other type of rabies virus vaccine and a documented history of rabies virus neutralizing antibody response to the prior vaccination.

of RIG and vaccine is recommended for both bite and non-bite exposures, regardless of the interval between exposure and initiation of PEP, within reason. If PEP has begun and laboratory diagnostic testing (i.e. direct fluorescent antibody test; http://www.cdc.gov/rabies/pdf/RabiesDFASPv2.pdf) demonstrates that the animal was not rabid, prophylaxis may be discontinued. The RIG is administered once at the start of PEP to provide immediate, specific antibody coverage until the active induction of antibody after vaccination. If not given when PEP was begun, RIG can be administered up to day 7. The RIG is not indicated thereafter as an antibody response to CCV has occurred. The RIG can suppress active production of antibody, so the amount administered should not exceed the recommended dose of 20 International Units (IU)/kg of estimated lean body mass (WHO, 2005). If feasible, the full dose of RIG should be thoroughly infiltrated into and around all wounds. After local infiltration (or if there is no discernible exposure site), the RIG can be injected intramuscularly at a site distant from vaccine administration. The RIG should not be administered in the same syringe or the same anatomical site as the first vaccine dose. Subsequent doses of vaccine in the series can be administered in the same anatomic location where the human RIG dose was administered. The first dose of vaccine should be administered as soon as possible after exposure. This date is then considered day 0 of the series. Additional doses should then be administered on days 3, 7 and 14 after the first vaccination (Briggs, 2011).

Deviations from recommended PEP practices

Obviously, every attempt should be made to adhere to the current PEP recommendations (Rupprecht and Gibbons, 2004). Delays of a few days appear unimportant, but the effect of longer lapses is unknown. Most interruptions in CCV schedules do not require re-initiation of PEP. For example, if a patient misses a dose scheduled for day 3, and presents for vaccination on day 7, the day 3 dose could be given on the day of presentation and the schedule resumed, maintaining the same interval between doses. When substantial deviations occur, immune status could be assessed by serological testing. Besides local anomalies, persons exposed abroad in localities where rabies is enzootic might receive PEP with regimens or biologics not used in their own country. If PEP is initiated with foreign or unknown products or regimens, or biologics of nerve tissue origin, additional prophylaxis may be needed once the patient presents at home. Failures have occurred if PEP was delayed, when less potent biologics were used, or if deviations were made from recommended protocols (Wilde and Lumlertdacha, 2011). Patients who contract rabies after PEP may have lacked adequate local wound cleansing, may not have received appropriate infiltration of RIG around the wound sites, or received vaccine administered in inappropriate anatomical sites. Delays between exposure and initiation of PEP are of great concern, especially with severe and multiple wounds to the face and head, which provide shorter access to the CNS.

Health economics of PEP

In a recent analysis, the cost-effectiveness of PEP was calculated for different contact scenarios, with the outcome the cost per life saved, primarily extrapolated from data in developed countries (Vaidya *et al.*, 2010). Cost-effective ratios were calculated using assigned transmission risk values and average costs of PEP. Direct medical costs associated with prophylaxis included a dose of RIG, multiple doses of CCV, and hospital and physician charges. Indirect costs included travel, lost wages, alternative medicine, and other costs. Human life was valued using the average present value of expected future lifetime earnings and housekeeping services. Rabies PEP, when given per recommendations, was assumed to be 100% effective in preventing a case of rabies. The analysis showed cost savings to give PEP if a patient was bitten by a laboratory-confirmed rabid animal or by a wild reservoir species (e.g. fox bite), even if the animal was not available. For other transmission risk situations, the average net cost effectiveness ratio ranged from nearly \$3 million to \$8 billion per life saved following lower risk situations. Similar economic savings are obtained when intradermal vaccination is employed in developing country scenarios (Hampson *et al.*, 2011).

Pre-exposure vaccination

Since the time of Pasteur, most rabies biologics have been used for naïve individuals in PEP settings (Wu *et al.*, 2011). However, rabies vaccination is also used in certain persons prior to exposure. As to rationale, pre-exposure vaccination is used to simplify future medical management by eliminating the need for RIG, decreasing the number of vaccine doses needed after exposure, providing a window of time when PEP may be delayed, and offering an element of immunity to persons at risk for unknown exposures (Briggs, 2011). Pre-exposure vaccination is considered for high-risk groups, such as veterinarians and staff, animal handlers, *Lyssavirus* researchers, laboratory diagnostic workers, and persons whose activities bring them into contact with virus or potentially rabid wildlife. International travellers may be candidates for vaccination if they are likely to come in contact with animals in areas where rabies is enzootic and immediate access to appropriate medical care and biologics might be limited (Blanton and Rupprecht, 2008). Routine pre-exposure vaccination for the general public or the routine tourist is not recommended. Recognizing the burden of rabies in children, pre-exposure vaccination has also been considered in paediatric populations in canine enzootic countries until such time that animal rabies prevention and control becomes a reality (Lang *et al.*, 2009). In addition, in parts of the Neotropics, where indigenous populations of Amazonia are under high risk of vampire bat predation, pre-exposure vaccination is one strategy to decrease the burden of rabies where PEP is impractical owing to unpredictability and timeliness of PEP application in remote settings (Carvalho-Costa *et al.*, 2012).

Primary vaccination and of previously vaccinated persons

For those at risk, three applications of CCV should be administered on days 0, 7 and 21 or 28 (Table 13.5). After exposure, local wound care remains an extremely important, but often overlooked part of prophylaxis, even for previously vaccinated persons, who should receive two doses of vaccine, one immediately and one three days later (WHO, 2005; CDC, 2008). Previously vaccinated persons are those who have received pre-exposure or PEP regimens of modern CCV, or who received another vaccine, with a documented rabies VNA titre. Administration of RIG is unnecessary and might inhibit the degree or rapidity of an expected anamnestic response. For previously vaccinated persons who are exposed to rabies, it is inappropriate to simply determine the current rabies virus neutralizing antibody titre (Briggs, 2011). First, there will be several days before an anamnestic response may be observed. Secondly, there is no known 'protective' titre, so appropriate management of the previously vaccinated person would still consist of the two doses of vaccine, one on day 0 (as soon as the exposure is recognized and administration of vaccine can be arranged) and the second on day 3. Lastly, although rabies VNA are important components, other immune effectors are also operative in disease prevention (Moore *et al.*, 2006).

Routine booster doses of vaccine and serological monitoring

Healthy persons tested serologically 2–4 weeks after completion of rabies vaccination demonstrate an adequate immunological response, currently defined by an arbitrary international level of VNA detection of 0.5 IU/ml (WHO, 2005). Application of PEP in any previously

Table 13.5 Conventional rabies pre-exposure intramuscular prophylaxis schedule (adapted from current ACIP/WHO recommendations)

Type of vaccination	Route	Regimen
Primary	Intramuscular	CCV; 1.0 ml (deltoid area), one each on days 0[a], 7, and 21 or 28
Booster	Intramuscular	CCV; 1.0 ml (deltoid area), day 0[a] only

CCV, modern cell culture vaccine from the WHO pre-approved list.
[a]Day 0 is the day the first dose of vaccine is administered.

vaccinated individuals occurs regardless of VNA titre, because sero-protective levels have not been defined, and owing to delays in serum collection and reporting of the test results (Briggs, 2011). However, unrecognized exposures may occur in those persons who work with *Lyssavirus* and suspect rabid animals. Absolute rabies VNA levels do not determine definitive susceptibility or protection related to virus exposure, but measurements are useful for monitoring immune status over time of those in various populations, as immunity wanes (Table 13.6). Among persons at occupational risk of exposure, titres should be checked periodically, with booster vaccine doses administered as needed when titres decline. For example, approximately two years after pre-exposure vaccination, adequate titres were found among 93–98% of persons who received a three-dose pre-exposure intramuscular (IM) series and 83–95% of persons after intradermal (ID) vaccination. If titres fall below a minimum acceptable antibody level, a single booster dose of vaccine results in a rapid anamnestic response (Briggs and Schwenke, 1992).

Management of adverse reactions to rabies biologics

Compared to older nerve tissue-based products, modern CCV and RIG are extremely safe (WHO, 2005). Considering potential acquisition of a fatal zoonosis, rabies PEP should not be discontinued because of mild adverse reactions once initiated. Such events can be managed with non-steroidal anti-inflammatory, antihistaminic, and antipyretic agents. When a history of hypersensitivity to a biologic is recognized, pre-treatment with antihistamines may be considered. Although serious systemic, anaphylactic, or neuroparalytic reactions are rare after the administration of rabies biologics, such events pose a serious dilemma (Briggs, 2011). The risk of rabies must be carefully weighed before deciding to discontinue

Table 13.6 Rabies pre-exposure prophylaxis guide by risk group (adapted from current ACIP/WHO recommendations)

Risk category	Nature of risk	Typical populations	Pre-exposure recommendations
Continuous	Virus present continuously, often in high concentrations. Specific viral exposures may go unrecognized. Bite, non-bite, or aerosol exposure	Rabies research laboratory workers[a]; rabies biologics production workers	Primary course Serological testing every 6 months; booster vaccination if antibody titre is below acceptable levels[b]
Frequent	Exposures usually episodic, with source recognized, but exposure might also be unrecognized. Bite, non-bite, or aerosol exposure	Rabies diagnostic lab workers[a], spelunkers, veterinarians and staff, and animal-control and wildlife workers in rabies-enzootic areas. All persons who frequently handle bats	Primary course Serological testing every ~2 years; booster vaccination if antibody titre is below acceptable levels[b]
Infrequent (but greater than the population at large)	Exposure nearly always episodic, with source recognized. Bite or non-bite exposure	Veterinarians and animal-control staff working with animals in areas where rabies is uncommon to rare. Veterinary students. Travellers visiting areas where rabies is enzootic and immediate access to appropriate medical care, including biologics is limited	Primary course No serological testing or booster vaccinations
Rare (population at large)	Exposure typically episodic with source recognized. Bite, or non-bite exposures.	Resident populations at large, including persons in rabies-epizootic areas	No vaccination necessary

[a]Judgement of risk and extra monitoring of vaccination status of workers at risk is the responsibility of the laboratory supervisor.
[b]Minimum arbitrary international acceptable antibody level is 0.5 IU/ml measured by the rapid fluorescent focus inhibition test. A booster dose should be administered if the value falls below this level.

vaccination. Assistance on management of serious adverse reactions should be sought from appropriate biomedical and public health officials. All clinically significant adverse events occurring following administration of rabies biologics should be reported to national regulatory authorities. Although such notification is subject to limitations common to any passive surveillance system (including underreporting and reporting bias), such analysed data may be valuable tools for characterizing safety profiles and identifying risk factors for rare serious adverse reactions. Health care providers are encouraged to report electronically to national centres to promote timeliness and quality of safety data, especially as may be related to the need for product recalls (CDC, 2008).

Precautions and contra-indications in rabies vaccination

There are no true contraindications to rabies PEP, but caution should be exercised in a variety of scenarios (WHO, 2005). Drugs and existing medical conditions interfere with development of adequate immunity after rabies vaccination, including corticosteroids, other immunosuppressive agents, some antimalarials, and many immunosuppressive illnesses. Patients who are immunosuppressed by disease or medications should postpone voluntary vaccinations and avoid activities for which pre-exposure vaccination is recommended. Monitoring of rabies VNA is one management option. Failure to seroconvert should be managed in consultation with knowledgeable public health officials. To date, cases of rabies PEP failure have not been documented among immunosuppressed persons, such as those due to HIV (Briggs, 2011). Immunosuppressive agents should not be administered during PEP unless absolutely essential for the treatment of other conditions (CDC, 2008). Pregnancy is not considered a contraindication to PEP. No increased incidence of abortion, premature births or fetal abnormalities has been associated with rabies vaccination (Abazeed and Cinti, 2007; Sudarshan *et al.*, 2007; Briggs, 2011). If the risk of exposure is substantial, pre-exposure might be indicated during pregnancy. Viral exposure or a diagnosis of rabies in the pregnant patient is not an indication for pregnancy termination.

Patients with a serious hypersensitivity to components of one rabies vaccine should be revaccinated with caution, and alternative rabies vaccines may be considered.

Rabies treatment

Rabies has the highest case fatality rate of any conventional infectious disease, and is an ideal biomedical application of the aphorism that a gram of prevention is worth much more than a kilogram of cure, because none has been proven as of yet, after the onset of clinical signs. However, with documentation of naturally occurring antibodies in a variety of mammals, and observations of animal recovery after experimental infection, a new thesis is emerging that rabies is not unique but may be a continuum, like other viral diseases (Rupprecht *et al.*, 2006). Historically, human treatment has included administration of a combination of intensive supportive care, anti-viral drugs, CCV, RIG, cytokines, steroids, and other measures. To date, since 1970, survival has been well documented for less than a dozen cases (Jackson and Wunner, 2007). In most examples, patients received rabies vaccination before disease onset. The first patient to recover from rabies without a history of prior vaccination occurred in the United States during 2004, after a Wisconsin teenager was bitten on her finger by a bat. Her acute medical management included ventilatory support, drug-induced coma, neuro-protective measures, and the use of ketamine, midazolam, ribavirin, and amantadine (Willoughby, 2007). After more than 10 weeks of hospitalization, she was discharged to intensive rehabilitation, and went on to a dramatic recovery, including graduation from college. Clearly, treatment of rabies is an extreme clinical and ethical challenge. Rapid ante mortem diagnosis is necessary to confirm a suspected diagnosis before starting therapy. Rather than be treated as a pariah, or be discharged to unknown home settings, comfort care and sedation of the patient should occur at a minimum in an appropriate medical facility. Sedation is necessary because patients can become extremely agitated, especially in the presence of excitatory stimuli. Patient and familial stress may be compounded by the trauma resulting from a sense of hopelessness related to the prognosis. As new modalities become available, tertiary care centres could specialize in an aggressive approach to experimental therapies in confirmed cases at an early stage of disease. Informed consent considerations should include the realistic probability for failure, anticipated expenses, lengthy rehabilitation, and quality of life issues. A better understanding of basic rabies pathogenesis and the applied development of relevant animal models will provide insights towards future improvement. To this effect, in a 'One Health' capacity, besides the use of traditional laboratory animals alone as potential surrogates for humans, breakthroughs in the treatment of viral encephalitis in domestic animals may also be illuminating. In the interim, efforts focusing upon the elimination of viral exposure and application of appropriate PEP remain the most effective public health measures to prevent human rabies (WHO, 2005; CDC, 2008; Briggs, 2011).

Considerations during clinical management of human rabies patients

Human-to-human transmission has occurred among transplant recipients from corneas, solid organs, and vascular tissue (Jackson and Wunner, 2007). Investigations revealed each of the donors had died of an illness compatible with or proven to be rabies. The cases occurred in the United States (five cases: one corneal transplant transmission, three solid organ transmissions, and one vascular graft transmission), Germany (four cases), Thailand (two cases), India (two cases), Iran (two cases), and France (one case). Apart from transplantation of infected organs and tissues, bite and non-bite exposures inflicted by infected humans could transmit rabies virus, in theory, but no laboratory-diagnosed cases occurring under such situations have been documented. At least two cases of human-to-human rabies transmission in Ethiopia have been suggested, but rabies as the cause of death was not confirmed by laboratory testing. The reported route of exposure in both cases was direct salivary contact from another human. Rabies transmission has not been reported in a health care setting. Beyond understandable fear, rabid patients do not pose any greater infection risk to health care workers than do persons affected with other more common bacterial and viral infections. Medical staff can adhere to standard precautions and use personal protection equipment, including outer disposable gowns, adequate eye protection, respiratory protection and gloves, particularly during procedures such as intubation and suctioning. Prophylaxis of staff or patient contacts is indicated only when the patient has bitten another person or when the patient's saliva or other potentially infectious material, such as neural tissue, has contaminated an open wound or mucous membrane.

Applications to domestic animals

Domestic animals are key parts of the dynamic web of rabies, both as reservoirs and as victims (Fig. 13.2). Rather than culling and population reduction, creation of herd immunity in susceptible populations by vaccination is a proven twenty-first century mainstay. Reflecting the diversity of domestic taxa, pure, potent, safe, and efficacious commercial vaccines have been licensed for 1- to 4-year parenteral application in most major groups, including dogs, cats, ferrets, cattle, sheep, and horses, in addition to other effective rabies prevention and control strategies (NASPHV, 2011).

Livestock

Several major organizations contribute to prevention and control outreach with a focus on the global agricultural aspects of rabies, particularly related to international health regulations, outbreak news, and applied aspects of the veterinary profession for domestic animal operations, including responsible units of WHO (http://www.who.int/rabies/en/) and its regional offices, such as PAHO, (http://www.paho.org/english/ad/dpc/vp/rabia.htm), and OIE's web-based rabies portal (http://www.oic.int/en/animal-health-in-the-world/rabies-portal/). In addition, the Animal Production and Health Division of the FAO is another international source for information on events and activities related to rabies (http://www.fao.org/ag/againfo/home/en/news_archive/AGA_in_action/2010_rabies.html).

The death of even a single animal to rabies not

only means the complete loss of food and fibre, but can be devastating to the economy of a local farmer. Epidemiologically similar to humans, affected hoofed-stock are typically single, dead-end cases, and are usually infected by rabid carnivores, with a few notable exceptions. For example, in Latin America, tens of thousands of livestock succumb to rabies annually after bite exposures by rabid vampire bats (Jackson and Wunner, 2007). Vaccination would prevent rabies in cattle and other ruminants, but not the direct predation by these hematophagous bats, which is detrimental to the substantial production of meat and milk, in regions from Mexico to Argentina. Historically, non-specific destruction of bats and roosts has not been successful in controlling rabies in vampire bats, plus impacts extend to many beneficial non-target species. Rather, selective targeting of vampire bats at farms and ranches has met with some success, by application of an anti-coagulant paste, which is spread in a bat colony as members mutually groom each other and die. Concerns about environmental contamination with pesticides and irregular use of such control, with eventual repopulation by vampires from non-affected areas, limit this tactic as a long-term solution. Suggestions of substituting oral vaccines in the paste to immunize vampire populations to prevent rabies would not change the destructive effects of the bat's blood meal upon livestock (Almeida *et al.*, 2008). Unlike most cases of rabies in domestic livestock, which after carnivore exposures tend to be isolated, or spatial collections of events associated with vampire bat- exposures, transmission has been suggested in several instances within farmed wild hoofed stock herds, notably kudu and deer (Mansfield *et al.*, 2006; Petersen *et al.*, 2012). Off-label use of domestic livestock vaccines could prevent rabies in these populations, as well as institution of other management practices to minimize the opportunity of viral introduction into such ungulates (i.e. hoofed animals) from local carnivore reservoirs (i.e. jackals, raccoons).

Dogs

Over the past ~15,000–30,000 years, humans and dogs have co-evolved significantly, due in no small part to the benefits early settlements gained through canine companionship, hauling, herding, hunting, sanitation, security, and other co-dependent activities (Vonholdt *et al.*, 2010; Galibert *et al.*, 2011; Skoglund *et al.*, 2011). As populations spread from centres of civilization over the centuries, dogs accompanied humans, representing complex genetic mosaics of founding migrants and newer arrivals throughout the world (Boyko *et al.*, 2009). Not surprisingly, historical and current human and canine translocations had a profound effect upon the cosmopolitan introduction, spread and perpetuation of rabies (Talbi *et al.*, 2010).

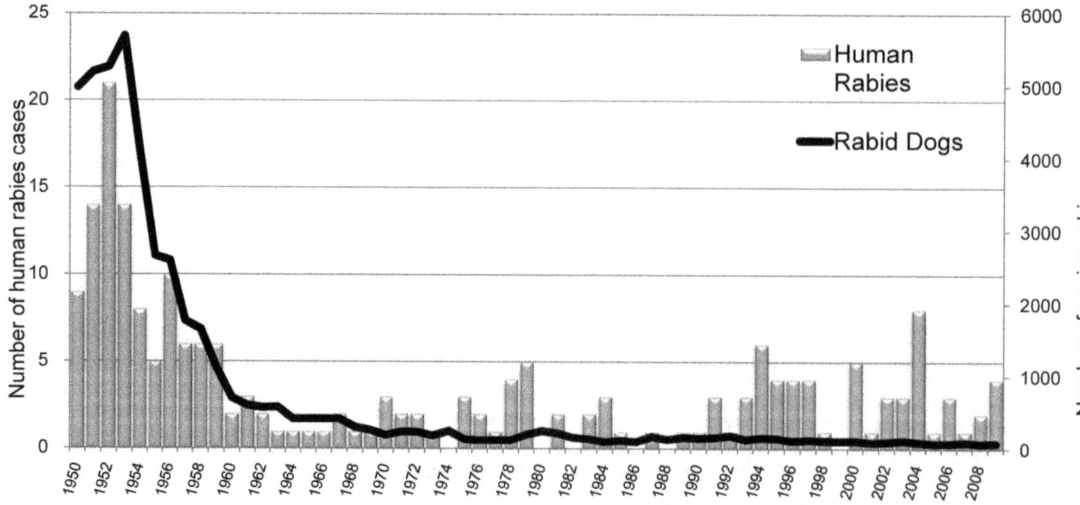

Figure 13.5 Comparison of human and animal rabies cases – United States of America.

As a result of focused large-scale canine vaccination and stray animal control programmes, marked decreases in domestic animal rabies cases have occurred. Such declines lead to a substantial decrease in indigenously acquired rabies among humans (Fig. 13.5). For example, in the United States during 1946, more than 8000 indigenous rabies cases were described among dogs, and 33 cases in humans, all reported from domestic animal exposure. In contrast, shortly after the turn of the twenty-first century, enzootic dog-to-dog transmission had been eliminated, and in 2011, none of the six reported human rabies cases were acquired from indigenous domestic animals. Moreover, typically less than 100 cases of rabies are reported in domestic dogs, all with sources representative of infection by indigenous wildlife reservoirs (Blanton *et al.*, 2011). Despite such dramatic success, annual rabies cases in cats now exceed those in dogs, because of much less attention focused upon feline vaccination. Unlike the situation in developing countries, nearly all human rabies cases in the United States are caused by variants of rabies viruses associated with wildlife and not dogs, although the risk from reintroduction remains, as exemplified by three imported human rabies cases in 2011, acquired abroad from infected dogs. Thus, the likelihood of human exposure to a rabid domestic animal in the United States and other developed countries has decreased greatly, as demonstrated from the notable success in canine rabies elimination (Table 13.7). However, the human rabies cases diagnosed in 2011 involved infections that were acquired outside the United States in countries where canine rabies is enzootic. Nevertheless, international travellers to areas where canine rabies remains endemic are at risk of exposure to rabies from infected dogs. Despite the creation of international guidelines for importation of dogs into so-called 'rabies-free' areas (including obligatory rabies vaccination, VNA serology, veterinary health certification, micro-chip identification), rabid dogs from enzootic areas continue to pose a risk for viral introduction into areas that have eliminated the disease (Johnson *et al.*, 2011).

Clearly, in developing countries, canine vaccination programmes face severe constraints, from maintaining seasonal supply of cold chains of biologics and obtaining trained, dedicated personnel, to the frequency, timing, and distribution of service to remote locations. Considering that the spatial distribution and relative abundance of dogs has an important influence on the local dynamics and spread of rabies, recent metapopulation 'patch-occupancy' applications, modelled on Tanzanian districts, have suggested methods to improve the effectiveness of vaccination deployment. Optimization of limited resources, and large, single-pulse campaigns yielded the best short-term protection, but higher frequencies of smaller applications were more effective at reducing long-term disease occurrence (Beyer *et al.*, 2011).

Recognizing the importance of canine rabies, and the dual central role of objective science and public advocacy in the prevention and control of any disease, several major routine international meetings (e.g. Rabies in the Americas Conference, Southern and Eastern African Rabies Group) and professional groups (e.g. Global Alliance for Rabies Control, Partners for Rabies Prevention) are promoting a greater awareness of this neglected zoonosis. Relevant web links are http://www.rabiescontrol.net; http://www.

Table 13.7 Selected examples of areas with a reported absence of circulation of canine rabies viruses

Australia	Comoros Islands	Panama
Canada	Europe (primarily north, south and west portions)	São Tomé and Príncipe
Cape Verde	Hong Kong	Seychelles
Caribbean islands (except Cuba, Grenada, Hispaniola, Puerto Rico)	Japan	Taiwan
Costa Rica	Mauritius	United States
Chile	Pacific Oceania (including New Zealand)	Uruguay

rabiescontrol.net/about-us/partners/partners-for-rabies-prevention.html; http://www.afroreb.info; http://www.areb.info/website/; http://www.meereb.info/and http://www.searg.info/doku.php. These groups and activities seek to focus global laboratory and field research towards relevant, cost-effective collaborative solutions (Lembo *et al.*, 2011). Notably, World Rabies Day was created in 2007 as an annual event to galvanize international attention upon such efforts and in part to publicize the creation of a blueprint for use by individuals and countries towards eventual canine rabies elimination (http://www.rabiesblueprint.com/). This document was developed collectively by international scientists with expertise in rabies surveillance, diagnosis, epidemiology, pathogenesis, prevention, and control, and provides a comprehensive outline of the roles, responsibilities, infrastructure, legislative framework, costs, communication plans, operational activities, references, and related components for a successful programme (Lembo and Partners for Rabies Prevention, 2012). Increasingly, developing countries are repeating the progress that Europe and North America have achieved, such as the regional programme in Latin America, where political support for improved public health through mass dog vaccination campaigns was first gained during 1983, leading to dramatic declines in both human and canine rabies cases (Fig. 13.6). Such progress in domestic animal programmes holds similar theoretical and applied opportunities for free-ranging wildlife reservoirs.

Domestic cats

Pet domestic cats fill beneficial roles in the lives of many people. Contact between people and cats, which have represented the primary domestic animal reported rabid in the United States since 1987, underscore the need for effective rabies prevention. In 2010, a total of 303 rabid cats was reported, mostly from eastern states with endemic raccoon rabies (Blanton *et al.*, 2011).

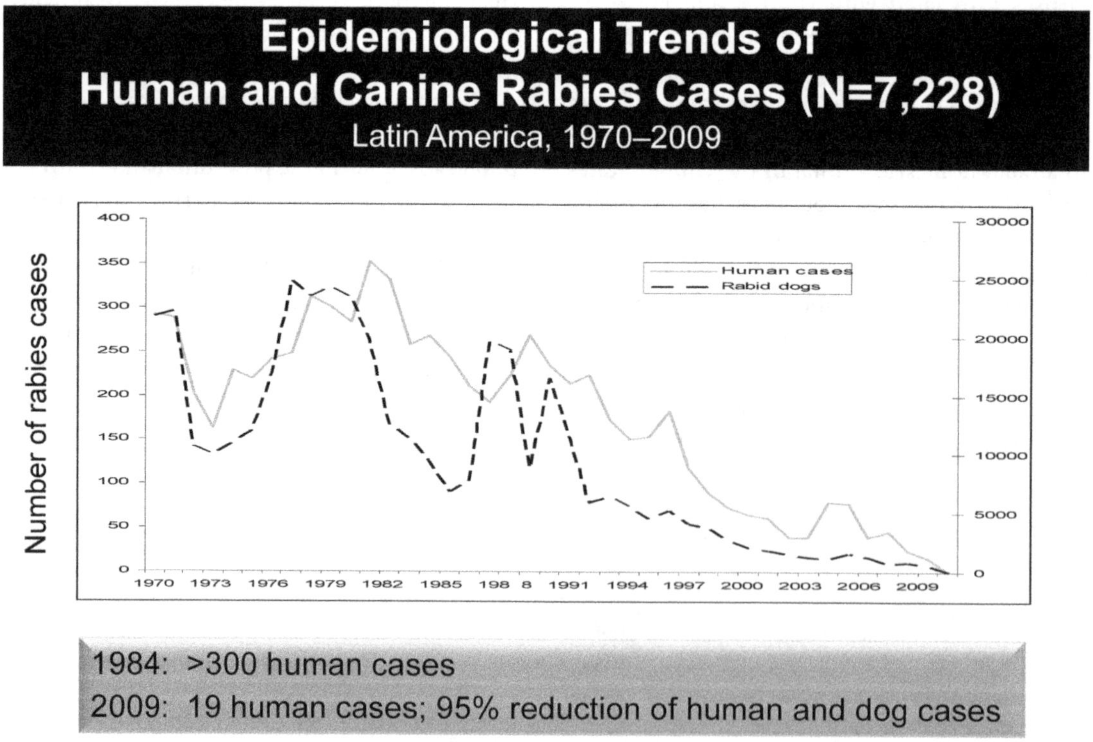

Figure 13.6 Decreases in human and canine rabies cases in Latin America (data kindly provided by Dr F. Leanes, Pan American Health Organization).

Domestic animals account for the majority of potential human rabies exposures that result in persons receiving PEP; cats account for around 50% of exposures involving domestic animals (Blanton *et al.*, 2005; O'Bell *et al.*, 2006). The increase in the association of cats with rabies among domestic species has largely been attributed to lower adherence to public health recommendations and to laws that cats, as well as dogs, should be vaccinated against rabies and prevented from roaming freely. While domestic cats have not been identified as a reservoir for rabies, *per se*, they do act as a primary vector for human exposures owing to their contact with wildlife rabies reservoirs such as the raccoon and skunk (Blanton *et al.*, 2011). Thus, public health recommendations to prevent rabies in domestic animals have focused on the vaccination and registration of all owned animals and the removal of stray populations to promote the maintenance of a sanitary environment and prevent the transmission of infectious disease among feral animals and from feral animals to humans (Suzan and Ceballos, 2005; Blanton *et al.*, 2011; NASPHV, 2011).

The emergence of trap–neuter–vaccinate–release (TNVR) programmes for cats is not expected to mitigate rabies public health issues associated with free roaming cats that are often cared for in colonies. Providing rabies vaccination and booster schedules recommended by vaccine manufacturers to achieve and maintain population immunity (Foley *et al.*, 2005; AAFP, 2007; Frymus *et al.*, 2009); identifying specific cats in the context of vaccination certification in the event of a human or pet exposure; achieving effective levels of contraception (Castillo and Clarke, 2003; Foley *et al.*, 2005; Natoli *et al.*, 2006; Schmidt *et al.*, 2009; Gunther *et al.*, 2011); and, eliminating contact with wildlife at supplemental feeding sites (Olson *et al.*, 2000) that could result in the transmission of rabies or other diseases (Gomot *et al.*, 1997; Maruyama *et al.*, 2003; Nutter *et al.*, 2004; Salant *et al.*, 2009; McElroy *et al.*, 2010; Al-Kappany *et al.*, 2011; Little, 2011) are among the serious concerns. Public health is therefore best served through adherence to the minimum current recommendations in the National Association of State Public Health Veterinarians' Compendium of Animal Rabies Prevention and Control, endorsed by CDC, APHA, AVMA, APHL, CSTE and NACA, which states that 'local governments should initiate and maintain effective programmes to ensure vaccination of all dogs, cats, and ferrets and to remove strays and unwanted animals' (NASPHV, 2011).

Management of rabies in wild carnivores

Focus

Wildlife rabies control has undergone a marked change over the past four decades. The historical limitations associated with population reduction as an independent approach to rabies control, and the accomplishments of oral rabies vaccination (ORV), since this novel method of disease prevention was conceived by the late George Baer during the 1960s, were central to such a transition. Intervention by ORV demonstrated the potential to produce lasting rabies management effects. The elimination of rabies in red foxes (*Vulpes vulpes*) over broad geographic areas in Europe (Blancou *et al.*, 2008; Wandeler, 2008) and much of southern Ontario, Canada (MacInnes *et al.*, 2001; Rosatte *et al.*, 2007a), are compelling examples of ORV success. While population reduction alone has historically been credited with eliminating fox rabies in Demark, by gassing fox dens and through hunting (Müller *et al.*, 2004) and suppressing rabies or dampening epizootic peaks (Rosatte *et al.*, 1986; Pybus 1988), in general such effects have been considered transient at best.

During its relatively short existence for field use since the late 1970s, ORV programmes have produced a noteworthy rabies management track record in specific wild meso-carnivore reservoirs, such as the red fox (MacInnes *et al.*, 2001; Pastoret *et al.*, 2004), grey fox (*Urocyon cinereoargenteus*) (Sidwa *et al.*, 2005), raccoon (*Procyon lotor*) (Rosatte *et al.*, 2009; Slate *et al.*, 2009) and coyote (*Canis latrans*) (Sidwa *et al.*, 2005). Successes represent key milestones to argue for continued ORV intervention, as a basis to seek enhancements for current methods, and to explore promising new tools and strategies. Yet, in today's harsh economic realities, there are no guarantees that all current ORV programmes will be maintained to achieve stated goals, let alone contemplate expansion of programmes on a continental basis. To better ensure that ORV programmes remain available to prevent and control rabies at its source in diverse reservoir species, cost-effective advances are needed to address complex technical issues, such as achieving rabies control in meso-carnivores like the striped skunk (*Mephitis mephitis*), and incorporating immuno-contraception into specific control strategies. Science-based advances in new vaccines, baits, biomarkers, delivery systems and strategies that meet the expectations of enhanced effectiveness and affordability in rabies management programmes will better serve as rallying points for marshalling focused programme support. Failure to meet these basic expectations increases the risk of ORV being relegated to 'luxury status' in the context of competing priorities for local, federal and international funding for disease abatement.

Several book chapters have been published over the past decade, which provide partial and comprehensive details and histories on wildlife rabies management (Rosatte *et al.*, 2007b; Rosatte, 2011; Slate and Rupprecht, 2011). This section will focus on recent opportunities, challenges and accomplishments in rabies management in wild meso-carnivores, with an emphasis on the past five years, and a look ahead towards the next five years. In so doing, our hope is to convey that 'the future is now' and that while we can celebrate some successes, we cannot ignore the fact that holistic management of rabies in wild carnivores is in full view of many funding decision-makers and remains 'on the clock' to produce measurable successes in the near term. If we are to achieve goals such as true elimination of specific rabies virus variants in key meso-carnivore reservoirs, timely, science-based results offer the best hope to continue to garner support for wildlife rabies management programmes, both now and in the future.

Collaborations and wildlife rabies management

Clearly, collaboration is the 'general rule' if effective wildlife rabies management is to be achieved. Examples span from the global level, as illustrated by the WHO Expert Panel (WHO, 1960), nationally in the United States by the Compendium of Animal Rabies Prevention and Control (NASPHV, 2011), the National Rabies Management Team (Slate *et al.*, 2009), and locally by state rabies task forces. These are typically composed of state health, agriculture and wildlife agencies, counties and municipalities, and may include federal representation, such as CDC and USDA, Wildlife Services (WS), private sector Nuisance Wildlife Control Operators, and others. The need for collaboration is clear in view of the diverse disciplines that are often brought to bear on the complexities associated with wildlife rabies management. The list of disciplines is exhaustive and may include: veterinary and human medicine, virology, epidemiology, ecology, economics, molecular biology, genetics, ecology, animal husbandry, wildlife biology and management, and applied mathematical modelling, among others.

The signing of the North American Rabies Management Plan (NARMP) at CDC during 2008 represented a recent milestone to foster such multisector collaboration (USDA, 2008a). Through this international plan, representatives from Canada, Mexico, the Navajo Nation and the United States have established an inter-jurisdictional and multidisciplinary framework that extends collaboration in rabies management to the North American continent. This plan may also serve as a model for other regions contemplating broader international rabies surveillance and control initiatives. Communications, surveillance, control, and research represent key areas for collaboration under the NARMP. A field comparison of sero-conversion of ONRAB® in New Brunswick, Canada, to Raboral V-RG® in Maine, United States (Fehlner-Gardiner *et al.*, 2012) and a study evaluating the GnRH immuno-contraceptive GonaCon™ in captive dogs in Mexico are examples of recent research initiatives undertaken through the NARMP. Both projects have immediate relevance in providing scientific support to pursue broader North American field trials of ONRAB® and other oral rabies vaccine–bait combinations and additional studies in Latin America to determine the feasibility of integrating

GonaCon™ (Miller *et al.*, 2000) into mass dog rabies vaccination campaigns (USDA, 2008a).

Enhanced surveillance – advances in a critical rabies management component

The ability for objective assessment of ORV success hinges on adequate surveillance. While the presence of biomarker and vaccine-induced rabies virus antibody represent commonly evaluated and important indices of ORV effectiveness (Johnston *et al.*, 1999, 2005; Rosatte *et al.*, 2009; Slate *et al.*, 2009), the absence of rabid target species infected with specific viral variants is the ultimate measure of rabies management success.

Enhanced rabies surveillance is conducted to complement public health surveillance, which is typically based on human or domestic animal exposure events brought to the attention of public health officials (Blanton *et al.*, 2007). Enhanced surveillance is designed to increase the geographic scope and intensity of sampling through the collection of specific high value samples beyond those normally tested to protect human and domestic animal health. Samples often include ill or strange acting animals, where there has been no human or domestic animal exposure, fresh road kills, animals found dead in addition to road kills, animals with injuries or lesions suggestive of aggressive behaviour, and animal samples from focal trapping at sites where rabid animals were recently confirmed. Consequently, enhanced surveillance information provides greater confidence in real time ORV decisions. Moreover, such enhanced rabies surveillance is at its highest value when applied along the leading edge ORV zones, where the risk of spread beyond existing ORV barriers exist, such as the western and northern distribution for raccoon rabies in North America.

Effective enhanced surveillance is typically the product of a collaborative process, where government and private sector individuals who often have access to suspect animals participate as a part of their routine professional activities to provide samples for diagnosis from designated areas. Health officials, veterinarians, wildlife biologists, conservation officers, police, and nuisance wildlife control officers are among the disciplines that often have contact with suspect animals of value to enhanced surveillance. For example, focal trapping has been conducted by WS to collect samples in real time where animals have recently been confirmed rabid in relation to a critical ORV area. Both trap–vaccinate–release (TVR) programmes and pre- and post-ORV sampling for serum often result in the capture of suspect animals that are submitted for rabies diagnosis.

An increase in the number of suspect rabid animal samples from enhanced surveillance can create a burden on some local rabies diagnostic laboratories for timely diagnosis, particularly because the recent economic downturn has impacted state budgets. A direct immunohistochemistry test (dRIT) developed at CDC (Lembo *et al.*, 2006) has largely mitigated this sample burden, while facilitating timely analysis under a mutually agreed upon protocol among WS, CDC and participating states. After primary training at CDC, WS personnel participate in routine dRIT proficiency testing. All suspected rabid animal samples are confirmed and characterized by CDC. As an example of activities to date, in the period 2005 to 21 September 2011, WS submitted 62,118 enhanced rabies surveillance samples, and 48,605 have been tested by the dRIT, resulting in at least 897 confirmed rabid animals that would not likely have been available for rabies management decisions through public health surveillance alone (Table 13.8).

Rabies management in wild carnivore reservoirs

In discussing international wildlife rabies management accomplishments for the last five years, and peering ahead at opportunities and challenges on a projected 5-year horizon, one major lesson learned is that achieving stated rabies management goals is technically, ecologically, logistically, and economically complex irrespective of the mesocarnivore species. To date, disease prevention, control, and elimination has proven less difficult in foxes than in raccoons and skunks. Moreover, often project managers face many more obstacles and uncertainties than with humans and domestic animals, such as those presented by environmental perturbations, weather patterns, climate change, long-term funding to sustain programmes in spite of notable accomplishments, and the effects of

Table 13.8 Summary of enhanced wildlife rabies surveillance and direct rapid immunohistochemistry testing (dRIT) conducted in the United States by USDA, Wildlife Services, 2005–2011[a]

Year	Number of enhanced surveillance samples	Number dRIT tested	Number positive by dRIT	Per cent positive by dRIT
2005	7623	2738	91	3.3
2006	7281	6400	87	1.4
2007	10,812	9037	189	2.1
2008	11,004	8790	142	1.6
2009	12,254	10,533	160	1.5
2010	9264	7324	145	2.0
2011[b]	4008	3783	83	2.2
Total	62,246	48,605	897	1.8

[a]Information subject to change based on revised data entry to Wildlife Service Management Information System; accessed September 2011.
[b]Incomplete estimated total for 2011.

human behaviour (e.g. accidental and intentional translocations) that may significantly affect success (Chipman *et al.*, 2008). Nevertheless, rabies management in the reservoir hosts is pursued primarily because of the potential for appreciable long-term benefits of improved public and animal health, and financial cost savings, with lasting returns on investments if actions can prevent the spread of specific rabies virus variants to naïve areas, or lead to their elimination.

Foxes – a case for rabies elimination

Elimination of rabies in red foxes has been possible in several European countries through strategic use of ORV to geographically compartmentalize the virus (Wandeler *et al.*, 1988; Cliquet and Aubert, 2004). A similar result appears within reach in southern Ontario where an Arctic fox rabies virus variant had emerged and spilled over to perpetuation in red foxes (MacInnes, 1987). However, the need for special programmes to address the recent re-emergence of fox rabies in Italy (Freuling and Müller, 2009) illustrates the need for adequate coordination among neighbouring countries to ensure that sufficient levels of surveillance and control occur where rabies remains a threat. In Ontario, persistence of residual rabies foci of Arctic fox virus in striped skunks illustrates the challenge associated with virus spill-over into a species that has proven to be refractory to ORV to date. Early favourable results with one recombinant adenovirus vaccine (ONRAB®), contained in an ultralite bait (Artemis Technologies, Guelph, Ontario, Canada), with narrow flight line spacing for bait distribution and high bait densities (300 baits/km^2) (Rosatte *et al.*, 2011) suggest that elimination of this virus variant in skunks may be possible. If achieved, southern Ontario, where much of Canada's human population resides, would become free of rabies in mesocarnivores.

Since 1996, ORV has been conducted to control a unique rabies virus variant in grey foxes in west Texas. This programme has had interim objectives shaped in large part by reduced state funding by 2000, but the ultimate goal was to eliminate this viral variant from Texas and consequently from the United States (Sidwa *et al.*, 2005). An approximately 50% reduction in the geographical area occupied by this grey fox rabies variant represents a significant accomplishment during the last five years. In fact, with enhanced surveillance in place, there have not been reported cases of this unique variant in Texas since 2009 (Blanton *et al.*, 2011). While elimination of this virus appears achievable, the long United States border with Mexico, where the status of this variant has not been adequately characterized, presents a challenge for future incursion and spread. The potential for long range grey fox movements (Deyoung *et al.*, 2009) and the lack of appreciable natural barriers along the

Rio Grande river, raises the issue of implementing the ideal strategy to prevent re-introduction. Maintenance of a continuous ORV zone along the Texas-Mexico border, development of contingency plans to augment focused ORV in strategic areas to swiftly address re-emergence, and other options will likely deserve consideration to optimize resource availability for wildlife rabies management in this example. Selection of management alternatives will largely depend on the ability to obtain meaningful enhanced wildlife rabies surveillance along the border, particularly south of the Rio Grande, which is currently hampered by continued illegal drug trafficking-related violence and, in part, by an apparent lack of infrastructure and local coordination among agencies within Mexico to conduct extensive surveillance in remote areas common to northern Mexican border states.

A distinct grey fox rabies virus variant originally described in Arizona has recently spread east to the United States Continental Divide in New Mexico (Blanton *et al.*, 2011). In this region, there are no obvious geographic or other impediments to prevent this variant from spreading farther east among wildlife, perhaps into the areas recently freed of grey fox rabies in west-central Texas. This event underscores the significance of stepping up enhanced surveillance and virus characterization in the expansive south-western region to better understand the spatial dynamics of grey fox rabies in general, and of this rabies variant specifically, before contemplating the commitment of resources for ORV. Again, an overall lack of surveillance south of the United States border in Mexico presents a complicating factor that will ultimately need to be addressed to consider if elimination is a practical, economically feasible long-term goal.

Arctic foxes, rabies and climate change

The Arctic fox (*Vulpes lagopus*) represents an important circumpolar reservoir for variants of rabies virus (Johnston and Fong, 1992; Krebs *et al.*, 2003a, 2004). The risk of rabies exposure may be elevated for indigenous peoples because segments of the Arctic fox population often habituate to human settlements during winter in response to easy access to anthropogenic foods (Rausch 1958). In addition, the effects of Arctic fox rabies have previously extended well into the temperate region by virtue of rabies virus spill-over along the Arctic–red fox interface. Through its broad geographic distribution, the red fox in turn served as a conduit for spread of Arctic fox virus into southern Ontario, Quebec and the north-eastern United States, where high numbers of people reside and are at risk. Instituting surveillance and preparing effective contingency action plans to prevent Arctic fox variant from spreading back into southern Ontario are not difficult to envisage, given the precedence for effective control and access to enhanced surveillance tools such as dRIT (Lembo *et al.*, 2006) and real-time GIS mapping (Blanton *et al.*, 2006a) capabilities. However, developing meaningful intervention strategies for Arctic environments requires a greater understanding of the dynamics of rabies virus variants and interspecific interactions along the Arctic–red fox interface (Pamperin *et al.*, 2006). How to most effectively intervene in the vast Arctic raises several near term and longer term questions. Some of these issues include the relatively short lifespan of the Arctic fox compared with other meso-carnivores; the apparent nomadic behaviour displayed in some Arctic foxes; the need for a cost-effective oral rabies vaccine with ideal deliverability and stability in Arctic environments (e.g. lyophilized or other); and the effects climate change will likely exert on fox species and rabies dynamics (Pamperin *et al.*, 2008). The specialized Arctic fox occupies a relatively narrow niche in comparison to the red fox, which is more of an ecological generalist. A warming trend that may be expected to have its greatest effects towards the poles could have a substantial negative impact on the Arctic fox, with uncertain consequences on rabies virus dynamics. Clearly, this rabies reservoir in an Arctic environment stressed by a warming climate is ripe for further study from an ecological, epizootiological, and public and animal health perspective.

Bats, carnivores, virus host shifts and intervention

In general, evolutionary rates of RNA viruses are faster than those of DNA viruses, particularly owing to a lack of proofreading ability by their polymerases. Lyssaviruses have a relatively simple genome compared to some other rhabdoviruses, and have constraints imposed in part by adaptation to the mammalian CNS. Genetic bottlenecks imposed by their existence in such a fairly stable intraspecific niche may be offset by the chances for new host invasion presented by ecological conditions and inter-specific social opportunities. Such postulated host switches have occurred historically and recently. For example, molecular evidence supports the hypothesis that lyssaviruses evolved in Chiroptera (likely via an insect route) and emerged later in carnivores, as a result of a viral host shifts from bats (Badrane and Tordo, 2001). Emergence of bat rabies viruses in grey foxes and skunks provides compelling real time support for this hypothesis (Leslie *et al.*, 2006). Given that Raboral V-RG® (Merial, Athens, Georgia, USA) has been highly effective in grey fox rabies control in Texas (Sidwa *et al.*, 2005), the intervention goal with ORV was to prevent radiation of this host shift in rabies virus from Flagstaff, Arizona to a broader geographic area in the south-western United States. There have been no reported cases in grey foxes since ORV and TVR began near Flagstaff in 2009, with enhanced surveillance in place, but forest fires, drought, canine distemper and other diseases cannot be ruled out as contributing factors in suppressing local carnivore populations and hence rabies perpetuation. Sporadic occurrence of bat rabies viruses in grey fox and skunk populations near Cave Junction, Oregon, or among coatis (*Nasua narica*) in Mexico represent other potential viral host shifts, but too little information is available at this time to define these events with confidence (Aréchiga-Ceballos *et al.*, 2010; Blanton *et al.*, 2011). Such incidents underscore the value of continued vigilance, enhanced surveillance, rapid viral characterization, and the near real time application of contingency actions to minimize disease reintroduction that would confound ORV programme success.

Rabies along the dog–wild carnivore interface

Where canine rabies control has not been completely achieved (i.e. through acquired herd immunity from vaccination), the dog-wild carnivore interface remains an active and viable route for two-way viral transmission. Movement of rabies via this interface brings with it public health risks and the potential for profound ecological ramifications for rarer species.

In North America, the epizootic that emerged in coyotes in south Texas in the 1980s was the likely result of spillover from infected dogs in Mexico (Clark *et al.*, 1994; Fearneyhough *et al.*, 1998) in spite of Mexico's remarkable efforts and accomplishments through annual mass parenteral rabies vaccination campaigns (Mexican Ministry of Health, 2011). An under-vaccinated domestic dog population combined with an abundant, susceptible coyote population in south Texas led to a rapid increase in cases in the early 1990s and at least two human deaths attributed to canine rabies (Sidwa *et al.*, 2005). An integrated approach using ORV to control rabies in coyotes in tandem with conventional dog rabies prevention and control (Sidwa *et al.*, 2005) led to the elimination of canine rabies from south Texas, allowing the United States to be declared canine rabies free in 2007 (based on conventional international OIE/WHO standards; Blanton *et al.*, 2007; Velasco-Villa *et al.*, 2008).

Effective dog rabies control is woefully inadequate or lacking throughout much of the developing world, including the majority of Africa. Public health impacts associated with dog rabies in Africa rank second only to Asia, with a continental estimate of 24,000 annual human deaths (Knobel *et al.*, 2005) that is generally viewed as conservative given that the etiologic agent for many human deaths is unknown. The Ethiopian wolf (*Canis simensis*) and African wild dog (*Lycaon pictus*) are examples of wildlife species profoundly impacted by spill-over of dog rabies (Gascoyne *et al.*, 1993). Both are on the International Union of Conservation of Nature (IUCN) Red List of Threatened Species (McNutt *et al.*, 2008; Marino and Sillero-Zubiri, 2011).

The Ethiopian wolf is the world's most threatened wild carnivore, with the total estimated number of adult wolves between 300 and 500 in seven fragmented populations in the Afroalpine habitats of Ethiopia (Stewart *et al.*, 2011). Spill-over of canine rabies is an additive source of mortality that places the already fragmented populations of the Ethiopian wolf in further jeopardy. Moreover, the Ethiopian wolf represents a sentinel species for the 'health' of the Afroalpine ecosystem and the biodiversity it supports. By extension, rabies from domestic dogs represents a direct impact and contributes to the cumulative impact on wolves and the Afroalpine ecosystem. Previous TVR efforts have had positive results on wolf survivorship (Stewart *et al.*, 2011), but ORV is needed in the short-term to more efficiently vaccinate wolves to establish an immune buffer to protect against spill-over of dog rabies into wolves. A long-term goal may be met if programmes are established to effectively vaccinate dogs to reduce the risk of canine rabies spill-over from this source.

Similar impacts have also occurred as a result of virus spill over into African wild dog populations in Kenya from domestic dog sources, with 21 of 23 wild dogs in packs monitored succumbing to rabies (Kat *et al.*, 1995). Rabies was also confirmed in one African wild dog in the Serengeti region, Tanzania. Inactivated rabies vaccine was administered intramuscularly to African wild dogs in the remaining two packs, by dart or by hand following anaesthesia. Three of 12 African wild dogs sampled at the time of vaccination had rabies VNA titres prior to vaccination. African wild dog recovery attempts in Madikwe Game Reserve, South Africa, have been impacted twice since their reintroduction in 1995 by spill-over of a canine rabies virus variant described in the black-backed jackal (*Canis mesomelas*) (Hofmeyr *et al.*, 2004). Wild dogs that had received parenterally administered multiple rabies vaccinations survived during the second outbreak in 2000. Clearly, ORV intervention strategies should be considered to protect this endangered species and enhance recovery efforts.

Challenges in skunk and raccoon rabies control

Skunk (primarily the striped skunk) rabies occupies the broadest geographic distribution in the United States, occurring as distinct variants in California and the north and south central plains (Blanton *et al.*, 2011). Much of this area is within the rural heartland of the United States, while skunks infected as a result of presumed spill-over of raccoon rabies virus occur within the enzootic area of raccoon rabies in the eastern US (Krebs *et al.*, 2003a). In addition, skunks were the mesocarnivore host in which big brown bat sources of rabies first appeared near Flagstaff, Arizona, in 2001 (Leslie *et al.*, 2006).

Annually, skunks are frequently infected with raccoon rabies virus variants (Krebs *et al.*, 2003a; Blanton *et al.*, 2011). To date, remarkable changes have not been detected in the raccoon variant of the rabies virus sequenced from rabid skunks in New England (Steve Rich, Tufts University, unpublished report) or elsewhere where raccoon rabies is sympatric with the striped skunk and samples from rabid skunks have been genetically evaluated (Guerra *et al.*, 2003). Because skunks could potentially serve as a maintenance host for raccoon rabies virus (Guerra *et al.*, 2003), the inability to effectively orally vaccinate skunks through campaigns targeting raccoons presents a management dilemma.

One captive study with Raboral V-RG® resulted in sero-conversion and survival of rabies challenge in four of six skunks that were orally vaccinated by syringe per os. However, only one of 11 skunks fed either single or multiple baits containing Raboral V-RG® survived a challenge, and none produced rabies VNA (Grosenbaugh *et al.*, 2007). Field evaluation of sera collected from ORV zones have been an even more convincing indication of the difficulty associated with orally vaccinating skunks in the wild. Baits [fishmeal polymer (FMP) and coated sachet (CS)] containing Raboral V-RG® that were distributed evenly, mostly at 75 baits/km^2 for raccoon rabies control in the eastern United States have not resulted in the detection of rabies VNA above 5.6% in skunks from 2001 to 2008, with the overall rabies VNA sero-positive rate of 2.2%, 11/506 (USDA, 2011a). In 2005 and 2006, ORV hand baiting (uneven distribution at ~75 baits/km^2) specifically targeting skunks in the vicinity of Flagstaff, Arizona resulted in only 1 of 66 skunks with rabies VNA (I.U. ≥ 0.06) that did not have a history of vaccination in TVR; although encapsulated tetracycline was added to the exterior coating of CS baits to aid in evaluation, this animal was not tetracycline-positive. Seven of the 66 skunks had rabies VNA in association with intramuscular vaccination in TVR (USDA, 2007, 2008b).

As a component of ORV to control a unique variant of rabies virus in grey foxes in west Texas, skunks and other non-target species were collected for post-ORV evaluation. From these serum collections, the highest recorded rabies VNA presence in skunks to date at 51% was reported. This percentage was based on 33 of 45 individuals captured from west Texas, where ORV with Raboral V-RG® (dog food polymer baits) had been conducted to control grey fox rabies (Sidwa *et al.*, 2005). However, all but four of the 45 skunks were collected in 2002 and 2003, coincidental with a reported peak for skunk rabies in Texas (Oertli *et al.*, 2008). Although the presence of rabies VNA was derived from collections in west Texas, most animal rabies cases were reported from central to east Texas, where more exposures to skunk rabies might be expected as a function of a higher human population. However, cases were also reported from central to west Texas in both hog-nosed (*Conepatus leuconotus*) and striped skunks (Oertli *et al.*, 2008). While the 51% sero-conversion rate may reflect response to Raboral V-RG®, the presence of rabies VNA from sublethal exposures to south central skunk rabies virus or other sources should not be ruled out as a potential contributing factor for the elevated presence of antibody in skunks. A similar antibody response in the absence of ORV has been reported for raccoons near rabies outbreaks in the south-eastern United States (McLean, 1971) and in areas west of the current known distribution of raccoon rabies and ORV in Ohio (Ramey *et al.*, 2008).

Considerations include that ORV bait densities for grey foxes averaged much lower at 27 to 39 baits/km^2 (Sidwa *et al.*, 2005) than those used for raccoon rabies control, which are typically at ~75 baits/km^2 and occasionally ~150 baits/km^2 (Slate *et al.*, 2005). Nevertheless, bait availability cannot be deduced in the absence of comparative data on target species densities and non-target species competition for baits. Bait preference is also unknown in the context of other foods available during ORV. Although tetracycline biomarker was not reported in association with sero-positive skunks (Sidwa *et al.*, 2005), other data suggest that ORV was a contributing factor to the observed rabies VNA titres (Sidwa, personal communication). In 2002, 7 of 12 skunks had rabies VNA titres, two of seven were tetracycline positive, and one of seven had a rabies VNA titre that was also biomarker positive. In 2003, 16 of 29 skunks had rabies VNA titres, 14 of 25 were tetracycline positive (biomarker samples were not available for four skunks), and 10 of 25 had rabies VNA titres that were also biomarker positive. Clearly, antibody presence reported in Texas skunks, in contrast to all other available field skunk sero-conversion data, points to the need to further evaluate antibody responses from ORV areas as well as naïve areas with similar south central skunk rabies dynamics. Unfortunately, the absence of tetracycline in commercially available CS baits eliminates the use of this biomarker as a correlate of sero-conversion that would add confidence to the interpretation of results. Anti-orthopox titres from poxvirus-vectored rabies vaccines may serve as a surrogate biomarker to aid in the determination of vaccine-induced sero-conversion, but specificity and sensitivity are currently lacking in tests to conclusively rule out a vaccine-induced antibody response (Weyer *et al.*, 2009).

Considering other biologics, the commercially produced modified live SAG2 (Virbac, France) and an experimental recombinant canine adenovirus (serotype 2)-rabies glycoprotein (CAV2-RG) oral vaccine have shown promise in protecting skunks and raccoons in captive studies. Among five skunks and five raccoons orally vaccinated by syringe per os with SAG2, two skunks and three raccoons developed rabies VNA pre-challenge and four skunks and all raccoons had rabies VNA titres 7 days post challenge. All vaccinated skunks and raccoons survived (Hanlon *et al.*, 2002). Eleven skunks in two experimental groups vaccinated with a 10-fold difference in dilution of CAV2-RG vaccine and all seven raccoons produced rabies VNA and survived virus challenge (Henderson *et al.*, 2009). Given that the results from oral vaccine captive studies, where delivery by oral installation have not been realized through bait delivery of oral rabies vaccine under field conditions, captive skunk studies should, where practical, also incorporate bait delivery of oral vaccine as an experimental treatment as an interim step prior to field trials. The presence of CAV2 antibody (mean 35.4%, $n = 89$ skunks, collected in one to five states from 2006 to 2008) in wild skunks suggests that additional samples should be evaluated from diverse locations to determine if naturally acquired population immunity levels would interfere with CAV2-RG in ORV campaigns (Slate *et al.*, 2011).

Application of ONRAB®, a recombinant human adenovirus [serotype 5] – rabies glycoprotein oral vaccine (Yarosh *et al.*, 1996), designated HAdRG1.3, produced inhibition levels based on a competitive ELISA (Elmgren and Wandeler, 1996) that ranged from 28% to 60% for strong to suspect positives, respectively, among skunks sampled from study areas with narrow flight line spacing (0.25 km) treated at 300 baits/km^2 in southern Ontario (Rosatte *et al.*, 2011). While these results are encouraging and strategies using high bait densities appear suitable to eliminate the remaining foci of Arctic fox rabies in skunks in southern Ontario, it is doubtful that bait densities approaching 300 baits/km^2 could be sustained over large tracts of land to achieve skunk rabies elimination. However, sero-conversion rates between skunks sampled from ONRAB® and Raboral V-RG® established ORV zones at 'traditional' raccoon bait densities (i.e. 75 baits/km^2) along the Maine–New Brunswick border were not significantly different [the proportion of antibody-positive skunks was higher in New Brunswick (15.2–18.2%) but significantly different from rates in Maine (2.8–11.1%), $\chi^2 = 7.34$, df = 5, $P = 0.197$] (Fehlner-Gardiner *et al.* 2012).

A clear dichotomy exists between strategies that mandate extremely high bait densities as a part of contingency actions to achieve a short term intervention goal such as the elimination of residual rabies foci maintained by skunks in Ontario, Canada and long-term strategies that may be formulated in the future for broader geographic application of ORV for skunk rabies control. Current bait pricing alone (~$1.25/CS unit, Merial, Athens, Georgia) would likely make it difficult to justify the costs of skunk bait densities that are not on par with or less than those generally used for raccoons (~75 baits/km^2). Benefits and costs will weigh heavily in the application of rabies control specifically targeting skunks in North America (Sterner *et al.*, 2008; Shwiff *et al.*, 2009).

Coordinated ORV among states along the leading western and northern edge of raccoon rabies has prevented this variant from gaining a broader geographic distribution in the United States (Slate *et al.*, 2009). However, this accomplishment has required dependence on local contingency actions that may integrate special ORV baiting (e.g. targeting specific areas representing higher risk for spread with high bait densities or multiple baiting/year) and TVR to bolster raccoon sero-conversion. Incursions of raccoon rabies into southern Ontario (Rosatte *et al.*, 2009), western New Brunswick (Badcock and Goltz, 2005), and southern Quebec have necessitated contingency actions as well. Point infection control (PIC) or modified versions of PIC have been applied with success (Rosatte *et al.*, 2009), but the use of ONRAB® in place of Raboral V-RG® apparently has contributed to accelerated success in Quebec where raccoon rabies virus was first detected in 2006. Since 2007, ONRAB® has been used as a part of contingency actions in Quebec and there have been no raccoon rabies cases detected in the province since 2009 (Canadian Food Inspection Agency, 2011).

Contingency actions have and will remain an integral part of rabies management in wild carnivores in North America. However, such actions are labour intensive, costing at least 2.5 times more than ORV alone (Rosatte *et al.*, 2001) and, therefore, are not sustainable over larger tracts of land or for extended periods of time. Often these special management actions are employed to preserve or restore existing ORV zones or to prevent rabies from spreading where no ORV zone has been established (Slate *et al.*, 2009). Frequent occurrence of raccoon rabies virus in striped skunks in the eastern United States (Krebs *et al.*, 2003a) also may necessitate TVR in contingency actions due to the limited effectiveness of Raboral V-RG®. Skunks may maintain the raccoon rabies virus variant (Guerra *et al.*, 2003) and confound local elimination of raccoon rabies. Enhanced surveillance cases from contingency action zones in northeast Ohio suggest numerically that the skunk may play a role in maintaining raccoon rabies, where after intensive control with ORV and TVR beginning in 2004 and 2005, two to eight skunks infected with raccoon rabies virus have been detected since 2009 (Fig. 13.7).

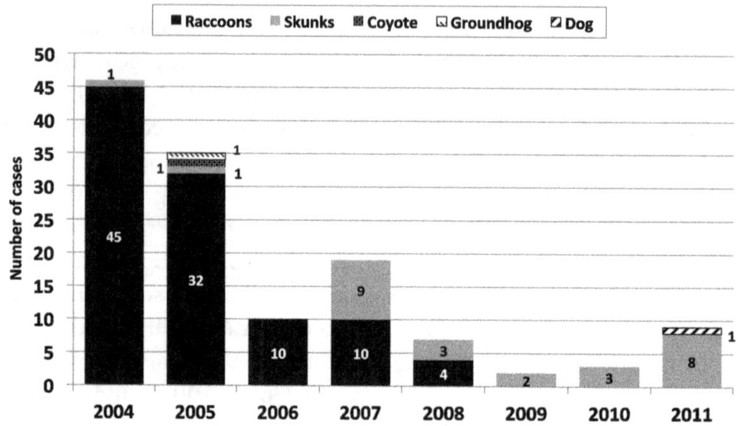

Figure 13.7 Raccoon rabies cases in the contingency action oral rabies vaccination zone of north-eastern Ohio, United States, 2004–2011 (through 10 October 2011).

While this example underscores the need for an effective vaccine bait to orally vaccinate skunks to better ensure success in raccoon rabies control efforts, a clearer understanding of the functional rabies dynamic between the skunk and raccoon is needed. One is still left to muse if the frequent occurrence of raccoon rabies in skunks reflects a genuine host shift or is simply spill-over and a dead-end infection.

To move beyond a 'hold the line' strategy, that has successfully prevented raccoon rabies from spreading, towards true elimination efforts, modelling (Thulke and Eisinger, 2008) and empirical evidence from Ontario, where elimination appears to have been achieved (Rosatte *et al.*, 2009), point to the need for sero-conversion rates higher than the 30% to 40% often observed in areas vaccinated with Raboral V-RG® (Slate *et al.*, 2009). In a comparative ORV study along the Maine-New Brunswick border, where covariates were similar (i.e. raccoon density, time of baiting, bait density, aerial delivery), sero-conversion for ONRAB® was greater than twice (73.5%) that observed for Raboral V-RG® (Fehlner-Gardiner *et al.*, 2012). The outcome of that study, early trends from a similar study along the Vermont-Quebec border (J. Mainguy, personal communication), and lack of access to other candidate vaccines that had cleared the regulatory and environmental compliance process, led to the first field trial using ONRAB® in the United States in September 2011 (USDA, 2011b). The results from this field trial will have a direct bearing on future trials with ONRAB® to address additional field safety and immunogenicity questions in strategic management locations elsewhere. Other ORV candidates will continue to merit consideration for testing as they become available.

Rabies in introduced species

Introduction for fur resources

Raccoons are a New World species. The raccoon was successfully introduced to Germany in 1934 (Müller-Using, 1956), and the former Soviet Union beginning in 1936, for its fur value (Aliev and Sanderson, 1966). Populations have prospered and the raccoon range has expanded to include several European countries with the attendant human-raccoon conflicts and disease risks. Raccoons have achieved their highest reported densities around residential suburbs in Germany, with Hesse supporting the highest known raccoon densities at ~ 50–150 raccoons/km^2. To date, the raccoon has not emerged as an important rabies spill-over host or reservoir. In fact, only 15 rabid raccoons were reported in association with fox rabies epizootics between 1960 and 1975, most from the northern part of Hesse (Rojahn, 1977). This epizootiologic pattern indicates that the raccoon did not play an important role in the emergence or spread of fox rabies in Germany (Tschirch, 2001).

Nevertheless, high raccoon abundance, particularly in residential areas, poses a potential risk if rabies should be introduced into these populations. Among the important lessons learned from rabies management in American raccoons is that the suburban interface supports both high raccoon and human population densities and is the most challenging environment in which to conduct control activities. Consequently, effective rabies management has required attention to unique bait distribution patterns, high bait densities, contingency actions that integrate several control tactics (Rosatte *et al.*, 2001; Slate *et al.*, 2009) and special attention to mitigate against vaccine-bait contacts by pets and people and the ability for a public health response in real time should such contacts occur (Roess *et al.*, 2012).

Introductions of the raccoon dog (*Nyctereutes procyonoides*) as a furbearer primarily to the European portion of the former Soviet Union from 1929–1955, including releases close to Finland (Helle and Kauhala, 1991), have resulted in expanding populations that have become established across much of northern and eastern Europe (Holmala and Kauhlala, 2006). Spill-over of rabies from Arctic sources in Finland during the late 1980s resulted in an epizootic in which raccoon dogs accounted for 73% of observed cases (18% in foxes) (Westerling, 1991). Thereafter, ORV has been applied with success in eliminating rabies in raccoon dogs from Finland (Brookes *et al.*, 2004). However, the potential contribution of hunting and winter dormancy towards this success are not clear (Nyberg *et al.*, 1992). Through 2003, the trend was towards an increasing proportion of rabies cases in raccoon dogs in the Baltic States, Poland and the Russian Federation (WHO, 2003; Holmala and Kauhlala, 2006). Greater numbers of cases in the red fox suggest that the raccoon dog is a spill-over host rather than a reservoir. However, 64% of rabies cases during the first quarter of 2004 in Estonia (WHO, 2004) and 52% of cases in Lithuania (WHO, 2006) were reported in raccoon dogs, raising suspicion about its status as a reservoir. Nevertheless, rabies in this exotic species has increased complexity to the meso-carnivore community level given that control strategies must be adapted to address both fox and raccoon dog ecological and behavioural differences and that a better understanding of the ultimate impact the raccoon dog may have in rabies dynamics in northern and eastern Europe is required (Holmala and Kauhlala, 2006, 2009). The effects of such a meso-carnivore introduction in the early to mid-twentieth century, such as the risk of combined threshold host density in relation to rabies dynamics (Holmala and Kauhlala, 2006) and added cost of control, are only now beginning to be more fully realized.

Introduction for presumed biological control

The small Asian mongoose (*Herpestes javanicus*) has been introduced to many sugarcane growing regions for the intended purpose of biological control of rodents [primarily rats (*Rattus* sp.)] that cause substantial losses in sugarcane production (Garlough, 1938; Martin *et al.*, 2007). These introductions have proven ineffective for a variety of reasons, including differences in predator-prey activity patterns, the mongoose being an ecological generalist, and the prolific reproductive capacity of rats (Hoagland *et al.*, 1989). Nevertheless, the mongoose has emerged as an important rabies reservoir and public health threat in Cuba, Grenada (Everard and Everard, 1985), Puerto Rico (Tierkel *et al.*, 1952; Krebs *et al.*, 2003b; Blanton *et al.*, 2007), the Dominican Republic, and likely in portions of Haiti, by extension (Everard and Everard, 1992).

The risk of rabies translocation (Chipman *et al.*, 2008) to established mongoose populations on the other 26 islands in the Caribbean (Horst *et al.*, 2001), currently free of rabies, persists in the absence of effective control. Translocation is a risk to mainland areas in southern North America, and Central and South America as well.

Sources of rabies viruses circulate among mongooses in the Caribbean, with distinct lineages in Puerto Rico closely related to a canine-originated north central skunk variant in North America (Nadin-Davis *et al.*, 2008) and a variant in Cuba closely related to Mexican dog rabies (Nadin-Davis *et al.*, 2006). Additional sequencing of rabies virus isolates from insular mongoose populations are required to better understand regional dynamics and establish baselines for evaluating potential translocations or virus evolutionary events. Recent molecular analyses relate mongoose rabies to historic dog lineages (Nadin-Davis *et al.*, 2008; Velasco-Villa *et al.*, 2008), which predated the introduction of mongoose to the Caribbean in the 1870s. Consequently, control of dog rabies in affected Caribbean countries cannot ignore rabies in the exotic mongoose.

A reverse genetics oral rabies virus vaccine with proven effectiveness in captive mongooses (Blanton *et al.*, 2006b) and documented acceptance of a variety of baits under field conditions (Linhart *et al.*, 1993; Creekmore *et al.*, 1994) provide the rudimentary tools to contemplate ORV field trials. Effective tools and strategies applied to insular mongoose populations provide an unparalleled opportunity to realistically consider rabies elimination in this exotic species to ultimately supplant lethal approaches to rabies control that produce transient local effectiveness (Everard and Everard, 1992). Given that baits are the most significant cost driver in ORV campaigns (Kemere *et al.*, 2002) and that mongooses can achieve high densities in specific habitats on some islands (Horst *et al.*, 2001), spatially temporally compartmentalized strategies will likely be required to conduct effective phased control. For example, strategies that reduce competition for baits by non-target domestic and feral cats and rats will likely be also critical, as preliminary results for Puerto Rico illustrate (Slate and Rupprecht, 2010) (Fig. 13.8). Bait distribution after daylight to mid-day may be one of several viable approaches to reduce competition from nocturnal rats. Non-fishy baits may reduce bait consumption by cats.

The small Asian mongoose has been listed among the world's 100 worst invasive species (IUCN, 2000). Consequently, profound adverse impacts on native fauna by the exotic mongoose, including species extirpations and extinctions (Seaman, 1952; Vilella, 1998; Roy *et al.*, 2002; Engeman *et al.*, 2006) pose a challenge and an opportunity in contemplating strategies for eliminating rabies in insular mongoose populations. In spite of substantial public health risks (Everard and Everard, 1985), the challenge is to provide convincing science-based arguments to decision makers that removal of rabies mortality would not result in additive adverse effects on rare and endangered native fauna (i.e. rabies is not an important mortality factor that regulates insular mongoose populations). Low levels of rabies mortality (0.6–3.5%) and proportionately high levels of naturally acquired rabies VNA (22–52%) during a declining rabies cycle on Grenada from 1971 to 1974 (Everard and Everard, 1985) suggest a different dynamic than occurs in foxes in Europe, where high levels of fox mortality occur through a rabies 'cycle' in the absence of control (Wandeler, 2004). Clearly, additional studies of mongoose and rabies dynamics are required, but

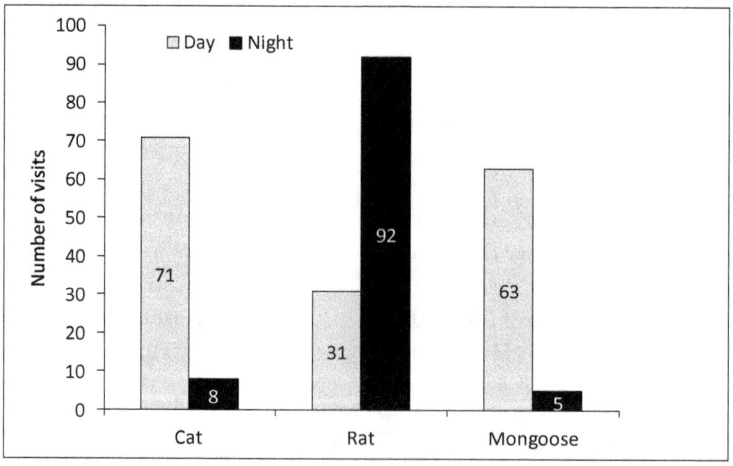

Figure 13.8 Estimation of mongoose, cat and rat visits to camera sites with placebo oral rabies vaccine baits (fishmeal polymer and coated sachet bait types) in Puerto Rico, July 2004.

in the interim period mongoose impacts continue to take their toll. In view of these potential conflicts, the mongoose 'problem' presents a potential opportunity to integrate immuno-contraception into oral rabies vaccine for dual effects. Encouraging research by Wu *et al.* (2009) that demonstrated that gonadotropin-releasing hormone (GnRH)-carrying rabies viruses induced antibodies against GnRH in immunized mice and protected 100% of vaccinated animals after rabies virus challenge should logically be extended to captive testing in mongoose. The absence of native terrestrial mammalian fauna on small islands distant from mainland in general and on Puerto Rico specifically (Turvey *et al.*, 2007), for example, eliminates non-target wildlife vaccine risks associated with contraception. This approach could significantly reduce the cost of baiting by suppressing fecundity to enhance rabies elimination that in turn could have measurable positive effects on native species, such as the recovery of the highly endangered Puerto Rican parrot. Benefit – cost analyses should be considered to evaluate scenarios that may be considered over the next 5 years.

Reflections on wildlife rabies prevention and control

Controlling rabies at its source in reservoir species offers the most exciting prospect of long-term benefits in human and animal health, and reduced expenditures for coexistence with rabies. However, domestic dog rabies has not been controlled over large parts of the developing world. Fox rabies control efforts in Europe and North America illustrate that the promise of elimination over large land masses is achievable in wild carnivore hosts through the strategic application of ORV. The appearance of raccoon rabies in Central Park, New York City, frame some of the challenges faced by wildlife rabies control in urban settings, particularly with species such as the raccoon that are capable of adapting and prospering from the abundance of anthropogenic energy subsidies available in such environments. Advances in modelling that can be applied to enhance control efforts will continue to be 'value added' given the time and resources required to collect and evaluate empirical data from control strategies applied in the field (Recuenco *et al.*, 2012). Rabies in insular populations of the small Asian mongoose has been largely ignored in the development of novel methods such as ORV. Ironically, this may be an opportunity to demonstrate true intervention for an insular population and contribute to the elimination of dog lineages of rabies viruses in the Caribbean. Wildlife control may be at a crossroads with respect to government funding, as any protracted economic downturn could result in a break in critical funding that could compromise accomplishments targeting certain reservoirs species to date. Therefore, it is necessary to remain focused and not become distracted from the primary mission of demonstrating successes and building advocacy from the platform of science-based results. Economic modelling will become increasingly vital to illustrate benefits relative to costs, rather than a sole preoccupation with programme costs outside of the context of the societal and environmental benefits they provide in the larger arena of a 'One Health' concept.

Acknowledgements

We thank all of our respective staff of the Rabies Program at CDC and at USDA, Wildlife Services for their dedication and contributions to date, as well as the academic, agricultural, public health and wildlife agencies and institutions in the United States that have contributed to the national programme, our partners in the North American Rabies Management Plan for their cooperation towards a continental approach of disease intervention, and our many colleagues around the world that have contributed to a global One Health vision towards modern, humane, economical, and effective disease prevention and control efforts.

The findings and conclusions in this report are those of the authors and do not necessarily represent the views of their institutions.

The use of any trade names or commercial products is for comparison purposes only and does not constitute endorsement by the United States government.

References

Abazeed, M.E., and Cinti, S. (2007). Rabies prophylaxis for pregnant women. Emerg. Infect. Dis. *13*, 1966–1967.

Aliev, F.F., and Sanderson, G.C. (1966). Distribution and status of the raccoon in the Soviet Union. J. Wildl. Manage. *30*, 497–502.

Al-Kappany, Y.M., Lappin, M.R., Kwok, O.C.H., Abu-Elwafa, S.A., Hilali, M., and Dubey, J.P., (2011). Seroprevalence of *Toxoplasma gondii* and concurrent *Bartonella* spp., feline immunodeficiency virus, feline leukemia virus, and *Dirofilaria imitis* infection in Egyptian cats. J. Parasit. *97*, 256–258.

Almeida, M.F., Martorelli, L.F., Aires, C.C., Barros, R.F., and Massad, E. (2008). Vaccinating the vampire bat *Desmodus rotundus* against rabies. Virus Res. *137*, 275–277.

Aréchiga-Ceballos, N., Velasco-Villa, A., Shi, M., Flores-Chávez, S., Barrón, B., Cuevas-Domínguez, E., González-Origel, A., and Aguilar-Setién, A. (2010). New rabies virus variant found during an epizootic in white-nosed coatis from the Yucatan Peninsula. Epidemiol. Infect. *138*, 1586–1589.

AAFP. (2007). American Association of Feline Practitioners: feline vaccination guidelines. Available at: http://catvets.com/professionals/guidelines/publications/?Id=176.

Badrane, H., and Tordo, N. (2001). Host switching in *Lyssavirus* history from the Chiroptera to the Carnivora orders. J. Virol. *75*, 8096–8104.

Badcock, J., and Goltz, J. (2005). Elimination of raccoon rabies from New Brunswick, Canada. Proceedings of the 16th Annual Conference on Rabies in the Americas. Ottawa, Canada, 16–21 October 2005.

Banyard, A.C., Hayman, D., Johnson, N., McElhinney, L., and Fooks, A.R. (2011). Bats and lyssaviruses. Adv. Virus Res. *79*, 239–289.

Beyer, H.L., Hampson, K., Lembo, T., Cleaveland, S., Kaare, M., and Haydon, D.T. (2011). The implications of metapopulation dynamics on the design of vaccination campaigns. Vaccine *30*, 1014–1022.

Blancou, J. (2008). The control of rabies in Eurasia: overview, history and background. Dev. Biol. *131*, 3–15.

Blanton, J.D., and Rupprecht, C.E. (2008). Travel vaccination for rabies. Expert Rev. Vaccines *7*, 613–620.

Blanton, J.D., Bowden, N.Y., Eidson, M., Wyatt, J.D., and Hanlon, C.A. (2005). Rabies postexposure prophylaxis, New York, 1995–2000. J. Emerg. Infect. Dis. *11*, 1921–1927.

Blanton, J.D., Manangan, A., Manangan, J., Hanlon, C.A., Slate, D., and Rupprecht. C.E. (2006a). Development of a GIS-based, real-time internet mapping tool for rabies surveillance. Int. J. Health Geographics *5*, 47.

Blanton, J.D., Meadows, A., Murphy, S.M., Manangan, J., Hanlon, C.A., Faber, M.-L., Dietzschold, B., and Rupprecht, C.E. (2006b). Vaccination of Small Asian Mongoose (*Herpestes javanicus*) against rabies. J. Wildl. Dis. *42*, 663–666.

Blanton, J.D., Hanlon, C.A., and Rupprecht, C.E. (2007). Rabies surveillance in the United States during 2006. J. Amer. Vet. Med. Assoc. *231*, 540–556.

Blanton, J.D., Palmer, D., Dyer, J., and Rupprecht, C.E. (2011). Rabies surveillance in the United States during 2010. J. Amer. Vet. Med. Assoc. *231*, 540–556.

Both, L., Banyard, A.C., van Dolleweerd, C., Horton, D.L., Ma., J.K., and Fooks, A.R. (2012). Passive immunity in the prevention of rabies. Lancet Infect. Dis. *12*, 397–407.

Boyko, A.R., Boyko, R.H., Boyko, C.M., Parker, H.G., Castelhano, M., Corey, L., Degenhardt, J.D., Auton, A., Hedimbi, M., Kityo, R., *et al.* (2009). Complex population structure in African village dogs and its implications for inferring dog domestication history. Proc. Natl. Acad. Sci. U.S.A. *106*, 13903–13908.

Briggs, D.J. (2011). Rabies: the biological basis for immunization, module 17 (WHO, Geneva, Switzerland), pp. 23.

Briggs, D.J., and Schwenke, J.R. (1992). Longevity of rabies antibody titre in recipients of human diploid cell rabies vaccine. Vaccine *10*, 125–129.

Brito, M.G., Chamone, T.L., da Silva, F.J., Wada, M.Y., Miranda, A.B., Castilho, J.G., Carrieri, M.L., Kotait, I., and Lemos, F.L. (2011). Ante mortem diagnosis of human rabies in a veterinarian infected when handling a herbivore in Minas Gerais, Brazil. Rev. Inst. Med. Trop. Sao Paulo *53*, 39–44.

Brookes, S.M., McElbinney, L.M., Johnson, N., Bradane, H., Tordo, N., and Fooks, A.R. (2004). Rabies virus variants and molecular epidemiology in Europe. In Historical Perspective of Rabies in Europe and the Mediterranean Basin, King, A.A., Fooks, A.R., Aubert, M., and Wandeler, A.I., eds (World Organization for Animal Health (OIE), Paris), pp. 243–258.

Canadian Food Inspection Agency. (2011). Positive rabies in Canada. Available at: http://www.inspection.gc.ca/english/anima/disemala/rabrag/statse.shtml (accessed October 2011).

Carroll, H.J., McCall, B.J., Looke, D., and Fraser, B. (2011). Challenges in postexposure prophylaxis of a vaccinated bat carer. Med. J. Aust. *195*, 323–324.

Carvalho-Costa, F.A., Tedesqui, V.L., de Jesus Nascimento Monteiro, M., and Bóia, M.N. (2012). Outbreaks of attacks by hematophagous bats in isolated riverine communities in the Brazilian Amazon: a challenge to rabies control. Zoonoses Public Health. *59*, 272–277.

Castillo, D., and Clarke, A.L. (2003). Trap/neuter/release methods ineffective in controlling domestic cat 'colonies' on public lands. Nat. Areas J. *23*, 247–253.

CDC. (1999). Mass treatment of humans who drank unpasteurized milk from rabid cows – Massachusetts, 1996–1998. MMWR Morb. Mortal. Wkly Rep.*48*, 228–229.

CDC. (2008). Human rabies prevention, Advisory Committee on Immunization Practices. MMWR Morb. Mortal. Wkly Rep.*57(R&R 3)*, 1–28.

CDC. (2011). Human rabies – Wisconsin, 2010. MMWR Morb. Mortal. Wkly Rep.*60*, 1164–1166.

CDC. (2012). Imported human rabies – New Jersey, 2011. MMWR Morb. Mortal. Wkly Rep.*60*, 1734–1736.

Childs, J.E., Colby, L., Krebs, J.W., Strine, T., Feller, M., Noah, D., Drenzek, C., Smith, J.S., and Rupprecht, C.E. (1997). Surveillance and spatiotemporal associations of rabies in rodents and lagomorphs in the United States, 1985–1994. J. Wildl. Dis. *33*, 20–27.

Chipman, R.B., Slate, D., Rupprecht, C., and Mendoza, M., (2008). Downside risk of wildlife translocation. Dev. Biol. (Basel) *131*, 223–232.

Clark, K.A., Neill, S.U., Smith, J.S., Wilson, P.J., Whadford, V.W., and McKirahan, G.W. (1994). Epizootic canine rabies transmitted by coyotes in south Texas. J. Amer. Vet. Med. Assoc. *204*, 536–540.

Cliquet, F., and Aubert, M. (2004). Elimination of terrestrial rabies in western European countries. Dev. Biol. *119*, 185–204.

Corn, J.L., and Conroy, M.J. (1998). Estimation of density of mongooses with capture-recapture and distance sampling. J. Mammal. *79*, 1009–1015.

Creekmore, T.E., Linhart, S.B., Corn, J.L., Whitney, M.D., Snyder, B.D., and Nettles, V.F. (1994). Field evaluation of baits and baiting strategies for delivering oral vaccine to mongooses in Antigua, West Indies. J. Wildl. Dis. *30*, 497–505.

David, D., Bellaiche, M., and Yakobson, B.A. (2010). Rabies in two vaccinated dogs in Israel. Vet. Rec. *167*, 907–908.

Deyoung, R.W., Zamorano, A., Mesenbrink, B.T., Campbell, T.A., Leland, B.R., Moore, G.M., Honeycutt, R.L., and Root, J.J. (2009). Landscape-genetic analysis of population structure in the Texas gray fox oral rabies vaccination zone. J. Wildl. Manage. *73*, 1292–1299.

Dietzschold, B., Morimoto, K., Hooper, D.C., Smith, J.S., Rupprecht, C.E., and Koprowski, H. (2000). Genotypic and phenotypic diversity of rabies virus variants involved in human rabies: implications for postexposure prophylaxis. J. Hum. Virol. 3, 50–57.

Dietzschold, B., Li, J., Faber, M., and Schnell, M. (2008). Concepts in the pathogenesis of rabies. Future Virol. 3, 481–490.

Dimaano, E.M., Scholand, S.J., Alera, M.T., and Belandres, D.B. (2011). Clinical and epidemiological features of human rabies cases in the Philippines: a review from 1987 to 2006. Int. J. Infect. Dis. *15*, e495–499.

Drolet, B.S., Stuart, M.A., and Derner, J.D. (2009). Infection of *Melanoplus sanguinipes* grasshoppers following ingestion of rangeland plant species harboring vesicular stomatitis virus. Appl. Environ. Microbiol. *75*, 3029–3033.

Dürr, S., Naïssengar, S., Mindekem, R., Diguimbye, C., Niezgoda, M., Kuzmin, I., Rupprecht, C.E., and Zinsstag, J. (2008). Rabies diagnosis for developing countries. PLoS Negl. Trop. Dis. 2, e206.

Eidson, M., Matthews, S.D., Willsey, A.L., Cherry, B., Rudd, R.J., and Trimarchi, C.V. (2005). Rabies virus infection in a pet guinea pig and seven pet rabbits. J. Am. Vet. Med. Assoc. *227*, 932–935.

Elmgren, L.D., and Wandeler, A.I. (1996). Competitive ELISA for the detection of rabies virus neutralizing antibodies. In Laboratory Techniques in Rabies, 4th edn, Meslin, F.-X., Kaplan, M.M., and Koprowski, H., eds (WHO, Geneva, Switzerland), pp. 200–208.

Engeman, R., Whisson, D.A., Quinn, J.H., Cano, F., Quinonez, P., and White, T.H., Jr. (2006). Monitoring invasive mammalian predator populations sharing habitat with the critically endangered Puerto Rican parrot (*Amazona vittata*). Oryx *40*, 95–102.

Enzmann, P.J., Castric, J., Bovo, G., Thiery, R., Fichtner, D., Schütze, H., and Wahli, T. (2010). Evolution of infectious hematopoietic necrosis virus (IHNV), a fish rhabdovirus, in Europe over 20 years: implications for control. Dis. Aquat. Organ. *89*, 9–15.

Everard, C.O.R., and Everard, J.D. (1985). Mongoose rabies in Grenada. In Population Dynamics of Rabies in Wildlife, Bacon, P.J., ed. (Academic Press, London, UK), pp. 43–69.

Everard, C.O.R., and Everard, J.D. (1992). Mongoose rabies in the Caribbean. Ann. N.Y. Acad. Sci. *653*, 356–366.

Fayaz, A., Simani, S., Janani, A., Farahtaj, F., Biglari, P., Howeizi, N., and Eslami, N. (2011). Antibody persistence, 32 years after post-exposure prophylaxis with human diploid cell rabies vaccine (HDCV). Vaccine *29*, 3742–3745.

Fearneyhough, M.G., Wilson, P.J., Clark, K.A., Smith, D.R., Johnston, D.H., Hicks, B.H., and Moore, G.M. (1998). Results of an oral rabies vaccination program for coyotes. J. Amer. Vet. Med. Assoc. *212*, 498–502.

Fehlner-Gardiner, C., Rudd, R., Donovan, D., Slate, D., Kempf, L., and Badcock, J. (2012). Comparison of ONRAB® and Raboral V-RG® oral rabies vaccine field performance in raccoons and striped skunks in New Brunswick, Canada and Maine, USA. J. Wildl. Dis. *48*, 157–167.

Foley, P., Foley, J.E., Levy, J.K., and Paik, T. (2005). Analysis of the impact of trap-neuter-return programs on populations of feral cats. J. Amer. Vet. Med. Assoc. *227*, 1775–1781.

Freuling, C., and Müller, T. (2009). Rabies Bulletin Europe 33, 3rd Quarter. Available at: http://www.who-rabies-bulletin.org/ (accessed March 2010).

Frymus, T., Addie, D., Belak, S., Boucraut-Baralon, C., Egberink, H., Gruffydd-Jones, T., Hartmann, K., Hosie, M.J., Lloret, A., Lutz, H., *et al.* (2009). Feline Rabies: ACBD guidelines on prevention and management. J. Feline Med. Surg. *11*, 585–593.

Galibert, F., Quignon, P., Hitte, C., and André, C. (2011). Toward understanding dog evolutionary and domestication history. CR Biol. *334*, 190–196.

Garlough, F.E. (1938). Rodents in relation to sugarcane growing. Proc. Int. Soc. Sugarcane Technol. *6*, 98–106.

Gascoyne, S.C., Laurenson, M.K., Lelo, S., and Borner, M. (1993). Rabies in African wild dogs (*Lycaon pictus*) in the Serengeti region, Tanzania. J. Wildl. Dis. *29*, 396–402.

Gibbons, R.V. (2002). Cryptogenic rabies, bats, and the question of aerosol transmission. Ann. Emerg. Med. *39*, 528–536.

Gomot, F., Artois, M., Langlais, M., Courchamp, F., and Pontier, D. (1997). Modelling the feline leukemia virus (FeLV) in natural populations of cats (*Felis catus*). Theor. Pop. Bio. *52*, 60–70.

Grosenbaugh, D.A., Maki, J.L., Rupprecht, C.E., and Wall, D.K. (2007). Rabies challenge of captive striped skunks (*Mephitis mephitis*) following oral administration of a live vaccinia-vectored rabies vaccine. J. Wildl. Dis. *43*, 124–128.

Guerra, M.A., Curns, A.T., Rupprecht, C.E., Hanlon, C.A., Krebs, J.W., and Childs, J.E. (2003). Skunk and raccoon rabies in the eastern United States: temporal and spatial analysis. Emerg. Infect. Dis. 9: 1143–1150.

Gunther, I., Finkler, H., and Terkel, J. (2100). Demographic differences between urban feeding groups of neutered and sexually intact free-roaming cats following a trap-neuter-return procedure. J. Am. Vet. Med. Assoc. *238*, 1134–1140.

Hampson, K., Cleaveland, S., and Briggs, D. (2011). Evaluation of cost-effective strategies for rabies post-exposure vaccination in low-income countries. PLoS Negl. Trop. Dis. *5*, e982.

Hanlon, C.A., Niezgoda, M., Morrill, P., and Rupprecht, C.E. (2002). Oral efficacy of an attenuated rabies virus vaccine in skunks and raccoons. J. Wildl. Dis. *38*, 420–427.

Helle, E., and Kauhala, K. (1991). Distribution history and present status of the raccoon dog in Finland. Holarctic Ecol. *14*, 278–286.

Henderson, H., Jackson, F., Bean, K., Panasuk, B., Niezgoda, M., Slate, D., Li, J., Dietzschold, B., Mattis, J., and Rupprecht, C. (2009). Oral immunization of raccoons and skunks with a canine adenovirus recombinant rabies vaccine. Vaccine *27*, 7194–7197.

Hoagland, D.B., Horst, G.R., and Kilpatrick, C.W. (1989). Biogeography and population biology of the mongoose in the West Indies. In Biogeography of the West Indies Past, Present and Future, Woods, C.A., ed. (Sandhill Crane Press, Inc., Gainsville, FL), pp. 611–634.

Hofmeyr, M., Hofmeyr, D., Nel, L., and Bingham, J. (2004). A second outbreak of rabies in African wild dogs (*Lycaon pictus*) in Madikwe Game Reserve, South Africa, demonstrating the efficacy of vaccination against natural rabies challenge. Animal Cons. *7*, 193–198.

Holmala, K., and Kauhlala, K. (2006). Ecology of wildlife rabies in Europe. Mammal Rev. *36*, 17–36.

Holmala, K., and Kauhlala, K. (2009). Habitat use of medium sized carnivores in southeast Finland – key habitats for rabies spread? Ann. Zool. Fenn. *46*, 223–246.

Horst, G.R., Hoagland, D.B., and Kilpatrick, C.W. (2001). The mongoose in the West Indies: the biogeography of an introduced species. In Biogeography of the West Indies: Patterns and Perspectives, 2nd edn, Woods, C.A., and Sergile, F.E., eds (CRC Press, Boca Raton, FL), pp. 409–424.

Hossain, M., Bulbul, T., Ahmed, K., Ahmed, Z., Salimuzzaman, M., Haque, M.S., Ali, A., Hossain, S., Yamada, K., Moji, K., *et al.* (2011). Five-year (January 2004–December 2008) surveillance on animal bite and rabies vaccine utilization in the Infectious Disease Hospital, Dhaka, Bangladesh. Vaccine *29*, 1036–1040.

IUCN. (2000). 100 of the world's worst invasive alien species. Aliens *12*, 11.

Jackson, A.C., and Wunner, W.H. (2007). Rabies, 2nd edn (Academic Press, New York).

Johnson, N., Freuling, C., Horton, D., Muller, T., and Fooks, A.R. (2011). Imported rabies, European Union and Switzerland, 2001–2010. Emerg. Infect. Dis. *17*, 753–754.

Johnston, D.H., and Fong, D.W. (1992). Epidemiology of arctic fox rabies. In Wildlife Rabies Control, Bogel, K., Meslin, F.-X., and Kaplan, M., eds (Wells Medical Ltd, Kent, UK), pp. 45–49.

Johnston, D.H., Joachim, D., Bachmann, P., Kardong, K., Stewart, R., Dix, L., Strickland, M., and Watt, I. (1999). Aging furbearers using tooth structure and biomarkers. In Wild Furbearer Management and Conservation in North America, Novak (CD edn), M., Baker, J., Obbard, M., and Malloch, B., eds (Ontario Fur Managers Federation, Sault Ste Marie, Canada), pp. 228–243.

Johnston, J.J., Primus, T.M., Buettgenbach, T., Furcolow, C.A., Goodall, M.J., Slate, D., Chipman, R.B., Snow, L.J., and DeLiberto, T.J. (2005). Evaluation and significance of tetracycline stability in rabies vaccine baits. J. Wildl. Dis. *41*, 549–558.

Kat, P.W., Alexander, K.A., Smith, J.S., and Munson, L. (1995). Rabies and African wild dogs in Kenya. Proc. Royal Soc. B: Biol. Sci. *262*, 229–233.

Kemere, P., Liddel, M.K., Evangelou, P., Slate, D., and Osmek, S. (2002). Economic analysis of a large-scale oral vaccination program to control raccoon rabies. In Human Conflicts with Wildlife: Economic Considerations, Clark, L., ed. (National Wildlife Research Center, Fort Collins, CO), pp. 109–116.

Kikuti, M., Paploski, I.A., Silva, M.C., de Oliveira, E.A., da Silva, A.W., and Biondo, A.W. (2011). Prevention educational program of human rabies transmitted by bats in rain forest preserved area of southern Brazilian coast. Zoonoses Public Health *58*, 529–532.

Kim, M.S., Kim, D.S., and Kim, K.H. (2011). Oral immunization of olive flounder (*Paralichthys olivaceus*) with recombinant live viral hemorrhagic septicemia virus (VHSV) induces protection against VHSV infection. Fish Shellfish Immunol. *31*, 212–216.

Knobel, D.L., Cleaveland, S., Coleman, P.G., Fèvre, E.M., Meltzer, M.I., Elizabeth, M., Miranda, G., Shaw, A., Zinsstag, J., and Meslin, F-X. (2005). Re-evaluating the burden of rabies in Africa and Asia. Bull. World Health Organ. *83*, 360–368.

Krebs, J.W., Wheeling, J.T., and Childs, J.E. (2003a). Rabies surveillance in the United States during 2002. J. Amer. Vet. Med. Assoc. *223*, 1736–1748.

Krebs, J.W., Williams, S.M., Smith, J.S., Rupprecht, C.E., and Childs, J.E. (2003b). Rabies among infrequently reported mammalian carnivores in the United States, 1960–2000. J. Wildl. Dis. *39*, 253–261.

Krebs, J.W., Mandel, E.J., Swerdlow, D.L., and Rupprecht, C.E. (2004). Rabies surveillance in the United States during 2003. J. Amer. Vet. Med. Assoc. *225*, 1837–1849.

Kuzmin, I.V., Mayer, A.E., Niezgoda, M., Markotter, W., Agwanda, B., Breiman, R.F., and Rupprecht, C.E. (2010). Shimoni bat virus, a new representative of the *Lyssavirus* genus. Virus Res. *149*, 197–210.

Lang, J., Feroldi, E., and Vien, N.C. (2009). Pre-exposure purified Vero cell rabies vaccine and concomitant routine childhood vaccinations: 5-year post-vaccination follow-up study of an infant cohort in Vietnam. J. Trop. Pediatr. *55*, 26–31.

Lembo, T., and Partners for Rabies Prevention (2012). The blueprint for rabies prevention and control: a novel operational toolkit for rabies elimination. PLoS Negl. Trop. Dis. *6*, e1388.

Lembo, T., Niezgoda, M., Velasco-Villa, A., Cleaveland, S., Ernest, E., and Rupprecht, C.E. (2006). Evaluation of a direct, rapid immuno-histochemical test for rabies diagnosis. Emerg. Infect. Dis. *12*, 310–313.

Lembo, T., Attlan, M., Bourhy, H., Cleaveland, S., Costa, P., de Balogh, K., Dodet, B., Fooks, A.R., Hiby, E., Leanes, F., *et al.* (2011). Renewed global partnerships and redesigned roadmaps for rabies prevention and control. Vet. Med. Int. *2011*, 923149.

Little, S. (2011). A review of feline leukemia virus and feline immunodeficiency virus seroprevalence in cats in Canada. Vet. Immunol. Immunopathol. *143*, 243–245.

Leslie, M.J., Messenger, S., Rohde, R.E., Smith, J., Cheshier, R., Hanlon, C., and Rupprecht, C.E. (2006). Bat-associated rabies virus in skunks. Emerg. Infect. Dis. *12*, 1274–1277.

Linhart, S.B., Creekmore, T.E., Corn, J.L., Whitney, M.D., Snyder, B.D., and Nettles, V.F. (1993). Evaluation of baits for oral rabies vaccination of mongooses: pilot field trials in Antigua, West Indies. J. Wildl. Dis. *29*, 290–294.

McElroy, K.M., Blagburn, B.L., Breitschwerdt, E.B., Mead, P.S., and McQuiston, J.H. (2010). Flea-associated zoonotic diseases of cats in the USA: bartonellosis, flea-borne rickettsioses, and plague. Trends Parasitol. *26*, 197–204.

MacInnes, C.D. (1987). Rabies. In Wild Furbearer Management and Conservation in North America, Novak, M., Baker, J.A., Obbard, M.E., and Malloch, B., eds (Ontario Trappers Association, North Bay, Canada), pp. 910–929.

MacInnes, C.D., Smith, S.M., Tinline, R.R., Ayers, N.R., Bachmann, P., Ball, D.G.A., Calder, L.A., Crosgrey, S.J., Fielding, C., Hauschildt, P., *et al.* (2001). Elimination of rabies from red foxes in eastern Ontario. J. Wildl. Dis. *37*, 119–132.

McLean, R.G. (1971). Rabies in raccoons in the southeastern United States. J. Infect. Dis. *123*, 680–681.

McNutt, J.W., Mills, M.G.L., McCreery, K., Rasmussen, G., Robbins, R., and Woodroffe, R. (2008). *Lycaon pictus*. In IUCN 2011 – red list of threatened species: version 2011.1. Available at: http://www.iucnredlist.org (accessed October 2011).

Maier, T., Schwarting, A., Mauer, D., Ross, R.S., Martens, A., Kliem, V., Wahl, J., Panning, M., Baumgarte, S., Müller, T., *et al.* (2010). Management and outcomes after multiple corneal and solid organ transplantations from a donor infected with rabies virus. Clin. Infect. Dis. *50*, 1112–1119.

Mansfield, K., McElhinney, L., Hübschle, O., Mettler, F., Sabeta, C., Nel, L.H., and Fooks, A.R. (2006). A molecular epidemiological study of rabies epizootics in kudu (*Tragelaphus strepsiceros*) in Namibia. BMC Vet. Res. *2*, 2.

Marino, J., and Sillero-Zubiri, C. (2011). *Canis simensis*. In IUCN 2011 – red list of threatened species: version 2011.1. Available at: http://www.iucnredlist.org (accessed October 2011).

Marston, D.A., Horton, D.L., Ngeleja, C., Hampson, K., McElhinney, L.M., Banyard, A.C., Haydon, D., Cleaveland, S., Rupprecht, C.E., Bigambo, M., *et al.* (2012). Ikoma lyssavirus, highly divergent novel lyssavirus in an African civet. Emerg. Infect. Dis. *18*, 664–667.

Martin, J.M., Raid, R.N., and Branch, L.C. (2007). Quantifying damage potential of three rodent species on sugarcane. J. Amer. Soc. Sugarcane Technol. *27*, 48–54.

Maruyama, S., Kabeya, H., Kakao, R., Tanaka, S., Sakai, T., Xuan, X., Katsube, T., and Mikama, T. (2003). Seroprevalence of *Bartonella henselae, Toxoplasma gondii*, FIV and FeLV infections in domestic cats in Japan. Microbiol. Immunol. *47*, 147–153.

Messenger, S.L., Smith, J.S., Orciari, L.A., Yager, P.A., and Rupprecht, C.E. (2003). Emerging pattern of rabies deaths and increased viral infectivity. Emerg Infect. Dis. *9*, 151–154.

Mexican Ministry of Health. (2011). Guidelines for the National Week of Canine Dog Vaccination in Mexico (National Center of Prevention and Disease Control, Mexico City, Mexico).

Miller, L.A., Johns, B.E., and Killian, G.J. (2000). Immunocontraception of white-tailed deer with GnRH vaccine. Amer. J. Reprod. Immunol. *44*, 266–274.

Moore, S.M., Wilkerson, M.J., Davis, R.D., Wyatt, C.R., and Briggs, D.J. (2006). Detection of cellular immunity to rabies antigens in human vaccines. J. Clin. Immunol. *26*, 533–545.

Müller, W., Cox, J., and Müller, T. (2004). Rabies in Germany, Denmark, and Austria. In Historical Perspective of Rabies in Europe and the Mediterranean Basin, King, A.A., Fooks, A.R., Aubert, M., and Wandeler, A.I., eds (World Organization for Animal Health (OIE), Paris), pp. 79–92.

Müller-Using, D. (1956). Die ausbreitung des waschbären in Westdeutschland. Z. Jagdwissensch. *5*, 108–109.

Murray, K.O., Holmes, K.C., and Hanlon, C.A. (2009). Rabies in vaccinated dogs and cats in the United States, 1997–2001. J. Amer. Vet. Med. Assoc. *235*, 691–695.

Nadin-Davis, S.A., Torres, G., De Los Angeles Ribas, M., Guzman, M., Cruz-De La Paz, R., Morales, M., and Wandeler, A.I. (2006). A molecular epidemiological study of rabies in Cuba. Epidemiol. Infect. *134*, 1313–1324.

Nadin-Davis, S.A., Velez, J., Malaga, C., and Wandeler, A.I. (2008). A molecular epidemiological study of rabies in Puerto Rico. Virus Res. *131*, 8–15.

NASPHV. (2011). Compendium of animal rabies prevention and control. MMWR Morb. Mortal. Wkly Rep.*60(R&R 6)*, 1–15.

Natoli, E., Maragliano, L., Cariola, G., Faini, A., Bonanni, R., Cafazzo, S., and Fantini, C. (2006). Management of feral domestic cats in the urban environment of Rome (Italy). Prev. Vet. Med. *77*, 180–185.

Niezgoda, M., Briggs, D.J., Shaddock, J., and Rupprecht, C.E. (1998). Viral excretion in domestic ferrets (*Mustela putorius furo*) inoculated with a raccoon rabies isolate. Amer. J. Vet. Res. *59*, 1629–1632.

Nutter, F.B., Dubey, J.P., Levine, J.F., Breitschwerdt, E.B., Ford, R.B., and Stoskopf, M.K. (2004). Seroprevalences of antibodies against *Bartonella henselae* and *Toxoplasma gondii* and fecal shedding of *Crptosporidium* spp, *Giardia* spp, and *Toxocara cati* in feral and pet domestic cats. J. Am. Vet. Med. Assoc. *225*, 1394–1398.

Nyberg, M., Kulonen, K., Neuvonen, E., Ek-Kommonen, C., Nuorgam, M., and Westerling, B. (1992). An epidemic of sylvatic rabies in Finland – descriptive epidemiology and results of oral vaccination. Acta Vet. Scand. *33*, 43–57.

O'Bell, S.A., McQuiston, J., Bell, L.J., Ferguson, S.C., and Williams, L.A. (2006). Human rabies exposures and post exposure prophylaxis in South Carolina, 1993–2002. Public Health Rep. *121*, 197–202.

Oertli, E.H., Wilson, P.J., Hunt, P.R., Sidwa, T.J., and Rohde, R.E. (2008). Epidemiology of rabies in skunks in Texas. J. Amer. Vet. Med. Assoc. *234*, 616–620.

Olson, C.A., Mitchell, K.D., and Werner, P.A. (2000). Bait ingestion by free-ranging raccoons and nontarget species in an oral rabies vaccine field trial in Florida. J. Wildl. Dis. *36*, 764–743.

Pamperin, N.J., Follmann, E.H., and Peterson, B. (2006). Interspecific killing of an arctic fox by a red fox at Prudhoe Bay, Alaska. Arctic *59*, 361–364.

Pamperin, N.J., Follmann, E.H., and Person, B.T. (2008). Sea ice use by arctic foxes in northern Alaska. Polar Biol *31*, 1421–1426.

Pastoret, P.-P., Kappeler, A., and Aubert, M. (2004). European rabies control and its history. In Historical Perspective of Rabies in Europe and the Mediterranean Basin, King, A.A., Fooks, A.R., Aubert, M., and Wandeler, A.I., eds (World Organization for Animal Health (OIE), Paris), pp. 337–350.

Petersen, B.W., Tack, D.M., Longenberger, A., Simeone, A., Moll, M.E., Deasy, M.P., Blanton, J.D., and Rupprecht, C.E. (2012). Rabies in captive deer, Pennsylvania, USA, 2007–2010. Emerg. Infect. Dis. *18*, 138–141.

Pybus, M.J. (1988). Rabies and rabies control in striped skunks (*Mephitis mephitis*) in three prairie regions of western North America. J. Wildl. Dis. *24*, 434–449.

Ramey, P.C., Blackwell, B.F., Gates, R.J., and Slemons, R.D. (2008). Oral rabies vaccination of a northern Ohio raccoon population: relevance of population density and prebait serology. J. Wildl. Dis. *44*, 553–568.

Rausch, R.L. (1958). Some observations on rabies in Alaska, with special reference to wild Canidae. J. Wildl. Manage. *22*, 246–260.

Recuenco, S., Blanton, J.D., and Rupprecht, C.E. (2012). A spatial model to forecast raccoon rabies emergence. Vector Borne Zoonotic Dis. *12*, 126–137.

Robertson, K., Lumlertdacha, B., Franka, R., Petersen, B., Bhengsri, S., Henchaichon, S., Peruski, L.F., Baggett, H.C., Maloney, S.A., and Rupprecht, C.E. (2011). Rabies-related knowledge and practices among persons at risk of bat exposures in Thailand. PLoS Negl. Trop. Dis. *5*, e1054.

Rojahn, A. (1977). Vorkommen der tollwut in der Bundesrepublik Deutschland (Occurrence of rabies in the Federal Republic of Germany). Berl. Münch. Tierärztl. *90*, 269–273.

Rosatte, R. (2011). Evolution of wildlife rabies control tactics. Adv. Virus Res. *79*, 397–419.

Rosatte, R.C., Pybus, M.J., and Gunson, J.R. (1986). Population reduction as a factor in the control of skunk rabies in Alberta. J. Wildl. Dis. *22*, 459–467.

Rosatte, R., Donovan, D., Allan, M., Howes, L., Silver, A., Bennett, K., MacInnes, C., Davies, C., Wandeler, A., and Radford, B. (2001). Emergency response to raccoon rabies introduction into Ontario. J. Wildl. Dis. *37*, 265–279.

Rosatte, R.C., Power, M.J., Donovan, D., Davies, J.C., Allan, M., Bachmann, P., Stevenson, B., Wandeler, A., and Muldoon, F. (2007a). Elimination of arctic fox variant rabies in red foxes, metropolitan Toronto. Emerg. Infect. Dis. *13*, 25–27.

Rosatte, R.C., Tinline, R.R., and Johnston, D.H. (2007b). Rabies control in wild carnivores. In Rabies, 2nd edn, Jackson, A.C., ed. (Elsevier Academic Press, Amsterdam, the Netherlands), pp. 595–624.

Rosatte, R.C., Donovan, D., Allan, M., Bruce, L., Buchanan, T., Sobey, K., Stevenson, B., Gibson, M., MacDonald, T., Whalen, M., *et al.* (2009). The control of raccoon rabies in Ontario Canada: proactive and reactive tactics, 1994–2007. J. Wildl. Dis. *45*, 772–784.

Rosatte, R.C., Donovan, D., Davies, J.C., Brown, L., Allan, M., von Zuben, V., Bachmann, P., Sobey, K., Silver, A., Bennett, K., *et al.* (2011). High-density baiting with ONRAB rabies vaccine baits to control arctic-variant rabies in striped skunks in Ontario, Canada. J. Wildl. Dis. *47*, 459–465.

Roess, A.A., Rea, N., Lederman, E., Dato, V., Chipman, R., Slate, D., Reynolds, M.G., Damon, I.K., and Rupprecht, C.E. (2012). National surveillance for human and pet contact with oral rabies vaccine baits, 2001–2009. J. Amer. Vet. Med. Assoc. *240*, 163–168.

Roy, S.S., Jones, C.G., and Harris, S. (2002). An ecological basis for control of the mongoose in Mauritius: is eradication possible? In Turning the Tide: The Eradication of Invasive Species, Veitch, C., and Clout, M., eds (World Conservation Union (IUCN), Gland, Switzerland), pp. 266–273.

Rupprecht, C.E., and Gibbons, R.V. (2004). Clinical practice: prophylaxis against rabies. N. Engl. J. Med. *351*, 2626–2635.

Rupprecht, C.E., Willoughby, R., and Slate, D. (2006). Current and future trends in the prevention, treatment and control of rabies. Expert Rev. Anti. Infect. Ther. *4*, 1021–1038.

Rupprecht, C.E., Barrett, J., Briggs, D., Cliquet, F., Fooks, A.R., Lumlertdacha, B., Meslin, F-X., Müler, T., Nel, L.H., Schneider, C., *et al.* (2008). Can rabies be eradicated? Dev. Biol. (Basel) *131*, 95–121.

Rupprecht, C.E., Turmelle, A., and Kuzmin, I.V. (2011). A perspective on lyssavirus emergence and perpetuation. Curr. Opin. Virol. *1*, 662–670.

Salahuddin, N., Jamali, S., Ibraheem, K., and Sardar, S. (2011). Awareness about rabies postexposure prophylaxis in Pakistan among patients and health care workers: results from an Asian Rabies Expert Bureau study. J. Coll. Physicians Surg. Pak. *21*, 491–494.

Salant, H., Weingram, T., Spira, D.T., and Eizenberg, T. (2009). An outbreak of Toxoplasmosis amongst squirrel monkeys in an Israeli monkey colony. Vet. Parasitol. *159*, 24–29.

Schmidt, P.M., Sawnnack, T.M., Lopez, R.R., and Slater, M.R. (2009). Use of matrix population models to estimate the efficacy of euthanasia versus trap-neuter-return for management of free-roaming cats. Wildl. Res. *36*, 117–125.

Seaman, G.A. (1952). The mongoose and Caribbean wildlife. Transactions 17th North American Wildlife Conference, pp. 188–197.

Shwiff, S.A., Sterner, R.T., Hale, R., Jay, M.T., Sun, B., and Slate, D. (2009). Benefit cost scenarios of potential oral rabies vaccination for skunks in California. J. Wildl. Dis. *45*, 227–233.

Sidwa, T.J., Wilson, P.J., Moore, G.M., Oertli, E.H., Hicks, B.H., Rohde, R.E., and Johnston, D.H. (2005). Evaluation of oral rabies vaccination programs for control of rabies epizootics in coyotes and gray foxes. J. Amer. Vet. Med. Assoc. *227*, 785–792.

Skoglund, P., Götherström, A., and Jakobsson, M. (2011). Estimation of population divergence times from non-overlapping genomic sequences: examples from dogs and wolves. Mol. Biol. Evol. *28*, 1505–1517.

Slate, D., and Rupprecht, C.E. (2010). Rabies management options for the small Asian mongoose. Unnumbered White Paper, Pan-American Health Organization, 2010 REIPRA, Buenos Aries, Argentina.

Slate, D., and Rupprecht, C.E. (2011). Rabies management in wild carnivores. In Fowler's Zoo and Wild Animal Medicine: Current Therapy 7, Miller, R.E., and Fowler, M.E., eds (W.B. Saunders Co., St. Louis, MO), pp. 366–375.

Slate, D., Rupprecht, C.E., Rooney, J.A., Donovan, D., Lein, D.H., and Chipman, R.B. (2005). Status of oral rabies vaccination in wild carnivores in the United States. Virus Res. *111*, 68–76.

Slate, D., Algeo, T.P., Nelson, K.M., Chipman, R.B., Donovan, D., Blanton, J.D., Niezgoda, M., and Rupprecht, C.E. (2009). Oral rabies vaccination in North America: opportunities, complexities, and challenges. PLoS Negl. Trop. Dis. *3*, e549.

Slate, D., Oertli, E.H., Nelson, K.M., Chipman, R.B., Algeo, T.P., Dubovi, E.J., and Rupprecht, C.E. (2011). Canine adenovirus antibody presence in meso-carnivores in the United States: Implications for oral rabies vaccination. Poster. The World Organization for Animal Health (OIE), Global Conference on Rabies Control: Towards Sustainable Prevention at the Source, Incheon, Republic of Korea, 7–9 September 2011.

Smith, T.G., Wu, X., Franka, R., and Rupprecht, C.E. (2011). Design of future rabies biologics and antiviral drugs. Adv. Virus Res. *79*, 345–363.

Sterner, R.T., Sun, B., Bourassa, J.B., Hale, R.L., Shwiff, S.A., Jay, M.T., and Slate, D. (2008). Skunk rabies in California (1992–2003) – implications for oral rabies vaccination. J. Wildl. Dis. *44*, 1008–1013.

Stewart, A.-M., Gordon, C., and Sillero-Zubiri, C. (2011). EWCP Annual Report. Unpublished annual report. 19 pp.

Sudarshan, M.K., Giri, M.S., Mahendra, B.J., Venkatesh, G.M., Sanjay, T.V., Narayana, D.H., and Ravish, H.S. (2007). Assessing the safety of post-exposure rabies immunization in pregnancy. Hum. Vaccin. *3*, 87–89.

Suzan, G., and Ceballos, G. (2005). The role of feral mammals on wildlife infectious disease prevalence in two nature reserves within Mexico City limits. J. Zool. Wildl. Med. *36*, 479–484.

Talbi, C., Lemey, P., Suchard, M.A., Abdelatif, E., Elharrak, M., Nourlil, J., Faouzi, A., Echevarría, J.E., Vazquez Morón, S., Rambaut, A., *et al.* (2010). Phylodynamics and human-mediated dispersal of a zoonotic virus. PLoS Pathog. *6*, e1001166.

Thiptara, A., Atwill, E.R., Kongkaew, W., and Chomel, B.B. (2011). Epidemiologic trends of rabies in domestic animals in southern Thailand, 1994–2008. Amer. J. Trop. Med. Hyg. *85*, 138–145.

Thulke, H.-H., and Eisinger, D. (2008). The strength of 70%: Revision of a standard threshold of rabies control. Dev. Biol. *131*, 291–298.

Tierkel, E.S., Arbona, G., Rivera, A., and De Juan, A. (1952). Mongoose rabies in Puerto Rico. Public Health Rep. *67*, 274–278.

Tschirch, W. (2001). Die Bedeutung von Luchs, Wildkatze, Waschbär und Marderhund in der Tollwut-Epidemiologie. Beitr. Jagd- Wildforsch. *26*, 281–298.

Turvey, S.T., Oliver, J.R., Narganes Storde, Y.M., and Rye, P. (2007). Late Holocene extinction of Puerto Rican native land mammals. Biol. Lett. *3*, 193–196.

USDA (2007). Cooperative Rabies Management Program National Report 2005. USDA, APHIS, Wildlife Services. Washington, DC (unnumbered report), p. 137.

USDA (2008a). North American Rabies Management Plan. Available at: http://www.aphis.usda.gov/wildlife_damage/oral_rabies/downloads/Final%20NARMP%209-30-2008%20(ENGLISH).pdf (accessed February 2009).

USDA (2008b). Cooperative Rabies Management Program National Report 2006. USDA, APHIS, Wildlife Services. Washington, DC (unnumbered report), p. 139.

USDA (2011a). Cooperative Rabies Management Program National Summary Report 2008. USDA, APHIS, Wildlife Services. Washington, DC (unnumbered report), p. 12.

USDA (2011b). Animal and Plant Health Inspection Service, Wildlife Services. Environmental Assessment (EA) and decision/finding of no significant impact (FONSI) – Field trial of an experimental rabies vaccine, human adenovirus type 5 vector in West Virginia.

Vaidya, S.A., Manning, S.E., Dhankhar, P., Meltzer, M.I., Rupprecht, C., Hull, H.F., and Fishbein, D.B. (2010). Estimating the risk of rabies transmission to humans in the U.S.: a Delphi analysis. BMC Public Health *10*, 278.

Van Thiel, P.P., de Bie, R.M., Eftimov, F., Tepaske, R., Zaaijer, H.L., van Doornum, G.J., Schutten, M.,

Osterhaus, A.D., Majoie, C.B., Aronica, E., *et al.* (2009). Fatal human rabies due to Duvenhage virus from a bat in Kenya: failure of treatment with coma-induction, ketamine, and antiviral drugs. PLoS Negl. Trop. Dis. *3*, e428.

Velasco-Villa, A., Reeder, S.A., Orciari, L.A., Yager, P.A., Franka, R., Blanton, J.D., Zuckero, L., Hunt, P., Oertli, E.H., Robinson, L.E., *et al.* (2008). Enzootic rabies elimination from dogs and reemergence in wild terrestrial carnivores, United States. Emerg. Infect. Dis. *14*, 1849–1854.

Vilella, F.J. (1998). Biology of the mongoose (*Herpestes javanicus*) in a rain forest of Puerto Rico. Biotropica *30*, 120–125.

Vonholdt, B.M., Pollinger, J.P., Lohmueller, K.E., Han, E., Parker, H.G., Quignon, P., Degenhardt, J.D., Boyko, A.R., Earl, D.A., Auton, A., *et al.* (2010). Genome-wide SNP and haplotype analyses reveal a rich history underlying dog domestication. Nature *464*, 898–902.

Vos, A., Nunan, C., Bolles, D., Müller, T., Fooks, A.R., Tordo, N., and Baer, G.M. (2011). The occurrence of rabies in pre-Columbian Central America: an historical search. Epidemiol. Infect. *139*, 1445–1452.

Walker, P.J. (2005). Bovine ephemeral fever in Australia and the world. Curr. Top. Microbiol. Immunol. *292*, 57–80.

Wandeler, A. (2004). Epidemiology and ecology of fox rabies in Europe. In Historical Perspective of Rabies in Europe and the Mediterranean Basin, King, A.A., Fooks, A.R., Aubert, M., and Wandeler, A.I., eds (World Organization for Animal Health (OIE), Paris), pp. 201–214.

Wandeler, A.I., Capt, S., Kappler, A., and Hauser, R. (1988). Oral immunization of wildlife against rabies: concept and first experiments. Rev. Infect. Dis. *10(Suppl. 4)*, S649–S653.

Wandeler, A.I. (2008). The rabies situation in Western Europe. Dev. Biol. *131*, 19–25.

Westerling, B. (1991). Rabies in Finland and its control 198 8–90. Suomen Riista *37*, 93–100. (in Finnish with English summary).

Weyer, J., Rupprecht, C.E., and Nel, L.H. (2009). Poxvirus-vectored vaccines for rabies – a review. Vaccine *27*, 7198–7201.

WHO (1960). Expert Committee on Rabies. Fourth Report, Geneva, Switzerland. Available at: http://whqlibdoc.who.int/trs/WHO_TRS_201.pdf (accessed October 2011).

WHO (2003). Country summaries of rabies cases, 2003 total. Rabies Bull. Europe *27*, 9–10.

WHO (2004). Country summaries of rabies cases, 1st quarter. Rabies Bull. Europe *28*, 4–21.

WHO (2005). Expert consultation on rabies. Tech. Report Series, *931*, 1–88.

WHO (2006). Country summaries of rabies cases, 2006 total. Rabies Bull. Europe *30*, 10.

Wilde, H., and Lumlertdacha, B. (2011). Rabies research in resource-poor countries. Adv. Virus Res. *79*, 449–455.

Willoughby, R.E., Jr. (2007). A cure for a rabies? Sci. Amer. *296*, 88–95.

Wu, X., Franka, R., Svoboda, P., Pohl, J., and Rupprecht, C.E. (2009). Development of combined vaccines for rabies and immunocontraception. Vaccine *27*, 7202–7209.

Wu, X., Smith, T.G., and Rupprecht, C.E. (2011). From brain passage to cell adaptation: the road of human rabies vaccine development. Expert Rev. Vaccines *10*, 1597–1608.

Yarosh, O.K., Wandeler, A.I., Graham, F.L., Campbell, J.B., and Prevec, L. (1996). Human adenovirus type 5 vectors expressing rabies glycoprotein. Vaccine *14*, 1257–1264.

Zhang, Y.Z., Fu, Z.F., Wang, D.M., Zhou, J.Z., Wang, Z.X., Lv, T.F., Xiong, C.L., Zou, Y., Yao, W.R., Li, M.H., *et al.* (2008). Investigation of the role of healthy dogs as potential carriers of rabies virus. Vector Borne Zoonotic Dis. *8*, 313–319.

Current Approaches in Lyssavirus Vaccine Development and Future Challenges

14

Xianfu Wu, Rongliang Hu and Todd G. Smith

Abstract

Current approaches to human rabies vaccine are focused on delivering safe, potent vaccine to exposed people in as few, cost-effective doses as possible. Given the current paradigm of inactivated rabies vaccines for humans, using more attenuated rabies virus seeds combined with industrialized cell fermentation could be one way to decrease manufacturing costs. Plasmid-based DNA vaccines, protein-based vaccines including virus-like particles and transgenic plants, and recombinant vaccines including rabies virus-vectored and other DNA or RNA virus-vectored recombinants, all comprise possible future rabies vaccine development pathways. For animals, a wider diversity of vaccines has been developed in the past and can be developed in the future including live attenuated, oral, and recombinant rabies vaccines. The main challenges in rabies control and prevention are not only in vaccine development but also in vaccine application especially in interrupting canine rabies transmission in dogs. New, more effective, strategies are needed to make human rabies vaccines more affordable and to simplify vaccine regimens. Improved strategies to vaccinate and humanely manage free-roaming dog populations can be implemented in the future. Finally, developing new rabies biologics to prevent emerging lyssavirus infections is necessary but challenging due to the lack of incentive and applicable strategies in the field.

Introduction

Efforts to improve potency, safety, delivery and cost-effectiveness are driving current rabies vaccine development. From a disease control perspective, rabies can be prevented in humans and domestic animals and can be managed in wild animal species to limit cross species transmission. Coordinated rabies control is often referred as the original one medicine concept (or one health: collective activities to promote human, animal and environmental health). The formalization of One Health is often attributed to Claude Bourgelat, the founder of the first veterinary school in 1761 in Lyon, France (Day, 2011). Unfortunately, in the early twentieth century, veterinary and medical professional activities began to diverge. Consequently, rabies vaccine research at the seminal stage followed a 'two medicine' approach: one for human rabies vaccines and another for animal rabies vaccines.

Vaccination is one of the most efficient ways to prevent infectious diseases and is usually administered to healthy populations before exposure to infectious agents. However, the first rabies vaccine was developed empirically in 1885 before the age of virology, and owing to the unknown aetiology of the disease or mechanisms of prophylaxis at that time, an aggressive post-exposure prophylaxis (PEP) method was designed (Pasteur *et al.*, 1885). This has set human rabies vaccine apart from other vaccines with regards to vaccine utilization; when, in fact, rabies vaccine is not unique. For example, rabies vaccine is similar to Japanese encephalitis vaccine (JEV), but JEV has since been recommended and included in extended immunization programmes for children in endemic countries while rabies vaccine has not.

Many current public health threats arise from zoonotic infectious diseases. With a broad spectrum of susceptible hosts, aetiological agents can cross species barriers from animals to humans, and the origin of the agent is not always clear. Global disease eradication has achieved two successes so far: smallpox for humans (WHO, 1980) and rinderpest for cattle (OIE, 2011). These diseases were caused by agents adapted to one host, and did not have other animal reservoirs. Thus, eradication of rabies, given the current tools, does not appear feasible because of its manifestation in bats as well as human factors, such as globalization and animal translocation (Constantine, 2003; Marano *et al.*, 2007; Castrodale *et al.*, 2008; Meng *et al.*, 2011). Vigilantly maintaining and increasing vaccination rates in domestic and wild animals is an ideal strategy to prevent rabies in humans and animals.

Assorted human and animal rabies vaccines have been developed by use of various substrates but a limited number of rabies virus seed strains (Table 14.1). The cost-effectiveness of human rabies vaccines might be greatly improved by industrialized fermentation techniques using more attenuated rabies virus seeds and cell substrates. From the very early Fermi and Semple type vaccines live attenuated rabies virus vaccines have been used but were abandoned in favour of inactivated rabies vaccines to ensure the safety

Table 14.1 Human and animal rabies vaccines

Vaccine substrate	Virus strain	Human vaccine	Animal vaccine
Rabbit brain	PAS	Pasteur, live	–
Goat/sheep brain	PAS	Fermi, live and killed mixture	1920s, Japan Dog vaccination
Goat/sheep brain	PAS	Semple, inactivated	+
Chicken embryo	LEP; HEP	–	+, live
Duck embryo	CVS; PM	+, inactivated	–
Neonatal mouse brain	PAS; PV	+, inactivated	–
PHKC cells	CL-60; Vnukovo-32; 3aG; CTN	+ Concentrated/inactivated	–
WI-38 cells	PM	+ Purified, inactivated	–
MRC-5 cells	PM	+ Purified, inactivated	–
FRhL-2 cells	CVS	+ Purified, inactivated	–
CEF cells	LEP; HEP	+ Purified, inactivated	+ Live or inactivated
DEF cells	PM; PV	+ Purified, inactivated	–
Primary dog or bovine kidney cells	PM; PV	+ Purified, inactivated	–
Vero cell line	PM; PV	+ Purified, inactivated	+ Inactivated
BHK cell line	ERA; SAD-Bern, B19, P5/88; SAG-1, SAG-2	–	+/– ERA, live/inactivated Kelev, inactivated

'+', available; '–', unavailable; PV, Pasteur virus; PAS, Louis Pasteur virus; LEP, Low egg passage; HEP, High egg passage; CVS, challenge virus standard; PHKC, primary hamster kidney cell; FRhL-2, a fetal rhesus monkey lung cell; CEF, chicken embryo fibroblast cells; DEF, duck embryo fibroblast cells; Vero, a cell line derived from African green monkey kidney cells; BHK, a baby hamster kidney cell line; SAD, street Alabama Dufferin; PM, Pitman–Moore.

WI-38 and MRC-5 cells are derived from human foetal lung cell strains. PM and CVS are derived from PV; ERA, SAD-Bern, SAD-B19, SAD-P5/88, Vnukovo-32, SAG-1 and SAG-2 are derived from SAD.

of the vaccine supply. Despite the accumulated knowledge of over a century of research on rabies virus pathogenicity and revitalized techniques in molecular virology, new approaches to human rabies vaccine development have been limited. Current research approaches include plasmid-based DNA vaccines, protein-vaccines (virus-like particles and transgenic plants), and recombinants (rabies virus- and other DNA and RNA virus-vectored recombinants).

In contrast to human rabies vaccines, many different strategies have been employed in animal rabies vaccine development (Table 14.2). For example, live attenuated and recombinant rabies vaccines have been licensed for oral vaccination of wildlife, and many different animal rabies vaccines, some containing adjuvant, have been targeted for different domestic species. Overall, rabies vaccine development has been stymied by underutilization or uncertain utilization strategies, resulting from large populations of unvaccinated dogs, cats, and susceptible wildlife, including emerging threats from bats and resulting in human deaths.

Seed viruses for rabies vaccines

Despite a long history of vaccine development, rabies vaccines have used a limited number of seed viruses for vaccine production. The residual virulence of the historical seeds is the main reason for the switch from live to inactivated vaccines in human rabies prophylaxis. As previously reviewed, these seed viruses were maintained by passage in laboratory animals or avian embryos and only briefly adapted to cell culture substrates (Wu *et al.*, 2011a). Using recently developed more attenuated seed viruses that are better adapted to cell culture, could increase the viral yield during

Table 14.2 New approaches to rabies vaccines

Virus-vector vaccines	Protein-based vaccines	DNA vaccines
Poxvirus-based:	*Expressed G protein:*	*G gene*
Vaccinia (V-RG, Licensed)	Virus-like particles	*G and N gene*
Canary pox (cats, Licensed)	Baculovirus system	*Chimeric G gene*
Raccoon pox	Transgenic plants	*Bi-cistronic*
Adenovirus-based:	Salmonella strains	*Lyssavirus G genes*
Human adenovirus (field trials)	Avian adenovirus system	
Chimpanzee adenovirus		
Canine adenovirus		
Herpesvirus-based:		
Pseudorabies		
Newcastle disease virus-based:		
NDV LaSota		
Rabies virus-based:		
Gene mutation:		
SAG1, 2; ERAg333 (trials)		
Extra copies of G		
Gene rearrangement:		
G before M, G before P		
Gene deletion:		
G, P, or M deletion		

Most of the investigations on recombinant virus vectors are targeting oral rabies vaccination of wildlife. No experimental human rabies vaccines, such as replication-deficient viruses (with gene deletions), have resulted in any clinical trials.

vaccine production leading to a more efficient manufacturing process and possibly negate the need for costly virus concentration. While many factors contribute to vaccine costs and some are unrelated to manufacturing, currently available high titre seed viruses can be used to generate vaccines of lower production cost and improve vaccine access in rabies endemic countries. An alternative strategy, using reverse genetics to manipulate traditionally attenuated vaccine strains such as the Street Alabama Dufferin (SAD) B19 and Evelyn–Rokitnicki–Abelseth (ERA) strains, results in greater attenuation and higher virus yield in growth (Bankovskiy *et al.*, 2008; Wu and Rupprecht, 2008). Unlike previous empirical attenuation where the mechanism and stability of the attenuations were unknown, attenuation engineered using reverse genetics is characterized, and can be more easily evaluated for stability. These newly developed virus strains grow to very high titres in cell culture, and because these viruses lack the pathogenicity of traditional seed viruses, the biological hazards of growing large amounts of virus for vaccine production are reduced. A shift in such seed viruses is not completely unprecedented. For example, the purified chick embryo cell rabies vaccine (PCECV) that was originally developed in Japan using one seed virus, was redeveloped later using a different seed virus (Barth *et al.*, 1984). The seed virus used to redevelop this vaccine, Flury low egg passage (LEP)-c25, was derived from a veterinary seed virus. This vaccine was licensed in the USA in 1997 (Centers for Disease Control and Prevention, 1998). This successful example could be used as a model for redevelopment of rabies vaccines using different seed viruses in the future.

Cell substrates for rabies vaccines

The path of rabies vaccine preparations has been reviewed previously from the historical perspective (Wu *et al.*, 2011a). Increased demand for domestically produced rabies vaccines has renewed interest in various cell substrates. Currently, some countries still use adult mammalian or neonatal rodent nerve-tissue vaccines (Ayele *et al.*, 2001). Given the lower potency and safety risks associated with these vaccines, alternative cell culture-based vaccines should be developed as recommended by WHO. Any country interested in manufacturing rabies vaccines should pursue cell culture vaccines, which are safe and effective as well as more humane, affordable and scalable than using animals. Vaccines derived from some cell substrates, such as primary hamster kidney cell and human diploid cell vaccines, are restricted in geographic distribution while others, such as PCECV and purified Vero cell vaccines (PVRV), have a wider distribution. In addition, some previously discontinued cell substrates, such as purified duck embryo cell vaccine, have been revived recently (Ashwathnarayana *et al.*, 2009). In contrast to the seed virus, the cell substrate may not contribute as much towards viral yield. Theoretically, any qualified cell substrate, meeting the criteria of residue DNA and carcinogenicity, when paired with an appropriate seed virus could be used for vaccine production. Moreover, all purified cell culture vaccines developed in the last 30 years have comparable qualities (Ajjan and Pilet, 1989; Arai and Montalban, 1993; Dietzschold and Hooper, 1998; Sabchareon *et al.*, 1999; Sudarshan *et al.*, 1999; Jones *et al.*, 2001; Madhusudana *et al.*, 2006; Ashwathnarayana *et al.*, 2009).

Vaccine potency

Vaccine potency testing is designed to assess and prevent lot-to-lot variation in vaccine manufacturing. Despite the utility of the NIH test over the past several decades as a basic measure of rabies vaccine potency, newer, more relevant *in vitro* assays and appropriate *in vivo* models are necessary for refinement and replacement of this historical measure, which requires large numbers of laboratory mice, and often exhibits wide confidence intervals. Replacement of the NIH test is a long standing issue that continues to plague vaccine manufactures around the world (Jivapaisarnpong *et al.*, 2009). A national panel of rabies experts met in 2011 to discuss the possible alternatives to the long standing issues of the NIH test. Use of alternative *in vitro* tests, such as a single radial immunodiffusion (SRID) test was originally proposed in the 1980s (Fitzgerald and Needy, 1986). Various enzyme-linked immunosorbent assays

(ELISA) have also been developed to measure levels of rabies glycoprotein (G) which directly correlates to vaccine potency. The most recent example is an immunocapture or sandwich ELISA developed in India (Nimmagadda *et al.*, 2010). Recently, retaining an *in vivo* test but using less animals, measuring immune response without challenge, or using peripheral challenge to reduce variability have been suggested as possible solutions (Wunderli *et al.*, 2006; de Moura *et al.*, 2009; Kramer *et al.*, 2010).

DNA vaccines

The first rabies DNA vaccine was developed in the early 1990s shortly after the first proof of concept of nucleic acid vaccination (Xiang *et al.*, 1994). Unfortunately, success in laboratory animals has not been translated into a vaccine for target species, including humans. DNA vaccines have been touted as less expensive, more accessible alternatives to current cell culture vaccines. However, additional regulations for DNA vaccine production and fears of a ground-breaking failure have slowed development.

In an effort to improve immunogenicity, the intradermal (ID) and intranasal (IN) routes were also tested. One study found that a rabies DNA vaccine delivered ID protected cats, while the same vaccine delivered intramuscular (IM) did not protect against a lethal challenge (Tesoro-Cruz *et al.*, 2008a). When compared to ID, IN administration resulted in delayed induction of virus neutralizing antibodies (VNA), but additional studies found that sufficient VNA titres (> 0.6 IU/ml) were generated in mice 30 days after a single IN dose (Tesoro-Cruz *et al.*, 2008b). When tested in a PEP model, this vaccine also protected mice and rabbits using a four-dose IN regimen (Feria-Romero *et al.*, 2011). Other strategies to improve immunogenicity of rabies DNA vaccines include modifying the antigen for increased expression, targeting antigen expression, and addition of adjuvant. Using the G gene from a genetically modified highly attenuated vaccine seed strain resulted in an effective DNA vaccine that delivered IM protected 80% of mice from lethal challenge (Osinubi *et al.*, 2009). The mutation in the G gene most likely increased antigen expression, as all mice succumbed when the parental G gene was given as a DNA vaccine. Similarly, using signal peptides, such as lysosomal-associated membrane protein-1 (LAMP-1) for targeting antigen expression to a particular subcellular compartment, increased the effectiveness of DNA vaccination compared to the unmodified G gene (Kaur *et al.*, 2009). The DNA rabies vaccine with LAMP-1 combined with oil-in-water adjuvant completely protected mice in a five-dose IM post-exposure regimen (Kaur *et al.*, 2010). Recombinant nucleic acid vaccines based on Sindbis virus expressing rabies virus G have been tested in both laboratory animals and target species. The Sindbis DNA rabies vaccine given IM protected all mice from lethal challenge, more than a conventional rabies DNA vaccine (80%), and the same as traditional cell culture vaccine (Saxena *et al.*, 2008). While similar to the conventional DNA vaccine, a Sindbis RNA rabies vaccine protected only 80% of mice (Saxena *et al.*, 2009). The Sindbis DNA vaccine administered ID in the ear of dogs induced VNA titres > 0.5 IU/ml after 21 days (Gupta *et al.*, 2009).

Virus-like particles

Virus-like particle (VLP) technology is one concept that is potentially more cost-effective than conventional rabies vaccines which require a costly concentration step. Such VLPs are collections of viral proteins that self-assemble into a structure similar to native virus and represent an emerging vaccine technology for generating effective immune protection (Vicente *et al.*, 2011). The three-dimensional structure of VLPs is readily recognized by the immune system, but VLPs lack core genetic material, making them unable to replicate and rendering them non-infectious. Unlike current conventional rabies vaccines that are manufactured with concentrated agents, VLP vaccines do not require a replication-competent aliquot of the virus to begin vaccine production. In human clinical trials, some VLPs activate key aspects of the immune response to achieve potent immune stimulation and long-term immunological memory (Emeny *et al.*, 2002; Pinto *et al.*, 2003). In addition, multiple expression systems exist. Transfection of insect cells with baculovirus is one of the most common systems, but yeast, mammalian cells and plants have been used to produce VLPs (Roldão *et al.*, 2010).

Although other rabies G-based vaccines have been proposed (see below), only one rabies VLP vaccine has been published so far. This VLP is a recombinant with fragments of the rabies G gene cloned into canine parvovirus capsid gene VP2 (Feng *et al.*, 2011). The recombinant VLPs expressing linear G epitopes (23 amino acids long) were recognized *in vitro* by polyclonal anti-rabies virus antibodies. Further work is needed to determine if this VLP induces rabies VNA *in vivo*, but this report suggests that a dual rabies and canine parvovirus VLP vaccine may be developed in the future. Interestingly, a close relative of rabies virus, vesicular stomatitis virus, was shown to be an efficient vector for recombinant VLPs and could be a feasible option for formation of rabies VLPs in the future (Ma and Li, 2011). Other VLP vaccine strategies include chimerias with proteins from two different viruses, such as coronavirus glycoprotein and influenza matrix protein or multiple proteins from the same virus, for example, a licensed hepatitis B VLP vaccine and the recently reported nipah VLP (Shapira *et al.*, 2001; Liu *et al.*, 2011; Walpita *et al.*, 2011).

Plant-based approaches

Transgenic plants are one alternative that offer low production costs and agricultural scale-up. Rabies virus G has been produced in various plant systems, including tobacco, maize, tomato and carrot (McGarvey *et al.*, 1995; Ashraf *et al.*, 2005; Loza-Rubio *et al.*, 2008; Rojas-Anaya *et al.*, 2009). These systems have failed to yield a promising vaccine candidate owing to low antigenic load. Advances in the field of plant-based biopharmaceuticals, such as variants that express human glycosylases, may warrant renewed attempts for a plant-based rabies vaccine. The most recent plant-based approach involved expressing recombinant rabies G fused to cholera toxin B in tobacco (Roy *et al.*, 2010). Cholera toxin B not only acts as an adjuvant enhancing antigen presentation by binding to receptors on the cell surface, but also acts as a scaffold for G protein expression. The recombinant protein formed pentamers that reacted *in vitro* with anti-rabies virus antibodies. Further *in vivo* studies, especially in target species, are still needed to validate this study and additional plant-based approaches.

Other subunit vaccines

A variety of systems has been used to express rabies virus G. In general, rabies virus G subunit vaccines have not stimulated sufficient immune responses to justify further vaccine development. For example, a synthetic peptide displaying a linear G epitope elicited a strong immune response in goats, but no VNA (Niederhauser *et al.*, 2008). However, new approaches may overcome these shortfalls. Ramya and co-authors describe conditions for expression and purification of soluble G from insect cells. Protein produced using this system induced VNA and protected mice from a lethal challenge (Ramya *et al.*, 2011). Rabies VLPs (discussed above) were produced in silkworm pupae which incorporates the benefits of agricultural scalability similar to plant-based approaches (Feng *et al.*, 2011). Recombinant G fused with *Escherichia coli* heat-labile enterotoxin B was expressed in a live attenuated *Salmonella* strain LH430 and used to orally immunize mice (Wang *et al.*, 2011). This approach could be used to develop a low-cost, oral veterinary vaccine in the near future. Finally, while production of rabies virus in embryonated chicken eggs is a time-proven concept, at least one biopharmaceutical company has expressed recombinant G at high levels in eggs using an avian adenovirus system (Grabko, 2002).

Animal rabies vaccines

Animal rabies vaccine development has a highly diverse history of using a variety of molecular techniques. In general, the tools for animal rabies vaccine development are more readily accepted than for human vaccines (Table 14.2). Current live rabies virus-vectored vaccines do not retain significant residual pathogenicity, since field applications of the constructs over the last three decades have validated their acceptability. Vaccinia virus and human adenovirus have the potential to recombine, and VR-G had caused infections in immune-compromised subjects in the past (Centers for Disease Control and Prevention, 2009).

Early animal rabies vaccines: Fermi- and Semple-type vaccines

Unlike human PEP practices, management of animal rabies focuses on prevention by active pre-exposure immunization. Early animal rabies vaccines did not differ much from human rabies vaccines. For example, Fermi-type vaccine was used for mass dog vaccination campaigns in Japan during the 1920s (Umeno and Doi, 1921). This classic model targeting dog rabies for elimination of human rabies set a strategic example for the rest of the world. Furthermore, Umeno's method, for the first time, applied a one-shot regimen for rabies pre-exposure prophylaxis (PrEP) in dogs.

Before cell-culture derived rabies vaccines became available in the 1960s, no qualitative differences were recognized in manufacturing human and animal rabies vaccines with regard to substrates and seed viruses. Semple-type vaccine was used widely for human PEP and preventative vaccination in dogs (Johnson, 1954). Phenolized and inactivated nerve tissue vaccines (NTVs) from sheep or goat brains, infected with CVS virus, were licensed for humans in the USA in the 1960s. Inactivated animal rabies vaccines, treated by carbolic acid, phenol, formalin, beta propiolactone, or ultraviolet light exposure, were in-parallel investigated during that time period (Webster and Casals, 1942).

Inactivated animal rabies vaccines

The historical rabies virus strains after empirical attenuation, including Pasteur virus (PV), LEP and ERA, possess residual virulence that causes rabies in certain animals. Fermi-type vaccine using the PV strain in Umeno's method was not recommended for puppies younger than four months old, owing to severe adverse effects caused by residual live virus in the vaccine preparation (Hata, 1924). Similar cases were observed with live LEP vaccine, which cannot be administrated in dogs less than five months of age, and live ERA vaccine-induced rabies was reported in cats after injection (Esh *et al.*, 1982). While Semple-type vaccines were commonly used for animal vaccinations, chicken embryo origin LEP and Flury high egg passage (HEP) vaccines were developed in the 1950s and 1960s, and were publicly recommended for animals. Like human rabies vaccine development when techniques for cell cultivation matured, inactivated animal rabies vaccines produced in cell substrates, such as the BHK-21 cell line, Vero cell line, and various primary animal cells including primary hamster kidney cells (PHKC), dog kidney cells, pig kidney cells and chicken embryo fibroblast cells (CEF) were also developed. In addition, aluminium hydroxide and oil-in-water adjuvants were included in inactivated animal rabies vaccines. However, injection-site sarcoma in cats has been linked to aluminium hydroxide adjuvant in rabies vaccine (Hendrick *et al.*, 1992).

Live animal rabies vaccines

Because of the long-standing debates over administration of live rabies vaccine in humans, rabies vaccines became unilaterally inactivated for human PEP (Chakrabarti, 2010). However, live LEP chick embryo-derived vaccine was produced and applied widely in canine rabies control campaigns despite reported vaccine-associated rabies cases. This scenario continues to repeat itself in some canine rabies endemic countries. One of the advantages of a live rabies vaccine is the quick and long-lasting immunity after a single injection. Live HEP vaccine produced in dog kidney cells was administrated in dogs and cats in the U.S. Live ERA (or it derivative Vnukovo-32) vaccines produced in dog, bovine, hamster, or pig kidney cells were used for dogs, cats and domestic herbivores in Canada, the former Soviet Union and Russia. More importantly, for more than three decades, ERA and it associated derivatives (SAG-1 or SAG-2), have been widely distributed as an oral vaccine for wildlife rabies control in Europe and North America. Other live animal vaccines using less characterized viruses, such as Kelev and Kissling strains, were not widely recommended.

Recombinant rabies vaccines

The milestones of recombinant rabies vaccines include oral vaccination of wildlife using live attenuated rabies virus, vaccinia-rabies virus glycoprotein recombinant (VR-G) and recombinant human adenovirus (Baer *et al.*, 1971; Rupprecht *et al.*, 1988; Charlton *et al.*, 1992). From the complete list of licensed human vaccines in the USA (www.fda.gov), only two recombinant vaccines are approved for immunization and distribution: hepatitis B and human papillomavirus. Experimental human rabies vaccines as discussed above have not successfully reached human clinical trials (Table 14.2; Dietzschold *et al.*, 2003; Sugiyama and Ito, 2007; McGettigan, 2010). Only one recently developed human rabies vaccine has been tested in human subjects. This vaccine was made using a veterinary vaccine substrate, BHK 21/c13 cells, and rabies virus PV 2061/Vero 15, but has not been licensed (Lalosevic *et al.*, 1998; Lalosevic *et al.*, 2008).

Orthopoxvirus-based recombinants

The drive for recombinant rabies vaccine research comes from safety concerns and the observed ineffectiveness of ERA oral vaccine in some wildlife species, such as raccoons and skunks (Tolson *et al.*, 1988; Rupprecht *et al.*, 1990). The initiative of rabies oral vaccination was owing to endemic fox rabies in Western Europe, Canada and the USA during the 1960s and 1970s. Rabies virus ERA strain tested in 1971 was a valid candidate as an oral vaccine for foxes (Baer *et al.*, 1971). Since 1978, ERA and its closely related derivatives (SAD-Bern, B19, P5/88, SAG-1 and SAG-2) have been widely distributed as baits for oral vaccination against rabies in wildlife. Nevertheless, the ERA oral vaccine was ineffective in skunks and raccoons prompting research for other alternatives, from which a recombinant vaccinia virus expressing the ERA G (V-RG) was constructed (Kieny *et al.*, 1984; Wiktor *et al.*, 1984; Blancou *et al.*, 1986; Rupprecht *et al.*, 1988). The V-RG vaccine was introduced in small field trials in 1989, and was conditionally licensed in 1995 in the USA for oral vaccination of wildlife. However, VR-G still does not induce sufficient rabies VNA responses in skunks, or dogs by a single dose (Rupprecht *et al.*, 2005; Blanton *et al.*, 2006; Grosenbaugh *et al.*, 2007).

Adenovirus-based recombinants

The initial adenovirus-vectored rabies vaccine (AdRG1.0) was constructed based on human adenovirus type 5 (Prevec *et al.*, 1990). Expression of rabies virus G was later improved by replacing the regulatory elements of the gene in the recombinant virus (Yarosh *et al.*, 1996). Thus, the constructed recombinant rabies vaccine, AdRG1.3, was examined in raccoons, skunks, and foxes under both laboratory and field conditions. Reasonable vaccine stability and efficacy was demonstrated. The vaccine after oral delivery also triggered humoral immune response in skunks and foxes (Charlton *et al.*, 1992). Subsequently, starting in 2006, a commercial product AdRG1.03 (trade name ONRAB®) has been approved for field trials in Canada. However, the recombinant adenovirus AdRG1.03 is replication-competent and can be isolated from excretes of vaccinated animals, although the negative impact of this virus on wildlife is presumed to be low (Lutze-Wallace *et al.*, 1995; Knowles *et al.*, 2009). Alternatively, a replication-deficient human adenovirus-vectored rabies vaccine was constructed by deletion of the E1 region, which was reported to be superior to the VR-G and safer than AdRG1.0 (Xiang *et al.*, 1996). The vaccine induced effective immune responses by IM and per os administrations in mice and foxes, but was not so efficacious for oral vaccination of dogs (Vos *et al.*, 2001). Another two adenovirus-vectored recombinant rabies vaccines were derived from chimpanzee adenovirus serotype 68 (Xiang *et al.*, 2002; Zhou *et al.*, 2006), and canine adenovirus type 2 (Hu *et al.*, 2006; Li *et al.*, 2006). After IM inoculation the vaccines induced effective protection against rabies in mice, swine and sheep, and in several target animals, such as dogs, cats, and skunks (Tims *et al.*, 2000; Hu *et al.*, 2007; Liu *et al.*, 2008; Bouet-Cararo *et al.*, 2011). The replication competent canine adenovirus type 2-vectored recombinant rabies vaccine also stimulated a protective immune response in dogs, raccoons and skunks after oral administration (Zhang *et al.*, 2008; Henderson *et al.*, 2009). Overall, all adenovirus-vectored rabies recombinants elicited high immune responses in animals, probably due in part to the non-cytopathic features of the virus, which permits sustained infection after delivery (Xiang *et al.*, 1996). In addition, adenoviruses primarily enter cells through interaction with the coxsackie-adenovirus receptor, and efficiently transduce dendritic cells, the main cell population presenting antigens to a naïve immune system (Bergelson *et al.*, 1997; Zhong *et al.*, 2000). Therefore, adenovirus-vectored recombinant rabies vaccines, especially replication-deficient adenovirus-vectored vaccines, are promising for future field applications.

Rabies virus-based recombinants

Traditionally, street rabies virus was attenuated empirically, with undefined genetic changes, through switching propagation in a variety of animal hosts or cell substrates. For example, strains of PV, Pitman-Moore (PM), LEP, HEP, ERA, Vnukovo-32, SAD-Bern, SAD-B19, P5/88, Kelev, CTN, 3aG, etc. were passaged in this way and applied as seed viruses for human and animal rabies vaccine manufacturing. The later developed strains SAG-1, SAG-2 and some other mutants (AvO1) were selected by rabies virus monoclonal antibody neutralization escape, which identified single or double nucleotides in the G gene corresponding to changes in amino acid residue 333. With the advent of reverse genetics techniques, rabies virus was manipulated in a recombinant manner with clear and defined genetic changes (Schnell *et al.*, 1994). Thereafter, a number of genes from other pathogenic agents (including human immunodeficiency virus HIV-1, simian immunodeficiency virus, hepatitis C virus, anthrax, severe acute respiratory syndrome and Ebola virus), marker genes (chloramphenicol acetyl-transferase, green fluorescent protein, beta-galactosidase and luciferase), cytochrome c, inflammatory cytokine and chemokine genes, or additional copies of the G have been incorporated into the rabies viral genome as recombinants for various purposes (McGettigan *et al.*, 2001; Faber *et al.*, 2002, 2005a, 2009; Foley *et al.*, 2002; McKenna *et al.*, 2004; Zhao *et al.*, 2009; Wen *et al.*, 2010; Gomme *et al.*, 2011). Meanwhile, site-directed mutagenesis mainly in the G gene and nucleoprotein (N) gene or limited deletions in the phosprotein (P) gene have also been introduced into rabies viruses (Mebatsion, 2001; Wu *et al.*, 2002; Faber *et al.*, 2005b). Full-length gene deletions in the G, P or matrix protein (M) gene resulted in replication-deficient mutants (Mebatsion *et al.*, 1996; Ito *et al.*, 2005; Morimoto *et al.*, 2005). Gene rearrangements that relocated the G gene closer to 3′ terminus in the viral genome led to further attenuation of the ERA strain (Wu and Rupprecht 2008; Wu *et al.*, 2011b). Limiting the fusion capacity of G by inserting small peptides has also been investigated, and the resulting recombinant served as a candidate for immuno-contraceptive studies (Wu *et al.*, 2009).

Other virus-based recombinants

From prokaryotic to eukaryotic systems, the rabies virus G protein has been expressed in different formulations, including insoluble and denatured forms in *Escherichia coli*, VLPs, and in transgenic plants. Similarly, the G gene has also been incorporated into other viruses for expression, besides the above-discussed V-RG and adenovirus recombinants.

Pseudorabies, a misnomer based on the phylogeny of the two viruses, was first reported in the USA in 1813. Pseudorabies virus (PRV), isolated in 1902 by Aladár Aujeszky, naturally infects adult pigs, but cross-species transmission can occur through contagious contact including cattle, sheep, goats, horses, rats, cats, dogs, skunks, raccoons and other mammals. The genetically engineered PRV (named OMNIVAC-PRV) as a gene-deleted live vaccine, first licensed by the US Department of Agriculture in 1986, is a pioneering symbol in animal disease control (Beardsley 1986; Kit 1990). By using a licensed PRV vaccine strain in China, a recombinant rPRV/eGFP/rgp, expressing rabies virus G was constructed (Yuan *et al.*, 2008). The recombinant induced neutralizing antibodies against rabies virus and PRV in dogs after oral delivery and intramuscular injections.

Newcastle disease (ND) was discovered in Newcastle upon Tyne, England in 1926, and the aetiological agent ND virus (NDV) primarily infects domestic and wild avian species. The NDV is a negative-stranded RNA virus in the genus *Avulavirus*, family *Paramyxoviridae*. Usually, NDV does not infect mammals, but NDV as a vaccine vector has been recently investigated for expression of antigens from respiratory syncytial virus (Martinez-Sobrido *et al.*, 2006), Rift Valley fever virus (Kortekaas *et al.*, 2010), measles virus (Kim *et al.*, 2011), HIV-1 (Khattar *et al.*, 2011), and rabies virus (Ge *et al.*, 2011). The NDV LaSota, a lentogenic vaccine strain, with ERA G incorporated, grew as well as the parental viruses. This rL-RVG induced high titres of neutralizing antibodies against rabies virus and NDV in a dose dependent manner after intramuscular injection in mice, cats and dogs.

Lyssaviruses other than classic rabies virus vaccines

Before the full-fledged diversity in the *Lyssavirus* genus was well recognized, chimeric G proteins or chimeric DNA vaccines were tested hoping to cover a broad spectrum of lyssavirus infections (Mebatsion *et al.*, 1995; Jallet *et al.*, 1999). Chimeric lyssavirus G protein (cLVG) containing the amino terminal half of rabies virus PV and carboxyl terminal half of a different lyssavirus genotypes (GT), such as pGMOK-PV (Mokola virus, MOKV plus PV) and pGEBL1-PV (European bat lyssavirus 1, EBLV-1 and PV), broadened the spectrum of protection against viruses from GT1 (classic rabies viruses), GT5 (EBLV-1) and GT6 (EBLV-2) (Perrin *et al.*, 2000). Using a recombinant vaccinia virus expressing the G from MOKV, Lagos bat virus (LBV) or West Caucasian bat virus (WCBV), either alone or in dual combinations, MOKV glycoprotein was shown to trigger cross neutralizing antibodies against LBV, but not against WCBV (Weyer *et al.*, 2008).

Challenges

Affordable human rabies vaccines, accompanied by a simplified PEP regimen, are critical in rabies prevention. To deliver efficacious vaccines to target animals in the field is of paramount importance in interrupting transmission to humans. Lyssavirus infections other than classic rabies have been discovered mostly in bats, with the exception for MOKV, which has only been identified in terrestrial animals (Nel 2005; Nandin-Davis and Fehlner-Gardiner 2008). The impact of these potential infections to humans has not attracted serious public health attention. Both the incentive and technical approaches to develop novel vaccines against lyssaviruses are limited at this moment. Currently, rabies vaccine availability and affordability, not novel lyssavirus exposures, are the main challenges in rabies prevention and control in endemic regions. However, any new tool aiming at improving current vaccine status deserves consideration. Other concerns include:

- To understand vaccine licensing regulations when developing new biologics. Vaccine development and regulatory review processes should be better integrated; ideally this should begin as early as the initiation of research on a new biologic. The basic standards of vaccine efficacy, safety (including targets, non-targets and environment), purity and delivery will eventually determine the products that can be licensed for clinical applications.
- To produce sufficient, affordable and safe human and animal rabies vaccines. This concern has been repeatedly discussed, but no solution has been consistently deployed. Rabies virus strains that are highly attenuated by reverse genetics, being able to adapt to industrial scale cell cultivation, may be useful in meeting this challenge. These highly attenuated strains based on traditional vaccine seed viruses, such as ERA or SAD derivatives, can be produced in large quantities in substrates, such as Vero, CEF and BHK cells, to render rabies vaccines more affordable in many underserved regions of the world.
- To standardize the rabies vaccine potency tests. Rabies vaccine potency testing is a problematic, long-standing issue from the early stages of vaccine development. To develop standardized, simple and quantitative methods for rabies vaccine potency is a necessity in the quality control of rabies vaccines, especially in the reduction, refinement and replacement in the use of animals.
- To simplify human rabies PEP based on scientific evidence. The current human PEP regimen is largely arbitrary and based on historical legacy. Without proper risk assessment after exposure to animal bites, human rabies PEP will continue to be misapplied and overused, which results in shortages of vaccine and rabies immune globulin (RIG). Consequently, rabies biologics may become inaccessible to people at real risk. Simplifying the PEP regimen is dauntingly difficult owing to historical dogma, but is a fundamental strategy in rabies prevention.

- To increase dog vaccination coverage, and manage stray and feral animals in a humane manner. Culling dog population for rabies control has proved to be ineffective and is no longer acceptable in modern civilization. Combined rabies vaccination and sterilization for dog population management is an ideal solution to this challenge.
- To develop efficient vaccine delivery methods to vaccinate stray or feral animals. Currently, most licensed animal vaccines are only efficacious by injection. However, development of other delivery methods to vaccinate free-ranging animals will be a significant step in broader rabies control.
- To develop novel vaccines against new-emerging lyssavirus infections. The spectrum of the *Lyssavirus* genus is steadily expanding as new members are being discovered. Available rabies biologicals are effective for phylogroup 1 lyssavirus infections, but do not cross-protect against phylogroup 2 lyssaviruses (Badrane *et al.*, 2001; Hanlon *et al.*, 2005). Close associations between humans, domestic animals and wildlife, coupled with increased deforestation and global travel, provide opportunities to contract emerging infectious diseases, such as bat lyssaviruses (van Thiel *et al.*, 2009). Having preventative tools (vaccines and immune globulins) available will be crucial in dealing with any potential exposures and future disease emergence.

Conclusions

Market and regulatory factors are the primary forces that determine rabies vaccine availability. However, to produce a cost-effective vaccine and to deliver an efficacious vaccine to free-ranging animals in the field are technical challenges. As rabies is neglected globally, the motivation for new biologics may be weaker than what was vigorously pursued over half a century ago. Furthermore, no reliable therapy for rabies exists. Timely administration of PEP is still the key to successful intervention after exposure to rabies virus. Efficacious vaccines and delivery to the targeted animals ensure effective herd immunity to interrupt rabies virus transmission. Experience gleaned from over a century has confirmed that rabies virus-based vaccines (inactivated or live) are efficacious in inducing protective immune responses after proper administration.

After Pasteur's pioneering achievements, technical modifications in the following years mainly increased the safety of the vaccine preparation, but did not clearly improve potency until the introduction of cell culture vaccines in the 1970s. Vaccine safety was compromised by insufficient potency, such as the initial avian-embryo tissue vaccines. Manufacturing of cell culture-derived rabies vaccines can be scaled up to meet current PEP needs by switching to cell substrates, such as Vero and CEF, and using more diverse and attenuated rabies virus strains. Replacing RIG with monoclonal antibody cocktails is one current strategy to produce adequate passive immunity to accompany rabies vaccines for human PEP. However, the cost of human PEP (mainly associated with RIG) in developed countries and deviation from PEP recommendations in rabies endemic regions where RIG is not available remain as long-term concerns. Current investigations focusing on approaches other than simplifying the PEP regimen, such as DNA vaccines, viral recombinant, protein vaccines (VLP or other subunit formulations) and transgenic plants, will have limited advantages over currently licensed rabies vaccines. Infections with lyssaviruses other than classic rabies should be a public health concern, but the overall impact in comparison with traditional rabies may remain small. At most, for some professionals who contact bats or suspicious animals on a regular basis, the tools for such prevention should be available in dealing with possible incidents. Therefore, development of new lyssavirus vaccines for pre-exposure prophylaxis is necessary.

In contrast to the inactivated vaccine scenario for human rabies vaccine development, animal rabies vaccine development remains an active area for research: from rabies virus, human virus-vectored, animal virus-vectored, bird virus-vectored vaccines, DNA vaccines, transgenic plant vaccines to adjuvant vaccines. However, no recombinant or purified protein vaccine provides better or longer-lasting protection than rabies virus-based vaccines. To deliver a safe and efficacious vaccine to the target animal should be the focus of animal rabies vaccine research. Oral delivery of an effective vaccine bait, with proven stability and safety profiles, is a valid method to immunize free-ranging animals, but no single recombinant construct works in all targeted animal species to date. Vaccines, such as V-RG and AdRG1.3-ONRAB, have been developed for wildlife species in North America. Adaption of these constructs to stray and free-ranging dogs, cats, and wildlife throughout the world needs careful evaluation, including efficacy, safety and acceptability for public health concerns. Production of sufficient, efficacious human and animal rabies biologics should be an immediate priority in rabies prevention and control.

Acknowledgements

We would like to thank Charles Rupprecht and Inger Damon for critical comments in preparation of the manuscript. Use of trade names and commercial sources are for identification only and does not imply endorsement by the US Department of Health and Human Services. The findings and conclusions in this report are those of the authors and do not necessarily represent the official position of the Centers for Disease Control and Prevention.

References

Ajjan, N., and Pilet, C. (1989). Comparative study of the safety and protective value, in pre-exposure use, of rabies vaccine cultivated on human diploid cells (HDCV) and of the new vaccine grown on Vero cells. Vaccine 7, 125–128.

Arai, Y.T., and Montalban, C.S. (1993). Comparative studies on chick embryo cell rabies vaccine and human diploid cell vaccine: pre-exposure use and purity of the vaccines. Southeast Asian J. Trop Med. Public Health *24*, 747–750.

Ashraf, S., Singh, P.K., Yadav, D.K., Shahnawaz, M., Mishra, S., Sawant, S.V., and Tuli, R. (2005). High level expression of surface glycoprotein of rabies virus in tobacco leaves and its immunoprotective activity in mice. J. Biotechnol. *119*, 1–14.

Ashwathnarayana, D.H., Madhusudana, S.N., Sampath, G., Sathpathy, D.M., Mankeshwar, R., Ravish, H.H., Ullas, P.T., Behra, T.R., Sudarshan, M.K., Gangaboraiah, *et al.* (2009). A comparative study on the safety and immunogenicity of purified duck embryo cell vaccine (PDEV, Vaxirab) with purified chick embryo cell vaccine (PCEC, Rabipur) and purifed vero cell rabies vaccine (PVRV, Verorab). Vaccine *28*, 148–151.

Ayele, W., Fekadu, M., Zewdie, B., Beyene, M., Bogale, Y., Mocha, K., and Egziabher, F.G. (2001). Immunogenicity and efficacy of Fermi-type nerve tissue rabies vaccine in mice and in humans undergoing post-exposure prophylaxis for rabies in Ethiopia. Ethiop. Med. J. *39*, 313–321.

Badrane, H., Bahloul, C., Perrin, P., and Tordo, N. (2001). Evidence of two *Lyssavirus* phylogroups with distinct pathogenicity and immunogenicity. J. Virol. *75*, 3268–3276.

Baer, G.M., Abelseth, M.K., and Debbie, J.G. (1971). Oral vaccination of foxes against rabies. Am. J. Epidemiol. *98*, 487–490.

Bankovskiy, D., Safonov, G., and Kurilchuk, Y. (2008). Immunogenicity of the ERA G 333 rabies virus strain in foxes and raccoon dogs. Dev. Biol. (Basel) *131*, 461–466.

Barth, R., Gruschkau, H., Bijok, U., Hilfenhaus, J., Hinz, J., Milcke, L., Moser, H., Jaeger, O., Ronneberger, H., and Weinmann, E. (1984). A new inactivated tissue culture rabies vaccine for use in man. Evaluation of PCEC-vaccine by laboratory tests. J. Biol. Stand. *12*, 29–46.

Beardsley, T. (1986). USDA goes too public too quickly. Nature *320* (6062), 473.

Bergelson, J.M., Cunningham. J.A., Droguett, G., Kurt-Jones, E.A., Krithivas, A., Hong, J.S., Horwitz, M.S., Crowell, R.L., and Finberg, R.W. (1997). Isolation of a common receptor for coxsackie B viruses and adenoviruses 2 and 5. Science *275*, 1320–1323.

Blancou, J., Kieny, M.P., Lathe, R., Lecocq, J.P., Pastoret, P.P., Soulebot, J.P., and Desmettre, P. (1986). Oral vaccination of the fox against rabies using a live recombinant vaccinia virus. Nature *322*, 373–375.

Blanton, J.D., Meadow, A., Murphy, S.M., Manangan, J., Hanlon, C.A., Faber, M.L., Dietzschold, B., and Rupprecht, C.E. (2006). Vaccination of small Asian mongoose (*Herpestes javanicus*) against rabies. J. Wildl. Dis. *42*, 663–666.

Bouet-Cararo, C., Contreras, V., Fournier, A., Jallet, C., Guibert, J.M., Dubois, E., Thiery, R., Bréard, E., Tordo, N., Richardson, J., *et al.* (2011). Canine adenoviruses elicit both humoral and cell-mediated immune responses against rabies following immunisation of sheep. Vaccine *29*, 1304–1310.

Castrodale, L., Walker, V., Baldwin, J., Hofmann, J., and Hanlon, C. (2008). Rabies in a puppy imported from India to the USA, March 2007. Zoonoses Public Health *55*, 427–430.

Centers for Disease Control and Prevention (1998). Availability of new rabies vaccine for human use. MMWR Morb. Mortal. Wkly Rep. *47*, 12, 19.

Centers for Disease Control and Prevention (2009). Human vaccinia infection after contact with a raccoon rabies vaccine bait – Pennsylvania, 2009. MMWR Morb. Mortal. Wkly Rep. *58*, 1204–1207.

Chakrabarti, P. (2010). 'Living versus dead': The Pasteurian paradigm and imperial vaccine research. Bull. Hist. Med. *84*, 387–423.

Charlton, K.M., Artois, M., Prevec, L., Campbell, J.B., Casey, G.A., Wandeler, A.I., and Armstrong, J. (1992). Oral rabies vaccination of skunks and foxes with a recombinant human adenovirus vaccine. Arch. Virol. *123*, 169–179.

Constantine, D.G. (2003). Geographic translocation of bats: known and potential problems. Emerg. Infect. Dis. *9*, 17–21.

Day, M.J. (2011). One health: the importance of companion animal vector-borne diseases. Parasit. Vectors *4*, 49.

Dietzschold, B., and Hooper, D.C. (1998). Human diploid cell culture rabies vaccine (HDCV) and purified chick embryo cell culture rabies vaccine (PCECV) both confer protective immunity against infection with the silver-haired bat rabies virus strain (SHBRV). Vaccine *16*, 1656–1659.

Dietzschold, B., Faber, M., and Schnell, M.J. (2003). New approaches to the prevention and eradication of rabies. Expert Rev. Vaccines *2*, 399–406.

Emeny, R.T., Wheeler, C.M., Jansen, K.U., Hunt, W.C., Fu, T.M., Smith, J.F., MacMullen, S., Esser, M.T., and Paliard, X. (2002). Priming of human papillomavirus type 11-specific humoral and cellular immune responses in college-aged women with a virus-like particle vaccine. J.Virol. *76*, 7832–7842.

Esh, J.B., Cunningham, J.G., and Wiktor, T.J. (1982). Vaccine-induced rabies in four cats. J. Am. Vet. Med. Assoc. *180*, 1336–1339.

Faber, M., Pulmanausahakul, R., Hodawakekar, S.S., Spitsin, S., McGettigan, J.P., Schnell, M.J., and Dietzschold, B. (2002). Overexpression of the rabies virus glycoprotein results in enhencement of apoptosis and antiviral immune response. J. Virol. *76*, 3374–3381.

Faber, M., Faber, M.L., Papaneri, A., Bette, M., Weihe, R., Dietzschold, B., and Schnell, M.J. (2005a). A single amino acid change in rabies virus glycoprotein increases virus spread and enhences virus pathogenicity. J. Virol. *79*, 14141–14148.

Faber, M., Lamirande, E.W., Roberts, A., Rice, A.B., Koprowski, D., Dietzschold, B., and Schnell, M.J. (2005b). A single immunization with a rhabdovirus-based vector expressing severe acute respiratory syndrome coronavirus (SARS-CoV) S protein results in the production of high levels of SARS-CoV-neutralizing antibodies. J. Gen. Virol. *86*, 1435–1440.

Faber, M., Li, J., Kean, R.B., Hooper, D.C., Alugupalli, K.R., and Dietzschold, B. (2009). Effective preexposure and postexposure prophylaxis of rabies with a highly attenuated recombinant rabies virus. Proc. Natl. Acad. Sci. USA. *106*, 11300–11305.

Feng, H., Liang, M., Wang, H.L., Zhang, T., Zhao, P.S., Shen, X.J., Zhang, R.Z., Hu, G.Q., Gao, Y.W., Wang, C.Y., *et al.* (2011). Recombinant canine parvovirus-like particles express foreign epitopes in silkworm pupae. Vet. Microbiol. *154*, 49–57.

Feria-Romero, I.A., Chavez-Rueda, K., Orozco-Suarez, S., Blanco-Favela, F., Calzada-Bermejo, F., Chavez-Sanchez, L., Manuel-Apolinar, L., Hernandez-Gonzalez, R., Aguilar-Setien, A., and Tesoro-Cruz, E. (2011). Intranasal Anti-rabies DNA Immunization Promotes Th1-related Cytokines Associated with Plasmid Survival Time. Arch. Med. Res. *42*, 563–571.

Fitzgerald, E.A., and Needy, C.F. (1986). Use of the single radial immunodiffusion test as a replacement for the NIH mouse potency test for rabies vaccine. Dev. Biol. Stand. *64*, 73–79.

Foly, H.D., Otero, M., Orenstein, J.M., Pomerantz, R.J., and Schnell, M.J. (2002). Rhabdovirus-based vectors with human immunodeficiency virus type 1 (HIV-1) envelope display HIV-1 like tropism and target human dentritic cells. J. Virol. *76*, 19–31.

Ge, J., Wang, X., Tao, L., Wen, Z., Feng, N., Yang, S., Xia, X., Yang, C., Chen, X., and Bu, Z. (2011). Newcastle disease virus-vectored rabies vaccine is safe, highly immunogenic, and provides long-lasting protection in dogs and cats. J. Virol. *85*, 8241–8252.

Gomme, E.A., Wanjalla, C.N., Wirblich, C., and Schnell, M.J. (2011). Rabies virus as a research tool and viral vaccine vector. Adv. Virus Res. *79*, 139–164.

Grabko, V.I., and Blyden, E.R. (2002). Recombinant Eggs And Gene Cloning And Expression Vectors Based On Avian Adenoviruses (Chemogen Inc (US), Blyden, Eluemuno, R. (US), Grabko, Vladimir, I. (US)).

Grosenbaugh, D.A., Maki, J.L., Rupprecht, C.E., and Wall, D.K. (2007). Rabies challenge of captive striped skunks (*Mephitis mephitis*) following oral administration of a live vaccinia-vectored rabies vaccine. J. Wild. Dis. *43*, 124–128.

Gupta, P.K., Dahiya, S.S., Kumar, P., Rai, A., Patel, C.L., Sonwane, A.A., and Saini, M. (2009). Sindbis virus replicon-based DNA vaccine encoding Rabies virus glycoprotein elicits specific humoral and cellular immune response in dogs. Acta. Virol. *53*, 83–88.

Hata, S. (1924). The Protection of Dogs Against Rabies by Umeno's Method of Preventive Inoculation. J. Immunol. *9*, 89–93.

Hanlon, C.A., Kuzmin, I.V., Blanton, J.D., Weldon, W.C., Manangan, J.S., and Rupprecht, C.E. (2005). Efficacy of rabies biologics against new lyssaviruses from Eurasia. Virus Res. *111*, 44–54.

Henderson, H., Jackson, F., Bean, K., Panasuk, B., Niezgoda, M., Slate, D., Li, J., Dietzschold, B., Mattis, J., and Rupprecht, C.E. (2009). Oral immunization of raccoons and skunks with a canine adenovirus recombinant rabies vaccine. Vaccine *27*, 7194–7197.

Hendrick, M.J., Goldschmidt, M.H., Shofer, F.S., Wang, Y.Y., and Somlyo, A.P. (1992). Postvaccinal sarcomas in the cat: epidemiology and electron probe microanalytical identification of aluminum. Cancer Res. *52*, 5391–5394.

Hu, R.L., Zhang, S., Fooks, A.R., Yuan, H., Liu, Y., Li, H., Tu, C., Xia, X., and Xiao, Y. (2006). Prevention of rabies virus infection in dogs by a recombinant canine adenovirus type-2 encoding the rabies virus glycoprotein. Microbes. Infect. *8*, 1090–1097.

Hu, R.L., Liu, Y., Zhang, S.F., Zhang, F., and Fooks, A.R. (2007). Experimental immunization of cats with a recombinant rabies-canine adenovirus vaccine elicits a long-lasting neutralizing antibody response against rabies. Vaccine *25*, 5301–5307.

Ito, N., Sugiyama, M., Yamada, K., Shimizu, K., Takayama-Ito, M., Hosokawa, J., and Minamoto, N. (2005). Characterization of M gene-deficient rabies virus with advantages of effective immunization and safety as a vaccine strain. Microbiol. Immunol. *49*, 971–979.

Jallet, C., Jacob, Y., Bahloul, C., Drings, A., Desmezieres, E., Tordo, N., and Perrin, P. (1999). Chimeric *Lyssavirus* glycoproteins with increased immunological potential. J. Virol. *73*, 225–233.

Jivapaisarnpong, T., Schofield, T., and Krause, P.R. (2009). A vaccine measured with a highly variable assay: rabies. Biologicals *37*, 412–415.

Johnson, H.N. (1954). Experimental studies on the duration of immunity in dogs vaccinated against rabies. Bull. World Health Organ. *10*, 725–729.

Jones, R.L., Froeschle, J.E., Atmar, R.L., Matthews, J.S., Sanders, R., Pardalos, J., Moeller, L., Chin, J.E., Famula, M., Briggs, D.J., *et al.* (2001). Immunogenicity, safety and lot consistency in adults of a chromatographically purified Vero-cell rabies vaccine: a randomized, double-blind trial with human diploid cell rabies vaccine. Vaccine *19*, 4635–4643.

Kaur, M., Rai, A., and Bhatnagar, R. (2009). Rabies DNA vaccine: no impact of MHC class I and class II targeting sequences on immune response and protection against lethal challenge. Vaccine *27*, 2128–2137.

Kaur, M., Saxena, A., Rai, A., and Bhatnagar, R. (2010). Rabies DNA vaccine encoding lysosome-targeted glycoprotein supplemented with Emulsigen-D confers complete protection in preexposure and postexposure studies in BALB/c mice. FASEB J. *24*, 173–183.

Khattar, S.K., Samal, S., Devico, A.L., Collins, P.L., and Samal, S.K. (2011). Newcastle disease virus expressing human immunodeficiency virus type 1 envelope glycoprotein induces strong mucosal and serum antibody responses in Guinea pigs. J. Virol. *85*, 10529–10541.

Kieny, M.P., Lathe, R., Drillien, R., Spehner, D., Skory, S., Schmitt, D., Wiktor, T., Koprowski, H., and Lecocq, J.P. (1984). Expression of rabies virus glycoprotein from a recombinant vaccinia virus. Nature *312*, 163–166.

Kim, D., Martinez-Sobrido, L., Choi, C., Petroff, N., García-Sastre, A., Niewiesk, S., and Carsillo, T. (2011). Induction of type 1 interferon secretion through recombinant Newcastle disease virus expressing measles virus hemagglutinin stimulates antibody secretion in the presence of maternal antibodies. J. Virol. *85*, 200–207.

Kit, S. (1990). Genetically engineered vaccines for control of Aujeszky's disease (pseudorabies). Vaccine *8*, 420–424.

Knowles, M.K., Nadin-Davis, S.A., Sheen, M., Rosatte, R., Mueller, R., and Beresford, A. (2009). Safety studies on an adenovirus recombinant vaccine for rabies (AdRG1.3-ONRAB) in target and non-target species. Vaccine *27*, 6619–6626.

Kortekaas, J., Dekker, A., de Boer, S.M., Weerdmeester, K., Vloet, R.P., de Wit, A.A., Peeters, B.P., and Moormann, R.J. (2010). Intramuscular inoculation of calves with an experimental Newcastle disease virus-based vector vaccine elicits neutralizing antibodies against Rift Valley fever virus. Vaccine *28*, 2271–2276.

Kramer, B., Bruckner, L., Daas, A., and Milne, C. (2010). Collaborative study for validation of a serological potency assay for rabies vaccine (inactivated) for veterinary use. Pharmeur. Bio. Sci. Notes *2010*, 37–55.

Lalosevic, D., Stankov, S., Lazarevic-Ivanc, L., Lalosevic, V., and Knezevic, I. (1998). Immunogenicity of BHK-rabies vaccine in human volunteers. Med. Pregl. *51*, 17–19.

Lalosevic, D., Lalosevic, V., Lazarevic-Ivanc, L., and Knezevic, I. (2008). BHK-21 cell culture rabies vaccine: immunogenicity of a candidate vaccine for humans. Dev. Biol. (Basel) *131*, 421–429.

Li, J., Faber, M., Papaneri, A., Faber, M.L., McGettigan, J.P., Schnell, M.J., and Dietzschold, B. (2006). A single immunization with a recombinant canine adenovirus expressing the rabies virus G protein confers protective immunity against rabies in mice. Virology *356*, 147–154.

Liu, Y., Zhang, S., Ma, G., Zhang, F., and Hu, R. (2008). Efficacy and safety of a live canine adenovirus-vectored rabies virus vaccine in swine. Vaccine *26*, 5368–5372.

Liu, Y.V., Massare, M.J., Barnard, D.L., Kort, T., Nathan, M., Wang, L., and Smith, G. (2011). Chimeric severe acute respiratory syndrome coronavirus (SARS-CoV) S glycoprotein and influenza matrix 1 efficiently form virus-like particles (VLPs) that protect mice against challenge with SARS-CoV. Vaccine *29*, 6606–6613.

Loza-Rubio, E., Rojas, E., Gomez, L., Olivera, M.T., and Gomez-Lim, M.A. (2008). Development of an edible rabies vaccine in maize using the Vnukovo strain. Dev. Biol. (Basel) *131*, 477–482.

Lutze-Wallace, C., Wandeler, A., Prevec, L., Sidhu, M., Sapp, T., and Armstrong, J. (1995). Characterization of a human adenovirus 5: rabies glycoprotein recombinant vaccine reisolated from orally vaccinated skunks. Biologicals *23*, 271–277.

Ma, Y., and Li, J. (2011). Vesicular stomatitis virus as a vector to deliver virus-like particles of human norovirus: a new vaccine candidate against an important noncultivable virus. J. Virol. *85*, 2942–2952.

McGarvey, P.B., Hammond, J., Dienelt, M.M., Hooper, D.C., Fu, Z.F., Dietzschold, B., Koprowski, H., and Michaels, F.H. (1995). Expression of the rabies virus glycoprotein in transgenic tomatoes. Biotechnology *13*, 1484–1487.

McGettigan, J.P., Foley, H.D., Belyakov, I.M., Berzofsky, J.A., Pomerantz, R.J., and Schnell, M.J. (2001). Rabies virus-based vectors expressing human immunodeficiency virus type 1 (HIV-1) envelope protein induce a strong, cross-reactive cytotoxic T-lymphocyte response against envelope protein from different HIV-1 isolates. J. Virol. *75*, 4430–4434.

McGettigan, J.P. (2010). Experimental rabies vaccines for humans. Expert. Rev. Vaccines *9*, 1177–1186.

McKenna, P.M., Aye, P.P., Dietzschold, B., Montefiori, D.C., Martin, L.N., Marx, P.A., Pomerantz, R.J., Lackner, A., and Schnell, M.J. (2004). Immunogenicity study of glycoprotein-deficient rabies virus expressing simian/human immunodeficiency virus SHIV89.6P envelope in a rhesus macaque. J. Virol. *78*, 13455–13459.

Madhusudana, S.N., Sanjay, T.V., Mahendra, B.J., Sudarshan, M.K., Narayana, D.H., Giri, A.,

Muhamuda, K., Ravi, V., Vakil, H.B., and Malerczyk, C. (2006). Comparison of saftey and immunogenicity of purified chick embryo cell rabies vaccine (PCECV) and purified vero cell rabies vaccine (PVRV) using the Thai Red Cross intradermal regimen at a dose of 0.1 ML. Hum. Vaccin. *2*, 200–204.

Marano, N., Arguin, P.M., and Pappaioanou, M. (2007). Impact of globalization and animal trade on infectious disease ecology. Emerg. Infect. Dis. *13*, 1807–1809.

Martinez-Sobrido, L., Gitiban, N., Fernandez-Sesma, A., Cros, J., Mertz, S.E., Jewell, N.A., Hammond, S., Flano, E., Durbin, R.K., García-Sastre, A., *et al.* (2006). Protection against respiratory syncytial virus by a recombinant Newcastle disease virus vector. J. Virol. *80*, 1130–1139.

Mebatsion, T. (2001). Extensive attenuation of rabies virus by simutaneously modifying the dynein light chain binding site in the P protein and replacing Arg 333 in the G protein. J. Virol. *75*, 11496–11502.

Mebatsion, T., Schnell, M.J., and Conzelmann, K.K. (1995). Mokola virus glycoprotein and chimeric proteins can replace rabies virus glycoprotein in the rescue of infectious defective rabies virus particles. J. Virol. *69*,144–1451.

Mebatsion, T., Konig, M., and Conzelmann, K.K. (1996). Budding of rabies virus particles in the absence of the spike glycoprotein. Cell *84*, 941–951.

Meng, S., Sun, Y., Wu, X., Tang, J., Xu, G., Lei, Y., Wu, J., Yan, J., Yang, X., and Rupprecht, C.E. (2011). Evolutionary dynamics of rabies viruses highlights the importance of China rabies transmission in Asia. Virology *410*, 403–409.

Morimoto, K., Shoji, Y., and Inoue, S. (2005). Characterization of P gene-deficient rabies virus: propagation, pathogenicity and antigenicity. Virus Res. *111*, 61–67.

de Moura, W.C., de Araujo, H.P., Cabello, P.H., Romijn, P.C., and Leite, J.P. (2009). Potency evaluation of rabies vaccine for human use: the impact of the reduction in the number of animals per dilution. J. Virol. Methods. *158*, 84–92.

Nadin-Davis, S.A., and Fehlner-Gardiner, C. (2008). Lyssaviruses-current trends. Adv. Virus Res. *71*, 207–249.

Nel, L.H. (2005). Vaccines for lyssaviruses other than rabies. Expert Rev. Vaccines *4*, 533–540.

Niederhauser, S., Bruegger, D., Zahno, M.L., Vogt, H.R., Peterhans, E., Zanoni, R., and Bertoni, G. (2008). A synthetic peptide encompassing the G5 antigenic region of the rabies virus induces high avidity but poorly neutralizing antibody in immunized animals. Vaccine *26*, 6749–6753.

Nimmagadda, S.V., Aavula, S.M., Biradhar, N., Rao, V.S., Shanmugham, R., Chandran, D., Thirumeni, N., Singanallur, N.B., and Villuppanoor, S.A. (2010). Recombinant diabody-based immunocapture enzyme-linked immunosorbent assay for quantification of rabies virus glycoprotein. Clin. Vaccine Immunol. *17*, 1261–1268.

Osinubi, M.O., Wu, X., Franka, R., Niezgoda, M., Nok, A.J., Ogunkoya, A.B., and Rupprecht, C.E. (2009). Enhancing comparative rabies DNA vaccine effectiveness through glycoprotein gene modifications. Vaccine 27, 7214–7218.

Pasteur, L., Chamberland, C.E., and Roux, E. (1885). Methode pour prevenir la rage après morsue. C. R. Acad. Sci. *101*, 765–774.

Perrin, P., Jacob, Y., Desmezieres, E., and Tordo, N. (2000). DNA-based immunisation against rabies and rabies-related viruses: towards multivalent vaccines. Dev. Biol. (Basel). *104*, 151–157.

Pinto, L.A., Edwards, J., Castle, P.E., Harro, C.D., Lowy, D.R., Schiller, J.T., Wallace, D., Kopp, W., Adelsberger, J.W., Baseler, M.W., *et al.* (2003). Cellular immune responses to human papillomavirus (HPV)-16 L1 in healthy volunteers immunized with recombinant HPV-16 L1 virus-like particles. J. Infect. Dis. *188*, 327–338.

Prevec, L., Campbell, J.B., Christie, B.S., Belbeck, L., and Graham, F.L. (1990). A recombinant human adenovirus vaccine against rabies. J. Infect. Dis. *161*, 27–30.

Ramya, R., Mohana Subramanian, B., Sivakumar, V., Senthilkumar, R.L., Sambasiva Rao, K.R., and Srinivasan, V.A. (2011). Expression and solubilization of insect cell-based rabies virus glycoprotein and assessment of its immunogenicity and protective efficacy in mice. Clin. Vaccine Immunol. *18*, 1673–1679.

Rojas-Anaya, E., Loza-Rubio, E., Olivera-Flores, M.T., and Gomez-Lim, M. (2009). Expression of rabies virus G protein in carrots (*Daucus carota*). Transgenic Res. *18*, 911–919.

Roldão, A., Mellado, M.C.M., Castilho, L.R., Carrondo, M.J.T., and Alves, P.M. (2010). Virus-like particles in vaccine development. Expert Rev. Vaccines *9*, 1149–1176.

Roy, S., Tyagi, A., Tiwari, S., Singh, A., Sawant, S.V., Singh, P.K., and Tuli, R. (2010). Rabies glycoprotein fused with B subunit of cholera toxin expressed in tobacco plants folds into biologically active pentameric protein. Protein Expr. Purif. *70*, 184–190.

Rupprecht, C.E., Hamir, A.N., Johnston, D.H., and Koprowski, H. (1988). Efficacy of a vaccinia-rabies glycoprotein recombinant virus vaccine in raccoons (*Procyon lotor*). Rev. Infect. Dis. *10*, S803–809.

Rupprecht, C.E., Charlton, K.M., Artois, M., Casey, G.A., Webster, W.A., Campbell, J.B., Lawson, K.F., and Schneider, L.G. (1990). Ineffectiveness and comparative pathogenicity of attenuated rabies virus vaccines for the striped skunk (*Mephitis mephitis*). J. Wildl. Dis. *26*, 99–102.

Rupprecht, C.E., Hanlon, C.A., Blanton, J., Managan, J., Morrill, P., Murphy, S., Niezgoda, M., Orciari, L.A., Schumacher, C.L., and Dietzschold, B. (2005). Oral vaccination of dogs with recombinant rabies virus vaccines. Virus Res. *111*, 101–105.

Sabchareon, A., Lang, J., Attanath, P., Sirivichayakul, C., Pengsaa, K., Le Mener, V., Chantavanich, P., Prarinyanuphab, V., Pojjaroen-Anant, C., Nimnual, S., *et al.* (1999). A new Vero cell rabies vaccine: results of a comparative trial with human diploid cell rabies vaccine in children. Clin. Infect. Dis. *29*, 141–149.

Saxena, S., Dahiya, S.S., Sonwane, A.A., Patel, C.L., Saini, M., Rai, A., and Gupta, P.K. (2008). A sindbis virus replicon-based DNA vaccine encoding the rabies virus glycoprotein elicits immune responses and complete protection in mice from lethal challenge. Vaccine *26*, 6592–6601.

Saxena, S., Sonwane, A.A., Dahiya, S.S., Patel, C.L., Saini, M., Rai, A., and Gupta, P.K. (2009). Induction of immune responses and protection in mice against rabies using a self-replicating RNA vaccine encoding rabies virus glycoprotein. Vet. Microbiol. *136*, 36–44.

Schnell, M.J., Mebatsion, T., and Conzelmann, K.K. (1994). Infectious rabies viruses from cloned cDNA. EMBO J. *13*, 4195–4203.

Shapira, M.Y., Zeira, E., Adler, R., and Shouval, D. (2001). Rapid seroprotection against hepatitis B following the first dose of a Pre-S1/Pre-S2/S vaccine. J. Hepatol. *34*, 123–127.

Sudarshan, M.K., Madhusudana, S.N., and Mahendra, B.J. (1999). Post-exposure prophylaxis with purified vero cell rabies vaccine during pregnancy – safety and immunogenicity. J. Commun. Dis. *31*, 229–236.

Sugiyama, M., and Ito, N. (2007). Control of rabies: epidemiology of rabies in Asia and development of new-generation vaccines for rabies. Comp. Immunol. Microbiol. Infect. Dis. *30*, 273–286.

Tesoro-Cruz, E., Calderon-Rodriguez, R., Hernandez-Gonzalez, R., Blanco-Favela, F., and Aguilar-Setien, A. (2008a). Intradermal DNA vaccination in ear pinnae is an efficient route to protect cats against rabies virus. Vet. Res. *39*, 16.

Tesoro-Cruz, E., Feria Romero, I.A., Lopez Mendoza, J.G., Orozco Suarez, S., Hernandez Gonzalez, R., Favela, F.B., Perez Torres, A., and Jose Alvaro Aguilar, S. (2008b). Efficient post-exposure prophylaxis against rabies by applying a four-dose DNA vaccine intranasally. Vaccine *26*, 6936–6944.

van, Thiel, P.P., de Bie, R.M., Eftimov, F., Tepaske, R., Zaaijer, H.L., van Doornum, G.J., Schutten, M., Osterhaus, A.D., Majoie, C.B., Aronica, E., *et al.* (2009). Fatal human rabies due to Duvenhage virus from a bat in Kenya: failure of treatment with coma-induction, ketamine, and antiviral drugs. PLoS Negl. Trop. Dis. *3*, e428.

Tims, T., Briggs, D.J., Davis, R.D., Moore, S.M., Xiang, Z., Ertl, H.C., and Fu, Z.F. (2000). Adult dogs receiving a rabies booster dose with a recombinant adenovirus expressing rabies virus glycoprotein develop high titers of neutralizing antibodies. Vaccine *18*, 2804–2807.

Tolson, N.D., Charlton, K.M., Lawson, K.F., Campbell, J.B., and Stewart, R.B. (1988). Studies of ERA/BHK-21 rabies vaccine in skunks and mice. Can. J. Vet. Res. *52*, 58–62.

Umeno, S., and Doi, Y. (1921). The study on the anti-rabic inoculation of dogs and the results of its practical application. Kitasato Arch. Exp. Med. *4*, 89–108.

Vicente, T., Roldao, A., Peixoto, C., Carrondo, M.J., and Alves, P.M. (2011). Large-scale production and purification of VLP-based vaccines. J. Invertebr. Pathol. *107*, S42–48.

Vos, A., Neubert, A., Pommerening, E., Müller, T., Döhner, L., Neubert, L., and Hughes, K. (2001). Immunogenicity of an E1-deleted recombinant human adenovirus against rabies by different routes of administration. J. Gen. Virol. *82*, 2191–2197.

Vos, A., Conzelmann, K.K., Finke, S., Müller, T., Teifke, J., Fooks, A.R., and Neubert, A. (2011). Immunogenicity studies in carnivores using a rabies virus construct with a site-directed deletion in the phosphoprotein. Adv. Pre. Med. *2011*:898171.

Walpita, P., Barr, J., Sherman, M., Basler, C.F., and Wang, L. (2011). Vaccine potential of Nipah virus-like particles. PLoS One *6*, e18437.

Wang, X., Liu, J., Wu, X., Yu, L., Chen, H., Guo, H., Zhang, M., Li, H., Liu, X., Sun, S., *et al.* (2011). Oral immunisation of mice with a recombinant rabies virus vaccine incorporating the heat-labile enterotoxin B subunit of *Escherichia coli* in an attenuated *Salmonella* strain. Res. Vet. Sci. In Press. DOI: 10.1016/j.rvsc.2011.09.015.

Webster, L.T., and Casals, J. (1942). A non-virulent, single-dose rabies vaccine for prophylactic immunization of dogs. J. Exp. Med. *76*, 185–194.

Wen, Y., Wang, H., Wu, H., Yang, F., Tripp, R.A., Hogan, R.J., and Fu, Z.F. (2010). Rabies virus expressing dendritic cell-activating molecules enhances the innate and adaptive immune response to vaccination. J.Virol. *85*, 1634–1644.

Weyer, J., Kuzmin, I.V., Rupprecht, C.E., and Nel, L.H. (2008). Cross-protective and cross-reactive immune responses to recombinant vaccinia viruses expressing full-length *Lyssavirus* glycoprotein genes. Epidemiol. Infect. *136*, 670–678.

Wiktor, T.J., Macfarlan, R.I., Reagan, K.J., Dietzschold, B., Curtis, P.J., Wunner, W.H., Kieny, M.P., Lathe, R., Lecocq, J.P., and Mackett, M. (1984). Protection from rabies by vaccinia virus recombinant containing the rabies virus glycoprotein. Proc. Natl. Acad. Sci. USA. *81*, 7194–7198.

Wu, X., and Rupprecht, C.E. (2008). Glycoprotein gene relocation in rabies virus. Virus Res. *131*, 95–99.

Wu, X., Gong, X., Foley, H.D., Schnell, M.J., and Fu, Z.F.(2002). Both viral transcription and replication are reduced when the rabies virus nucleoprotein is not phosphorylated. J. Virol. *76*, 4153–4161.

Wu, X., Franka, R., Svoboda, P., Pohl, J., and Rupprecht, C.E. (2009). Development of combined vaccine for rabies and immunocontraception. Vaccine *27*, 7202–7209.

Wu, X., Smith, T.G., and Rupprecht, C.E. (2011a). From brain passage to cell adaptation: the road of human rabies vaccine development. Expert Rev. Vaccines *10*, 1597–1608.

Wu, X., Frank, R., Henderson, H., and Rupprecht, C.E. (2011b). Live attenuated rabies virus co-infected with street rabies virus protects animals against rabies. Vaccine *29*, 4195–4201.

Wunderli, P.S., Dreesen, D.W., Miller, T.J., and Baer, G.M. (2006). The rabies peripheral challenge test: more accurate determination of vaccine potency. Vaccine *24*, 7115–7123.

Xiang, Z., Gao, G., Reyes-Sandoval, A., Cohen, C.J., Li, Y., Bergelson, J.M., Wilson, J.M., and Ertl, H.C. (2002). Novel, chimpanzee serotype 68-based adenoviral vaccine carrier for induction of antibodies to a transgene product. J. Virol. *76*, 2667–2675.

Xiang, Z.Q., Spitalnik, S., Tran, M., Wunner, W.H., Cheng, J., and Ertl, H.C. (1994). Vaccination with a plasmid vector carrying the rabies virus glycoprotein gene induces protective immunity against rabies virus. Virology *199*, 132–140.

Xiang, Z.Q., Yang, Y., Wilson, J.M., and Ertl, H.C. (1996). A replication-defective human adenovirus recombinant serves as a highly efficacious vaccine carrier. Virology *219*, 220–227.

Yarosh, O.K., Wandeler, A.I., Graham, F.L., Campbell, J.B., and Prevec, L. (1996). Human adenovirus type 5 vectors expressing rabies glycoprotein. Vaccine *14*, 1257–1264.

Yuan, Z., Zhang, S., Liu, Y., Zhang, F., Fooks, A.R., Liu, Q., and Hu, R. (2008). A recombinant pseudorabies virus expressing rabies virus glycoprotein: safety and immunogenicity in dogs. Vaccine *26*, 1314–1321.

Zhang, S., Liu, Y., Fooks, A.R., Zhang, F., and Hu, R. (2008). Oral vaccination of dogs (*Canis familiaris*) with baits containing the recombinant rabies-canine adenovirus type-2 vaccine confers long-lasting immunity against rabies. Vaccine *26*, 345–350.

Zhao, L., Toriumi, H., Kuang, Y., Chen, H., and Fu, Z.F. (2009). The roles of chemokines in rabies virus infection: Overexpression may not always be beneficial. J. Virol. *83*, 11808–11818.

Zhong, L., Granelli-Piperno, A., Pope, M., Ignatius, R., Lewis, M.G., Frankel, S.S., and Steinman, R.M. (2000). Presentation of SIVgag to monkey T cells using dendritic cells transfected with a recombinant adenovirus. Eur. J. Immunol. *30*, 3281–3290.

Zhou, D., Cun, A., Li, Y., Xiang, Z., and Ertl, H.C. (2006). A chimpanzee-origin adenovirus vector expressing the rabies virus glycoprotein as an oral vaccine against inhalation infection with rabies virus. Mol. Ther. *14*, 662–672.

Index

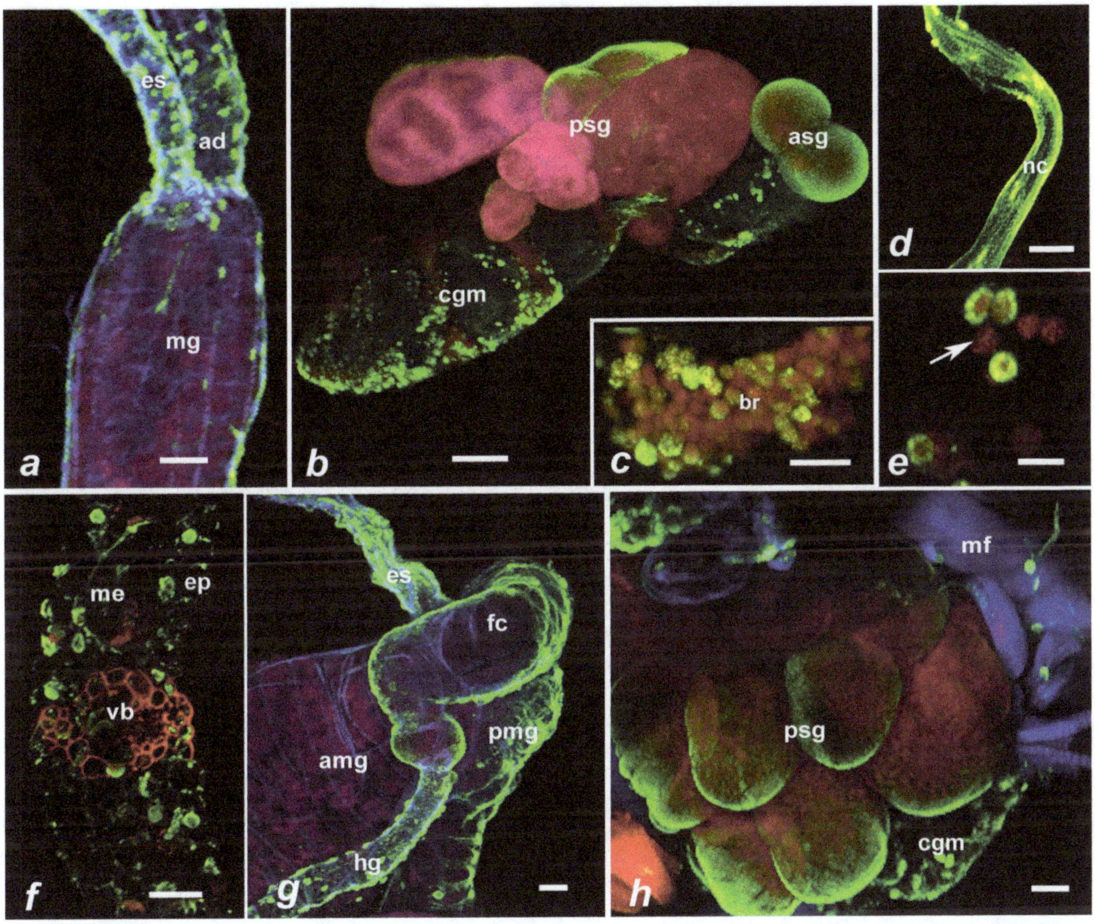

Figure 10.3 Immunofluorescence localization of maize-infecting rhabdoviruses in organs and tissues of their insect vectors. (a–e) Localization of maize mosaic virus (green fluorescence) in cells of the planthopper vector's (*Peregrinus maidis*) alimentary canal (a), salivary gland and compound ganglion (b) brain, (c) nerve cord (d) and hemocytes (e). (f–h) Localization of maize fine streak virus (green fluorescence) in cells of a maize leaf (f), and the leafhopper vector's (*Graminella nigrifrons*) alimentary canal (g), salivary gland and compound ganglion (h). Dissected organs or hemolymph smears were incubated with virus-specific antibodies, then with a secondary antibody, Alexa Fluor 488 (green), the nuclear stain propidium iodide (red), and (in a, g and h) with an actin stain, Phalloidin (blue/purple). Abbreviations: ad, anterior diverticulum; amg, anterior midgut; asg, accessory salivary gland; br, brain; cgm, compound ganglionic mass; ep, epidermis; es, oesophagus; fc, filter chamber; hg, hindgut; me, mesophyll; mf, muscle fibres; mg, midgut; nc; nerve cord; pmg, posterior midgut; psg, principal salivary gland; vb, vascular bundle. The arrow in (e) indicates the nucleus of an uninfected haemocyte. Scale bars = 40 μm, except in (c) and (e) (10 μm). Adapted from Ammar *et al.* (2009).

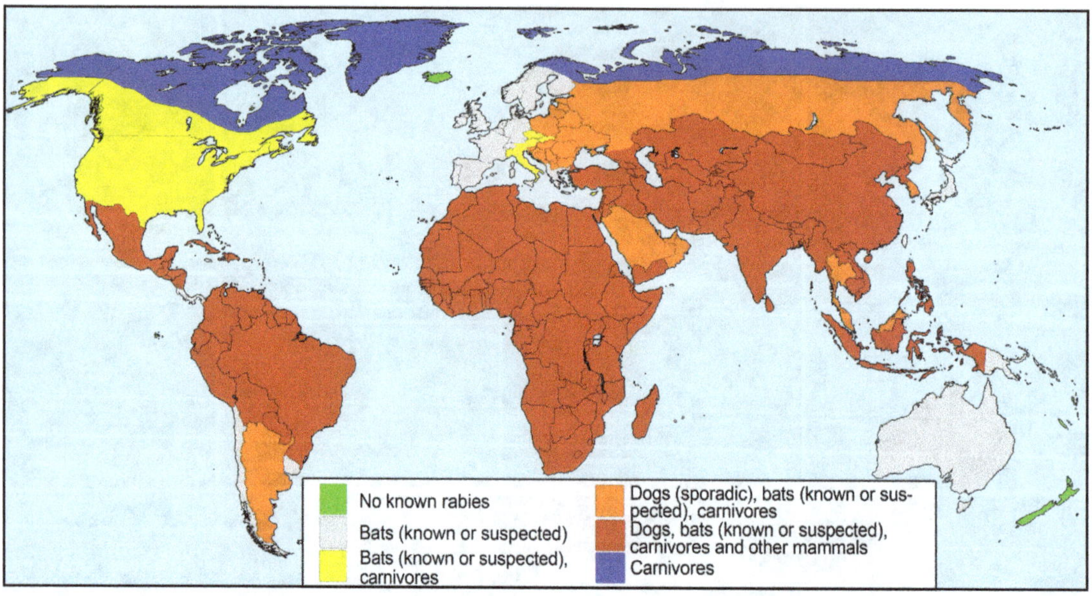

Figure 13.1 Global rabies distribution by reservoir. (Map redesigned courtesy of I.V. Kuzmin).

www.ingramcontent.com/pod-product-compliance
Lightning Source LLC
LaVergne TN
LVHW062226200926
848484LV00006B/1483